Praise for Brian Greene's

THE FABRIC OF THE COSMOS

"As pure intellectual adventure, this is about as good as it gets. . . . Even compared with *A Brief History of Time*, Greene's book stands out for its sweeping ambition . . . stripping down the mystery from difficult concepts without watering down the science." —*Newsday*

"Greene is as elegant as ever, cutting through the fog of complexity with insight and clarity. Space and time, you might even say, become putty in his hands." —*Los Angeles Times*

"Highly informed, lucid and witty. . . . There is simply no better introduction to the strange wonders of general relativity and quantum mechanics, the fields of knowledge essential for any real understanding of space and time." —*Discover*

"The author's informed curiosity is inspiring and his enthusiasm infectious." —*Kansas City Star*

"Mind-bending. . . . [Greene] is both a gifted theoretical physicist and a graceful popularizer [with] virtuoso explanatory skills." —*The Oregonian*

"Brian Greene is the new Hawking, only better." —*The Times* (London)

"Greene's gravitational pull rivals a black hole's." —*Newsweek*

"Greene is an excellent teacher, humorous and quick. . . . Read [to your friends] the passages of this book that boggle your mind. (You may find yourself reading them every single paragraph.)" —*The Boston Globe*

"Inexhaustibly witty . . . a must-read for the huge constituency of lay readers enticed by the mysteries of cosmology." —*The Sunday Times*

Brian Greene

The Fabric of the Cosmos

Brian Greene received his undergraduate degree from Harvard University and his doctorate from Oxford University, where he was a Rhodes Scholar. He joined the physics faculty at Cornell University in 1990, was appointed to a full professorship in 1995, and in 1996 joined Columbia University where he is professor of physics and mathematics. He has lectured at both a general and a technical level in more than twenty-five countries and is widely regarded for a number of groundbreaking discoveries in superstring theory. He lives in Andes, New York, and New York City.

ALSO BY BRIAN GREENE

The Elegant Universe:
Superstrings, Hidden Dimensions,
and the Quest for the Ultimate Theory

THE FABRIC
OF THE COSMOS

THE FABRIC
OF THE COSMOS

SPACE, TIME, AND THE TEXTURE OF REALITY

BRIAN GREENE

VINTAGE BOOKS

A DIVISION OF RANDOM HOUSE, INC. • NEW YORK

FIRST VINTAGE BOOKS EDITION, FEBRUARY 2005

Copyright © 2004 by Brian R. Greene

All rights reserved under International and Pan-American Copyright Conventions.
Published in the United States by Vintage Books, a division of Random House, Inc.,
New York. Originally published in hardcover in the United States by Alfred A. Knopf,
a division of Random House, Inc., New York, and in Canada by Random House
of Canada Limited, Toronto, in 2003.

Vintage and colophon are registered trademarks of Random House, Inc.

The Library of Congress has cataloged the Knopf edition as follows:
Greene, B. (Brian).
The fabric of the cosmos : space, time, and the texture of reality /
Brian Greene.
p. cm.
Includes bibliographical references (pp. 543–44).
1. Cosmology—Popular works.   I. Title.
QB982.G74 2004
523.1—dc22   2003058918

Vintage ISBN: 0-375-72720-5

Book design by Virginia Tan

www.vintagebooks.com

Printed in the United States of America
10 9 8 7 6

Contents

Preface

Space and time capture the imagination like no other scientific subject. For good reason. They form the arena of reality, the very fabric of the cosmos. Our entire existence—everything we do, think, and experience—takes place in some region of space during some interval of time. Yet science is still struggling to understand what space and time actually are. Are they real physical entities or simply useful ideas? If they're real, are they fundamental, or do they emerge from more basic constituents? What does it mean for space to be empty? Does time have a beginning? Does it have an arrow, flowing inexorably from past to future, as common experience would indicate? Can we manipulate space and time? In this book, we follow three hundred years of passionate scientific investigation seeking answers, or at least glimpses of answers, to such basic but deep questions about the nature of the universe.

Our journey also brings us repeatedly to another, tightly related question, as encompassing as it is elusive: What *is* reality? We humans only have access to the internal experiences of perception and thought, so how can we be sure they truly reflect an external world? Philosophers have long recognized this problem. Filmmakers have popularized it through story lines involving artificial worlds, generated by finely tuned neurological stimulation that exist solely within the minds of their protagonists. And physicists such as myself are acutely aware that the reality we observe—matter evolving on the stage of space and time—may have little to do with the reality, if any, that's out there. Nevertheless, because observations are all we have, we take them seriously. We choose hard data and the framework of mathematics as our guides, not unrestrained imagination or unrelenting skepticism, and seek the simplest yet most wide-reaching theories capable of explaining and predicting the outcome of today's and future experiments. This severely restricts the theories we pursue. (In this book, for example, we won't find a hint that I'm floating in a tank, connected to thousands of brain-stimulating wires, making me merely *think* that I'm now writing this text.) But during the last hundred years, discoveries in physics have suggested revisions to our everyday sense of reality that are as dramatic, as mind-bending, and as paradigm-shaking as the most imaginative science fiction. These revolutionary upheavals will frame our passage through the pages that follow.

Many of the questions we explore are the same ones that, in various guises, furrowed the brows of Aristotle, Galileo, Newton, Einstein, and countless others through the ages. And because this book seeks to convey science in the making, we follow these questions as they've been declared answered by one generation, overturned by their successors, and refined and reinterpreted by scientists in the centuries that followed.

For example, on the perplexing question of whether completely empty space is, like a blank canvas, a real entity or merely an abstract idea, we follow the pendulum of scientific opinion as it swings between Isaac Newton's seventeenth-century declaration that space is real, Ernst Mach's conclusion in the nineteenth century that it isn't, and Einstein's twentieth-century dramatic reformulation of the question itself, in which he merged space and time, and largely refuted Mach. We then encounter subsequent discoveries that transformed the question once again by redefining the meaning of "empty," envisioning that space is unavoidably suffused with what are called quantum fields and possibly a diffuse uniform energy called a cosmological constant—modern echoes of the old and discredited notion of a space-filling aether. What's more, we then describe how upcoming space-based experiments may confirm particular features of Mach's conclusions that happen to agree with Einstein's general relativity, illustrating well the fascinating and tangled web of scientific development.

In our own era we encounter inflationary cosmology's gratifying insights into time's arrow, string theory's rich assortment of extra spatial dimensions, M-theory's radical suggestion that the space we inhabit may be but a sliver floating in a grander cosmos, and the current wild speculation that the universe we see may be nothing more than a cosmic hologram. We don't yet know if the more recent of these theoretical proposals are right. But outrageous as they sound, we investigate them thoroughly because they are where our dogged search for the deepest laws of the universe leads. Not only can a strange and unfamiliar reality arise from the fertile imagination of science fiction, but one may also emerge from the cutting-edge findings of modern physics.

The Fabric of the Cosmos is intended primarily for the general reader who has little or no formal training in the sciences but whose desire to understand the workings of the universe provides incentive to grapple with a number of complex and challenging concepts. As in my first book, *The Elegant Universe*, I've stayed close to the core scientific ideas throughout, while stripping away the mathematical details in favor of metaphors, analogies, stories, and illustrations. When we reach the book's most difficult sections, I forewarn the reader and provide brief summaries for those who decide to skip or skim these more involved discussions. In this way, the reader should be able to walk the path of discovery and gain not just knowledge of physics' current worldview, but an understanding of how and why that worldview has gained prominence.

Students, avid readers of general-level science, teachers, and professionals should also find much of interest in the book. Although the initial chapters cover the necessary but standard background material in relativity and quantum mechanics, the focus on the corporeality of space and time is somewhat unconventional in its approach. Subsequent chapters cover a wide range of topics—Bell's theorem, delayed choice experiments, quantum measurement, accelerated expansion, the possibility of producing black holes in the next generation of particle accelerators, fanciful wormhole time machines, to name a few—and so will bring such readers up to date on a number of the most tantalizing and debated advances.

Some of the material I cover is controversial. For those issues that remain up in the air, I've discussed the leading viewpoints in the main text. For the points of contention that I feel have achieved more of a consensus, I've relegated differing viewpoints to the notes. Some scientists, especially those holding minority views, may take exception to some of my judgments, but through the main text and the notes, I've striven for a balanced treatment. In the notes, the particularly diligent reader will also find more complete explanations, clarifications, and caveats relevant to points I've simplified, as well as (for those so inclined) brief mathematical counterparts to the equation-free approach taken in the main text. A short glossary provides a quick reference for some of the more specialized scientific terms.

Even a book of this length can't exhaust the vast subject of space and time. I've focused on those features I find both exciting and essential to forming a full picture of the reality painted by modern science. No doubt, many of these choices reflect personal taste, and so I apologize to those who feel their own work or favorite area of study is not given adequate attention.

While writing *The Fabric of the Cosmos*, I've been fortunate to receive valuable feedback from a number of dedicated readers. Raphael Kasper, Lubos Motl, David Steinhardt, and Ken Vineberg read various versions of the entire manuscript, sometimes repeatedly, and offered numerous, detailed, and insightful suggestions that substantially enhanced both the clarity and the accuracy of the presentation. I offer them heartfelt thanks. David Albert, Ted Baltz, Nicholas Boles, Tracy Day, Peter Demchuk, Richard Easther, Anna Hall, Keith Goldsmith, Shelley Goldstein, Michael Gordin, Joshua Greene, Arthur Greenspoon, Gavin Guerra, Sandra Kauffman, Edward Kastenmeier, Robert Krulwich, Andrei Linde, Shani Offen, Maulik Parikh, Michael Popowits, Marlin Scully, John Stachel, and Lars Straeter read all or part of the manuscript, and their comments were extremely useful. I benefited from conversations with Andreas Albrecht, Michael Bassett, Sean Carrol, Andrea Cross, Rita Greene, Wendy Greene, Susan Greene, Alan Guth, Mark Jackson, Daniel Kabat, Will Kinney, Justin Khoury, Hiranya Peiris, Saul Perlmutter, Koenraad Schalm, Paul Steinhardt, Leonard Susskind, Neil Turok, Henry Tye, William Warmus, and Erick Weinberg. I owe special thanks to Raphael Gunner, whose keen sense of the genuine argument and whose willingness to critique various of my attempts proved invaluable. Eric Martinez provided critical and tireless assistance in the production phase of the book, and Jason Severs did a stellar job of creating the illustrations. I thank my agents, Katinka Matson and John Brockman. And I owe a great debt of gratitude to my editor, Marty Asher, for providing a wellspring of encouragement, advice, and sharp insight that substantially improved the quality of the presentation.

During the course of my career, my scientific research has been funded by the Department of Energy, the National Science Foundation, and the Alfred P. Sloan Foundation. I gratefully acknowledge their support.

I

REALITY'S
ARENA

1

Roads to Reality

SPACE, TIME, AND WHY THINGS ARE AS THEY ARE

None of the books in my father's dusty old bookcase were forbidden. Yet while I was growing up, I never saw anyone take one down. Most were massive tomes—a comprehensive history of civilization, matching volumes of the great works of western literature, numerous others I can no longer recall—that seemed almost fused to shelves that bowed slightly from decades of steadfast support. But way up on the highest shelf was a thin little text that, every now and then, would catch my eye because it seemed so out of place, like Gulliver among the Brobdingnagians. In hindsight, I'm not quite sure why I waited so long before taking a look. Perhaps, as the years went by, the books seemed less like material you read and more like family heirlooms you admire from afar. Ultimately, such reverence gave way to teenage brashness. I reached up for the little text, dusted it off, and opened to page one. The first few lines were, to say the least, startling.

"There is but one truly philosophical problem, and that is suicide," the text began. I winced. "Whether or not the world has three dimensions or the mind nine or twelve categories," it continued, "comes afterward"; such questions, the text explained, were part of the game humanity played, but they deserved attention only after the one true issue had been settled. The book was *The Myth of Sisyphus* and was written by the Algerian-born philosopher and Nobel laureate Albert Camus. After a moment, the iciness of his words melted under the light of comprehension. Yes, of course, I thought. You can ponder this or analyze that till the cows come home, but the real question is whether all your ponderings and analyses will convince you that life is worth living. That's what it all comes down to. Everything else is detail.

My chance encounter with Camus' book must have occurred during an especially impressionable phase because, more than anything else I'd read, his words stayed with me. Time and again I'd imagine how various people I'd met, or heard about, or had seen on television would answer this primary of all questions. In retrospect, though, it was his second assertion—regarding the role of scientific progress—that, for me, proved particularly challenging. Camus acknowledged value in understanding the structure of the universe, but as far as I could tell, he rejected the possibility that such understanding could make any difference to our assessment of life's worth. Now, certainly, my teenage reading of existential philosophy was about as sophisticated as Bart Simpson's reading of Romantic poetry, but even so, Camus' conclusion struck me as off the mark. To this aspiring physicist, it seemed that an informed appraisal of life absolutely required a full understanding of life's arena—the universe. I remember thinking that if our species dwelled in cavernous outcroppings buried deep underground and so had yet to discover the earth's surface, brilliant sunlight, an ocean breeze, and the stars that lie beyond, or if evolution had proceeded along a different pathway and we had yet to acquire any but the sense of touch, so everything we knew came only from our tactile impressions of our immediate environment, or if human mental faculties stopped developing during early childhood so our emotional and analytical skills never progressed beyond those of a five-year-old—in short, if our experiences painted but a paltry portrait of reality—our appraisal of life would be thoroughly compromised. When we finally found our way to earth's surface, or when we finally gained the ability to see, hear, smell, and taste, or when our minds were finally freed to develop as they ordinarily do, our collective view of life and the cosmos would, of necessity, change radically. Our previously compromised grasp of reality would have shed a very different light on that most fundamental of all philosophical questions.

But, you might ask, what of it? Surely, any sober assessment would conclude that although we might not understand everything about the universe—every aspect of how matter behaves or life functions—we are privy to the defining, broad-brush strokes gracing nature's canvas. Surely, as Camus intimated, progress in physics, such as understanding the number of space dimensions; or progress in neuropsychology, such as understanding all the organizational structures in the brain; or, for that matter, progress in any number of other scientific undertakings may fill in important details, but their impact on our evaluation of life and reality would be minimal. Surely, reality is what we think it is; reality is revealed to us by our experiences.

To one extent or another, this view of reality is one many of us hold, if only implicitly. I certainly find myself thinking this way in day-to-day life; it's easy to be seduced by the face nature reveals directly to our senses. Yet, in the decades since first encountering Camus' text, I've learned that modern science tells a very different story. *The* overarching lesson that has emerged from scientific inquiry over the last century is that human experience is often a misleading guide to the true nature of reality. Lying just beneath the surface of the everyday *is* a world we'd hardly recognize. Followers of the occult, devotees of astrology, and those who hold to religious principles that speak to a reality beyond experience have, from widely varying perspectives, long since arrived at a similar conclusion. But that's not what I have in mind. I'm referring to the work of ingenious innovators and tireless researchers—the men and women of science—who have peeled back layer after layer of the cosmic onion, enigma by enigma, and revealed a universe that is at once surprising, unfamiliar, exciting, elegant, and thoroughly unlike what anyone ever expected.

These developments are anything but details. Breakthroughs in physics have forced, and continue to force, dramatic revisions to our conception of the cosmos. I remain as convinced now as I did decades ago that Camus rightly chose life's value as the ultimate question, but the insights of modern physics have persuaded me that assessing life through the lens of everyday experience is like gazing at a van Gogh through an empty Coke bottle. Modern science has spearheaded one assault after another on evidence gathered from our rudimentary perceptions, showing that they often yield a clouded conception of the world we inhabit. And so whereas Camus separated out physical questions and labeled them secondary, I've become convinced that they're primary. For me, physical reality both sets the arena and provides the illumination for grappling with Camus' question. Assessing existence while failing to embrace the insights of modern physics would be like wrestling in the dark with an unknown opponent. By deepening our understanding of the true nature of physical reality, we profoundly reconfigure our sense of ourselves and our experience of the universe.

The central concern of this book is to explain some of the most prominent and pivotal of these revisions to our picture of reality, with an intense focus on those that affect our species' long-term project to understand space and time. From Aristotle to Einstein, from the astrolabe to the Hubble Space Telescope, from the pyramids to mountaintop observatories, space and time have framed thinking since thinking began. With the advent of the modern scientific age, their importance has only been heightened. Over the last three centuries, developments in physics have revealed space and time as the most baffling and most compelling concepts, and as those most instrumental in our scientific analysis of the universe. Such developments have also shown that space and time top the list of age-old scientific constructs that are being fantastically revised by cutting-edge research.

To Isaac Newton, space and time simply were—they formed an inert, universal cosmic stage on which the events of the universe played themselves out. To his contemporary and frequent rival Gottfried Wilhelm von Leibniz, "space" and "time" were merely the vocabulary of relations between where objects were and when events took place. Nothing more. But to Albert Einstein, space and time were the raw material underlying reality. Through his theories of relativity, Einstein jolted our thinking about space and time and revealed the principal part they play in the evolution of the universe. Ever since, space and time have been the sparkling jewels of physics. They are at once familiar and mystifying; fully understanding space and time has become physics' most daunting challenge and sought-after prize.

The developments we'll cover in this book interweave the fabric of space and time in various ways. Some ideas will challenge features of space and time so basic that for centuries, if not millennia, they've seemed beyond questioning. Others will seek the link between our theoretical understanding of space and time and the traits we commonly experience. Yet others will raise questions unfathomable within the limited confines of ordinary perceptions.

We will speak only minimally of philosophy (and not at all about suicide and the meaning of life). But in our scientific quest to solve the mysteries of space and time, we will be unrestrained. From the universe's smallest speck and earliest moments to its farthest reaches and most distant future, we will examine space and time in environments familiar and far-flung, with an unflinching eye seeking their true nature. As the story of space and time has yet to be fully written, we won't arrive at any final assessments. But we will encounter a series of developments—some intensely strange, some deeply satisfying, some experimentally verified, some thoroughly speculative—that will show how close we've come to wrapping our minds around the fabric of the cosmos and touching the true texture of reality.

Classical Reality

Historians differ on exactly when the modern scientific age began, but certainly by the time Galileo Galilei, René Descartes, and Isaac Newton had had their say, it was briskly under way. In those days, the new scientific mind-set was being steadily forged, as patterns found in terrestrial and astronomical data made it increasingly clear that there is an order to all the comings and goings of the cosmos, an order accessible to careful reasoning and mathematical analysis. These early pioneers of modern scientific thought argued that, when looked at the right way, the happenings in the universe not only are explicable but predictable. The power of science to foretell aspects of the future—consistently and quantitatively—had been revealed.

Early scientific study focused on the kinds of things one might see or experience in everyday life. Galileo dropped weights from a leaning tower (or so legend has it) and watched balls rolling down inclined surfaces; Newton studied falling apples (or so legend has it) and the orbit of the moon. The goal of these investigations was to attune the nascent scientific ear to nature's harmonies. To be sure, physical reality was the stuff of experience, but the challenge was to hear the rhyme and reason behind the rhythm and regularity. Many sung and unsung heroes contributed to the rapid and impressive progress that was made, but Newton stole the show. With a handful of mathematical equations, he synthesized everything known about motion on earth and in the heavens, and in so doing, composed the score for what has come to be known as *classical physics*.

In the decades following Newton's work, his equations were developed into an elaborate mathematical structure that significantly extended both their reach and their practical utility. Classical physics gradually became a sophisticated and mature scientific discipline. But shining clearly through all these advances was the beacon of Newton's original insights. Even today, more than three hundred years later, you can see Newton's equations scrawled on introductory-physics chalkboards worldwide, printed on NASA flight plans computing spacecraft trajectories, and embedded within the complex calculations of forefront research.

Newton brought a wealth of physical phenomena within a single theoretical framework.

But while formulating his laws of motion, Newton encountered a critical stumbling block, one that is of particular importance to our story (Chapter 2). Everyone knew that things could move, but what about the arena within which the motion took place? Well, that's space, we'd all answer. But, Newton would reply, what *is* space? Is space a real physical entity or is it an abstract idea born of the human struggle to comprehend the cosmos? Newton realized that this key question had to be answered, because without taking a stand on the meaning of space and time, his equations describing motion would prove meaningless. Understanding requires context; insight must be anchored.

And so, with a few brief sentences in his *Principia Mathematica*, Newton articulated a conception of space and time, declaring them absolute and immutable entities that provided the universe with a rigid, unchangeable arena. According to Newton, space and time supplied an invisible scaffolding that gave the universe shape and structure.

Not everyone agreed. Some argued persuasively that it made little sense to ascribe existence to something you can't feel, grasp, or affect. But the explanatory and predictive power of Newton's equations quieted the critics. For the next two hundred years, his absolute conception of space and time was dogma.

Relativistic Reality

The classical Newtonian worldview was pleasing. Not only did it describe natural phenomena with striking accuracy, but the details of the description—the mathematics—aligned tightly with experience. If you push something, it speeds up. The harder you throw a ball, the more impact it has when it smacks into a wall. If you press against something, you feel it pressing back against you. The more massive something is, the stronger its gravitational pull. These are among the most basic properties of the natural world, and when you learn Newton's framework, you see them represented in his equations, clear as day. Unlike a crystal ball's inscrutable hocus-pocus, the workings of Newton's laws were on display for all with minimal mathematical training to take in fully. Classical physics provided a rigorous grounding for human intuition.

Newton had included the force of gravity in his equations, but it was not until the 1860s that the Scottish scientist James Clerk Maxwell extended the framework of classical physics to take account of electrical and magnetic forces. Maxwell needed additional equations to do so and the mathematics he employed required a higher level of training to grasp fully. But his new equations were every bit as successful at explaining electrical and magnetic phenomena as Newton's were at describing motion. By the late 1800s, it was evident that the universe's secrets were proving no match for the power of human intellectual might.

Indeed, with the successful incorporation of electricity and magnetism, there was a growing sense that theoretical physics would soon be complete. Physics, some suggested, was rapidly becoming a finished subject and its laws would shortly be chiseled in stone. In 1894, the renowned experimental physicist Albert Michelson remarked that "most of the grand underlying principles have been firmly established" and he quoted an "eminent scientist"—most believe it was the British physicist Lord Kelvin—as saying that all that remained were details of determining some numbers to a greater number of decimal places.[1] In 1900, Kelvin himself did note that "two clouds" were hovering on the horizon, one to do with properties of light's motion and the other with aspects of the radiation objects emit when heated,[2] but there was a general feeling that these were mere details, which, no doubt, would soon be addressed.

Within a decade, everything changed. As anticipated, the two problems Kelvin had raised were promptly addressed, but they proved anything but minor. Each ignited a revolution, and each required a fundamental rewriting of nature's laws. The classical conceptions of space, time, and reality—the ones that for hundreds of years had not only worked but also concisely expressed our intuitive sense of the world—were overthrown.

The relativity revolution, which addressed the first of Kelvin's "clouds," dates from 1905 and 1915, when Albert Einstein completed his special and general theories of relativity (Chapter 3). While struggling with puzzles involving electricity, magnetism, and light's motion, Einstein realized that Newton's conception of space and time, the cornerstone of classical physics, was flawed. Over the course of a few intense weeks in the spring of 1905, he determined that space and time are not independent and absolute, as Newton had thought, but are enmeshed and relative in a manner that flies in the face of common experience. Some ten years later, Einstein hammered a final nail in the Newtonian coffin by rewriting the laws of gravitational physics. This time, not only did Einstein show that space and time are part of a unified whole, he also showed that by warping and curving they participate in cosmic evolution. Far from being the rigid, unchanging structures envisioned by Newton, space and time in Einstein's reworking are flexible and dynamic.

The two theories of relativity are among humankind's most precious achievements, and with them Einstein toppled Newton's conception of reality. Even though Newtonian physics seemed to capture mathematically much of what we experience physically, the reality it describes turns out not to be the reality of our world. Ours is a relativistic reality. Yet, because the deviation between classical and relativistic reality is manifest only under extreme conditions (such as extremes of speed and gravity), Newtonian physics still provides an approximation that proves extremely accurate and useful in many circumstances. But utility and reality are very different standards. As we will see, features of space and time that for many of us are second nature have turned out to be figments of a false Newtonian perspective.

Quantum Reality

The second anomaly to which Lord Kelvin referred led to the quantum revolution, one of the greatest upheavals to which modern human understanding has ever been subjected. By the time the fires subsided and the smoke cleared, the veneer of classical physics had been singed off the newly emerging framework of quantum reality.

A core feature of classical physics is that if you know the positions and velocities of all objects at a particular moment, Newton's equations, together with their Maxwellian updating, can tell you their positions and velocities at any other moment, past or future. Without equivocation, classical physics declares that the past and future are etched into the present. This feature is also shared by both special and general relativity. Although the relativistic concepts of past and future are subtler than their familiar classical counterparts (Chapters 3 and 5), the equations of relativity, together with a complete assessment of the present, determine them just as completely.

By the 1930s, however, physicists were forced to introduce a whole new conceptual schema called *quantum mechanics*. Quite unexpectedly, they found that only quantum laws were capable of resolving a host of puzzles and explaining a variety of data newly acquired from the atomic and subatomic realm. But according to the quantum laws, even if you make the most perfect measurements possible of how things are today, the best you can ever hope to do is predict the *probability* that things will be one way or another at some chosen time in the future, or that things were one way or another at some chosen time in the past. The universe, according to quantum mechanics, is *not* etched into the present; the universe, according to quantum mechanics, participates in a game of chance.

Although there is still controversy over precisely how these developments should be interpreted, most physicists agree that probability is deeply woven into the fabric of quantum reality. Whereas human intuition, and its embodiment in classical physics, envision a reality in which things are always definitely one way *or* another, quantum mechanics describes a reality in which things sometimes hover in a haze of being partly one way *and* partly another. Things become definite only when a suitable observation forces them to relinquish quantum possibilities and settle on a specific outcome. The outcome that's realized, though, cannot be predicted—we can predict only the odds that things will turn out one way or another.

This, plainly speaking, is weird. We are unused to a reality that remains ambiguous until perceived. But the oddity of quantum mechanics does not stop here. At least as astounding is a feature that goes back to a paper Einstein wrote in 1935 with two younger colleagues, Nathan Rosen and Boris Podolsky, that was intended as an attack on quantum theory.[3] With the ensuing twists of scientific progress, Einstein's paper can now be viewed as among the first to point out that quantum mechanics— if taken at face value—implies that something you do over here can be *instantaneously* linked to something happening over there, regardless of distance. Einstein considered such instantaneous connections ludicrous and interpreted their emergence from the mathematics of quantum theory as evidence that the theory was in need of much development before it would attain an acceptable form. But by the 1980s, when both theoretical and technological developments brought experimental scrutiny to bear on these purported quantum absurdities, researchers confirmed that there *can* be an instantaneous bond between what happens at widely separated locations. Under pristine laboratory conditions, what Einstein thought absurd really happens (Chapter 4).

The implications of these features of quantum mechanics for our picture of reality are a subject of ongoing research. Many scientists, myself included, view them as part of a radical quantum updating of the meaning and properties of space. Normally, spatial separation implies physical independence. If you want to control what's happening on the other side of a football field, you have to go there, or, at the very least, you have to send someone or something (the assistant coach, bouncing air molecules conveying speech, a flash of light to get someone's attention, etc.) across the field to convey your influence. If you don't—if you remain spatially isolated—you will have no impact, since intervening space ensures the absence of a physical connection. Quantum mechanics challenges this view by revealing, at least in certain circumstances, a capacity to transcend space; long-range quantum connections can bypass spatial separation. Two objects can be far apart in space, but as far as quantum mechanics is concerned, it's as if they're a single entity. Moreover, because of the tight link between space and time found by Einstein, the quantum connections also have temporal tentacles. We'll shortly encounter some clever and truly wondrous experiments that have recently explored a number of the startling spatio-temporal interconnections entailed by quantum mechanics and, as we'll see, they forcefully challenge the classical, intuitive worldview many of us hold.

Despite these many impressive insights, there remains one very basic feature of time—that it seems to have a direction pointing from past to future—for which neither relativity nor quantum mechanics has provided an explanation. Instead, the only convincing progress has come from research in an area of physics called *cosmology*.

Cosmological Reality

To open our eyes to the true nature of the universe has always been one of physics' primary purposes. It's hard to imagine a more mind-stretching experience than learning, as we have over the last century, that the reality we experience is but a glimmer of the reality that is. But physics also has the equally important charge of explaining the elements of reality that we actually do experience. From our rapid march through the history of physics, it might seem as if this has already been achieved, as if ordinary experience is addressed by pre–twentieth-century advances in physics. To some extent, this is true. But even when it comes to the everyday, we are far from a full understanding. And among the features of common experience that have resisted complete explanation is one that taps into one of the deepest unresolved mysteries in modern physics—the mystery that the great British physicist Sir Arthur Eddington called the *arrow of time*.[4]

We take for granted that there is a direction to the way things unfold in time. Eggs break, but they don't unbreak; candles melt, but they don't unmelt; memories are of the past, never of the future; people age, but they don't unage. These asymmetries govern our lives; the distinction between forward and backward in time is a prevailing element of experiential reality. If forward and backward in time exhibited the same symmetry we witness between left and right, or back and forth, the world would be unrecognizable. Eggs would unbreak as often as they broke; candles would unmelt as often as they melted; we'd remember as much about the future as we do about the past; people would unage as often as they aged. Certainly, such a time-symmetric reality is not our reality. But where does time's asymmetry come from? What is responsible for this most basic of all time's properties?

It turns out that the known and accepted laws of physics show no such asymmetry (Chapter 6): each direction in time, forward and backward, is treated by the laws without distinction. *And that's the origin of a huge puzzle.* Nothing in the equations of fundamental physics shows any sign of treating one direction in time differently from the other, and that is totally at odds with everything we experience.[5]

Surprisingly, even though we are focusing on a familiar feature of everyday life, the most convincing resolution of this mismatch between fundamental physics and basic experience requires us to contemplate the most unfamiliar of events—the beginning of the universe. This realization has its roots in the work of the great nineteenth-century physicist Ludwig Boltzmann, and in the years since has been elaborated on by many researchers, most notably the British mathematician Roger Penrose. As we will see, special physical conditions at the universe's inception (a highly ordered environment at or just after the big bang) may have imprinted a direction on time, rather as winding up a clock, twisting its spring into a highly ordered initial state, allows it to tick forward. Thus, in a sense we'll make precise, the breaking—as opposed to the unbreaking—of an egg bears witness to conditions at the birth of the universe some 14 billion years ago.

This unexpected link between everyday experience and the early universe provides insight into why events unfold one way in time and never the reverse, but it does not fully solve the mystery of time's arrow. Instead, it shifts the puzzle to the realm of *cosmology*—the study of the origin and evolution of the entire cosmos—and compels us to find out whether the universe actually had the highly ordered beginning that this explanation of time's arrow requires.

Cosmology is among the oldest subjects to captivate our species. And it's no wonder. We're storytellers, and what story could be more grand than the story of creation? Over the last few millennia, religious and philosophical traditions worldwide have weighed in with a wealth of versions of how everything—the universe—got started. Science, too, over its long history, has tried its hand at cosmology. But it was Einstein's discovery of general relativity that marked the birth of modern scientific cosmology.

Shortly after Einstein published his theory of general relativity, both he and others applied it to the universe as a whole. Within a few decades, their research led to the tentative framework for what is now called the *big bang theory*, an approach that successfully explained many features of astronomical observations (Chapter 8). In the mid-1960s, evidence in support of big bang cosmology mounted further, as observations revealed a nearly uniform haze of microwave radiation permeating space—invisible to the naked eye but readily measured by microwave detectors—that was predicted by the theory. And certainly by the 1970s, after a decade of closer scrutiny and substantial progress in determining how basic ingredients in the cosmos respond to extreme changes in heat and temperature, the big bang theory secured its place as the leading cosmological theory (Chapter 9).

Its successes notwithstanding, the theory suffered significant shortcomings. It had trouble explaining why space has the overall shape revealed by detailed astronomical observations, and it offered no explanation for why the temperature of the microwave radiation, intently studied ever since its discovery, appears thoroughly uniform across the sky. Moreover, what is of primary concern to the story we're telling, the big bang theory provided no compelling reason why the universe might have been highly ordered near the very beginning, as required by the explanation for time's arrow.

These and other open issues inspired a major breakthrough in the late 1970s and early 1980s, known as *inflationary cosmology* (Chapter 10). Inflationary cosmology modifies the big bang theory by inserting an extremely brief burst of astoundingly rapid expansion during the universe's earliest moments (in this approach, the size of the universe increased by a factor larger than a million trillion trillion in less than a millionth of a trillionth of a trillionth of a second). As will become clear, this stupendous growth of the young universe goes a long way toward filling in the gaps left by the big bang model—of explaining the shape of space and the uniformity of the microwave radiation, and also of suggesting why the early universe might have been highly ordered—thus providing significant progress toward explaining both astronomical observations and the arrow of time we all experience (Chapter 11).

Yet, despite these mounting successes, for two decades inflationary cosmology has been harboring its own embarrassing secret. Like the standard big bang theory it modified, inflationary cosmology rests on the equations Einstein discovered with his general theory of relativity. Although volumes of research articles attest to the power of Einstein's equations to accurately describe large and massive objects, physicists have long known that an accurate theoretical analysis of small objects—such as the observable universe when it was a mere fraction of a second old—requires the use of quantum mechanics. The problem, though, is that when the equations of general relativity commingle with those of quantum mechanics, the result is disastrous. The equations break down entirely, and this prevents us from determining how the universe was born and whether at its birth it realized the conditions necessary to explain time's arrow.

It's not an overstatement to describe this situation as a theoretician's nightmare: the absence of mathematical tools with which to analyze a vital realm that lies beyond experimental accessibility. And since space and time are so thoroughly entwined with this particular inaccessible realm—the origin of the universe—understanding space and time fully requires us to find equations that can cope with the extreme conditions of huge density, energy, and temperature characteristic of the universe's earliest moments. This is an absolutely essential goal, and one that many physicists believe requires developing a so-called *unified theory*.

Unified Reality

Over the past few centuries, physicists have sought to consolidate our understanding of the natural world by showing that diverse and apparently distinct phenomena are actually governed by a single set of physical laws. To Einstein, this goal of unification—of explaining the widest array of phenomena with the fewest physical principles—became a lifelong passion. With his two theories of relativity, Einstein united space, time, and gravity. But this success only encouraged him to think bigger. He dreamed of finding a single, all-encompassing framework capable of embracing all of nature's laws; he called that framework a *unified theory*. Although now and then rumors spread that Einstein had found a unified theory, all such claims turned out to be baseless; Einstein's dream went unfulfilled.

Einstein's focus on a unified theory during the last thirty years of his life distanced him from mainstream physics. Many younger scientists viewed his single-minded search for the grandest of all theories as the ravings of a great man who, in his later years, had turned down the wrong path. But in the decades since Einstein's passing, a growing number of physicists have taken up his unfinished quest. Today, developing a unified theory ranks among the most important problems in theoretical physics.

For many years, physicists found that the central obstacle to realizing a unified theory was the fundamental conflict between the two major breakthroughs of twentieth-century physics: general relativity and quantum mechanics. Although these two frameworks are typically applied in vastly different realms—general relativity to big things like stars and galaxies, quantum mechanics to small things like molecules and atoms—each theory claims to be universal, to work in all realms. However, as mentioned above, whenever the theories are used in conjunction, their combined equations produce nonsensical answers. For instance, when quantum mechanics is used with general relativity to calculate the probability that some process or other involving gravity will take place, the answer that's often found is not something like a probability of 24 percent or 63 percent or 91 percent; instead, out of the combined mathematics pops an *infinite* probability. That doesn't mean a probability so high that you should put all your money on it because it's a shoo-in. Probabilities bigger than 100 percent are meaningless. Calculations that produce an infinite probability simply show that the combined equations of general relativity and quantum mechanics have gone haywire.

Scientists have been aware of the tension between general relativity and quantum mechanics for more than half a century, but for a long time relatively few felt compelled to search for a resolution. Instead, most researchers used general relativity solely for analyzing large and massive objects, while reserving quantum mechanics solely for analyzing small and light objects, carefully keeping each theory a safe distance from the other so their mutual hostility would be held in check. Over the years, this approach to détente has allowed for stunning advances in our understanding of each domain, but it does not yield a lasting peace.

A very few realms—extreme physical situations that are both massive and tiny—fall squarely in the demilitarized zone, requiring that general relativity and quantum mechanics simultaneously be brought to bear. The center of a black hole, in which an entire star has been crushed by its own weight to a minuscule point, and the big bang, in which the entire observable universe is imagined to have been compressed to a nugget far smaller than a single atom, provide the two most familiar examples. Without a successful union between general relativity and quantum mechanics, the end of collapsing stars and the origin of the universe would remain forever mysterious. Many scientists were willing to set aside these realms, or at least defer thinking about them until other, more tractable problems had been overcome.

But a few researchers couldn't wait. A conflict in the known laws of physics means a failure to grasp a deep truth and that was enough to keep these scientists from resting easy. Those who plunged in, though, found the waters deep and the currents rough. For long stretches of time, research made little progress; things looked bleak. Even so, the tenacity of those who had the determination to stay the course and keep alive the dream of uniting general relativity and quantum mechanics is being rewarded. Scientists are now charging down paths blazed by those explorers and are closing in on a harmonious merger of the laws of the large and small. The approach that many agree is a leading contender is *superstring theory* (Chapter 12).

As we will see, superstring theory starts off by proposing a new answer to an old question: what are the smallest, indivisible constituents of matter? For many decades, the conventional answer has been that matter is composed of particles—electrons and quarks—that can be modeled as dots that are indivisible and that have no size and no internal structure. Conventional theory claims, and experiments confirm, that these particles combine in various ways to produce protons, neutrons, and the wide variety of atoms and molecules making up everything we've ever encountered. Superstring theory tells a different story. It does not deny the key role played by electrons, quarks, and the other particle species revealed by experiment, but it does claim that these particles are not dots. Instead, according to superstring theory, every particle is composed of a tiny filament of energy, some hundred billion billion times smaller than a single atomic nucleus (much smaller than we can currently probe), which is shaped like a little string. And just as a violin string can vibrate in different patterns, each of which produces a different musical tone, the filaments of superstring theory can also vibrate in different patterns. These vibrations, though, don't produce different musical notes; remarkably, the theory claims that they produce different particle properties. A tiny string vibrating in one pattern would have the mass and the electric charge of an electron; according to the theory, such a vibrating string would *be* what we have traditionally called an electron. A tiny string vibrating in a different pattern would have the requisite properties to identify it as a quark, a neutrino, or any other kind of particle. All species of particles are unified in superstring theory since each arises from a different vibrational pattern executed by the same underlying entity.

Going from dots to strings-so-small-they-look-like-dots might not seem like a terribly significant change in perspective. But it is. From such humble beginnings, superstring theory combines general relativity and quantum mechanics into a single, consistent theory, banishing the perniciously infinite probabilities afflicting previously attempted unions. And as if that weren't enough, superstring theory has revealed the breadth necessary to stitch all of nature's forces and all of matter into the same theoretical tapestry. In short, superstring theory is a prime candidate for Einstein's unified theory.

These are grand claims and, if correct, represent a monumental step forward. But the most stunning feature of superstring theory, one that I have little doubt would have set Einstein's heart aflutter, is its profound impact on our understanding of the fabric of the cosmos. As we will see, superstring theory's proposed fusion of general relativity and quantum mechanics is mathematically sensible only if we subject our conception of spacetime to yet another upheaval. Instead of the three spatial dimensions and one time dimension of common experience, superstring theory requires *nine* spatial dimensions and one time dimension. And, in a more robust incarnation of superstring theory known as *M-theory*, unification requires *ten* space dimensions and one time dimension—a cosmic substrate composed of a total of eleven spacetime dimensions. As we don't see these extra dimensions, superstring theory is telling us that *we've so far glimpsed but a meager slice of reality.*

Of course, the lack of observational evidence for extra dimensions might also mean they don't exist and that superstring theory is wrong. However, drawing that conclusion would be extremely hasty. Even decades before superstring theory's discovery, visionary scientists, including Einstein, pondered the idea of spatial dimensions beyond the ones we see, and suggested possibilities for where they might be hiding. String theorists have substantially refined these ideas and have found that extra dimensions might be so tightly crumpled that they're too small for us or any of our existing equipment to see (Chapter 12), or they might be large but invisible to the ways we probe the universe (Chapter 13). Either scenario comes with profound implications. Through their impact on string vibrations, the geometrical shapes of tiny crumpled dimensions might hold answers to some of the most basic questions, like why our universe has stars and planets. And the room provided by large extra space dimensions might allow for something even more remarkable: other, nearby worlds—not nearby in ordinary space, but nearby in the extra dimensions—of which we've so far been completely unaware.

Although a bold idea, the existence of extra dimensions is not just theoretical pie in the sky. It may shortly be testable. If they exist, extra dimensions may lead to spectacular results with the next generation of atom smashers, like the first human synthesis of a microscopic black hole, or the production of a huge variety of new, never before discovered species of particles (Chapter 13). These and other exotic results may provide the first evidence for dimensions beyond those directly visible, taking us one step closer to establishing superstring theory as the long-sought unified theory.

If superstring theory is proven correct, we will be forced to accept that the reality we have known is but a delicate chiffon draped over a thick and richly textured cosmic fabric. Camus' declaration notwithstanding, determining the number of space dimensions—and, in particular, finding that there aren't just three—would provide far more than a scientifically interesting but ultimately inconsequential detail. The discovery of extra dimensions would show that the entirety of human experience had left us completely unaware of a basic and essential aspect of the universe. It would forcefully argue that even those features of the cosmos that we have thought to be readily accessible to human senses need not be.

Past and Future Reality

With the development of superstring theory, researchers are optimistic that we finally have a framework that will not break down under any conditions, no matter how extreme, allowing us one day to peer back with our equations and learn what things were like at the very moment when the universe as we know it got started. To date, no one has gained sufficient dexterity with the theory to apply it unequivocally to the big bang, but understanding cosmology according to superstring theory has become one of the highest priorities of current research. Over the past few years, vigorous worldwide research programs in superstring cosmology have yielded novel cosmological frameworks (Chapter 13), suggested new ways to test superstring theory using astrophysical observations (Chapter 14), and provided some of the first insights into the role the theory may play in explaining time's arrow.

The arrow of time, through the defining role it plays in everyday life and its intimate link with the origin of the universe, lies at a singular threshold between the reality we experience and the more refined reality cutting-edge science seeks to uncover. As such, the question of time's arrow provides a common thread that runs through many of the developments we'll discuss, and it will surface repeatedly in the chapters that follow. This is fitting. Of the many factors that shape the lives we lead, time is among the most dominant. As we continue to gain facility with superstring theory and its extension, M-theory, our cosmological insights will deepen, bringing both time's origin and its arrow into ever-sharper focus. If we let our imaginations run wild, we can even envision that the depth of our understanding will one day allow us to navigate spacetime and hence explore realms that, to this point in our experience, remain well beyond our ability to access (Chapter 15).

Of course, it is extremely unlikely that we will ever achieve such power. But even if we never gain the ability to control space and time, deep understanding yields its own empowerment. Our grasp of the true nature of space and time would be a testament to the capacity of the human intellect. We would finally come to know space and time—the silent, ever-present markers delineating the outermost boundaries of human experience.

Coming of Age in Space and Time

When I turned the last page of *The Myth of Sisyphus* many years ago, I was surprised by the text's having achieved an overarching feeling of optimism. After all, a man condemned to pushing a rock up a hill with full knowledge that it will roll back down, requiring him to start pushing anew, is not the sort of story that you'd expect to have a happy ending. Yet Camus found much hope in the ability of Sisyphus to exert free will, to press on against insurmountable obstacles, and to assert his choice to survive even when condemned to an absurd task within an indifferent universe. By relinquishing everything beyond immediate experience, and ceasing to search for any kind of deeper understanding or deeper meaning, Sisyphus, Camus argued, triumphs.

I was struck by Camus' ability to discern hope where most others would see only despair. But as a teenager, and only more so in the decades since, I found that I couldn't embrace Camus' assertion that a deeper understanding of the universe would fail to make life more rich or worthwhile. Whereas Sisyphus was Camus' hero, the greatest of scientists—Newton, Einstein, Niels Bohr, and Richard Feynman—became mine. And when I read Feynman's description of a rose—in which he explained how he could experience the fragrance and beauty of the flower as fully as anyone, but how his knowledge of physics enriched the experience enormously because he could also take in the wonder and magnificence of the underlying molecular, atomic, and subatomic processes—I was hooked for good. I wanted what Feynman described: to assess life and to experience the universe on all possible levels, not just those that happened to be accessible to our frail human senses. The search for the deepest understanding of the cosmos became my lifeblood.

As a professional physicist, I have long since realized that there was much naïveté in my high school infatuation with physics. Physicists generally do not spend their working days contemplating flowers in a state of cosmic awe. Instead, we devote much of our time to grappling with complex mathematical equations scrawled across well-scored chalkboards. Progress can be slow. Promising ideas, more often than not, lead nowhere. That's the nature of scientific research. Yet, even during periods of minimal progress, I've found that the effort spent puzzling and calculating has only made me feel a closer connection to the cosmos. I've found that you can come to know the universe not only by resolving its mysteries, but also by immersing yourself within them. Answers are great. Answers confirmed by experiment are greater still. But even answers that are ultimately proven wrong represent the result of a deep engagement with the cosmos—an engagement that sheds intense illumination on the questions, and hence on the universe itself. Even when the rock associated with a particular scientific exploration happens to roll back to square one, we nevertheless learn something and our experience of the cosmos is enriched.

Of course, the history of science reveals that the rock of our collective scientific inquiry—with contributions from innumerable scientists across the continents and through the centuries—does not roll down the mountain. Unlike Sisyphus, we don't begin from scratch. Each generation takes over from the previous, pays homage to its predecessors' hard work, insight, and creativity, and pushes up a little further. New theories and more refined measurements are the mark of scientific progress, and such progress builds on what came before, almost never wiping the slate clean. Because this is the case, our task is far from absurd or pointless. In pushing the rock up the mountain, we undertake the most exquisite and noble of tasks: to unveil this place we call home, to revel in the wonders we discover, and to hand off our knowledge to those who follow.

For a species that, by cosmic time scales, has only just learned to walk upright, the challenges are staggering. Yet, over the last three hundred years, as we've progressed from classical to relativistic and then to quantum reality, and have now moved on to explorations of unified reality, our minds and instruments have swept across the grand expanse of space and time, bringing us closer than ever to a world that has proved a deft master of disguise. And as we've continued to slowly unmask the cosmos, we've gained the intimacy that comes only from closing in on the clarity of truth. The explorations have far to go, but to many it feels as though our species is finally reaching childhood's end.

To be sure, our coming of age here on the outskirts of the Milky Way[6] has been a long time in the making. In one way or another, we've been exploring our world and contemplating the cosmos for thousands of years. But for most of that time we made only brief forays into the unknown, each time returning home somewhat wiser but largely unchanged. It took the brashness of a Newton to plant the flag of modern scientific inquiry and never turn back. We've been heading higher ever since. And all our travels began with a simple question.

What is space?

2

The Universe
and the Bucket

IS SPACE A HUMAN ABSTRACTION OR A PHYSICAL ENTITY?

It's not often that a bucket of water is the central character in a three-hundred-year-long debate. But a bucket that belonged to Sir Isaac Newton is no ordinary bucket, and a little experiment he described in 1689 has deeply influenced some of the world's greatest physicists ever since. The experiment is this: Take a bucket filled with water, hang it by a rope, twist the rope tightly so that it's ready to unwind, and let it go. At first, the bucket starts to spin but the water inside remains fairly stationary; the surface of the stationary water stays nice and flat. As the bucket picks up speed, little by little its motion is communicated to the water by friction, and the water starts to spin too. As it does, the water's surface takes on a concave shape, higher at the rim and lower in the center, as in Figure 2.1.

That's the experiment—not quite something that gets the heart racing. But a little thought will show that this bucket of spinning water is extremely puzzling. And coming to grips with it, as we have not yet done in over three centuries, ranks among the most important steps toward grasping the structure of the universe. Understanding why will take some background, but it is well worth the effort.

Figure 2.1 The surface of the water starts out flat and remains so as the bucket starts to spin. Subsequently, as the water also starts to spin, its surface becomes concave, and it remains concave while the water spins, even as the bucket slows and stops.

Relativity Before Einstein

"Relativity" is a word we associate with Einstein, but the concept goes much further back. Galileo, Newton, and many others were well aware that *velocity*—the speed and direction of an object's motion—is relative. In modern terms, from the batter's point of view, a well-pitched fastball might be approaching at 100 miles per hour. From the baseball's point of view, it's the *batter* who is approaching at 100 miles per hour. Both descriptions are accurate; it's just a matter of perspective. Motion has meaning only in a relational sense: An object's velocity can be specified only in relation to that of another object. You've probably experienced this. When the train you are on is next to another and you see relative motion, you can't immediately tell which train is actually moving on the tracks. Galileo described this effect using the transport of his day, boats. Drop a coin on a smoothly sailing ship, Galileo said, and it will hit your foot just as it would on dry land. From your perspective, you are justified in declaring that you are stationary and it's the water that is rushing by the ship's hull. And since from this point of view you are not moving, the coin's motion relative to your foot will be exactly what it would have been before you embarked.

Of course, there are circumstances under which your motion seems intrinsic, when you can feel it and you seem able to declare, without recourse to external comparisons, that you are definitely moving. This is the case with *accelerated* motion, motion in which your speed and/or your direction changes. If the boat you are on suddenly lurches one way or another, or slows down or speeds up, or changes direction by rounding a bend, or gets caught in a whirlpool and spins around and around, you know that you are moving. And you realize this without looking out and comparing your motion with some chosen point of reference. Even if your eyes are closed, you know you're moving, because you feel it. Thus, while you can't feel motion with constant speed that heads in an unchanging straight-line trajectory—*constant velocity motion*, it's called—you can feel *changes* to your velocity.

But if you think about it for a moment, there is something odd about this. What is it about changes in velocity that allows them to stand alone, to have intrinsic meaning? If velocity is something that makes sense only by comparisons—by saying that *this* is moving with respect to *that*—how is it that changes in velocity are somehow different, and don't also require comparisons to give them meaning? In fact, could it be that they actually *do* require a comparison to be made? Could it be that there is some implicit or hidden comparison that is actually at work every time we refer to or experience accelerated motion? This is a central question we're heading toward because, perhaps surprisingly, it touches on the deepest issues surrounding the meaning of space and time.

Galileo's insights about motion, most notably his assertion that the earth itself moves, brought upon him the wrath of the Inquisition. A more cautious Descartes, in his *Principia Philosophiae*, sought to avoid a similar fate and couched his understanding of motion in an equivocating framework that could not stand up to the close scrutiny Newton gave it some thirty years later. Descartes spoke about objects' having a resistance to changes to their state of motion: something that is motionless will stay motionless unless someone or something forces it to move; something that is moving in a straight line at constant speed will maintain that motion until someone or something forces it to change. But what, Newton asked, do these notions of "motionless" or "straight line at constant speed" really mean? Motionless or constant speed with respect to what? Motionless or constant speed from whose viewpoint? If velocity is not constant, with respect to what or from whose viewpoint is it not constant? Descartes correctly teased out aspects of motion's meaning, but Newton realized that he left key questions unanswered.

Newton—a man so driven by the pursuit of truth that he once shoved a blunt needle between his eye and the socket bone to study ocular anatomy and, later in life as Master of the Mint, meted out the harshest of punishments to counterfeiters, sending more than a hundred to the gallows—had no tolerance for false or incomplete reasoning. So he decided to set the record straight. This led him to introduce the bucket.[1]

The Bucket

When we left the bucket, both it and the water within were spinning, with the water's surface forming a concave shape. The issue Newton raised is, *Why* does the water's surface take this shape? Well, because it's spinning, you say, and just as we feel pressed against the side of a car when it takes a sharp turn, the water gets pressed against the side of the bucket as it spins. And the only place for the pressed water to go is upward. This reasoning is sound, as far as it goes, but it misses the real intent of Newton's question. He wanted to know what it *means* to say that the water is spinning: spinning with respect to what? Newton was grappling with the very foundation of motion and was far from ready to accept that accelerated motion such as spinning—is somehow beyond the need for external comparisons.*

A natural suggestion is to use the bucket itself as the object of reference. As Newton argued, however, this fails. You see, at first when we let the bucket start to spin, there is definitely *relative* motion between the bucket and the water, because the water does not immediately move. Even so, the surface of the water stays flat. Then, a little later, when the water is spinning and there *isn't* relative motion between the bucket and the water, the surface of the water *is* concave. So, with the bucket as our object of reference, we get exactly the opposite of what we expect: when there is relative motion, the water's surface is flat; and when there is no relative motion, the surface is concave.

In fact, we can take Newton's bucket experiment one small step further. As the bucket continues to spin, the rope will twist again (in the other direction), causing the bucket to slow down and momentarily come to rest, while the water inside continues to spin. At this point, the relative

---

*The terms *centrifugal* and *centripetal* force are sometimes used when describing spinning motion. But they are merely labels. Our intent is to understand why spinning motion gives rise to force.

motion between the water and the bucket is the *same* as it was near the very beginning of the experiment (except for the inconsequential difference of clockwise vs. counterclockwise motion), but the shape of the water's surface is *different* (previously being flat, now being concave); this shows conclusively that the relative motion cannot explain the surface's shape.

Having ruled out the bucket as a relevant reference for the motion of the water, Newton boldly took the next step. Imagine, he suggested, another version of the spinning bucket experiment carried out in deep, cold, completely empty space. We can't run exactly the same experiment, since the shape of the water's surface depended in part on the pull of earth's gravity, and in this version the earth is absent. So, to create a more workable example, let's imagine we have a huge bucket—one as large as any amusement park ride—that is floating in the darkness of empty space, and imagine that a fearless astronaut, Homer, is strapped to the bucket's interior wall. (Newton didn't actually use this example; he suggested using two rocks tied together by a rope, but the point is the same.) The telltale sign that the bucket is spinning, the analog of the water being pushed outward yielding a concave surface, is that Homer will *feel* pressed against the inside of the bucket, his facial skin pulling taut, his stomach slightly compressing, and his hair (both strands) straining back toward the bucket wall. Here is the question: in *totally* empty space—no sun, no earth, no air, no doughnuts, no anything—what could possibly serve as the "something" with respect to which the bucket is spinning? At first, since we are imagining space is completely empty except for the bucket and its contents, it looks as if there simply isn't anything else to serve as the something. Newton disagreed.

He answered by fixing on the ultimate container as the relevant frame of reference: *space itself.* He proposed that the transparent, empty arena in which we are all immersed and within which all motion takes place exists as a real, physical entity, which he called *absolute space.*[2] We can't grab or clutch absolute space, we can't taste or smell or hear absolute space, but nevertheless Newton declared that absolute space is a something. It's the something, he proposed, that provides the truest reference for describing motion. An object is truly at rest when it is at rest with respect to absolute space. An object is truly moving when it is moving with respect to absolute space. And, most important, Newton concluded, an object is truly accelerating when it is accelerating with respect to absolute space.

Newton used this proposal to explain the terrestrial bucket experiment in the following way. At the beginning of the experiment, the bucket is spinning with respect to absolute space, but the water is stationary with respect to absolute space. That's why the water's surface is flat. As the water catches up with the bucket, it is now spinning with respect to absolute space, and that's why its surface becomes concave. As the bucket slows because of the tightening rope, the water continues to spin—spinning with respect to absolute space—and that's why its surface continues to be concave. And so, whereas relative motion between the water and the bucket cannot account for the observations, relative motion between the water and absolute space can. Space itself provides the true frame of reference for defining motion.

The bucket is but an example; the reasoning is of course far more general. According to Newton's perspective, when you round the bend in a car, you feel the change in your velocity because you are accelerating with respect to absolute space. When the plane you are on is gearing up for takeoff, you feel pressed back in your seat because you are accelerating with respect to absolute space. When you spin around on ice skates, you feel your arms being flung outward because you are accelerating with respect to absolute space. By contrast, if someone were able to spin the entire ice arena while you stood still (assuming the idealized situation of frictionless skates)—giving rise to the same relative motion between you and the ice—you would not feel your arms flung outward, because you would not be accelerating with respect to absolute space. And, just to make sure you don't get sidetracked by the irrelevant details of examples that use the human body, when Newton's two rocks tied together by a rope twirl around in empty space, the rope pulls taut because the rocks are accelerating with respect to absolute space. Absolute space has the final word on what it means to move.

But what is absolute space, really? In dealing with this question, Newton responded with a bit of fancy footwork and the force of fiat. He first wrote in the *Principia* "I do not define time, space, place, and motion, as [they] are well known to all,"[3] sidestepping any attempt to describe these concepts with rigor or precision. His next words have become famous: "Absolute space, in its own nature, without reference to anything external, remains always similar and unmovable." That is, absolute space just is, and is forever. Period. But there are glimmers that Newton was not completely comfortable with simply declaring the existence and importance of something that you can't directly see, measure, or affect. He wrote,

It is indeed a matter of great difficulty to discover and effectually
to distinguish the true motions of particular bodies from the
apparent, because the parts of that immovable space in which
those motions are performed do by no means come under the
observations of our senses.[4]

So Newton leaves us in a somewhat awkward position. He puts
absolute space front and center in the description of the most basic and
essential element of physics—motion—but he leaves its definition vague
and acknowledges his own discomfort about placing such an important
egg in such an elusive basket. Many others have shared this discomfort.

Space Jam

Einstein once said that if someone uses words like "red," "hard," or "disappointed," we all basically know what is meant. But as for the word
"space," "whose relation with psychological experience is less direct, there
exists a far-reaching uncertainty of interpretation."[5] This uncertainty
reaches far back: the struggle to come to grips with the meaning of space
is an ancient one. Democritus, Epicurus, Lucretius, Pythagoras, Plato,
Aristotle, and many of their followers through the ages wrestled in one
way or another with the meaning of "space." Is there a difference between
space and matter? Does space have an existence independent of the presence of material objects? Is there such a thing as empty space? Are space
and matter mutually exclusive? Is space finite or infinite?

For millennia, the philosophical parsings of space often arose in tandem with theological inquiries. God, according to some, is omnipresent,
an idea that gives space a divine character. This line of reasoning was
advanced by Henry More, a seventeenth-century theologian/philosopher
who, some think, may have been one of Newton's mentors.[6] He believed
that if space were empty it would not exist, but he also argued that this is
an irrelevant observation because, even when devoid of material objects,
space is filled with spirit, so it is *never* truly empty. Newton himself took
on a version of this idea, allowing space to be filled by "spiritual substance" as well as material substance, but he was careful to add that such
spiritual stuff "can be no obstacle to the motion of matter; no more than if
nothing were in its way."[7] Absolute space, Newton declared, is the senso-
rium of God.

Such philosophical and religious musings on space can be compelling and provocative, yet, as in Einstein's cautionary remark above, they lack a critical sharpness of description. But there *is* a fundamental and precisely framed question that emerges from such discourse: should we ascribe an independent reality to space, as we do for other, more ordinary material objects like the book you are now holding, or should we think of space as merely a language for describing relationships between ordinary material objects?

The great German philosopher Gottfried Wilhelm von Leibniz, who was Newton's contemporary, firmly believed that space does not exist in any conventional sense. Talk of space, he claimed, is nothing more than an easy and convenient way of encoding where things are relative to one another. Without the objects *in* space, Leibniz declared, space itself has no independent meaning or existence. Think of the English alphabet. It provides an order for twenty-six letters—it provides relations such as *a* is next to *b*, *d* is six letters before *j*, *x* is three letters after *u*, and so on. But without the letters, the alphabet has no meaning—it has no "supra-letter," independent existence. Instead, the alphabet comes into being with the letters whose lexicographic relations it supplies. Leibniz claimed that the same is true for space: Space has no meaning beyond providing the natural language for discussing the relationship between one object's location and another. According to Leibniz, if all objects were removed from space—if space were completely empty—it would be as meaningless as an alphabet that's missing its letters.

Leibniz put forward a number of arguments in support of this so-called *relationist* position. For example, he argued that if space really exists as an entity, as a background substance, God would have had to choose where in this substance to place the universe. But how could God, whose decisions all have sound justification and are never random or haphazard, have possibly distinguished one location in the uniform void of empty space from another, as they are all alike? To the scientifically receptive ear, this argument sounds tinny. However, if we remove the theological element, as Leibniz himself did in other arguments he put forward, we are left with thorny issues: What is the location of the universe within space? If the universe were to move as a whole—leaving all relative positions of material objects intact—ten feet to the left or right, how would we know? What is the speed of the entire universe through the substance of space? If we are fundamentally unable to detect space, or changes within space, how can we claim it actually exists?

It is here that Newton stepped in with his bucket and dramatically changed the character of the debate. While Newton agreed that certain features of absolute space seem difficult or perhaps impossible to detect directly, he argued that the existence of absolute space does have consequences that are observable: accelerations, such as those at play in the rotating bucket, are accelerations with respect to absolute space. Thus, the concave shape of the water, according to Newton, is a consequence of the existence of absolute space. And Newton argued that once one has any solid evidence for something's existence, no matter how indirect, that ends the discussion. In one clever stroke, Newton shifted the debate about space from philosophical ponderings to scientifically verifiable data. The effect was palpable. In due course, Leibniz was forced to admit, "I grant there is a difference between absolute true motion of a body and a mere relative change of its situation with respect to another body."[8] This was not a capitulation to Newton's absolute space, but it was a strong blow to the firm relationist position.

During the next two hundred years, the arguments of Leibniz and others against assigning space an independent reality generated hardly an echo in the scientific community.[9] Instead, the pendulum had clearly swung to Newton's view of space; his laws of motion, founded on his concept of absolute space, took center stage. Certainly, the success of these laws in describing observations was the essential reason for their acceptance. It's striking to note, however, that Newton himself viewed all of his achievements in physics as merely forming the solid foundation to support what he considered his really important discovery: absolute space. For Newton, it was all about space.[10]

Mach and the Meaning of Space

When I was growing up, I used to play a game with my father as we walked down the streets of Manhattan. One of us would look around, secretly fix on something that was happening—a bus rushing by, a pigeon landing on a windowsill, a man accidentally dropping a coin—and describe how it would look from an unusual perspective such as the wheel of the bus, the pigeon in flight, or the quarter falling earthward. The challenge was to take an unfamiliar description like "I'm walking on a dark, cylindrical surface surrounded by low, textured walls, and an unruly bunch of thick white tendrils is descending from the sky," and figure out that it was the view of an ant walking on a hot dog that a street vendor was garnishing with sauerkraut. Although we stopped playing years before I took my first physics course, the game is at least partly to blame for my having a fair amount of distress when I encountered Newton's laws.

The game encouraged seeing the world from different vantage points and emphasized that each was as valid as any other. But according to Newton, while you are certainly free to contemplate the world from any perspective you choose, the different vantage points are by no means on an equal footing. From the viewpoint of an ant on an ice skater's boot, it is the ice and the arena that are spinning; from the viewpoint of a spectator in the stands, it is the ice skater that is spinning. The two vantage points seem to be equally valid, they seem to be on an equal footing, they seem to stand in the symmetric relationship of each spinning with respect to the other. Yet, according to Newton, one of these perspectives is more right than the other since if it *really* is the ice skater that is spinning, his or her arms will splay outward, whereas if it *really* is the arena that is spinning, his or her arms will not. Accepting Newton's absolute space meant accepting an absolute conception of acceleration, and, in particular, accepting an absolute answer regarding who or what is really spinning. I struggled to understand how this could possibly be true. Every source I consulted—textbooks and teachers alike—agreed that only relative motion had relevance when considering constant velocity motion, so why in the world, I endlessly puzzled, would accelerated motion be so different? Why wouldn't *relative* acceleration, like relative velocity, be the only thing that's relevant when considering motion at velocity that isn't constant? The existence of absolute space decreed otherwise, but to me this seemed thoroughly peculiar.

Much later I learned that over the last few hundred years many physicists and philosophers—sometimes loudly, sometimes quietly—had struggled with the very same issue. Although Newton's bucket seemed to show definitively that absolute space is what selects one perspective over another (if someone or something is spinning with respect to absolute space then they are *really* spinning; otherwise they are not), this resolution left many people who mull over these issues unsatisfied. Beyond the intuitive sense that no perspective should be "more right" than any other, and beyond the eminently reasonable proposal of Leibniz that only relative motion between material objects has meaning, the concept of absolute space left many wondering how absolute space can allow us to identify true accelerated motion, as with the bucket, while it cannot provide a way to identify true constant velocity motion. After all, if absolute space really exists, it should provide a benchmark for *all* motion, not just accelerated motion. If absolute space really exists, why doesn't it provide a way of identifying where we are located in an absolute sense, one that need not use our position relative to other material objects as a reference point? And, if absolute space really exists, how come it can affect us (causing our arms to splay if we spin, for example) while we apparently have no way to affect it?

In the centuries since Newton's work, these questions were sometimes debated, but it wasn't until the mid-1800s, when the Austrian physicist and philosopher Ernst Mach came on the scene, that a bold, prescient, and extremely influential new view about space was suggested—a view that, among other things, would in due course have a deep impact on Albert Einstein.

To understand Mach's insight—or, more precisely, one modern reading of ideas often attributed to Mach*—let's go back to the bucket for a moment. There is something odd about Newton's argument. The bucket experiment challenges us to explain why the surface of the water is flat in one situation and concave in another. In hunting for explanations, we examined the two situations and realized that the key difference between them was whether or not the water was spinning. Unsurprisingly, we tried to explain the shape of the water's surface by appealing to its state of motion. But here's the thing: before introducing absolute space, Newton focused solely on the bucket as the possible reference for determining the motion of the water and, as we saw, that approach fails. There are other references, however, that we could naturally use to gauge the water's motion, such as the laboratory in which the experiment takes place—its floor, ceiling, and walls. Or if we happened to perform the experiment on a sunny day in an open field, the surrounding buildings or trees, or the ground under our feet, would provide the "stationary" reference to determine whether the water was spinning. And if we happened to perform this experiment while floating in outer space, we would invoke the distant stars as our stationary reference.

---

*There is debate concerning Mach's precise views on the material that follows. Some of his writings are a bit ambiguous and some of the ideas attributed to him arose from subsequent interpretations of his work. Since he seems to have been aware of these interpretations and never offered corrections, some have suggested that he agreed with their conclusions. But historical accuracy might be better served if every time I write "Mach argued" or "Mach's ideas," you read it to mean "the prevailing interpretation of an approach initiated by Mach."

This leads to the following question. Might Newton have kicked the bucket aside with such ease that he skipped too quickly over the relative motion we are apt to invoke in real life, such as between the water and the laboratory, or the water and the earth, or the water and the fixed stars in the sky? Might it be that such relative motion *can* account for the shape of the water's surface, eliminating the need to introduce the concept of absolute space? That was the line of questioning raised by Mach in the 1870s.

To understand Mach's point more fully, imagine you're floating in outer space, feeling calm, motionless, and weightless. You look out and you can see the distant stars, and they too appear to be perfectly stationary. (It's a real Zen moment.) Just then, someone floats by, grabs hold of you, and sets you spinning around. You will notice two things. First, your arms and legs will feel pulled from your body and if you let them go they will splay outward. Second, as you gaze out toward the stars, they will no longer appear stationary. Instead, they will seem to be spinning in great circular arcs across the distant heavens. Your experience thus reveals a close association between feeling a force on your body and witnessing motion with respect to the distant stars. Hold this in mind as we try the experiment again but in a different environment.

Imagine now that you are immersed in the blackness of *completely* empty space: no stars, no galaxies, no planets, no air, nothing but total blackness. (A real existential moment.) This time, if you start spinning, will you feel it? Will your arms and legs feel pulled outward? Our experiences in day-to-day life lead us to answer yes: any time we change from not spinning (a state in which we feel nothing) to spinning, we feel the difference as our appendages are pulled outward. But the current example is unlike anything any of us has ever experienced. In the universe as we know it, there are always other material objects, either nearby or, at the very least, far away (such as the distant stars), that can serve as a reference for our various states of motion. In this example, however, there is absolutely no way for you to distinguish "not spinning" from "spinning" by comparisons with other material objects; there *aren't* any other material objects. Mach took this observation to heart and extended it one giant step further. He suggested that in this case there might also be no way to *feel* a difference between various states of spinning. More precisely, Mach argued that in an otherwise empty universe there is *no distinction* between spinning and not spinning—there is no conception of motion or acceleration if there are no benchmarks for comparison—and so spinning and not spinning are the same. If Newton's two rocks tied together by a rope were set spinning in an otherwise empty universe, Mach reasoned that the rope would remain slack. If you spun around in an otherwise empty universe, your arms and legs would not splay outward, and the fluid in your ears would be unaffected; you'd feel nothing.

This is a deep and subtle suggestion. To really absorb it, you need to put yourself into the example earnestly and fully imagine the black, uniform stillness of totally empty space. It's not like a dark room in which you feel the floor under your feet or in which your eyes slowly adjust to the tiny amount of light seeping in from outside the door or window; instead, we are imagining that there are *no* things, so there is no floor and there is absolutely no light to adjust to. Regardless of where you reach or look, you feel and see absolutely nothing at all. You are engulfed in a cocoon of unvarying blackness, with no material benchmarks for comparison. And without such benchmarks, Mach argued, the very concepts of motion and acceleration cease to have meaning. It's not just that you won't feel anything if you spin; it's more basic. In an otherwise empty universe, standing perfectly motionless and spinning uniformly are indistinguishable.*

Newton, of course, would have disagreed. He claimed that even completely empty space still has *space*. And, although space is not tangible or directly graspable, Newton argued that it still provides a something with respect to which material objects can be said to move. But remember how Newton came to this conclusion: He pondered rotating motion and *assumed* that the results familiar from the laboratory (the water's surface becomes concave; Homer feels pressed against the bucket wall; your arms splay outward when you spin around; the rope tied between two spinning rocks becomes taut) would hold true if the experiment were carried out in empty space. This assumption led him to search for something in empty space relative to which the motion could be defined, and the something he came up with was space itself. Mach strongly challenged the key

---

*While I like human examples because they make an immediate connection between the physics we're discussing and innate sensations, a drawback is our ability to move, volitionally, one part of our body relative to another—in effect, to use one part of our body as the benchmark for another part's motion (like someone who spins one of his arms relative to his head). I emphasize *uniform* spinning motion—spinning motion in which every part of the body spins together—to avoid such irrelevant complications. So, when I talk about your body's spinning, imagine that, like Newton's two rocks tied by a rope or a skater in the final moments of an Olympic routine, every part of your body spins at the same rate as every other.

assumption: He argued that what happens in the laboratory is not what would happen in completely empty space.

Mach's was the first significant challenge to Newton's work in more than two centuries, and for years it sent shock waves through the physics community (and beyond: in 1909, while living in London, Vladimir Lenin wrote a philosophical pamphlet that, among other things, discussed aspects of Mach's work[11]). But if Mach was right and there was no notion of spinning in an otherwise empty universe—a state of affairs that would eliminate Newton's justification for absolute space—that still leaves the problem of explaining the terrestrial bucket experiment, in which the water certainly does take on a concave shape. Without invoking absolute space—if absolute space is not a something—how would Mach explain the water's shape? The answer emerges from thinking about a simple objection to Mach's reasoning.

Mach, Motion, and the Stars

Imagine a universe that is not completely empty, as Mach envisioned, but, instead, one that has just a handful of stars sprinkled across the sky. If you perform the outer-space-spinning experiment now, the stars—even if they appear as mere pinpricks of light coming from enormous distance—provide a means of gauging your state of motion. If you start to spin, the distant pinpoints of light will appear to circle around you. And since the stars provide a visual reference that allows you to distinguish spinning from not spinning, you would expect to be able to feel it, too. But how can a few distant stars make such a difference, their presence or absence somehow acting as a switch that turns on or off the sensation of spinning (or more generally, the sensation of accelerated motion)? If you can feel spinning motion in a universe with merely a few distant stars, perhaps that means Mach's idea is just wrong—perhaps, as assumed by Newton, in an empty universe you *would* still feel the sensation of spinning.

Mach offered an answer to this objection. In an empty universe, according to Mach, you feel nothing if you spin (more precisely, there is not even a concept of spinning vs. nonspinning). At the other end of the spectrum, in a universe populated by all the stars and other material objects existing in our real universe, the splaying force on your arms and legs is what you experience when you actually spin. (Try it.) And—here is the point—in a universe that is not empty but that has less matter than ours, Mach suggested that the force you would feel from spinning would lie between nothing and what you would feel in our universe. That is, the force you feel is proportional to the amount of matter in the universe. In a universe with a single star, you would feel a minuscule force on your body if you started spinning. With two stars, the force would get a bit stronger, and so on and so on, until you got to a universe with the material content of our own, in which you feel the full familiar force of spinning. In this approach, the force you feel from acceleration arises as a collective effect, a collective influence of all the other matter in the universe.

Again, the proposal holds for all kinds of accelerated motion, not just spinning. When the airplane you are on is accelerating down the runway, when the car you are in screeches to a halt, when the elevator you are in starts to climb, Mach's ideas imply that the force you feel represents the combined influence of all the other matter making up the universe. If there were more matter, you would feel greater force. If there were less matter, you would feel less force. And if there were no matter, you wouldn't feel anything at all. So, in Mach's way of thinking, only relative motion and relative acceleration matter. *You feel acceleration only when you accelerate relative to the average distribution of other material inhabiting the cosmos.* Without other material—without any benchmarks for comparison—Mach claimed there would be no way to experience acceleration.

For many physicists, this is one of the most seductive proposals about the cosmos put forward during the last century and a half. Generations of physicists have found it deeply unsettling to imagine that the untouchable, ungraspable, unclutchable fabric of space is really a something—a something substantial enough to provide the ultimate, absolute benchmark for motion. To many it has seemed absurd, or at least scientifically irresponsible, to base an understanding of motion on something so thoroughly imperceptible, so completely beyond our senses, that it borders on the mystical. Yet these same physicists were dogged by the question of how else to explain Newton's bucket. Mach's insights generated excitement because they raised the possibility of a new answer, one in which space is not a something, an answer that points back toward the relationist conception of space advocated by Leibniz. Space, in Mach's view, is very much as Leibniz imagined—it's the language for expressing the relationship between one object's position and another's. But, like an alphabet without letters, space does not enjoy an independent existence.

Mach vs. Newton

I learned of Mach's ideas when I was an undergraduate, and they were a godsend. Here, finally, was a theory of space and motion that put all perspectives back on an equal footing, since only relative motion and relative acceleration had meaning. Rather than the Newtonian benchmark for motion—an invisible thing called absolute space—Mach's proposed benchmark is out in the open for all to see—the matter that is distributed throughout the cosmos. I felt sure Mach's had to be the answer. I also learned that I was not alone in having this reaction; I was following a long line of physicists, including Albert Einstein, who had been swept away when they first encountered Mach's ideas.

Is Mach right? Did Newton get so caught up in the swirl of his bucket that he came to a wishy-washy conclusion regarding space? Does Newton's absolute space exist, or had the pendulum firmly swung back to the relationist perspective? During the first few decades after Mach introduced his ideas, these questions couldn't be answered. For the most part, the reason was that Mach's suggestion was not a complete theory or description, since he never specified *how* the matter content of the universe would exert the proposed influence. If his ideas were right, how do the distant stars and the house next door contribute to your feeling that you are spinning when you spin around? Without specifying a physical mechanism to realize his proposal, it was hard to investigate Mach's ideas with any precision.

From our modern vantage point, a reasonable guess is that gravity might have something to do with the influences involved in Mach's suggestion. In the following decades, this possibility caught Einstein's attention and he drew much inspiration from Mach's proposal while developing his own theory of gravity, the general theory of relativity. When the dust of relativity had finally settled, the question of whether space is a something—of whether the absolutist or relationist view of space is correct—was transformed in a manner that shattered all previous ways of looking at the universe.

3

Relativity and the Absolute

Some discoveries provide answers to questions. Other discoveries are so deep that they cast questions in a whole new light, showing that previous mysteries were misperceived through lack of knowledge. You could spend a lifetime—in antiquity, some did—wondering what happens when you reach earth's edge, or trying to figure out who or what lives on earth's underbelly. But when you learn that the earth is round, you see that the previous mysteries are not solved; instead, they're rendered irrelevant.

During the first decades of the twentieth century, Albert Einstein made two deep discoveries. Each caused a radical upheaval in our understanding of space and time. Einstein dismantled the rigid, absolute structures that Newton had erected, and built his own tower, synthesizing space and time in a manner that was completely unanticipated. When he was done, time had become so enmeshed with space that the reality of one could no longer be pondered separately from the other. And so, by the third decade of the twentieth century the question of the corporeality of space was outmoded; its Einsteinian reframing, as we'll talk about shortly, became: Is *spacetime* a something? With that seemingly slight modification, our understanding of reality's arena was transformed.

Is Empty Space Empty?

Light was the primary actor in the relativity drama written by Einstein in the early years of the twentieth century. And it was the work of James Clerk Maxwell that set the stage for Einstein's insights. In the mid-1800s, Maxwell discovered four powerful equations that, for the first time, set out a rigorous theoretical framework for understanding electricity, magnetism, and their intimate relationship.[1] Maxwell developed these equations by carefully studying the work of the English physicist Michael Faraday, who in the early 1800s had carried out tens of thousands of experiments that exposed hitherto unknown features of electricity and magnetism. Faraday's key breakthrough was the concept of the *field*. Later expanded on by Maxwell and many others, this concept has had an enormous influence on the development of physics during the last two centuries, and underlies many of the little mysteries we encounter in everyday life. When you go through airport security, how is it that a machine that doesn't touch you can determine whether you're carrying metallic objects? When you have an MRI, how is it that a device that remains outside your body can take a detailed picture of your insides? When you look at a compass, how is it that the needle swings around and points north even though nothing seems to nudge it? The familiar answer to the last question invokes the earth's magnetic field, and the concept of magnetic fields helps to explain the previous two examples as well.

I've never seen a better way to get a visceral sense of a magnetic field than the elementary school demonstration in which iron filings are sprinkled in the vicinity of a bar magnet. After a little shaking, the iron filings align themselves in an orderly pattern of arcs that begin at the magnet's north pole and swing up and around, to end at the magnet's south pole, as in Figure 3.1. The pattern traced by the iron filings is direct evidence that the magnet creates an invisible something that permeates the space around it—a something that can, for example, exert a force on shards of metal. The invisible something is the *magnetic field* and, to our intuition, it resembles a mist or essence that can fill a region of space and thereby exert a force beyond the physical extent of the magnet itself. A magnetic field provides a magnet what an army provides a dictator and what auditors provide the IRS: influence beyond their physical boundaries, which allows force to be exerted out in the "field." That is why a magnetic field is also called a force field.

Figure 3.1 Iron filings sprinkled near a bar magnet trace out its magnetic field.

It is the pervasive, space-filling capability of magnetic fields that makes them so useful. An airport metal detector's magnetic field seeps through your clothes and causes metallic objects to give off their own magnetic fields—fields that then exert an influence back on the detector, causing its alarm to sound. An MRI's magnetic field seeps into your body, causing particular atoms to gyrate in just the right way to generate their own magnetic fields—fields that the machine can detect and decode into a picture of internal tissues. The earth's magnetic field seeps through the compass casing and turns the needle, causing it to point along an arc that, as a result of eons-long geophysical processes, is aligned in a nearly south–north direction.

Magnetic fields are one familiar kind of field, but Faraday also analyzed another: the *electric field*. This is the field that causes your wool scarf to crackle, zaps your hand in a carpeted room when you touch a metal doorknob, and makes your skin tingle when you're up in the mountains during a powerful lightning storm. And if you happened to examine a compass during such a storm, the way its magnetic needle deflected this way and that as the bolts of electric lightning flashed nearby would have given you a hint of a deep interconnection between electric and magnetic fields—something first discovered by the Danish physicist Hans Oersted and investigated thoroughly by Faraday through painstaking experimentation. Just as developments in the stock market can affect the bond market which can then affect the stock market, and so on, these scientists found that changes in an electric field can produce changes in a nearby magnetic field, which can then cause changes in the electric field, and so on. Maxwell found the mathematical underpinnings of these interrelationships, and because his equations showed that electric and magnetic fields are as entwined as the fibers in a Rastafarian's dreadlocks, they were eventually christened *electromagnetic* fields, and the influence they exert the *electromagnetic* force.

Today, we are constantly immersed in a sea of electromagnetic fields. Your cellular telephone and car radio work over enormous expanses because the electromagnetic fields broadcast by telephone companies and radio stations suffuse impressively wide regions of space. The same goes for wireless Internet connections; computers can pluck the entire World Wide Web from electromagnetic fields that are vibrating all around us—in fact, right through us. Of course, in Maxwell's day, electromagnetic technology was less developed, but among scientists his feat was no less recognized: through the language of fields, Maxwell had shown that electricity and magnetism, although initially viewed as distinct, are really just different aspects of a single physical entity.

Later on, we'll encounter other kinds of fields—gravitational fields, nuclear fields, Higgs fields, and so on—and it will become increasingly clear that the field concept is central to our modern formulation of physical law. But for now the critical next step in our story is also due to Maxwell. Upon further analyzing his equations, he found that changes or disturbances to electromagnetic fields travel in a wavelike manner at a particular speed: 670 million miles per hour. As this is precisely the value other experiments had found for the speed of light, Maxwell realized that light must be nothing other than an electromagnetic wave, one that has the right properties to interact with chemicals in our retinas and give us the sensation of sight. This achievement made Maxwell's already towering discoveries all the more remarkable: he had linked the force produced by magnets, the influence exerted by electrical charges, and the light we use to see the universe—but it also raised a deep question.

When we say that the speed of light is 670 million miles per hour, experience, and our discussion so far, teach us this is a meaningless statement if we don't specify relative to *what* this speed is being measured. The funny thing was that Maxwell's equations just gave this number, 670 million miles per hour, without specifying or apparently relying on any such reference. It was as if someone gave the location for a party as 22 miles north without specifying the reference location, without specifying north of *what*. Most physicists, including Maxwell, attempted to explain the speed his equations gave in the following way: Familiar waves such as ocean waves or sound waves are carried by a substance, a medium. Ocean waves are carried by water. Sound waves are carried by air. And the speeds of these waves are specified *with respect to the medium*. When we talk about the speed of sound at room temperature being 767 miles per hour (also known as Mach 1, after the same Ernst Mach encountered earlier), we mean that sound waves travel through otherwise still air at this speed. Naturally, then, physicists surmised that light waves — electromagnetic waves — must also travel through some particular medium, one that had never been seen or detected but that must exist. To give this unseen light-carrying stuff due respect, it was given a name: the *luminiferous aether*, or the *aether* for short, the latter being an ancient term that Aristotle used to describe the magical catchall substance of which heavenly bodies were imagined to be made. And, to square this proposal with Maxwell's results, it was suggested that his equations implicitly took the perspective of some-one at rest with respect to the aether. The 670 million miles per hour his equations came up with, then, was the speed of light relative to the stationary aether.

As you can see, there is a striking similarity between the luminiferous aether and Newton's absolute space. They both originated in attempts to provide a reference for defining motion; accelerated motion led to absolute space, light's motion led to the luminiferous aether. In fact, many physicists viewed the aether as a down-to-earth stand-in for the divine spirit that Henry More, Newton, and others had envisioned permeating absolute space. (Newton and others in his age had even used the term "aether" in their descriptions of absolute space.) But what actually *is* the aether? What is it made of? Where did it come from? Does it exist everywhere?

These questions about the aether are the same ones that for centuries had been asked about absolute space. But whereas the full Machian test for absolute space involved spinning around in a completely empty universe, physicists were able to propose doable experiments to determine whether the aether really existed. For example, if you swim through water toward an oncoming water wave, the wave approaches you more quickly; if you swim away from the wave, it approaches you more slowly. Similarly, if you move through the supposed aether toward or away from an oncom-ing light wave, the light wave's approach should, by the same reasoning, be faster or slower than 670 million miles per hour. In 1887, however, when Albert Michelson and Edward Morley measured the speed of light, time and time again they found exactly the same speed of 670 million miles per hour *regardless of their motion or that of the light's source*. All sorts of clever arguments were devised to explain these results. Maybe, some suggested, the experimenters were unwittingly dragging the aether along with them as they moved. Maybe, a few ventured, the equipment was being warped as it moved through the aether, corrupting the measurements. But it was not until Einstein had his revolutionary insight that the explanation finally became clear.

Relative Space, Relative Time

In June 1905, Einstein wrote a paper with the unassuming title "On the Electrodynamics of Moving Bodies," which once and for all spelled the end of the luminiferous aether. In one stroke, it also changed forever our understanding of space and time. Einstein formulated the ideas in the paper over an intense five-week period in April and May 1905, but the issues it finally laid to rest had been gnawing at him for over a decade. As a teenager, Einstein struggled with the question of what a light wave would look like if you were to chase after it at exactly light speed. Since you and the light wave would be zipping through the aether at exactly the same speed, you would be keeping perfect pace with the light. And so, Einstein concluded, from your perspective the light should appear as though it wasn't moving. You should be able to reach out and grab a handful of motionless light just as you can scoop up a handful of newly fallen snow.

But here's the problem. It turns out that Maxwell's equations do not allow light to appear stationary—to look as if it's standing still. And certainly, there is no reliable report of anyone's ever actually catching hold of a stationary clump of light. So, the teenage Einstein asked, what are we to make of this apparent paradox?

Ten years later, Einstein gave the world his answer with his special theory of relativity. There has been much debate regarding the intellectual roots of Einstein's discovery, but there is no doubt that his unshakable belief in simplicity played a critical role. Einstein was aware of at least some experiments that had failed to detect evidence for the existence of the aether.[2] So why dance around trying to find fault with the experiments? Instead, Einstein declared, take the simple approach: The experiments were failing to find the aether because there is no aether. And since Maxwell's equations describing the motion of light—the motion of electromagnetic waves—do not invoke any such medium, both experiment and theory would converge on the same conclusion: light, unlike any other kind of wave ever encountered, does not need a medium to carry it along. Light is a lone traveler. Light can travel through empty space.

But what, then, are we to make of Maxwell's equation giving light a speed of 670 million miles per hour? If there is no aether to provide the standard of rest, what is the *what* with respect to which this speed is to be interpreted? Again, Einstein bucked convention and answered with ultimate simplicity. If Maxwell's theory does not invoke any particular standard of rest, the most direct interpretation is that we don't need one. *The speed of light,* Einstein declared, *is 670 million miles per hour relative to anything and everything.*

Well, this is certainly a simple statement; it fit well a maxim often attributed to Einstein: "Make everything as simple as possible, but no simpler." The problem is that it also seems crazy. If you run after a departing beam of light, common sense dictates that from your perspective the speed of the departing light has to be less than 670 million miles per hour. If you run toward an approaching beam of light, common sense dictates that from your perspective the speed of the approaching light will be greater than 670 million miles per hour. Throughout his life, Einstein challenged common sense, and this time was no exception. He forcefully argued that regardless of how fast you move toward or away from a beam of light, you will always measure its speed to be 670 million miles per hour—not a bit faster, not a bit slower, no matter what. This would certainly solve the paradox that stumped him as a teenager: Maxwell's theory does not allow for stationary light because light *never* is stationary; regardless of your state of motion, whether you chase a light beam, or run from it, or just stand still, the light retains its one fixed and never changing speed of 670 million miles per hour. But, we naturally ask, how can light possibly behave in such a strange manner?

Think about speed for a moment. Speed is measured by how far something goes divided by how long it takes to get there. It is a measure of space (the distance traveled) divided by a measure of time (the duration of the journey). Ever since Newton, space had been thought of as absolute, as being out there, as existing "without reference to anything external." Measurements of space and spatial separations must therefore also be absolute: regardless of who measures the distance between two things in space, if the measurements are done with adequate care, the answers will always agree. And although we have not yet discussed it directly, Newton declared the same to be true of time. His description of time in the *Prin-*

cipia echoes the language he used for space: "Time exists in and of itself and flows equably without reference to anything external." In other words, according to Newton, there is a universal, absolute conception of time that applies everywhere and everywhen. In a Newtonian universe, regardless of who measures how much time it takes for something to happen, if the measurements are done accurately, the answers will always agree.

These assumptions about space and time comport with our daily experiences and for that reason are the basis of our commonsense conclusion that light should appear to travel more slowly if we run after it. To see this, imagine that Bart, who's just received a new nuclear-powered skateboard, decides to take on the ultimate challenge and race a beam of light. Although he is a bit disappointed to see that the skateboard's top speed is only 500 million miles per hour, he is determined to give it his best shot. His sister Lisa stands ready with a laser; she counts down from 11 (her hero Schopenhauer's favorite number) and when she reaches 0, Bart and the laser light streak off into the distance. What does Lisa see? Well, for every hour that passes, Lisa sees the light travel 670 million miles while Bart travels only 500 million miles, so Lisa rightly concludes that the light is speeding away from Bart at 170 million miles per hour. Now let's bring Newton into the story. His ideas dictate that Lisa's observations about space and time are absolute and universal in the sense that anyone else performing these measurements would get the same answers. To Newton, such facts about motion through space and time were as objective as two plus two equaling four. According to Newton, then, Bart will agree with Lisa and will report that the light beam was speeding away from him at 170 million miles per hour.

But when Bart returns, he doesn't agree at all. Instead, he dejectedly claims that no matter what he did—no matter how much he pushed the skateboard's limit—he saw the light speed away at 670 million miles per hour, not a bit less.[3] And if for some reason you don't trust Bart, bear in mind that thousands of meticulous experiments carried out during the last hundred years, which have measured the speed of light using moving sources and receivers, support his observations with precision.

How can this be?

Einstein figured it out, and the answer he found is a logical yet profound extension of our discussion so far. It must be that Bart's measurements of distances and durations, the input that he uses to figure out how fast the light is receding from him, are different from Lisa's measurements. Think about it. Since speed is nothing but distance divided by time, there is no other way for Bart to have found a different answer from Lisa's for how fast the light was outrunning him. So, Einstein concluded, Newton's ideas of absolute space and absolute time were wrong. Einstein realized that experimenters who are moving relative to each other, like Bart and Lisa, will not find identical values for measurements of distances and durations. The puzzling experimental data on the speed of light can be explained only if their perceptions of space and time are different.

Subtle but Not Malicious

The relativity of space and of time is a startling conclusion. I have known about it for more than twenty-five years, but even so, whenever I quietly sit and think it through, I am amazed. From the well-worn statement that the speed of light is constant, we conclude that *space and time are in the eye of the beholder*. Each of us carries our own clock, our own monitor of the passage of time. Each clock is equally precise, yet when we move relative to one another, these clocks do not agree. They fall out of synchroniza-tion; they measure different amounts of elapsed time between two chosen events. The same is true of distance. Each of us carries our own yardstick, our own monitor of distance in space. Each yardstick is equally precise, yet when we move relative to one another, these yardsticks do not agree; they measure different distances between the locations of two specified events. If space and time did not behave this way, the speed of light would not be constant and would depend on the observer's state of motion. But it *is* constant; space and time *do* behave this way. Space and time adjust themselves in an exactly compensating manner so that observations of light's speed yield the same result, regardless of the observer's velocity.

Getting the quantitative details of precisely how the measurements of space and time differ is more involved, but requires only high school alge-bra. It is not the depth of mathematics that makes Einstein's special rela-tivity challenging. It is the degree to which the ideas are foreign and apparently inconsistent with our everyday experiences. But once Einstein had the key insight—the realization that he needed to break with the more than two-hundred-year-old Newtonian perspective on space and time—it was not hard to fill in the details. He was able to show precisely how one person's measurements of distances and durations must differ from those of another in order to ensure that each measures an identical value for the speed of light.[4]

To get a fuller sense of what Einstein found, imagine that Bart, with heavy heart, has carried out the mandatory retrofitting of his skateboard, which now has a maximum speed of 65 miles per hour. If he heads due north at top speed—reading, whistling, yawning, and occasionally glancing at the road—and then merges onto a highway pointing in a northeasterly direction, his speed in the northward direction will be *less* than 65 miles per hour. The reason is clear. Initially, all his speed was devoted to northward motion, but when he shifted direction some of that speed was diverted into eastward motion, leaving a little less for heading north. This extremely simple idea actually allows us to capture the core insight of special relativity. Here's how:

We are used to the fact that objects can move through space, but there is another kind of motion that is equally important: objects also move through time. Right now, the watch on your wrist and the clock on the wall are ticking away, showing that you and everything around you are relentlessly moving through time, relentlessly moving from one second to the next and the next. Newton thought that motion through time was totally separate from motion through space—he thought these two kinds of motion had nothing to do with each other. But Einstein found that they are intimately linked. In fact, *the* revolutionary discovery of special relativity is this: When you look at something like a parked car, which from your viewpoint is stationary—not moving through space, that is—*all* of its motion is through time. The car, its driver, the street, you, your clothes are all moving through time in perfect synch: second followed by second, ticking away uniformly. But if the car speeds away, some of its motion through time is *diverted* into motion through space. And just as Bart's speed in the northward direction slowed down when he diverted some of his northward motion into eastward motion, the speed of the car through *time* slows down when it diverts some of its motion through time into motion through *space*. This means that the car's progress through time slows down and therefore *time elapses more slowly for the moving car and its driver than it elapses for you and everything else that remains stationary.*

That, in a nutshell, is special relativity. In fact, we can be a bit more precise and take the description one step further. Because of the retrofitting, Bart had no choice but to limit his top speed to 65 miles per hour. This is important to the story, because if he sped up enough when he angled northeast, he could have compensated for the speed diversion and thereby maintained the same net speed toward the north. But with the retrofitting, no matter how hard he revved the skateboard's engine, his total speed—the combination of his speed toward the north and his speed toward the east—remained fixed at the maximum of 65 miles per hour. And so when he shifted his direction a bit toward the east, he necessarily caused a decreased northward speed.

Special relativity declares a similar law for all motion: *the combined speed of any object's motion through space and its motion through time is always precisely equal to the speed of light.* At first, you may instinctively recoil from this statement since we are all used to the idea that nothing but light can travel at light speed. *But that familiar idea refers solely to motion through space.* We are now talking about something related, yet richer: an object's combined motion through space and time. The key fact, Einstein discovered, is that these two kinds of motion are always complementary. When the parked car you were looking at speeds away, what really happens is that some of its light-speed motion is diverted from motion through time into motion through space, *keeping their combined total unchanged.* Such diversion unassailably means that the car's motion through time slows down.

As an example, if Lisa had been able to see Bart's watch as he sped along at 500 million miles per hour, she would have seen that it was ticking about two-thirds as fast as her own. For every three hours that passed on Lisa's watch, she would see that only two had passed on Bart's. His rapid motion through space would have proved a significant drain on his speed through time.

Moreover, the maximum speed through space is reached when all light-speed motion through time is fully diverted into light-speed motion through space—one way of understanding why it is impossible to go through space at greater than light speed. Light, which always travels at light speed through space, is special in that it always achieves such total diversion. And just as driving due east leaves no motion for traveling north, moving at light speed through space leaves no motion for traveling through time! Time stops when traveling at the speed of light through space. A watch worn by a particle of light would not tick at all. Light realizes the dreams of Ponce de León and the cosmetics industry: it doesn't age.[5]

As this description makes clear, the effects of special relativity are most pronounced when speeds (through space) are a significant fraction of light speed. But the unfamiliar, complementary nature of motion through space and time always applies. The lesser the speed, the smaller the deviation from prerelativity physics—from common sense, that is—but the deviation is still there, to be sure.

Truly. This is not dexterous wordplay, sleight of hand, or psychological illusion. This is how the universe works.

In 1971, Joseph Hafele and Richard Keating flew state-of-the-art cesium-beam atomic clocks around the world on a commercial Pan Am jet. When they compared the clocks flown on the plane with identical clocks left stationary on the ground, they found that less time had elapsed on the moving clocks. The difference was tiny—a few hundred billionths of a second—but it was precisely in accord with Einstein's discoveries. You can't get much more nuts-and-bolts than that.

In 1908, word began to spread that newer, more refined experiments were finding evidence for the aether.[6] If that had been so, it would have meant that there was an absolute standard of rest and that Einstein's special relativity was wrong. On hearing this rumor, Einstein replied, "Subtle is the Lord, malicious He is not." Peering deeply into the workings of nature to tease out insights into space and time was a profound challenge, one that had gotten the better of everyone until Einstein. But to allow such a startling and beautiful theory to exist, and yet to make it irrelevant to the workings of the universe, that would be malicious. Einstein would have none of it; he dismissed the new experiments. His confidence was well placed. The experiments were ultimately shown to be wrong, and the luminiferous aether evaporated from scientific discourse.

But What About the Bucket?

This is certainly a tidy story for light. Theory and experiment agree that light needs no medium to carry its waves and that regardless of the motion of either the source of light or the person observing, its speed is fixed and unchanging. Every vantage point is on an equal footing with every other. There is no absolute or preferred standard of rest. Great. But what about the bucket?

Remember, while many viewed the luminiferous aether as the physical substance giving credibility to Newton's absolute space, it had nothing to do with *why* Newton introduced absolute space. Instead, after wrangling with accelerated motion such as the spinning bucket, Newton saw no option but to invoke some invisible background stuff with respect to which motion could be unambiguously defined. Doing away with the

aether did not do away with the bucket, so how did Einstein and his special theory of relativity cope with the issue?

Well, truth be told, in special relativity, Einstein's main focus was on a
special kind of motion: constant-velocity motion. It was not until 1915,
some ten years later, that he fully came to grips with more general, accelerated motion, through his general theory of relativity. Even so, Einstein
and others repeatedly considered the question of rotating motion using
the insights of special relativity; they concluded, like Newton and unlike
Mach, that even in an otherwise completely empty universe you would
feel the outward pull from spinning—Homer would feel pressed against
the inner wall of a spinning bucket; the rope between the two twirling
rocks would pull taut.[7] Having dismantled Newton's absolute space and
absolute time, how did Einstein explain this?

The answer is surprising. Its name notwithstanding, Einstein's theory
does not proclaim that everything is relative. Special relativity does claim
that *some* things are relative: velocities are relative; distances across space
are relative; durations of elapsed time are relative. But the theory actually
introduces a grand, new, sweepingly absolute concept: *absolute spacetime.* Absolute spacetime is as absolute for special relativity as absolute
space and absolute time were for Newton, and partly for this reason Einstein did not suggest or particularly like the name "relativity theory."
Instead, he and other physicists suggested *invariance theory,* stressing that
the theory, at its core, involves something that everyone agrees on, something that is *not* relative.[8]

Absolute spacetime is the vital next chapter in the story of the bucket,
because, even if devoid of all material benchmarks for defining motion,
the absolute spacetime of special relativity provides a something with
respect to which objects can be said to accelerate.

Carving Space and Time

To see this, imagine that Marge and Lisa, seeking some quality together-
time, enroll in a Burns Institute extension course on urban renewal. For
their first assignment, they are asked to redesign the street and avenue layout of Springfield, subject to two requirements: first, the street/avenue
grid must be configured so that the Soaring Nuclear Monument is
located right at the grid's center, at 5th Street and 5th Avenue, and, second, the designs must use streets 100 meters long, and avenues, which run

perpendicular to streets, that are also 100 meters long. Just before class, Marge and Lisa compare their designs and realize that something is terribly wrong. After appropriately configuring her grid so that the Monument lies in the center, Marge finds that Kwik-E-Mart is at 8th Street and 5th Avenue and the nuclear power plant is at 3rd Street and 5th Avenue, as shown in Figure 3.2a. But in Lisa's design, the addresses are completely different: the Kwik-E-Mart is near the corner of 7th Street and 3rd Avenue, while the power plant is at 4th Street and 7th Avenue, as in Figure 3.2b. Clearly, someone has made a mistake.

After a moment's thought, though, Lisa realizes what's going on. There are no mistakes. She and Marge are both right. They merely chose different orientations for their street and avenue grids. Marge's streets and avenues run at an angle relative to Lisa's; their grids are rotated relative to each other; they have sliced up Springfield into streets and avenues in two different ways (see Figure 3.2c). The lesson here is simple, yet important. There is freedom in how Springfield—a region of space—can be organized by streets and avenues. There are no "absolute" streets or "absolute" avenues. Marge's choice is as valid as Lisa's—or any other possible orientation, for that matter.

Hold this idea in mind as we paint time into the picture. We are used to thinking about space as the arena of the universe, but physical processes occur in some region of space *during some interval of time*. As an example, imagine that Itchy and Scratchy are having a duel, as illustrated in Figure 3.3a, and the events are recorded moment by moment in

(a)   (b)

Figure 3.2 (a) Marge's street design. **(b)** Lisa's street design.

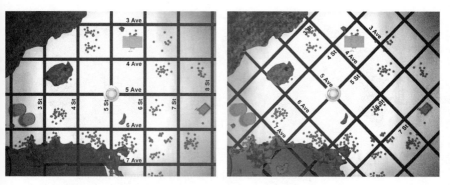

Figure 3.2 (c) Overview of Marge's and Lisa's street/avenue designs. Their grids differ by a rotation.

the fashion of one of those old-time flip books. Each page is a "time slice"—like a still frame in a filmstrip—that shows what happened in a region of space at one moment of time. To see what happened at a different moment of time you flip to a different page.* (Of course, space is three-dimensional while the pages are two-dimensional, but let's make this simplification for ease of thinking and drawing figures. It won't compromise any of our conclusions.) By way of terminology, a region of space considered over an interval of time is called a region of *spacetime*; you can think of a region of spacetime as a record of all things that happen in some region of space during a particular span of time.

Now, following the insight of Einstein's mathematics professor Hermann Minkowski (who once called his young student a lazy dog), consider the region of spacetime as an entity unto itself: consider the complete flip book as an object in its own right. To do so, imagine that, as in Figure 3.3b, we expand the binding of the flip-card book and then imagine that, as in Figure 3.3c, all the pages are completely transparent, so when you look at the book you see one continuous block containing all the events that happened during a given time interval. From this perspective, the pages should be thought of as simply providing a convenient way of organizing the content of the block—that is, of organizing the events of

*Like the pages in any flip book, the pages in Figure 3.3 only show representative moments of time. This may suggest to you the interesting question of whether time is discrete or infinitely divisible. We'll come back to that question later, but for now imagine that time is infinitely divisible, so our flip book really should have an infinite number of pages interpolating between those shown.

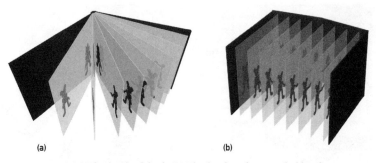

Figure 3.3 (a) Flip book of duel. **(b)** Flip book with expanded binding.

spacetime. Just as a street/avenue grid allows us to specify locations in a city easily, by giving their street and avenue address, the division of the spacetime block into pages allows us to easily specify an event (Itchy shooting his gun, Scratchy being hit, and so on) by giving the time when the event occurred—the page on which it appears—and the location within the region of space depicted on the pages.

Here is the key point: Just as Lisa realized that there are different, equally valid ways to slice up a region of space into streets and avenues,

Figure 3.3 (c) Block of spacetime containing the duel. Pages, or "time slices," organize the events in the block. The spaces between slices are for visual clarity only; they are not meant to suggest that time is discrete, a question we discuss later.

Einstein realized that there are different, equally valid ways to slice up a region of spacetime—a block like that in Figure 3.3c—into regions of space at moments of time. *The pages in Figures 3.3a, b, and c*—with, again, each page denoting one moment of time—*provide but one of the many possible slicings.* This may sound like only a minor extension of what we know intuitively about space, but it's the basis for overturning some of the most basic intuitions that we've held for thousands of years. Until 1905, it was thought that everyone experiences the passage of time identically, that everyone agrees on what events occur at a given moment of time, and hence, that everyone would concur on what belongs on a given page in the flip book of spacetime. But when Einstein realized that two observers in relative motion have clocks that tick off time differently, this all changed. Clocks that are moving relative to each other fall out of synchronization and therefore give different notions of simultaneity. Each page in Figure 3.3b is but one observer's view of the events in space taking place at a given moment of his or her time. Another observer, moving relative to the first, will declare that the events on a single one of these pages *do not* all happen at the same time.

This is known as the *relativity of simultaneity*, and we can see it directly. Imagine that Itchy and Scratchy, pistols in paws, are now facing each other on opposite ends of a long, moving railway car with one referee on the train and another officiating from the platform. To make the duel as fair as possible, all parties have agreed to forgo the three-step rule, and instead, the duelers will draw when a small pile of gunpowder, set midway between them, explodes. The first referee, Apu, lights the fuse, takes a sip of his refreshing Chutney Squishee, and steps back. The gunpowder flares, and both Itchy and Scratchy draw and fire. Since Itchy and Scratchy are the same distance from the gunpowder, Apu is certain that light from the flare reaches them simultaneously, so he raises the green flag and declares it a fair draw. But the second referee, Martin, who was watching from the platform, wildly squeals foul play, claiming that Itchy got the light signal from the explosion before Scratchy did. He explains that because the train was moving forward, Itchy was heading toward the light while Scratchy was moving away from it. This means that the light did not have to travel quite as far to reach Itchy, since he moved closer to it; moreover, the light had to travel farther to reach Scratchy, since he moved away from it. Since the speed of light, moving left or right from anyone's perspective, is constant, Martin claims that it took the light longer to reach Scratchy since it had to travel farther, rendering the duel unfair.

Who is right, Apu or Martin? Einstein's unexpected answer is that they both are. Although the conclusions of our two referees differ, the observations and the reasoning of each are flawless. Like the bat and the baseball, they simply have different perspectives on the same sequence of events. The shocking thing that Einstein revealed is that their different perspectives yield different but equally valid claims of what events happen at the same time. Of course, at everyday speeds like that of the train, the disparity is small—Martin claims that Scratchy got the light less than a trillionth of a second after Itchy—but were the train moving faster, near light speed, the time difference would be substantial.

Think about what this means for the flip-book pages slicing up a region of spacetime. Since observers moving relative to each other do not agree on what things happen simultaneously, the way each of them will slice a block of spacetime into pages—with each page containing all events that happen at a given moment from each observer's perspective—will not agree, either. Instead, observers moving relative to each other cut a block of spacetime up into pages, into time slices, in different but equally valid ways. What Lisa and Marge found for space, Einstein found for spacetime.

Angling the Slices

The analogy between street/avenue grids and time slicings can be taken even further. Just as Marge's and Lisa's designs differed by a rotation, Apu's and Martin's time slicings, their flip-book pages, also differ by a rotation, but one that involves both space and time. This is illustrated in Figures 3.4a and 3.4b, in which we see that Martin's slices are rotated relative to Apu's, leading him to conclude that the duel was unfair. A critical difference of detail, though, is that whereas the rotation angle between Marge's and Lisa's schemes was merely a design choice, the rotation angle between Apu's and Martin's slicings is determined by their relative speed. With minimal effort, we can see why.

Imagine that Itchy and Scratchy have reconciled. Instead of trying to shoot each other, they just want to ensure that clocks on the front and back of the train are perfectly synchronized. Since they are still equidistant from the gunpowder, they come up with the following plan. They agree to set their clocks to noon just as they see the light from the flaring gunpowder. From their perspective, the light has to travel the same dis-

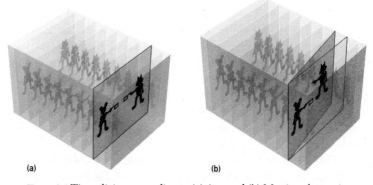

Figure 3.4 Time slicings according to (**a**) Apu and (**b**) Martin, who are in relative motion. Their slices differ by a rotation through space and time. According to Apu, who is on the train, the duel is fair; according to Martin, who is on the platform, it isn't. Both views are equally valid. In (**b**), the different angle of their slices through spacetime is emphasized.

tance to reach either of them, and since light's speed is constant, it will reach them simultaneously. But, by the same reasoning as before, Martin and anyone else viewing from the platform will say that Itchy is heading toward the emitted light while Scratchy is moving away from it, and so Itchy will receive the light signal a little before Scratchy does. Platform observers will therefore conclude that Itchy set his clock to 12:00 *before* Scratchy and will therefore claim that Itchy's clock is set a bit ahead of Scratchy's. For example, to a platform observer like Martin, when it's 12:06 on Itchy's clock, it may be only 12:04 on Scratchy's (the precise numbers depend on the length and the speed of the train; the longer and faster it is, the greater the discrepancy). Yet, from the viewpoint of Apu and everyone on the train, Itchy and Scratchy performed the synchronization perfectly. Again, although it's hard to accept at a gut level, there is no paradox here: *observers in relative motion do not agree on simultaneity—they do not agree on what things happen at the same time.*

This means that one page in the flip book as seen from the perspective of those on the train, a page containing events they consider simultaneous—such as Itchy's and Scratchy's setting their clocks—contains events that lie on *different* pages from the perspective of those observing from the platform (according to platform observers, Itchy set his clock *before* Scratchy, so these two events are on different pages from the platform observer's perspective). And there we have it. A single page from the perspective of those on the train contains events that lie on earlier and later pages of a platform observer. This is why Martin's and Apu's slices in Figure 3.4 are rotated relative to each other: what is a single time slice, from one perspective, cuts across many time slices, from the other perspective.

If Newton's conception of absolute space and absolute time were correct, everyone would agree on a single slicing of spacetime. Each slice would represent absolute space as viewed at a given moment of absolute time. This, however, is not how the world works, and the shift from rigid Newtonian time to the newfound Einsteinian flexibility inspires a shift in our metaphor. Rather than viewing spacetime as a rigid flip book, it will sometimes be useful to think of it as a huge, fresh loaf of bread. In place of the fixed pages that make up a book—the fixed Newtonian time slices— think of the variety of angles at which you can slice a loaf into parallel pieces of bread, as in Figure 3.5a. Each piece of bread represents space at one moment of time from one observer's perspective. But as illustrated in Figure 3.5b, another observer, moving relative to the first, will slice the spacetime loaf at a different angle. The greater the relative velocity of the two observers, the larger the angle between their respective parallel slices (as explained in the endnotes, the speed limit set by light translates into a maximum 45° rotation angle for these slicings[9]) and the greater the discrepancy between what the observers will report as having happened at the same moment.

The Bucket, According to Special Relativity

The relativity of time and space requires a dramatic change in our thinking. Yet there is an important point, mentioned earlier and illustrated now by the loaf of bread, which often gets lost: *not everything in relativity is relative.* Even if you and I were to imagine slicing up a loaf of bread in two different ways, there is still something that we would fully agree upon: the totality of the loaf itself. Although our slices would differ, if I were to imagine putting all of my slices together and you were to imagine doing the same for all of your slices, we would reconstitute the same loaf of bread. How could it be otherwise? We both imagined cutting up the same loaf.

Similarly, the totality of all the slices of space at successive moments

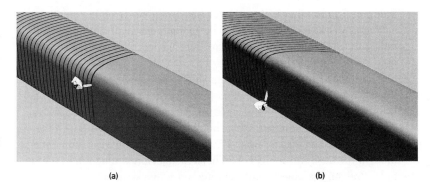

(a)   (b)

Figure 3.5 Just as one loaf of bread can be sliced at different angles, a block of spacetime is "time sliced" at different angles by observers in relative motion. The greater the relative speed, the greater the angle (with a maximum angle of 45° corresponding to the maximum speed set by light).

of time, from any single observer's perspective (see Figure 3.4), collectively yield the same region of spacetime. Different observers slice up a region of spacetime in different ways, but the region itself, like the loaf of bread, has an independent existence. Thus, although Newton definitely got it wrong, his intuition that there was something absolute, something that everyone would agree upon, was not fully debunked by special relativity. Absolute space does not exist. Absolute time does not exist. But according to special relativity, absolute spacetime does exist. With this observation, let's visit the bucket once again.

In an otherwise empty universe, with respect to *what* is the bucket spinning? According to Newton, the answer is absolute space. According to Mach, there is no sense in which the bucket can even be said to spin. According to Einstein's special relativity, the answer is absolute spacetime.

To understand this, let's look again at the proposed street and avenue layouts for Springfield. Remember that Marge and Lisa disagreed on the street and avenue address of the Kwik-E-Mart and the nuclear plant because their grids were rotated relative to each other. Even so, regardless of how each chose to lay out the grid, there are some things they definitely still agree on. For example, if in the interest of increasing worker efficiency during lunchtime, a trail is painted on the ground from the nuclear plant straight to the Kwik-E-Mart, Marge and Lisa will not agree on the streets and avenues through which the trail passes, as you can see in Figure 3.6. But they will certainly agree on the *shape* of the trail: they will agree that it is a straight line. The geometrical shape of the painted trail is independent of the particular street/avenue grid one happens to use.

Einstein realized that something similar holds for spacetime. Even though two observers in relative motion slice up spacetime in different ways, there are things they still agree on. As a prime example, consider a straight line not just through space, but through spacetime. Although the inclusion of time makes such a trajectory less familiar, a moment's thought reveals its meaning. For an object's trajectory through spacetime to be straight, the object must not only move in a straight line through space, but its motion must also be uniform through time; that is, both its speed and direction must be unchanging and hence it must be moving with constant velocity. Now, even though different observers slice up the spacetime loaf at different angles and thus will not agree on how much time has elapsed or how much distance is covered between various points on a trajectory, such observers will, like Marge and Lisa, still agree on whether a trajectory through spacetime is a straight line. Just as the geometrical shape of the painted trail to the Kwik-E-Mart is independent of the street/avenue slicing one uses, so the geometrical shapes of trajectories in spacetime are independent of the time slicing one uses.[10]

This is a simple yet critical realization, because with it special relativity provided an absolute criterion—one that all observers, regardless of their constant relative velocities, would agree on—for deciding whether or not something is accelerating. If the trajectory an object follows through spacetime is a straight line, like that of the gently resting astro-

Figure 3.6 Regardless of which street grid is used, everyone agrees on the shape of a trail, in this case, a straight line.

naut (a) in Figure 3.7, it is not accelerating. If the trajectory an object follows has any other shape but a straight line through spacetime, it *is* accelerating. For example, should the astronaut fire up her jetpack and fly around in a circle over and over again, like astronaut (b) in Figure 3.7, or should she zip out toward deep space at ever increasing speed, like astronaut (c), her trajectory through spacetime will be curved—the telltale sign of acceleration. And so, with these developments we learn that *geometrical shapes of trajectories in spacetime provide the absolute standard that determines whether something is accelerating.* Spacetime, not space alone, provides the benchmark.

In this sense, then, special relativity tells us that spacetime itself is the ultimate arbiter of accelerated motion. Spacetime provides the backdrop with respect to which something, like a spinning bucket, can be said to accelerate even in an otherwise empty universe. With this insight, the pendulum swung back again: from Leibniz the relationist to Newton the absolutist to Mach the relationist, and now back to Einstein, whose special relativity showed once again that the arena of reality—viewed as spacetime, not as space—*is* enough of a something to provide the ultimate benchmark for motion.[11]

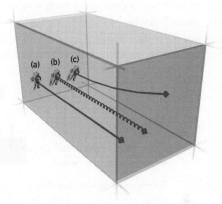

Figure 3.7 The paths through spacetime followed by three astronauts. Astronaut (**a**) does not accelerate and so follows a straight line through spacetime. Astronaut (**b**) flies repeatedly in a circle, and so follows a spiral through spacetime. Astronaut (**c**) accelerates into deep space, and so follows another curved trajectory in spacetime.

Gravity and the Age-old Question

At this point you might think we've reached the end of the bucket story, with Mach's ideas having been discredited and Einstein's radical updating of Newton's absolute conceptions of space and time having won the day. The truth, though, is more subtle and more interesting. But if you're new to the ideas we've covered so far, you may need a break before pressing on to the last sections of this chapter. In Table 3.1 you'll find a summary to refresh your memory when you've geared up to reengage.

Okay. If you're reading these words, I gather you're ready for the next major step in spacetime's story, a step catalyzed in large part by none other than Ernst Mach. Although special relativity, unlike Mach's theory, concludes that even in an otherwise empty universe you would feel pressed against the inside wall of a spinning bucket and that the rope tied between two twirling rocks would pull taut, Einstein remained deeply fascinated by Mach's ideas. He realized, however, that serious consideration of these ideas required significantly extending them. Mach never really specified a mechanism whereby distant stars and other matter in the universe might play a role in how strongly your arms splay outward when you spin or how forcefully you feel pressed against the inner wall of a spinning bucket. Einstein began to suspect that if there were such a mechanism it might have something to do with gravity.

This realization had a particular allure for Einstein because in special relativity, to keep the analysis tractable, he had completely ignored grav-

Newton	Space is an entity; accelerated motion is not relative; absolutist position.
Leibniz	Space is not an entity; all aspects of motion are relative; relationist position.
Mach	Space is not an entity; accelerated motion is relative to average mass distribution in the universe; relationist position.
Einstein (Special Relativity)	Space and time are individually relative; spacetime is an absolute entity.

Table 3.1 A summary of various positions on the nature of space and spacetime.

ity. Maybe, he speculated, a more robust theory, which embraced both special relativity and gravity, would come to a different conclusion regarding Mach's ideas. Maybe, he surmised, a generalization of special relativity that incorporated gravity would show that matter, both near and far, determines the force we feel when we accelerate.

Einstein also had a second, somewhat more pressing, reason for turning his attention to gravity. He realized that special relativity, with its central dictum that the speed of light is the fastest that anything or any disturbance can travel, was in direct conflict with Newton's universal law of gravity, the monumental achievement that had for over two hundred years predicted with fantastic precision the motion of the moon, the planets, comets, and all things tossed skyward. The experimental success of Newton's law notwithstanding, Einstein realized that according to Newton, gravity exerts its influence from place to place, from the sun to the earth, from the earth to the moon, from any-here to any-there, instantaneously, in no time at all, *much faster than light*. And that directly contradicted special relativity.

To illustrate the contradiction, imagine you've had a really disappointing evening (hometown ball club lost, no one remembered your birthday, someone ate the last chunk of Velveeta) and need a little time alone, so you take the family skiff out for some relaxing midnight boating. With the moon overhead, the water is at high tide (it's the moon's gravity pulling up on bodies of water that creates the tides), and beautiful moonlight reflections dance on its waving surface. But then, as if your night hadn't already been irritating enough, hostile aliens zap the moon and beam it clear across to the other side of the galaxy. Now, certainly, the moon's sudden disappearance would be odd, but if Newton's law of gravity was right, the episode would demonstrate something odder still. Newton's law predicts that the water would start to recede from high tide, because of the loss of the moon's gravitational pull, about a second and a half *before* you saw the moon disappear from the sky. *Like a sprinter jumping the gun, the water would seem to retreat a second and a half too soon.*

The reason is that, according to Newton, at the very moment the moon disappears its gravitational pull would *instantaneously* disappear too, and without the moon's gravity, the tides would immediately start to diminish. Yet, since it takes light a second and a half to travel the quarter million miles between the moon and the earth, you wouldn't immediately see that the moon had disappeared; for a second and a half, it would seem that the tides were receding from a moon that was still shining high overhead as usual. Thus, according to Newton's approach, gravity can affect us before light—gravity can outrun light—and this, Einstein felt certain, was wrong.[12]

And so, around 1907, Einstein became obsessed with the goal of formulating a new theory of gravity, one that would be at least as accurate as Newton's but would not conflict with the special theory of relativity. This turned out to be a challenge beyond all others. Einstein's formidable intellect had finally met its match. His notebook from this period is filled with half-formulated ideas, near misses in which small errors resulted in long wanderings down spurious paths, and exclamations that he had cracked the problem only to realize shortly afterward that he'd made another mistake. Finally, by 1915, Einstein emerged into the light. Although Einstein did have help at critical junctures, most notably from the mathematician Marcel Grossmann, the discovery of *general relativity* was the rare heroic struggle of a single mind to master the universe. The result is the crowning jewel of pre-quantum physics.

Einstein's journey toward general relativity began with a key question that Newton, rather sheepishly, had sidestepped two centuries earlier. How does gravity exert its influence over immense stretches of space? How does the vastly distant sun affect earth's motion? The sun doesn't touch the earth, so how does it do that? In short, how does gravity get the job done? Although Newton discovered an equation that described the effect of gravity with great accuracy, he fully recognized that he had left unanswered the important question of how gravity actually works. In his *Principia*, Newton wryly wrote, "I leave this problem to the consideration of the reader."[13] As you can see, there is a similarity between this problem and the one Faraday and Maxwell solved in the 1800s, using the idea of a magnetic field, regarding the way a magnet exerts influence on things that it doesn't literally touch. So you might suggest a similar answer: gravity exerts its influence by another field, the gravitational field. And, broadly speaking, this is the right suggestion. But realizing this answer in a manner that does not conflict with special relativity is easier said than done.

Much easier. It was this task to which Einstein boldly dedicated himself, and with the dazzling framework he developed after close to a decade of searching in the dark, Einstein overthrew Newton's revered theory of gravity. What is equally dazzling, the story comes full circle because Einstein's key breakthrough was tightly linked to the very issue Newton highlighted with the bucket: What is the true nature of accelerated motion?

The Equivalence of Gravity and Acceleration

In special relativity, Einstein's main focus was on observers who move with constant velocity—observers who feel no motion and hence are all justified in proclaiming that they are stationary and that the rest of the world moves by them. Itchy, Scratchy, and Apu on the train do not feel any motion. From their perspective, it's Martin and everyone else on the platform who are moving. Martin also feels no motion. To him, it's the train and its passengers that are in motion. Neither perspective is more correct than the other. But accelerated motion is different, because you *can* feel it. You feel squeezed back into a car seat as it accelerates forward, you feel pushed sideways as a train rounds a sharp bend, you feel pressed against the floor of an elevator that accelerates upward.

Nevertheless, the forces you'd feel struck Einstein as very familiar. As you approach a sharp bend, for example, your body tightens as you brace for the sideways push, because the impending force is inevitable. There is no way to shield yourself from its influence. The only way to avoid the force is to change your plans and not take the bend. This rang a loud bell for Einstein. He recognized that exactly the same features characterize the gravitational force. If you're standing on planet earth you are subject to planet earth's gravitational pull. It's inevitable. There is no way around it. While you can shield yourself from electromagnetic and nuclear forces, there is no way to shield yourself from gravity. And one day in 1907, Einstein realized that this was no mere analogy. In one of those flashes of insight that scientists spend a lifetime longing for, Einstein realized that gravity and accelerated motion are two sides of the same coin.

Just as by changing your planned motion (to avoid accelerating) you can avoid feeling squeezed back in your car seat or feeling pushed sideways on the train, Einstein understood that by suitably changing your motion you can also avoid feeling the usual sensations associated with gravity's pull. The idea is wonderfully simple. To understand it, imagine that Barney is desperately trying to win the Springfield Challenge, a monthlong competition among all belt-size-challenged males to see who can shed the greatest number of inches. But after two weeks on a liquid diet (Duff Beer), when he still has an obstructed view of the bathroom scale, he loses all hope. And so, in a fit of frustration, with the scale stuck to his feet, he leaps from the bathroom window. On his way down, just before plummeting into his neighbor's pool, Barney looks at the scale's reading and what does he see? Well, Einstein was the first person to realize, and realize fully, that Barney will see the scale's reading drop to zero. The scale falls at exactly the same rate as Barney does, so his feet don't press against it at all. *In free fall, Barney experiences the same weightlessness that astronauts experience in outer space.*

Indeed, if we imagine that Barney jumps out his window into a large shaft from which all air has been evacuated, then on his way down not only would air resistance be eliminated, but because every atom of his body would be falling at exactly the same rate, all the usual external bodily stresses and strains—his feet pushing up against his ankles, his legs pushing into his hips, his arms pulling down on his shoulders—would be eliminated as well.[14] By closing his eyes during the descent, Barney would feel exactly what he would if he were floating in the darkness of deep space. (And, again, in case you're happier with nonhuman examples: if you drop two rocks tied by a rope into the evacuated shaft, the rope will remain slack, just as it would if the rocks were floating in outer space.) Thus, by changing his state of motion—by fully "giving in to gravity"— Barney is able to simulate a gravity-free environment. (As a matter of fact, NASA trains astronauts for the gravity-free environment of outer space by having them ride in a modified 707 airplane, nicknamed the *Vomit Comet*, that periodically goes into a state of free fall toward earth.)

Similarly, by a suitable change in motion you can create a force that is essentially identical to gravity. For example, imagine that Barney joins astronauts floating weightless in their space capsule, with the bathroom scale still stuck to his feet and still reading zero. If the capsule should fire up its boosters and accelerate, things will change significantly. Barney will feel pressed to the capsule's floor, just as you feel pressed to the floor of an upward accelerating elevator. And since Barney's feet are now pressing against the scale, its reading is no longer zero. If the captain fires the boosters with just the right oomph, the reading on the scale will agree precisely with what Barney saw in the bathroom. Through appropriate acceleration, Barney is now experiencing a force that is indistinguishable from gravity.

The same is true of other kinds of accelerated motion. Should Barney join Homer in the outer space bucket, and, as the bucket spins, stand at a right angle to Homer—feet and scale against the inner bucket wall—the scale will register a nonzero reading since his feet will press against it. If the bucket spins at just the right rate, the scale will give the same reading Barney found earlier in the bathroom: the acceleration of the spinning bucket can also simulate earth's gravity.

All this led Einstein to conclude that the force one feels from gravity and the force one feels from acceleration are the same. They are equivalent. Einstein called this the *principle of equivalence*.

Take a look at what it means. Right now you feel gravity's influence. If you are standing, your feet feel the floor supporting your weight. If you are sitting, you feel the support somewhere else. And unless you are reading in a plane or a car, you probably also think that you are stationary—that you are not accelerating or even moving at all. But according to Einstein you actually are accelerating. Since you're sitting still this sounds a little silly, but don't forget to ask the usual question: Accelerating according to what benchmark? Accelerating from whose viewpoint?

With special relativity, Einstein proclaimed that absolute spacetime provides the benchmark, but special relativity does not take account of gravity. Then, through the equivalence principle, Einstein supplied a more robust benchmark that does include the effects of gravity. And this entailed a radical change in perspective. *Since gravity and acceleration are equivalent, if you feel gravity's influence, you must be accelerating.* Einstein argued that only those observers who feel no force at all—including the force of gravity—are justified in declaring that they are not accelerating. Such force-free observers provide the true reference points for discussing motion, and it's this recognition that requires a major turnabout in the way we usually think about such things. When Barney jumps from his window into the evacuated shaft, we would ordinarily describe him as accelerating down toward the earth's surface. But this is not a description Einstein would agree with. According to Einstein, Barney is *not* accelerating. *He* feels no force. *He* is weightless. *He* feels as he would floating in the deep darkness of empty space. *He* provides the standard against which all motion should be compared. And by this comparison, when you are calmly reading at home, *you* are accelerating. From Barney's perspective as he freely falls by your window—the perspective, according to Einstein, of a true benchmark for motion—you and the earth and all the other things we usually think of as stationary are *accelerating upward*. Einstein would argue that it was Newton's head that rushed up to meet the apple, not the other way around.

Clearly, this is a radically different way of thinking about motion. But it's anchored in the simple recognition that you feel gravity's influence only when you resist it. By contrast, when you fully give in to gravity you don't feel it. Assuming you are not subject to any other influences (such as air resistance), when you give in to gravity and allow yourself to fall freely, you feel as you would if you were freely floating in empty space—a perspective which, unhesitatingly, we consider to be unaccelerated.

In sum, only those individuals who are freely floating, regardless of whether they are in the depths of outer space or on a collision course with the earth's surface, are justified in claiming that they are experiencing no acceleration. If you pass by such an observer and there is relative acceleration between the two of you, then according to Einstein, *you* are accelerating.

As a matter of fact, notice that neither Itchy, nor Scratchy, nor Apu, nor Martin was truly justified in saying that he was stationary during the duel, since they all felt the downward pull of gravity. This has no bearing on our earlier discussion, because there, we were concerned only with horizontal motion, motion that was unaffected by the vertical gravity experienced by all participants. But as an important point of principle, the link Einstein found between gravity and acceleration means, once again, that we are justified only in considering stationary those observers who feel *no* forces whatsoever.

Having forged the link between gravity and acceleration, Einstein was now ready to take up Newton's challenge and seek an explanation of how gravity exerts its influence.

Warps, Curves, and Gravity

Through special relativity, Einstein showed that every observer cuts up spacetime into parallel slices that he or she considers to be all of space at successive instants of time, with the unexpected twist that observers moving relative to one another at constant velocity will cut through spacetime at different angles. If one such observer should start accelerating, you might guess that the moment-to-moment changes in his speed and/or direction of motion would result in moment-to-moment changes in the angle and orientation of his slices. Roughly speaking, this is what happens. Einstein (using geometrical insights articulated by Carl Friedrich Gauss, Georg Bernhard Riemann, and other mathematicians in the nineteenth century) developed this idea—by fits and starts—and showed that the differently angled cuts through the spacetime loaf smoothly merge into slices that are *curved* but fit together as perfectly as spoons in a silverware tray, as schematically illustrated in Figure 3.8. *An accelerated observer carves spatial slices that are warped.*

With this insight, Einstein was able to invoke the equivalence principle to profound effect. Since gravity and acceleration are equivalent, Einstein understood that gravity itself must be nothing but warps and curves in the fabric of spacetime. Let's see what this means.

If you roll a marble along a smooth wooden floor, it will travel in a straight line. But if you've recently had a terrible flood and the floor dried with all sorts of bumps and warps, a rolling marble will no longer travel along the same path. Instead, it will be guided this way and that by the warps and curves on the floor's surface. Einstein applied this simple idea to the fabric of the universe. He imagined that in the absence of matter or energy—no sun, no earth, no stars—spacetime, like the smooth wooden floor, has no warps or curves. It's flat. This is schematically illustrated in Figure 3.9a, in which we focus on one slice of space. Of course, space is really three dimensional, and so Figure 3.9b is a more accurate depiction, but drawings that illustrate two dimensions are easier to understand, so we'll continue to use them. Einstein then imagined that the presence of matter or energy has an effect on space much like the effect the flood had on the floor. Matter and energy, like the sun, cause space (and spacetime*) to warp and curve as illustrated in Figures 3.10a and 3.10b. And just as a marble rolling on the warped floor travels along a curved path, Einstein showed that anything moving through warped space—such as the earth moving in the vicinity of the sun—will travel along a curved trajectory, as illustrated in Figure 3.11a and Figure 3.11b.

It's as if matter and energy imprint a network of chutes and valleys along which objects are guided by the invisible hand of the spacetime fabric. That, according to Einstein, is how gravity exerts its influence. The same idea also applies closer to home. Right now, your body would like to slide down an indentation in the spacetime fabric caused by the earth's presence. But your motion is being blocked by the surface on which you're sitting or standing. The upward push you feel almost every moment of your life—be it from the ground, the floor of your house, the corner easy chair, or your kingsize bed—is acting to stop you from sliding

---

*It's easier to picture warped space, but because of their intimate connection, time is also warped by matter and energy. And just as a warp in space means that space is stretched or compressed, as in Figure 3.10, a warp in time means that time is stretched or compressed. That is, clocks experiencing different gravitational pulls—like one on the sun and another in deep, empty space—tick off time at different rates. In fact, it turns out that the warping of space caused by ordinary bodies like the earth and sun (as opposed to black holes) is far less pronounced than the warping they inflict on time.[15]

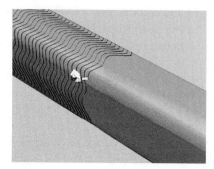

Figure 3.8 According to general relativity, not only will the spacetime loaf be sliced into space at moments of time at different angles (by observers in relative motion), but the slices themselves will be warped or curved by the presence of matter or energy.

down a valley in spacetime. By contrast, should you throw yourself off the high diving board, you are giving in to gravity by allowing your body to move freely along one of its spacetime chutes.

Figures 3.9, 3.10, and 3.11 schematically illustrate the triumph of Einstein's ten-year struggle. Much of his work during these years aimed at determining the precise shape and size of the warping that would be caused by a given amount of matter or energy. The mathematical result Einstein found underlies these figures and is embodied in what are called the *Einstein field equations*. As the name indicates, Einstein viewed the warping of spacetime as the manifestation—the geometrical embodiment—of a gravitational field. By framing the problem geometrically,

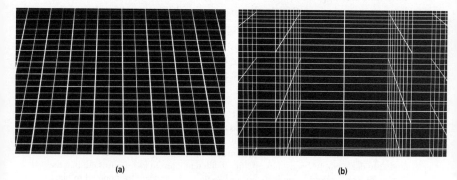

(a)   (b)

Figure 3.9 (a) Flat space (2-d version). **(b)** Flat space (3-d version).

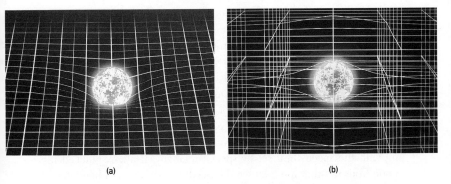

Figure 3.10 (a) The sun warping space (2-d version). (b) The sun warping space (3-d version).

Einstein was able to find equations that do for gravity what Maxwell's equations did for electromagnetism.[16] And by using these equations, Einstein and many others made predictions for the path that would be followed by this or that planet, or even by light emitted by a distant star, as it moves through curved spacetime. Not only have these predictions been confirmed to a high level of accuracy, but in head-to-head competition with the predictions of Newton's theory, Einstein's theory consistently matches reality with finer fidelity.

Of equal importance, since general relativity specifies the detailed mechanism by which gravity works, it provides a mathematical frame-

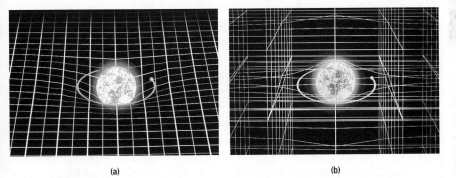

Figure 3.11 The earth stays in orbit around the sun because it follows curves in the spacetime fabric caused by the sun's presence. (a) 2-d version. (b) 3-d version.

work for determining how fast it transmits its influence. The speed of transmission comes down to the question of how fast the shape of space can change in time. That is, how quickly can warps and ripples—ripples like those on the surface of a pond caused by a plunging pebble—race from place to place through space? Einstein was able to work this out, and the answer he came to was enormously gratifying. He found that warps and ripples—gravity, that is—do not travel from place to place instantaneously, as they do in Newtonian calculations of gravity. Instead, *they travel at exactly the speed of light.* Not a bit faster or slower, fully in keeping with the speed limit set by special relativity. If aliens plucked the moon from its orbit, the tides would recede a second and a half later, at the exact same moment we'd see that the moon had vanished. Where Newton's theory failed, Einstein's general relativity prevailed.

General Relativity and the Bucket

Beyond giving the world a mathematically elegant, conceptually powerful, and, for the first time, fully consistent theory of gravity, the general theory of relativity also thoroughly reshaped our view of space and time. In both Newton's conception and that of special relativity, space and time provided an unchanging stage for the events of the universe. Even though the slicing of the cosmos into space at successive moments has a flexibility in special relativity unfathomable in Newton's age, space and time do not respond to happenings in the universe. Spacetime—the loaf, as we've been calling it—is taken as a given, once and for all. In general relativity, all this changes. Space and time become players in the evolving cosmos. They come alive. Matter here causes space to warp there, which causes matter over there to move, which causes space way over there to warp even more, and so on. General relativity provides the choreography for an entwined cosmic dance of space, time, matter, and energy.

This is a stunning development. But we now come back to our central theme: What about the bucket? Does general relativity provide the physical basis for Mach's relationist ideas, as Einstein hoped it would?

Over the years, this question has generated much controversy. Initially, Einstein thought that general relativity fully incorporated Mach's perspective, a viewpoint he considered so important that he christened it *Mach's principle.* In fact, in 1913, as Einstein was furiously working to put the final pieces of general relativity in place, he wrote Mach an enthusiastic letter in which he described how general relativity would confirm Mach's analysis of Newton's bucket experiment.[17] And in 1918, when Einstein wrote an article enumerating the three essential ideas behind general relativity, the third point in his list was Mach's principle. But general relativity is subtle and it had features that took many years for physicists, including Einstein himself, to appreciate completely. As these aspects were better understood, Einstein found it increasingly difficult to fully incorporate Mach's principle into general relativity. Little by little, he grew disillusioned with Mach's ideas and by the later years of his life came to renounce them.[18]

With an additional half century of research and hindsight, we can consider anew the extent to which general relativity conforms to Mach's reasoning. Although there is still some controversy, I think the most accurate statement is that in some respects general relativity has a distinctly Machian flavor, but it does not conform to the fully relationist perspective Mach advocated. Here's what I mean.

Mach argued[19] that when the spinning water's surface becomes concave, or when you feel your arms splay outward, or when the rope tied between the two rocks pulls taut, this has nothing to do with some hypothetical—and, in his view, thoroughly misguided—notion of absolute space (or absolute spacetime, in our more modern understanding). Instead, he argued that it's evidence of accelerated motion with respect to all the matter that's spread throughout the cosmos. Were there no matter, there'd be no notion of acceleration and none of the enumerated physical effects (concave water, splaying arms, rope pulling taut) would happen.

What does general relativity say?

According to general relativity, the benchmarks for all motion, and accelerated motion in particular, are freely falling observers—observers who have fully given in to gravity and are being acted on by no other forces. Now, a key point is that the gravitational force to which a freely falling observer acquiesces arises from all the matter (and energy) spread throughout the cosmos. The earth, the moon, the distant planets, stars, gas clouds, quasars, and galaxies all contribute to the gravitational field (in geometrical language, to the curvature of spacetime) right where you're now sitting. Things that are more massive and less distant exert a greater gravitational influence, but the gravitational field you feel represents the combined influence of the matter that's out there.[20] The path you'd take were you to give in to gravity fully and assume free-fall motion—the benchmark you'd become for judging whether some other object is accelerating—*would* be influenced by all matter in the cosmos, by the stars in the heavens and by the house next door. Thus, in general relativity, when an object is said to be accelerating, it means the object is accelerating with respect to a benchmark determined by matter spread throughout the universe. That's a conclusion which has the feel of what Mach advocated. So, in this sense, general relativity does incorporate some of Mach's thinking.

Nevertheless, general relativity does not confirm all of Mach's reasoning, as we can see directly by considering, once again, the spinning bucket in an otherwise empty universe. In an empty unchanging universe—no stars, no planets, no anything at all—there is no gravity.[21] And without gravity, spacetime is not warped—it takes the simple, uncurved shape shown in Figure 3.9b—and that means we are back in the simpler setting of special relativity. (Remember, Einstein ignored gravity while developing special relativity. General relativity made up for this deficiency by incorporating gravity, but when the universe is empty and unchanging there is no gravity, and so general relativity reduces to special relativity.) If we now introduce the bucket into this empty universe, it has such a tiny mass that its presence hardly affects the shape of space at all. And so the discussion we had earlier for the bucket in special relativity applies equally well to general relativity. In contradiction to what Mach would have predicted, general relativity comes to the same answer as special relativity, and proclaims that even in an otherwise empty universe, you *will* feel pressed against the inner wall of the spinning bucket; in an otherwise empty universe, your arms *will* feel pulled outward if you spin around; in an otherwise empty universe, the rope tied between two twirling rocks *will* become taut. The conclusion we draw is that even in general relativity, empty spacetime provides a benchmark for accelerated motion.

Hence, although general relativity incorporates some elements of Mach's thinking, it does not subscribe to the completely relative conception of motion Mach advocated.[22] Mach's principle is an example of a provocative idea that provided inspiration for a revolutionary discovery even though that discovery ultimately failed to fully embrace the idea that inspired it.

Spacetime in the Third Millennium

The spinning bucket has had a long run. From Newton's absolute space and absolute time, to Leibniz's and then Mach's relational conceptions, to Einstein's realization in special relativity that space and time are relative and yet in their union fill out absolute spacetime, to his subsequent discovery in general relativity that spacetime is a dynamic player in the unfolding cosmos, the bucket has always been there. Twirling in the back of the mind, it has provided a simple and quiet test for whether the invisible, the abstract, the untouchable stuff of space—and spacetime, more generally—is substantial enough to provide the ultimate reference for motion. The verdict? Although the issue is still debated, as we've now seen, the most straightforward reading of Einstein and his general relativity is that spacetime can provide such a benchmark: *spacetime is a something.*[23]

Notice, though, that this conclusion is also cause for celebration among supporters of a more broadly defined relationist outlook. In Newton's view and subsequently that of special relativity, space and then spacetime were invoked as entities that provide the reference for defining accelerated motion. And since, according to these perspectives, space and spacetime are absolutely unchangeable, this notion of acceleration is absolute. In general relativity, though, the character of spacetime is completely different. Space and time are dynamic in general relativity: they are mutable; they respond to the presence of mass and energy; they are not absolute. Spacetime and, in particular, the way it warps and curves, is an embodiment of the gravitational field. Thus, in general relativity, acceleration relative to spacetime is a far cry from the absolute, staunchly unrelational conception invoked by previous theories. Instead, as Einstein argued eloquently a few years before he died,[24] acceleration relative to general relativity's spacetime *is* relational. It is not acceleration relative to material objects like stones or stars, but it is acceleration relative to something just as real, tangible, and changeable: a field—the gravitational field.* In this sense, spacetime—by being the incarnation of gravity—is *so* real in general relativity that the benchmark it provides is one that many relationists can comfortably accept.

Debate on the issues discussed in this chapter will no doubt continue as we grope to understand what space, time, and spacetime actually are. With the development of quantum mechanics, the plot only thickens.

---

*In special relativity—the special case of general relativity in which the gravitational field is zero—this idea applies unchanged: a zero gravitational field is still a field, one that can be measured and changed, and hence provides a something relative to which acceleration can be defined.

The concepts of empty space and of nothingness take on a whole new meaning when quantum uncertainty takes the stage. Indeed, since 1905, when Einstein did away with the luminiferous aether, the idea that space is filled with invisible substances has waged a vigorous comeback. As we will see in later chapters, key developments in modern physics have reinstituted various forms of an aetherlike entity, none of which set an absolute standard for motion like the original luminiferous aether, but all of which thoroughly challenge the naïve conception of what it means for spacetime to be empty. Moreover, as we will now see, *the* most basic role that space plays in a classical universe—as the medium that separates one object from another, as the intervening stuff that allows us to declare definitively that one object is distinct and independent from another—is thoroughly challenged by startling quantum connections.

4

Entangling Space

WHAT DOES IT MEAN TO BE SEPARATE IN A QUANTUM UNIVERSE?

To accept special and general relativity is to abandon Newtonian absolute space and absolute time. While it's not easy, you can train your mind to do this. Whenever you move around, imagine your *now* shifting away from the *nows* experienced by all others not moving with you. While you are driving along a highway, imagine your watch ticking away at a different rate compared with timepieces in the homes you are speeding past. While you are gazing out from a mountaintop, imagine that because of the warping of spacetime, time passes more quickly for you than for those subject to stronger gravity on the ground far below. I say "imagine" because in ordinary circumstances such as these, the effects of relativity are so tiny that they go completely unnoticed. Everyday experience thus fails to reveal how the universe really works, and that's why a hundred years after Einstein, almost no one, not even professional physicists, feels relativity in their bones. This isn't surprising; one is hard pressed to find the survival advantage offered by a solid grasp of relativity. Newton's flawed conceptions of absolute space and absolute time work wonderfully well at the slow speeds and moderate gravity we encounter in daily life, so our senses are under no evolutionary pressure to develop relativistic acumen. Deep awareness and true understanding therefore require that we diligently use our intellect to fill in the gaps left by our senses.

While relativity represented a monumental break with traditional ideas about the universe, between 1900 and 1930 another revolution was also turning physics upside down. It started at the turn of the twentieth century with a couple of papers on properties of radiation, one by Max Planck and the other by Einstein; these, after three decades of intense research, led to the formulation of *quantum mechanics*. As with relativity, whose effects become significant under extremes of speed or gravity, the new physics of quantum mechanics reveals itself abundantly only in another extreme situation: the realm of the extremely tiny. But there is a sharp distinction between the upheavals of relativity and those of quantum mechanics. The weirdness of relativity arises because our personal experience of space and time differs from the experience of others. It is a weirdness born of comparison. We are forced to concede that our view of reality is but one among many—an infinite number, in fact—which all fit together within the seamless whole of spacetime.

Quantum mechanics is different. Its weirdness is evident without comparison. It is harder to train your mind to have quantum mechanical intuition, because quantum mechanics shatters our own personal, individual conception of reality.

The World According to the Quantum

Every age develops its stories or metaphors for how the universe was conceived and structured. According to an ancient Indian creation myth, the universe was created when the gods dismembered the primordial giant Purusa, whose head became the sky, whose feet became the earth, and whose breath became the wind. To Aristotle, the universe was a collection of fifty-five concentric crystalline spheres, the outermost being heaven, surrounding those of the planets, earth and its elements, and finally the seven circles of hell.[1] With Newton and his precise, deterministic mathematical formulation of motion, the description changed again. The universe was likened to an enormous, grand clockwork: after being wound and set into its initial state, the clockwork universe ticks from one moment to the next with complete regularity and predictability.

Special and general relativity pointed out important subtleties of the clockwork metaphor: there is no single, preferred, universal clock; there is no consensus on what constitutes a moment, what constitutes a *now*. Even so, you can still tell a clockworklike story about the evolving universe. The clock is your clock. The story is your story. But the universe unfolds with the same regularity and predictability as in the Newtonian framework. If by some means you know the state of the universe right now—if you know where *every* particle is and how fast and in what direction each is moving—then, Newton and Einstein agree, you can, in principle, use the laws of physics to predict everything about the universe arbitrarily far into the future or to figure out what it was like arbitrarily far into the past.[2]

Quantum mechanics breaks with this tradition. We *can't* ever know the exact location and exact velocity of even a single particle. We *can't* predict with total certainty the outcome of even the simplest of experiments, let alone the evolution of the entire cosmos. Quantum mechanics shows that the best we can ever do is predict the *probability* that an experiment will turn out this way or that. And as quantum mechanics has been verified through decades of fantastically accurate experiments, the Newtonian cosmic clock, even with its Einsteinian updating, is an untenable metaphor; it is demonstrably *not* how the world works.

But the break with the past is yet more complete. Even though Newton's and Einstein's theories differ sharply on the nature of space and time, they do agree on certain basic facts, certain truths that appear to be self-evident. If there is space between two objects—if there are two birds in the sky and one is way off to your right and the other is way off to your left—we can and do consider the two objects to be independent. We regard them as separate and distinct entities. Space, whatever it is fundamentally, provides the medium that separates and distinguishes one object from another. That is what space does. Things occupying different locations in space are different things. Moreover, in order for one object to influence another, it must in some way negotiate the space that separates them. One bird can fly to the other, traversing the space between them, and then peck or nudge its companion. One person can influence another by shooting a slingshot, causing a pebble to traverse the space between them, or by yelling, causing a domino effect of bouncing air molecules, one jostling the next until some bang into the recipient's eardrum. Being yet more sophisticated, one can exert influence on another by firing a laser, causing an electromagnetic wave—a beam of light—to traverse the intervening space; or, being more ambitious (like the extraterrestrial pranksters of last chapter) one can shake or move a massive body (like the moon) sending a gravitational disturbance speeding from one location to another. To be sure, if we are over here we can influence someone over there, but no matter how we do it, the procedure always involves someone or something traveling from here to there, and only when the someone or something gets there can the influence be exerted.

Physicists call this feature of the universe *locality*, emphasizing the point that you can directly affect only things that are next to you, that are local. Voodoo contravenes locality, since it involves doing something over here and affecting something over there without the need for anything to travel from here to there, but common experience leads us to think that verifiable, repeatable experiments would confirm locality.[3] And most do.

But a class of experiments performed during the last couple of decades has shown that something we do over here (such as measuring certain properties of a particle) *can* be subtly entwined with something that happens over there (such as the outcome of measuring certain properties of another distant particle), *without* anything being sent from here to there. While intuitively baffling, this phenomenon fully conforms to the laws of quantum mechanics, and was predicted using quantum mechanics long before the technology existed to do the experiment and observe, remarkably, that the prediction is correct. This sounds like voodoo; Einstein, who was among the first physicists to recognize—and sharply criticize—this possible feature of quantum mechanics, called it "spooky." But as we shall see, the long-distance links these experiments confirm are extremely delicate and are, in a precise sense, fundamentally beyond our ability to control.

Nevertheless, these results, coming from both theoretical and experimental considerations, strongly support the conclusion that the universe admits interconnections that are not local.[4] Something that happens over here can be entwined with something that happens over there even if nothing travels from here to there—and even if there isn't enough time for anything, even light, to travel between the events. This means that space cannot be thought of as it once was: intervening space, *regardless of how much there is*, does not ensure that two objects are separate, since quantum mechanics allows an entanglement, a kind of connection, to exist between them. A particle, like one of the countless number that make up you or me, can run but it can't hide. According to quantum theory and the many experiments that bear out its predictions, the quantum connection between two particles can persist even if they are on opposite sides of the universe. From the standpoint of their entanglement, notwithstanding the many trillions of miles of space between them, it's as if they are right on top of each other.

Numerous assaults on our conception of reality are emerging from modern physics; we will encounter many in the following chapters. But of those that have been experimentally verified, I find none more mind-boggling than the recent realization that our universe is not local.

The Red and the Blue

To get a feel for the kind of nonlocality emerging from quantum mechanics, imagine that Agent Scully, long overdue for a vacation, retreats to her family's estate in Provence. Before she's had time to unpack, the phone rings. It's Agent Mulder calling from America.

"Did you get the box—the one wrapped in red and blue paper?"

Scully, who has dumped all her mail in a pile by the door, looks over and sees the package. "Mulder, please, I didn't come all the way to Aix just to deal with another stack of files."

"No, no, the package is not from me. I got one too, and inside there are these little lightproof titanium boxes, numbered from 1 to 1,000, and a letter saying that you would be receiving an identical package."

"Yes, so?" Scully slowly responds, beginning to fear that the titanium boxes may somehow wind up cutting her vacation short.

"Well," Mulder continues, "the letter says that each titanium box contains an alien sphere that will flash red or blue the moment the little door on its side is opened."

"Mulder, am I supposed to be impressed?"

"Well, not yet, but listen. The letter says that *before* any given box is opened, the sphere has the capacity to flash either red or blue, and it *randomly* decides between the two colors at the moment the door is opened. But here's the strange part. The letter says that although your boxes work exactly the same way as mine—even though the spheres inside each one of our boxes *randomly* choose between flashing red or blue—our boxes somehow work in tandem. The letter claims that there is a mysterious connection, so that if there is a blue flash when I open my box 1, you will also find a blue flash when you open your box 1; if I see a red flash when I open box 2, you will also see a red flash in your box 2, and so on."

"Mulder, I'm really exhausted; let's let the parlor tricks wait till I get back."

"Scully, please. I know you're on vacation, but we can't just let this go. We'll only need a few minutes to see if it's true."

Reluctantly, Scully realizes that resistance is futile, so she goes along and opens her little boxes. And on comparing the colors that flash inside each box, Scully and Mulder do indeed find the agreement predicted in the letter. Sometimes the sphere in a box flashes red, sometimes blue, but on opening boxes with the same number, Scully and Mulder always see the same color flash. Mulder grows increasingly excited and agitated by the alien spheres but Scully is thoroughly unimpressed.

"Mulder," Scully sternly says into the phone, "*you* really need a vacation. This is silly. Obviously, the sphere inside each of our boxes has been programmed to flash red or it has been programmed to flash blue when the door to its box is opened. And whoever sent us this nonsense programmed our boxes identically so that you and I find the same color flash in boxes with the same number."

"But no, Scully, the letter says each alien sphere *randomly* chooses between flashing blue and red when the door is opened, *not* that the sphere has been preprogrammed to choose one color or the other."

"Mulder," Scully sighs, "my explanation makes perfect sense and it fits all the data. What more do you want? And look here, at the bottom of the letter. Here's the biggest laugh of all. The 'alien' small print informs us that not only will opening the door to a box cause the sphere inside to flash, but any other tampering with the box to figure out how it works—for example, if we try to examine the sphere's color composition or chemical makeup before the door is opened—will also cause it to flash. In other words, we can't analyze the supposed random selection of red or blue because any such attempt will contaminate the very experiment we are trying to carry out. It's as if I told you I'm really a blonde, but I become a redhead whenever you or anyone or anything looks at my hair or analyzes it in any way. How could you ever prove me wrong? Your tiny green men are pretty clever—they've set things up so their ruse can't be unmasked. Now, go and play with your little boxes while I enjoy a little peace and quiet."

It would seem that Scully has this one soundly wrapped up on the side of science. Yet, here's the thing. Quantum mechanicians—scientists, not aliens—have for nearly eighty years been making claims about how the universe works that closely parallel those described in the letter. And the rub is that there is now strong scientific evidence that a viewpoint along the lines of Mulder's—not Scully's—is supported by the data. For instance, according to quantum mechanics, a particle can hang in a state of limbo between having one or another particular property—like an "alien" sphere hovering between flashing red and flashing blue before the door to its box is opened—and only when the particle is looked at (measured) does it randomly commit to one definite property or another. As if this weren't strange enough, quantum mechanics also predicts that there can be connections between particles, similar to those claimed to exist between the alien spheres. Two particles can be so entwined by quantum effects that their random selection of one property or another is correlated: just as each of the alien spheres chooses randomly between red and blue and yet, somehow, the colors chosen by spheres in boxes with the same number are correlated (both flashing red or both flashing blue), the properties chosen randomly by two particles, even if they are far apart in space, can similarly be aligned perfectly. Roughly speaking, even though the two particles are widely separated, quantum mechanics shows that whatever one particle does, the other will do too.

As a concrete example, if you are wearing a pair of sunglasses, quantum mechanics shows that there is a 50–50 chance that a particular photon—like one that is reflected toward you from the surface of a lake or from an asphalt roadway—will make it through your glare-reducing polarized lenses: when the photon hits the glass, it randomly "chooses" between reflecting back and passing through. The astounding thing is that such a photon can have a partner photon that has sped miles away in the opposite direction and yet, when confronted with the same 50–50 probability of passing through another polarized sunglass lens, will somehow do whatever the initial photon does. *Even though each outcome is determined randomly and even though the photons are far apart in space, if one photon passes through, so will the other.* This is the kind of nonlocality predicted by quantum mechanics.

Einstein, who was never a great fan of quantum mechanics, was loath to accept that the universe operated according to such bizarre rules. He championed more conventional explanations that did away with the notion that particles randomly select attributes and outcomes when measured. Instead, Einstein argued that if two widely separated particles are observed to share certain attributes, this is not evidence of some mysterious quantum connection instantaneously correlating their properties. Rather, just as Scully argued that the spheres do not randomly choose between red and blue, but instead are programmed to flash one particular color when observed, Einstein claimed that particles do not randomly choose between having one feature or another but, instead, are similarly "programmed" to have one particular, definite feature when suitably measured. The correlation between the behavior of widely separated photons is evidence, Einstein claimed, that the photons were endowed with identical properties when emitted, not that they are subject to some bizarre long-distance quantum entanglement.

For close to five decades, the issue of who was right—Einstein or the supporters of quantum mechanics—was left unresolved because, as we shall see, the debate became much like that between Scully and Mulder: any attempt to disprove the proposed strange quantum mechanical connections and leave intact Einstein's more conventional view ran afoul of the claim that the experiments themselves would necessarily contaminate the very features they were trying to study. All this changed in the 1960s. With a stunning insight, the Irish physicist John Bell showed that the issue could be settled experimentally, and by the 1980s it was. The most straightforward reading of the data is that Einstein was wrong and there can be strange, weird, and "spooky" quantum connections between things over here and things over there.[5]

The reasoning behind this conclusion is so subtle that it took physicists more than three decades to appreciate fully. But after covering the essential features of quantum mechanics we will see that the core of the argument reduces to nothing more complex than a Click and Clack puzzler.

Casting a Wave

If you shine a laser pointer on a little piece of black, overexposed 35mm film from which you have scratched away the emulsion in two extremely close and narrow lines, you will see direct evidence that light is a wave. If you've never done this, it's worth a try (you can use many things in place of the film, such as the wire mesh in a fancy coffee plunger). The image you will see when the laser light passes through the slits on the film and hits a screen consists of light and dark bands, as in Figure 4.1, and the explanation for this pattern relies on a basic feature of waves. Water waves are easiest to visualize, so let's first explain the essential point with waves on a large, placid lake, and then apply our understanding to light.

A water wave disturbs the flat surface of a lake by creating regions where the water level is higher than usual and regions where it is lower than usual. The highest part of a wave is called its *peak* and the lowest part is called its *trough*. A typical wave involves a periodic succession: peak followed by trough followed by peak, and so forth. If two waves head toward

Figure 4.1 Laser light passing through two slits etched on a piece of black film yields an interference pattern on a detector screen, showing that light is a wave.

each other—if, for example, you and I each drop a pebble into the lake at nearby locations, producing outward-moving waves that run into each other—when they cross there results an important effect known as *interference*, illustrated in Figure 4.2a. When a peak of one wave and a peak of the other cross, the height of the water is even greater, being the sum of the two peak heights. Similarly, when a trough of one wave and a trough of the other cross, the depression in the water is even deeper, being the sum of the two depressions. And here is the most important combination: when a peak of one wave crosses the trough of another, they tend to cancel each other out, as the peak tries to make the water go up while the trough tries to drag it down. If the height of one wave's peak equals the depth of the other's trough, there will be perfect cancellation when they cross, so the water at that location will not move at all.

The same principle explains the pattern that light forms when it passes through the two slits in Figure 4.1. Light is an electromagnetic wave; when it passes through the two slits, it splits into two waves that head toward the screen. Like the two water waves just discussed, the two light waves interfere with each other. When they hit various points on the screen, sometimes both waves are at their peaks, making the screen bright; sometimes both waves are at their troughs, also making it bright; but sometimes one wave is at its peak and the other is at its trough and they cancel, making that point on the screen dark. We illustrate this in Figure 4.2b.

When the wave motion is analyzed in mathematical detail, including the cases of partial cancellations between waves at various stages between peaks and troughs, one can show that the bright and dark spots fill out the bands seen in Figure 4.1. The bright and dark bands are therefore a telltale sign that light is a wave, an issue that had been hotly debated ever since Newton claimed that light is not a wave but instead is made up of a

(a)   (b)

Figure 4.2 (a) Overlapping water waves produce an interference pattern. **(b)** Overlapping light waves produce an interference pattern.

stream of particles (more on this in a moment). Moreover, this analysis applies equally well to *any* kind of wave (light wave, water wave, sound wave, you name it) and thus, interference patterns provide the metaphorical smoking gun: you know you are dealing with a wave if, when it is forced to pass through two slits of the right size (determined by the distance between the wave's peaks and troughs), the resulting intensity pattern looks like that in Figure 4.1 (with bright regions representing high intensity and dark regions being low intensity).

In 1927, Clinton Davisson and Lester Germer fired a beam of electrons—particulate entities without any apparent connection to waves—at a piece of nickel crystal; the details need not concern us, but what does matter is that this experiment is equivalent to firing a beam of electrons at a barrier with two slits. When the experimenters allowed the electrons that passed through the slits to travel onward to a phosphor screen where their impact location was recorded by a tiny flash (the same kind of flashes responsible for the picture on your television screen), the results were astonishing. Thinking of the electrons as little pellets or bullets, you'd naturally expect their impact positions to line up with the two slits, as in Figure 4.3a. But that's not what Davisson and Germer found. Their experiment produced data schematically illustrated in Figure 4.3b: the electron impact positions filled out an interference pattern characteristic of waves. Davisson and Germer had found the smoking gun. *They had*

(a)   (b)

Figure 4.3 (a) Classical physics predicts that electrons fired at a barrier with two slits will produce two bright stripes on a detector. (b) Quantum physics predicts, and experiments confirm, that electrons will produce an interference pattern, showing that they embody wavelike features.

shown that the beam of particulate electrons must, unexpectedly, be some kind of wave.

Now, you might not think this is particularly surprising. Water is made of H_2O molecules, and a water wave arises when many molecules move in a coordinated pattern. One group of H_2O molecules goes up in one location, while another group goes down in a nearby location. Perhaps the data illustrated in Figure 4.3 show that electrons, like H_2O molecules, sometimes move in concert, creating a wavelike pattern in their overall, macroscopic motion. While at first blush this might seem to be a reasonable suggestion, the actual story is far more unexpected.

We initially imagined that a flood of electrons was fired continuously from the electron gun in Figure 4.3. But we can tune the gun so that it fires fewer and fewer electrons every second; in fact, we can tune it all the way down so that it fires, say, only one electron every ten seconds. With enough patience, we can run this experiment over a long period of time and record the impact position of each individual electron that passes through the slits. Figures 4.4a–4.4c show the resulting cumulative data after an hour, half a day, and a full day. In the 1920s, images like these rocked the foundations of physics. *We see that even individual, particulate electrons, moving to the screen independently, separately, one by one, build up the interference pattern characteristic of waves.*

This is as if an *individual* H_2O molecule could still embody something akin to a water wave. But how in the world could that be? Wave motion seems to be a collective property that has no meaning when applied to separate, particulate ingredients. If every few minutes individual spectators in the bleachers get up and sit down separately, independently, they are *not* doing the wave. More than that, wave interference seems to require a wave from *here* to cross a wave from *there*. So how can

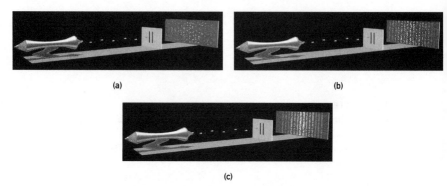

Figure 4.4 Electrons fired one by one toward slits build up an interference pattern dot by dot. In (a)–(c) we illustrate the pattern forming over time.

interference be at all relevant to single, individual, particulate ingredients? But somehow, as attested by the interference data in Figure 4.4, even though individual electrons are tiny particles of matter, each and every one also embodies a wavelike character.

Probability and the Laws of Physics

If an individual electron is also a wave, what is it that is waving? Erwin Schrödinger weighed in with the first guess: maybe the stuff of which electrons are made can be smeared out in space and it's this smeared electron essence that does the waving. An electron particle, from this point of view, would be a sharp spike in an electron mist. It was quickly realized, though, that this suggestion couldn't be correct because even a sharply spiked wave shape—such as a giant tidal wave—ultimately spreads out. And if the spiked electron wave were to spread we would expect to find part of a single electron's electric charge over here or part of its mass over there. But we never do. When we locate an electron, we always find all of its mass and all of its charge concentrated in one tiny, pointlike region. In 1927, Max Born put forward a different suggestion, one that turned out to be the decisive step that forced physics to enter a radically new realm. The wave, he claimed, is not a smeared-out electron, nor is it anything ever previously encountered in science. The wave, Born proposed, is a *probability* wave.

To understand what this means, picture a snapshot of a water wave that shows regions of high intensity (near the peaks and troughs) and regions of low intensity (near the flatter transition regions between peaks and troughs). The higher the intensity, the greater the potential the water wave has for exerting force on nearby ships or on coastline structures. The probability waves envisioned by Born also have regions of high and low intensity, but the meaning he ascribed to these wave shapes was unexpected: *the size of a wave at a given point in space is proportional to the probability that the electron is located at that point in space.* Places where the probability wave is large are locations where the electron is most likely to be found. Places where the probability wave is small are locations where the electron is unlikely to be found. And places where the probability wave is zero are locations where the electron will not be found.

Figure 4.5 gives a "snapshot" of a probability wave with the labels emphasizing Born's probabilistic interpretation. Unlike a photograph of water waves, though, this image could not actually have been made with a camera. No one has ever directly seen a probability wave, and conventional quantum mechanical reasoning says that no one ever will. Instead, we use mathematical equations (developed by Schrödinger, Niels Bohr, Werner Heisenberg, Paul Dirac, and others) to figure out what the probability wave should look like in a given situation. We then test such theoretical calculations by comparing them with experimental results in the following way. After calculating the purported probability wave for the electron in a given experimental setup, we carry out identical versions of

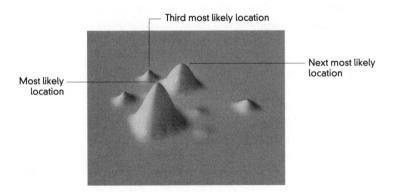

Figure 4.5 The probability wave of a particle, such as an electron, tells us the likelihood of finding the particle at one location or another.

the experiment over and over again from scratch, each time recording the measured position of the electron. *In contrast to what Newton would have expected, identical experiments and starting conditions do not necessarily lead to identical measurements.* Instead, our measurements yield a variety of measured locations. Sometimes we find the electron here, sometimes there, and every so often we find it *way* over there. If quantum mechanics is right, the number of times we find the electron at a given point should be proportional to the size (actually, the square of the size), at that point, of the probability wave that we calculated. Eight decades of experiments have shown that the predictions of quantum mechanics are confirmed to spectacular precision.

Only a portion of an electron's probability wave is shown in Figure 4.5: according to quantum mechanics, every probability wave extends throughout all of space, throughout the entire universe.[6] In many circumstances, though, a particle's probability wave quickly drops very close to zero outside some small region, indicating the overwhelming likelihood that the particle is in that region. In such cases, the part of the probability wave left out of Figure 4.5 (the part extending throughout the rest of the universe) looks very much like the part near the edges of the figure: quite flat and near the value zero. Nevertheless, so long as the probability wave somewhere in the Andromeda galaxy has a nonzero value, no matter how small, there is a tiny but genuine—nonzero—chance that the electron could be found there.

Thus, the success of quantum mechanics forces us to accept that the electron, a constituent of matter that we normally envision as occupying a tiny, pointlike region of space, also has a description involving a wave that, to the contrary, is spread through the entire universe. Moreover, according to quantum mechanics this particle-wave fusion holds for all of nature's constituents, not just electrons: protons are both particlelike and wavelike; neutrons are both particlelike and wavelike, and experiments in the early 1900s even established that light—which demonstrably behaves like a wave, as in Figure 4.1—can also be described in terms of particulate ingredients, the little "bundles of light" called photons mentioned earlier.[7] The familiar electromagnetic waves emitted by a hundred-watt bulb, for example, can equally well be described in terms of the bulb's emitting about a hundred billion billion photons each second. In the quantum world, we've learned that everything has both particlelike and wavelike attributes.

Over the last eight decades, the ubiquity and utility of quantum mechanical probability waves to predict and explain experimental results has been established beyond any doubt. Yet there is still no universally agreed-upon way to envision what quantum mechanical probability waves actually are. Whether we should say that an electron's probability wave *is* the electron, or that it's *associated* with the electron, or that it's a *mathematical device* for describing the electron's motion, or that it's the *embodiment of what we can know* about the electron is still debated. What is clear, though, is that through these waves, quantum mechanics injects probability into the laws of physics in a manner that no one had anticipated. Meteorologists use probability to predict the likelihood of rain. Casinos use probability to predict the likelihood you'll throw snake eyes. But probability plays a role in these examples because we haven't all of the information necessary to make definitive predictions. According to Newton, if we knew in complete detail the state of the environment (the positions and velocities of every one of its particulate ingredients), we would be able to predict (given sufficient calculational prowess) with certainty whether it will rain at 4:07 p.m. tomorrow; if we knew all the physical details of relevance to a craps game (the precise shape and composition of the dice, their speed and orientation as they left your hand, the composition of the table and its surface, and so on), we would be able to predict with certainty how the dice will land. Since, in practice, we can't gather all this information (and, even if we could, we do not yet have sufficiently powerful computers to perform the calculations required to make such predictions), we set our sights lower and predict only the probability of a given outcome in the weather or at the casino, making reasonable guesses about the data we don't have.

The probability introduced by quantum mechanics is of a different, more fundamental character. Regardless of improvements in data collection or in computer power, the best we can ever do, according to quantum mechanics, is predict the probability of this or that outcome. The best we can ever do is predict the probability that an electron, or a proton, or a neutron, or any other of nature's constituents, will be found here or there. Probability reigns supreme in the microcosmos.

As an example, the explanation quantum mechanics gives for individual electrons, one by one, over time, building up the pattern of light and dark bands in Figure 4.4, is now clear. Each individual electron is described by its probability wave. When an electron is fired, its probability wave flows through both slits. And just as with light waves and water waves, the probability waves emanating from the two slits interfere with each other. At some points on the detector screen the two probability waves reinforce and the resulting intensity is large. At other points the waves partially cancel and the intensity is small. At still other points the peaks and troughs of the probability waves completely cancel and the resulting wave intensity is exactly zero. That is, there are points on the screen where it is very likely an electron will land, points where it is far less likely that it will land, and places where there is no chance at all that an electron will land. Over time, the electrons' landing positions are distributed according to this probability profile, and hence we get some bright, some dimmer, and some completely dark regions on the screen. Detailed analysis shows that these light and dark regions will look exactly as they do in Figure 4.4.

Einstein and Quantum Mechanics

Because of its inherently probabilistic nature, quantum mechanics differs sharply from any previous fundamental description of the universe, qualitative or quantitative. Since its inception last century, physicists have struggled to mesh this strange and unexpected framework with the common worldview; the struggle is still very much under way. The problem lies in reconciling the macroscopic experience of day-to-day life with the microscopic reality revealed by quantum mechanics. We are used to living in a world that, while admittedly subject to the vagaries of economic or political happenstance, appears stable and reliable at least as far as its physical properties are concerned. You do not worry that the atomic constituents of the air you are now breathing will suddenly disband, leaving you gasping for breath as they manifest their quantum wavelike character by rematerializing, willy-nilly, on the dark side of the moon. And you are right not to fret about this outcome, because according to quantum mechanics the probability of its happening, while not zero, is absurdly small. But what makes the probability so small?

There are two main reasons. First, on a scale set by atoms, the moon is enormously far away. And, as mentioned, in many circumstances (although by no means all), the quantum equations show that a probability wave typically has an appreciable value in some small region of space and quickly drops nearly to zero as you move away from this region (as in Figure 4.5). So the likelihood that even a single electron that you expect to be in the same room as you—such as one of those that you just exhaled—will be found in a moment or two on the dark side of the moon, while not zero, is extremely small. So small, that it makes the probability that you will marry Nicole Kidman or Antonio Banderas seem enormous by comparison. Second, there are *a lot* of electrons, as well as protons and neutrons, making up the air in your room. The likelihood that *all* of these particles will do what is extremely unlikely even for one is so small that it's hardly worth a moment's thought. It would be like not only marrying your movie-star heartthrob but then also winning every state lottery every week for, well, a length of time that would make the current age of the universe seem a mere cosmic flicker.

This gives some sense of why we do not directly encounter the probabilistic aspects of quantum mechanics in day-to-day life. Nevertheless, because experiments confirm that quantum mechanics does describe fundamental physics, it presents a frontal assault on our basic beliefs as to what constitutes reality. Einstein, in particular, was deeply troubled by the probabilistic character of quantum theory. Physics, he would emphasize again and again, is in the business of determining with certainty what has happened, what is happening, and what will happen in the world around us. Physicists are not bookies, and physics is not the business of calculating odds. But Einstein could not deny that quantum mechanics was enormously successful in explaining and predicting, albeit in a statistical framework, experimental observations of the microworld. And so rather than attempting to show that quantum mechanics was wrong, a task that still looks like a fool's errand in light of its unparalleled successes, Einstein expended much effort on trying to show that quantum mechanics was not the final word on how the universe works. Even though he could not say what it was, Einstein wanted to convince everyone that there was a deeper and less bizarre description of the universe yet to be found.

Over the course of many years, Einstein mounted a series of ever more sophisticated challenges aimed at revealing gaps in the structure of quantum mechanics. One such challenge, raised in 1927 at the Fifth Physical Conference of the Solvay Institute,[8] concerns the fact that even though an electron's probability wave might look like that in Figure 4.5, whenever we measure the electron's whereabouts we always find it at one definite position or another. So, Einstein asked, doesn't that mean that the probability wave is merely a temporary stand-in for a more precise description—one yet to be discovered—that would predict the electron's position with certainty? After all, if the electron is found at X, doesn't that mean, in reality, it was *at* or *very near* X a moment before the measurement was carried out? And if so, Einstein prodded, doesn't quantum mechanics' reliance on the probability wave—a wave that, in this example, says the electron had some probability to have been far from X—reflect the theory's inadequacy to describe the true underlying reality?

Einstein's viewpoint is simple and compelling. What could be more natural than to expect a particle to be located at, or, at the very least, near where it's found a moment later? If that's the case, a deeper understanding of physics should provide *that* information and dispense with the coarser framework of probabilities. But the Danish physicist Niels Bohr and his entourage of quantum mechanics defenders disagreed. Such reasoning, they argued, is rooted in conventional thinking, according to which each electron follows a single, definite path as it wanders to and fro. And this thinking is strongly challenged by Figure 4.4, since if each electron did follow one definite path—like the classical image of a bullet fired from a gun—it would be extremely hard to explain the observed interference pattern: what would be interfering with what? Ordinary bullets fired one by one from a single gun certainly can't interfere with each other, so if electrons did travel like bullets, how would we explain the pattern in Figure 4.4?

Instead, according to Bohr and the Copenhagen interpretation of quantum mechanics he forcefully championed, *before one measures the electron's position there is no sense in even asking where it is.* It does not have a definite position. The probability wave encodes the likelihood that the electron, when examined suitably, will be found here or there, and that *truly* is all that can be said about its position. Period. The electron has a definite position in the usual intuitive sense only at the moment we "look" at it—at the moment when we measure its position—identifying its location with certainty. But before (and after) we do that, all it has are potential positions described by a probability wave that, like any wave, is subject to interference effects. It's not that the electron has a position and that we don't know the position before we do our measurement. Rather, contrary to what you'd expect, the electron simply *does not have* a definite position before the measurement is taken.

This is a radically strange reality. In this view, when we measure the electron's position we are not measuring an objective, preexisting feature of reality. Rather, the act of measurement is deeply enmeshed in creating the very reality it is measuring. Scaling this up from electrons to everyday life, Einstein quipped, "Do you really believe that the moon is not there unless we are looking at it?" The adherents of quantum mechanics responded with a version of the old saw about a tree falling in a forest: if no one is looking at the moon—if no one is "measuring its location by seeing it"—then there is no way for us to know whether it's there, so there is no point in asking the question. Einstein found this deeply unsatisfying. It was wildly at odds with his conception of reality; he firmly believed that the moon is there, whether or not anyone is looking. But the quantum stalwarts were unconvinced.

Einstein's second challenge, raised at the Solvay conference in 1930, followed closely on the first. He described a hypothetical device, which (through a clever combination of a scale, a clock, and a cameralike shutter) seemed to establish that a particle like an electron *must* have definite features—before it is measured or examined—that quantum mechanics said it couldn't. The details are not essential but the resolution is particularly ironic. When Bohr learned of Einstein's challenge, he was knocked back on his heels—at first, he couldn't see a flaw in Einstein's argument. Yet, within days, he bounced back and fully refuted Einstein's claim. And the surprising thing is that the key to Bohr's response was general relativity! Bohr realized that Einstein had failed to take account of his own discovery that gravity warps time—that a clock ticks at a rate dependent on the gravitational field it experiences. When this complication was included, Einstein was forced to admit that his conclusions fell right in line with orthodox quantum theory.

Even though his objections were shot down, Einstein remained deeply uncomfortable with quantum mechanics. In the following years he kept Bohr and his colleagues on their toes, leveling one new challenge after another. His most potent and far-reaching attack focused on something known as the *uncertainty principle*, a direct consequence of quantum mechanics, enunciated in 1927 by Werner Heisenberg.

Heisenberg and Uncertainty

The uncertainty principle provides a sharp, quantitative measure of how tightly probability is woven into the fabric of a quantum universe. To understand it, think of the prix-fixe menus in certain Chinese restaurants. Dishes are arranged in two columns, A and B, and if, for example, you order the first dish in column A, you are not allowed to order the first dish in column B; if you order the second dish in column A, you are not allowed to order the second dish in column B, and so forth. In this way, the restaurant has set up a dietary dualism, a culinary complementarity (one, in particular, that is designed to prevent you from piling up the most expensive dishes). On the prix-fixe menu you can have Peking Duck or Lobster Cantonese, but not both.

Heisenberg's uncertainty principle is similar. It says, roughly speaking, that the physical features of the microscopic realm (particle positions, velocities, energies, angular momenta, and so on) can be divided into two lists, A and B. And as Heisenberg discovered, knowledge of the first feature from list A fundamentally compromises your ability to have knowledge about the first feature from list B; knowledge of the second feature from list A fundamentally compromises your ability to have knowledge of the second feature from list B; and so on. Moreover, like being allowed a dish containing some Peking Duck and some Lobster Cantonese, but only in proportions that add up to the same total price, the more precise your knowledge of a feature from one list, the less precise your knowledge can possibly be about the corresponding feature from the second list. The fundamental inability to determine simultaneously all features from both lists—to determine with certainty all of these features of the microscopic realm—is the uncertainty revealed by Heisenberg's principle.

As an example, the more precisely you know where a particle is, the less precisely you can possibly know its speed. Similarly, the more precisely you know how fast a particle is moving, the less you can possibly know about where it is. Quantum theory thereby sets up its own duality: you can determine with precision certain physical features of the microscopic realm, but in so doing you eliminate the possibility of precisely determining certain other, complementary features.

To understand why, let's follow a rough description developed by Heisenberg himself, which, while incomplete in particular ways that we will discuss, does give a useful intuitive picture. When we measure the position of any object, we generally interact with it in some manner. If we search for the light switch in a dark room, we know we have located it when we touch it. If a bat is searching for a field mouse, it bounces sonar off its target and interprets the reflected wave. The most common instance of all is locating something by seeing it—by receiving light that has reflected off the object and entered our eyes. The key point is that these interactions not only affect us but also affect the object whose position is being determined. Even light, when bouncing off an object, gives it a tiny push. Now, for day-to-day objects such as the book in your hand or a clock on the wall, the wispy little push of bouncing light has no noticeable effect. But when it strikes a tiny particle like an electron it can have a big effect: as the light bounces off the electron, it changes the electron's speed, much as your own speed is affected by a strong, gusty wind that whips around a street corner. In fact, the more precisely you want to identify the electron's position, the more sharply defined and energetic the light beam must be, yielding an even larger effect on the electron's motion.

This means that if you measure an electron's position with high accuracy, you necessarily contaminate your own experiment: the act of precision position measurement disrupts the electron's velocity. You can therefore know precisely where the electron is, but you cannot also know precisely how fast, at that moment, it was moving. Conversely, you can measure precisely how fast an electron is moving, but in so doing you will contaminate your ability to determine with precision its position. Nature has a built-in limit on the precision with which such complementary features can be determined. And although we are focusing on electrons, the uncertainty principle is completely general: it applies to everything.

In day-to-day life we routinely speak about things like a car passing a particular stop sign (position) while traveling at 90 miles per hour (velocity), blithely specifying these two physical features. In reality, quantum mechanics says that such a statement has no precise meaning since you can't ever simultaneously measure a definite position and a definite speed. The reason we get away with such incorrect descriptions of the physical world is that on everyday scales the amount of uncertainty involved is tiny and generally goes unnoticed. You see, Heisenberg's principle does not just declare uncertainty, it also specifies—with complete certainty—the minimum *amount* of uncertainty in any situation. If we apply his formula to your car's velocity just as it passes a stop sign whose position is known to within a centimeter, then the uncertainty in speed turns out to be just shy of a billionth of a billionth of a billionth of a billionth of a mile per hour. A state trooper would be fully complying with the laws of quantum physics if he asserted that your speed was between 89.999999999999999999999999999999999 and 90.000000000000000000000000000000001 miles per hour as you blew past the stop sign; so much for a possible uncertainty-principle defense. But if we were to replace your massive car with a delicate electron whose position we knew to within a billionth of a meter, then the uncertainty in its speed would be a whopping 100,000 miles per hour.

Uncertainty is always present, but it becomes significant only on microscopic scales.

The explanation of uncertainty as arising through the unavoidable disturbance caused by the measurement process has provided physicists with a useful intuitive guide as well as a powerful explanatory framework in certain specific situations. However, it can also be misleading. It may give the impression that uncertainty arises only when we lumbering experimenters meddle with things. This is not true. Uncertainty is built into the wave structure of quantum mechanics and exists whether or not we carry out some clumsy measurement. As an example, take a look at a particularly simple probability wave for a particle, the analog of a gently rolling ocean wave, shown in Figure 4.6. Since the peaks are all uniformly moving to the right, you might guess that this wave describes a particle moving with the velocity of the wave peaks; experiments confirm that supposition. But where is the particle? Since the wave is uniformly spread throughout space, there is no way for us to say the electron is *here* or *there*. When measured, it literally could be found anywhere. So, while we know precisely how fast the particle is moving, there is huge uncertainty about its position. And as you see, this conclusion does not depend on our disturbing the particle. We never touched it. Instead, it relies on a basic feature of waves: they can be spread out.

Although the details get more involved, similar reasoning applies to all other wave shapes, so the general lesson is clear. In quantum mechanics, uncertainty just is.

Figure 4.6 A probability wave with a uniform succession of peaks and troughs represents a particle with a definite velocity. But since the peaks and troughs are uniformly spread in space, the particle's position is completely undetermined. It has an equal likelihood of being anywhere.

Einstein, Uncertainty, and a Question of Reality

An important question, and one that may have occurred to you, is whether the uncertainty principle is a statement about what we can know about reality or whether it is a statement about reality itself. Do objects making up the universe really have a position and a velocity, like our usual classical image of just about everything—a soaring baseball, a jogger on the boardwalk, a sunflower slowly tracking the sun's flight across the sky—although quantum uncertainty tells us these features of reality are forever beyond our ability to know simultaneously, even in principle? Or does quantum uncertainty break the classical mold completely, telling us that the list of attributes our classical intuition ascribes to reality, a list headed by the positions and velocities of the ingredients making up the world, is misguided? Does quantum uncertainty tell us that, at any given moment, particles simply do not possess a definite position and a definite velocity?

To Bohr, this issue was on par with a Zen koan. Physics addresses only things we can measure. From the standpoint of physics, that *is* reality. Trying to use physics to analyze a "deeper" reality, one beyond what we can know through measurement, is like asking physics to analyze the sound of one hand clapping. But in 1935, Einstein together with two colleagues, Boris Podolsky and Nathan Rosen, raised this issue in such a forceful and clever way that what had begun as one hand clapping reverberated over fifty years into a thunderclap that heralded a far greater assault on our understanding of reality than even Einstein ever envisioned.

The intent of the Einstein-Podolsky-Rosen paper was to show that quantum mechanics, while undeniably successful at making predictions and explaining data, could not be the final word regarding the physics of the microcosmos. Their strategy was simple, and was based on the issues just raised: they wanted to show that every particle does possess a definite position and a definite velocity at any given instant of time, and thus they wanted to conclude that the uncertainty principle reveals a fundamental limitation of the quantum mechanical approach. If every particle has a position and a velocity, but quantum mechanics cannot deal with these features of reality, then quantum mechanics provides only a partial description of the universe. Quantum mechanics, they intended to show, was therefore an incomplete theory of physical reality and, perhaps, merely a stepping-stone toward a deeper framework waiting to be discovered. In actuality, as we will see, they laid the groundwork for demonstrating something even more dramatic: the nonlocality of the quantum world.

Einstein, Podolsky, and Rosen (EPR) were partly inspired by Heisenberg's rough explanation of the uncertainty principle: when you measure where something is you necessarily disturb it, thereby contaminating any attempt to simultaneously ascertain its velocity. Although, as we have seen, quantum uncertainty is more general than the "disturbance" explanation indicates, Einstein, Podolsky, and Rosen invented what appeared to be a convincing and clever end run around *any* source of uncertainty. What if, they suggested, you could perform an indirect measurement of both the position and the velocity of a particle in a manner that never brings you into contact with the particle itself? For instance, using a classical analogy, imagine that Rod and Todd Flanders decide to do some lone wandering in Springfield's newly formed Nuclear Desert. They start back to back in the desert's center and agree to walk straight ahead, in opposite directions, at exactly the same prearranged speed. Imagine further that, nine hours later, their father, Ned, returning from his trek up Mount Springfield, catches sight of Rod, runs to him, and desperately asks about Todd's whereabouts. Well, by that point, Todd is far away, but by questioning and observing Rod, Ned can nevertheless learn much about Todd. If Rod is exactly 45 miles due east of the starting location, Todd must be exactly 45 miles due west of the starting location. If Rod is walking at exactly 5 miles per hour due east, Todd must be walking at exactly 5 miles per hour due west. So even though Todd is some 90 miles away, Ned can determine his position and speed, albeit indirectly.

Einstein and his colleagues applied a similar strategy to the quantum domain. There are well-known physical processes whereby two particles emerge from a common location with properties that are related in somewhat the same way as the motion of Rod and Todd. For example, if an initial single particle should disintegrate into two particles of equal mass that fly off "back-to-back" (like an explosive shooting off two chunks in opposite directions), something that is common in the realm of subatomic particle physics, the velocities of the two constituents will be equal and opposite. Moreover, the positions of the two constituent particles will also be closely related, and for simplicity the particles can be thought of as always being equidistant from their common origin.

An important distinction between the classical example involving Rod and Todd, and the quantum description of the two particles, is that although we can say with certainty that there is a definite relationship between the speeds of the two particles—if one were measured and found to be moving to the left at a given speed, then the other would necessarily be moving to the right at the same speed—we cannot predict the actual numerical value of the speed with which the particles move. Instead, the best we can do is use the laws of quantum physics to predict the probability that any particular speed is the one attained. Similarly, while we can say with certainty that there is a definite relationship between the positions of the particles—if one is measured at a given moment and found to be at some location, the other necessarily is located the same distance from the starting point but in the opposite direction—we cannot predict with certainty the actual location of either particle. Instead, the best we can do is predict the probability that one of the particles is at any chosen location. Thus, while quantum mechanics does not give definitive answers regarding particle speeds or positions, it does, in certain situations, give definitive statements regarding the *relationships* between the particle speeds and positions.

Einstein, Podolsky, and Rosen sought to exploit these relationships to show that each of the particles actually has a definite position and a definite velocity at every given instant of time. Here's how: imagine you measure the position of the right-moving particle and in this way learn, indirectly, the position of the left-moving particle. EPR argued that since you have done nothing, absolutely nothing, to the left-moving particle, it must have *had* this position, and all you have done is determine it, albeit indirectly. They then cleverly pointed out that you could have chosen instead to measure the right-moving particle's velocity. In that case you would have, indirectly, determined the velocity of the left-moving particle without at all disturbing it. Again, EPR argued that since you would have done nothing, absolutely nothing, to the left-moving particle, it must have *had* this velocity, and all you would have done is determine it. Putting both together—the measurement that you did and the measurement that you *could* have done—EPR concluded that the left-moving particle has a definite position and a definite velocity at any given moment.

As this is subtle and crucial, let me say it again. EPR reasoned that nothing in your act of measuring the right-moving particle could possibly have any effect on the left-moving particle, because they are separate and distant entities. The left-moving particle is totally oblivious to what you have done or could have done to the right-moving particle. The particles might be meters, kilometers, or light-years apart when you do your measurement on the right-moving particle, so, in short, the left-moving particle couldn't care less what you do. Thus, any feature that you actually learn or could in principle learn about the left-moving particle from studying its right-moving counterpart must be a *definite, existing* feature of the left-moving particle, totally independent of your measurement. And since if you had measured the position of the right particle you would have learned the position of the left particle, and if you had measured the velocity of the right particle you would have learned the velocity of the left particle, it must be that the left-moving particle actually has both a definite position and velocity. Of course, this whole discussion could be carried out interchanging the roles of left-moving and right-moving particles (and, in fact, before doing any measurement we can't even say which particle is moving left and which is moving right); this leads to the conclusion that both particles have definite positions and speeds.

Thus, EPR concluded that quantum mechanics is an incomplete description of reality. Particles have definite positions and speeds, but the quantum mechanical uncertainty principle shows that these features of reality are beyond the bounds of what the theory can handle. If, in agreement with these and most other physicists, you believe that a full theory of nature should describe every attribute of reality, the failure of quantum mechanics to describe both the positions and the velocities of particles means that it misses some attributes and is therefore not a complete theory; it is not the final word. That is what Einstein, Podolsky, and Rosen vigorously argued.

The Quantum Response

While EPR concluded that each particle has a definite position and velocity at any given moment, notice that if you follow their procedure you will fall short of actually determining these attributes. I said, above, that you could have chosen to measure the right-moving particle's velocity. Had you done so, you would have disturbed its position; on the other hand, had you chosen to measure its position you would have disturbed its velocity. If you don't have both of these attributes of the right-moving particle in hand, you don't have them for the left-moving particle either. Thus, *there is no conflict with the uncertainty principle*: Einstein and his collaborators fully recognized that they could not identify both the location and the velocity of any given particle. But, and this is key, even without determining both the position and velocity of either particle, EPR's reasoning shows that each *has* a definite position and velocity. To them, it was a question of reality. To them, a theory could not claim to be complete if there were elements of reality that it could not describe.

After a bit of intellectual scurrying in response to this unexpected observation, the defenders of quantum mechanics settled down to their usual, pragmatic approach, summarized well by the eminent physicist Wolfgang Pauli: "One should no more rack one's brain about the problem of whether something one cannot know anything about exists all the same, than about the ancient question of how many angels are able to sit on the point of a needle."[9] Physics in general, and quantum mechanics in particular, can deal only with the measurable properties of the universe. Anything else is simply not in the domain of physics. If you can't measure both the position and the velocity of a particle, then there is no sense in talking about whether it has both a position and a velocity.

EPR disagreed. Reality, they maintained, was more than the readings on detectors; it was more than the sum total of all observations at a given moment. When no one, absolutely no one, no device, no equipment, no anything at all is "looking" at the moon, they believed, the moon was still there. They believed that it was still part of reality.

In a way, this standoff echoes the debate between Newton and Leibniz about the reality of space. Can something be considered real if we can't actually touch it or see it or in some way measure it? In Chapter 2, I described how Newton's bucket changed the character of the space debate, suddenly suggesting that an influence of space could be observed directly, in the curved surface of spinning water. In 1964, in a single stunning stroke that one commentator has called "the most profound discovery of science,"[10] the Irish physicist John Bell did the same for the quantum reality debate.

In the following four sections, we will describe Bell's discovery, judiciously steering clear of all but a minimum of technicalities. All the same, even though the discussion uses reasoning less sophisticated than working out the odds in a craps game, it does involve a couple of steps that we must describe and then link together. Depending on your particular taste for detail, there may come a point when you just want the punch line. If this happens, feel free to jump to page 112, where you'll find a summary and a discussion of conclusions stemming from Bell's discovery.

Bell and Spin

John Bell transformed the central idea of the Einstein-Podolsky-Rosen paper from philosophical speculation into a question that could be answered by concrete experimental measurement. Surprisingly, all he needed to accomplish this was to consider a situation in which there were not just two features—for instance, position and velocity—that quantum uncertainty prevents us from simultaneously determining. He showed that if there are three or more features that simultaneously come under the umbrella of uncertainty—three or more features with the property that in measuring one, you contaminate the others and hence can't determine anything about them—then there *is* an experiment to address the reality question. The simplest such example involves something known as *spin*.

Since the 1920s, physicists have known that particles spin—they execute rotational motion akin to a soccer ball's spinning around as it heads toward the goal. Quantum particle spin, however, differs from this classical image in a number of essential ways, and foremost for us are the following two points. First, particles—for example, electrons and photons—can spin only clockwise or counterclockwise at one never-changing rate about any particular axis; a particle's spin axis can change directions but its rate of spin cannot slow down or speed up. Second, quantum uncertainty applied to spin shows that just as you can't simultaneously determine the position and the velocity of a particle, so also you can't simultaneously determine the spin of a particle about more than one axis. For example, if a soccer ball is spinning about a northeast-pointing axis, its spin is shared between a northward- and an eastward-pointing axis—and by a suitable measurement, you could determine the fraction of spin about each. But if you measure an electron's spin about any randomly chosen axis, you never find a fractional amount of spin. Ever. It's as if the measurement itself forces the electron to gather together all its spinning motion and direct it to be either clockwise or counterclockwise about the axis you happened to have focused on. Moreover, because of your measurement's influence on the electron's spin, you lose the ability to determine how it was spinning about a horizontal axis, about a back-and-forth axis, or about any other axis, prior to your measurement. These features of quantum mechanical spin are hard to picture fully, and the difficulty highlights the limits of classical images in revealing the true nature of the quantum world. Nevertheless, the mathematics of quantum theory, and decades of experiment, assure us that these characteristics of quantum spin are beyond doubt.

The reason for introducing spin here is not to delve into the intricacies of particle physics. Rather, the example of particle spin will, in just a moment, provide a simple laboratory for extracting wonderfully unexpected answers to the reality question. That is, does a particle simultaneously *have* a definite amount of spin about each and every axis, although we can never know it for more than one axis at a time because of quantum uncertainty? Or does the uncertainty principle tell us something else? Does it tell us, contrary to any classical notion of reality, that a particle simply does not and cannot possess such features simultaneously? Does it tell us that a particle resides in a state of quantum limbo, having no definite spin about any given axis, until someone or something measures it, causing it to snap to attention and attain—with a probability determined by quantum theory—one particular spin value or another (clockwise or counterclockwise) about the selected axis? By studying this question, essentially the same one we asked in the case of particle positions and velocities, we can use spin to probe the nature of quantum reality (and to extract answers that greatly transcend the specific example of spin). Let's see this.

As explicitly shown by the physicist David Bohm,[11] the reasoning of Einstein, Podolsky, and Rosen can easily be extended to the question of whether particles have definite spins about any and all chosen axes. Here's how it goes. Set up two detectors capable of measuring the spin of an incoming electron, one on the left side of the laboratory and the other on the right side. Arrange for two electrons to emanate back-to-back from a source midway between the two detectors, such that their spins—rather than their positions and velocities as in our earlier example—are correlated. The details of how this is done are not important; what is important is that it can be done and, in fact, can be done easily. The correlation can be arranged so that if the left and right detectors are set to measure the spins along axes pointing in the same direction, they will get the same result: if the detectors are set to measure the spin of their respective incoming electrons about a vertical axis and the left detector finds that the spin is clockwise, so will the right detector; if the detectors are set to measure spin along an axis 60 degrees clockwise from the vertical and the left detector measures a counterclockwise spin, so will the right detector; and so on. Again, in quantum mechanics the best we can do is predict the probability that the detectors will find clockwise or counterclockwise spin, but we can predict with 100 percent certainty that whatever one detector finds the other will find, too.*

Bohm's refinement of the EPR argument is now, for all intents and purposes, the same as it was in the original version that focused on position and velocity. The correlation between the particles' spins allows us to measure indirectly the spin of the left-moving particle about some axis by measuring that of its right-moving companion about that axis. Since this measurement is done far on the right side of the laboratory, it can't possibly influence the left-moving particle in any way. Hence, the latter must all along have had the spin value just determined; all we did was measure it, albeit indirectly. Moreover, since we could have chosen to perform this measurement about *any* axis, the same conclusion must hold for any axis: the left-moving electron must have a definite spin about each and every axis, even though we can explicitly determine it only about one axis at a time. Of course, the roles of left and right can be reversed, leading to the conclusion that each particle has a definite spin about any axis.[12]

At this stage, seeing no obvious difference from the position/velocity example, you might take Pauli's lead and be tempted to respond that there is no point in thinking about such issues. If you can't actually measure the spin about different axes, what is the point in wondering whether the particle nevertheless has a definite spin—clockwise versus counterclockwise—about each? Quantum mechanics, and physics more generally, is obliged only to account for features of the world that can be measured. And neither Bohm, Einstein, Podolsky, nor Rosen would have argued that the measurements can be done. Instead, they argued that the particles possess features forbidden by the uncertainty principle even though we can never explicitly know their particular values. Such features have come to be known as *hidden features*, or, more commonly, *hidden variables*.

Here is where John Bell changed everything. He discovered that even if you can't actually determine the spin of a particle about more than one axis, still, if in fact it *has* a definite spin about all axes, then there are testable, observable consequences of that spin.

---

*To avoid linguistic complications, I'm describing the electron spins as perfectly correlated, even though the more conventional description is one in which they're perfectly *anti*correlated: whatever result one detector finds, the other will find the opposite. To compare with the conventional description, imagine that I've interchanged all the clockwise and counterclockwise labels on one of the detectors.

Reality Testing

To grasp the gist of Bell's insight, let's return to Mulder and Scully and imagine that they've each received another package, also containing titanium boxes, but with an important new feature. Instead of having one door, each titanium box has three: one on top, one on the side, and one on the front.[13] The accompanying letter informs them that the sphere inside each box now randomly chooses between flashing red and flashing blue when any one of the box's three doors is opened. If a different door (top versus side versus front) on a given box were opened, the color randomly selected by the sphere might be different, but once one door is opened and the sphere has flashed, there is no way to determine what would have happened had another door been chosen. (In the physics application, this feature captures quantum uncertainty: once you measure one feature you can't determine anything about the others.) Finally, the letter tells them that there is again a mysterious connection, a strange entanglement, between the two sets of titanium boxes: Even though all the spheres *randomly* choose what color to flash when one of their box's three doors is opened, if both Mulder and Scully happen to open the *same* door on a box with the *same* number, the letter predicts that they will see the same color flash. If Mulder opens the top door on his box 1 and sees blue, then the letter predicts that Scully will also see blue if she opens the top door on her box 1; if Mulder opens the side door on his box 2 and sees red, then the letter predicts that Scully will also see red if she opens the side door on her box 2, and so forth. Indeed, when Scully and Mulder open the first few dozen boxes—agreeing by phone which door to open on each—they verify the letter's predictions.

Although Mulder and Scully are being presented with a somewhat more complicated situation than previously, at first blush it seems that the same reasoning Scully used earlier applies equally well here.

"Mulder," says Scully, "this is as silly as yesterday's package. Once again, there is no mystery. The sphere inside each box must simply be programmed. Don't you see?"

"But now there are three doors," cautions Mulder, "so the sphere can't possibly 'know' which door we'll choose to open, right?"

"It doesn't need to," explains Scully. "That's part of the programming. Look, here's an example. Grab hold of the next unopened box, box 37, and I'll do the same. Now, imagine, for argument's sake, that the sphere in my box 37 is programmed, say, to flash red if the top door is opened, to flash blue if the side door is opened, and to flash red if the front door is opened. I'll call this program *red, blue, red*. Clearly, then, if whoever is sending us this stuff has input this same program into your box 37, and if we both open the same door, we will see the same color flash. This explains the 'mysterious connection': if the boxes in our respective collections with the same number have been programmed with the same instructions, then we will see the same color if we open the same door. There is *no* mystery!"

But Mulder does not believe that the spheres are programmed. He believes the letter. He believes that the spheres are randomly choosing between red and blue when one of their box's doors is opened and hence he believes, fervently, that his and Scully's boxes *do* have some mysterious long-range connection.

Who is right? Since there is no way to examine the spheres before or during the supposed random selection of color (remember, any such tampering will cause the sphere instantly to choose randomly between red or blue, confounding any attempt to investigate how it really works), it seems impossible to prove definitively whether Scully or Mulder is right.

Yet, remarkably, after a little thought, Mulder realizes that there *is* an experiment that will settle the question completely. Mulder's reasoning is straightforward, but it does require a touch more explicit mathematical reasoning than most things we cover. It's definitely worth trying to follow the details—there aren't that many—but don't worry if some of it slips by; we'll shortly summarize the key conclusion.

Mulder realizes that he and Scully have so far only considered what happens if they each open the same door on a box with a given number. And, as he excitedly tells Scully after calling her back, there is much to be learned if they do not always choose the same door and, instead, randomly and independently choose which door to open on each of their boxes.

"Mulder, please. Just let me enjoy my vacation. What can we possibly learn by doing that?"

"Well, Scully, we can determine whether your explanation is right or wrong."

"Okay, I've got to hear this."

"It's simple," Mulder continues. "If you're right, then here's what I realized: if you and I separately and randomly choose which door to open on a given box and record the color we see flash, then, after doing this for many boxes we must find that we saw the same color flash *more* than 50 percent of the time. But if that isn't the case, if we find that we don't agree on the color for more than 50 percent of the boxes, then you can't be right."

"Really, how is that?" Scully is getting a bit more interested.

"Well," Mulder continues, "here's an example. Assume you're right, and each sphere operates according to a program. Just to be concrete, imagine the program for the sphere in a particular box happens to be *blue, blue, red*. Now since we both choose from among three doors, there are a total of nine possible door combinations that we might select to open for this box. For example, I might choose the top door on my box while you might choose the side door on your box; or I might choose the front door and you might choose the top door; and so on."

"Yes, of course," Scully jumps in. "If we call the top door 1, the side door 2, and the front door 3, then the nine possible door combinations are just (1,1), (1,2), (1,3), (2,1), (2,2), (2,3), (3,1), (3,2), (3,3)."

"Yes, that's right," Mulder continues. "Now here is the point: Of these nine possibilities notice that five door combinations—(1,1), (2,2), (3,3), (1,2), (2,1)—will result in us seeing the spheres in our boxes flash the same color. The first three door combinations are the ones in which we happen to choose the same door, and as we know, that *always* results in our seeing the same color. The other two door combinations, (1,2) and (2,1), result in the same color because the program dictates that the spheres will flash the same color—blue—if either door 1 or door 2 is opened. Now, since 5 is more than half of 9, this means that for more than half—more than 50 percent—of the possible combination of doors that we might select to open, the spheres will flash the same color."

"But wait," Scully protests. "That's just one example of a particular program: *blue, blue, red*. In my explanation, I proposed that differently numbered boxes can and generally will have different programs."

"Actually, that doesn't matter. The conclusion holds for all of the possible programs. You see, my reasoning with the *blue, blue, red* program only relied on the fact that two of the colors in the program are the same, and so an identical conclusion follows for any program: *red, red, blue*, or *red, blue, red*, and so on. Any program has to have at least two colors the same; the only programs that are really different are those in which all three colors are the same—*red, red, red* and *blue, blue, blue*. But for boxes with either of these programs, we'll get the same color to flash regardless of which doors we happen to open, and so the overall fraction on which we should agree will only increase. So, if your explanation is right and the boxes operate according to programs—even with programs that vary from one numbered box to another—we must agree on the color we see *more* than 50 percent of the time."

That's the argument. The hard part is now over. The bottom line is that there *is* a test to determine whether Scully is correct and each sphere operates according to a program that determines definitively which color to flash depending on which door is opened. If she and Mulder independently and randomly choose which of the three doors on each of their boxes to open, and then compare the colors they see—box by numbered box—they must find agreement for *more* than 50 percent of the boxes.

When cast in the language of physics, as it will be in the next section, Mulder's realization is nothing but John Bell's breakthrough.

Counting Angels with Angles

The translation of this result into physics is straightforward. Imagine we have two detectors, one on the left side of the laboratory and another on the right side, that measure the spin of an incoming particle like an electron, as in the experiment discussed in the section before last. The detectors require you to choose the axis (vertical, horizontal, back-forth, or one of the innumerable axes that lie in between) along which the spin is to be measured; for simplicity's sake, imagine that we have bargain-basement detectors that offer only three choices for the axes. In any given run of the experiment, you will find that the incoming electron is either spinning clockwise or counterclockwise about the axis you selected.

According to Einstein, Podolsky, and Rosen, each incoming electron provides the detector it enters with what amounts to a program: Even though it's hidden, even though you can't measure it, EPR claimed that each electron has a definite amount of spin—either clockwise or counter-clockwise—about each and every axis. Hence, when an electron enters a detector, the electron definitively determines whether you will measure its spin to be clockwise or counterclockwise about whichever axis you happen to choose. For example, an electron that is spinning clockwise about each of the three axes provides the program *clockwise, clockwise, clockwise*; an electron that is spinning clockwise about the first two axes and counterclockwise about the third provides the program *clockwise, clockwise, counterclockwise*, and so forth. In order to explain the correlation between the left-moving and right-moving electrons, Einstein, Podolsky, and Rosen simply claim that such electrons have identical spins and thus provide the detectors they enter with identical programs. Thus, if the same axes are chosen for the left and right detectors, the spin detectors will find identical results.

Notice that these spin detectors exactly reproduce everything encountered by Scully and Mulder, though with simple substitutions: instead of choosing a door on a titanium box, we are choosing an axis; instead of seeing a red or blue flash, we record a clockwise or counterclockwise spin. So, just as opening the same doors on a pair of identically numbered titanium boxes results in the same color flashing, choosing the same axes on the two detectors results in the same spin direction being measured. Also, just as opening one particular door on a titanium box prevents us from ever knowing what color would have flashed had we chosen another door, measuring the electron spin about one particular axis prevents us, via quantum uncertainty, from ever knowing which spin direction we would have found had we chosen a different axis.

All of the foregoing means that Mulder's analysis of how to learn who's right applies in exactly the same way to this situation as it does to the case of the alien spheres. If EPR are correct and each electron actually has a definite spin value about all three axes—if each electron provides a "program" that definitively determines the result of any of the three possible spin measurements—then we can make the following prediction. Scrutiny of data gathered from many runs of the experiment—runs in which the axis for each detector is randomly and independently selected—will show that *more than half the time, the two electron spins agree, being both clockwise or both counterclockwise.* If the electron spins do not agree more than half the time, then Einstein, Podolsky, and Rosen are wrong.

This is Bell's discovery. It shows that even though you can't actually measure the spin of an electron about more than one axis—even though you can't explicitly "read" the program it is purported to supply to the detector it enters—this does *not* mean that trying to learn whether it nonetheless has a definite amount of spin about more than one axis is tantamount to counting angels on the head of a pin. Far from it. Bell found that there is a bona fide, testable consequence associated with a particle having definite spin values. By using axes at three angles, Bell provided a way to count Pauli's angels.

No Smoke but Fire

In case you missed any of the details, let's summarize where we've gotten. Through the Heisenberg uncertainty principle, quantum mechanics claims that there are features of the world—like the position and the velocity of a particle, or the spin of a particle about various axes—that cannot simultaneously have definite values. *A particle, according to quantum theory, cannot have a definite position and a definite velocity; a particle cannot have a definite spin (clockwise or counterclockwise) about more than one axis; a particle cannot simultaneously have definite attributes for things that lie on opposite sides of the uncertainty divide.* Instead, particles hover in quantum limbo, in a fuzzy, amorphous, probabilistic mixture of all possibilities; only when measured is one definite outcome selected from the many. Clearly, this is a drastically different picture of reality than that painted by classical physics.

Ever the skeptic about quantum mechanics, Einstein, together with his colleagues Podolsky and Rosen, tried to use this aspect of quantum mechanics as a weapon against the theory itself. EPR argued that even though quantum mechanics does not allow such features to be simultaneously determined, particles nevertheless do have definite values for position and velocity; particles do have definite spin values about all axes; particles do have definite values for all things forbidden by quantum uncertainty. EPR thus argued that quantum mechanics cannot handle all elements of physical reality—it cannot handle the position and velocity of a particle; it cannot handle the spin of a particle about more than one axis—and hence is an incomplete theory.

For a long time, the issue of whether EPR were correct seemed more a question of metaphysics than of physics. As Pauli said, if you can't actually measure features forbidden by quantum uncertainty, what difference could it possibly make if they, nevertheless, exist in some hidden fold of reality? But, remarkably, John Bell found something that had escaped Einstein, Bohr, and all the other giants of twentieth-century theoretical physics: he found that the mere existence of certain things, even if they are beyond explicit measurement or determination, does make a difference—a difference that can be checked experimentally. Bell showed that if EPR were correct, the results found by two widely separated detectors measuring certain particle properties (spin about various randomly chosen axes, in the approach we have taken) would have to agree more than 50 percent of the time.

Bell had this insight in 1964, but at that time the technology did not exist to undertake the required experiments. By the early 1970s it did. Beginning with Stuart Freedman and John Clauser at Berkeley, followed by Edward Fry and Randall Thompson at Texas A&M, and culminating in the early 1980s with the work of Alain Aspect and collaborators working in France, ever more refined and impressive versions of these experiments were carried out. In the Aspect experiment, for example, the two detectors were placed 13 meters apart and a container of energetic calcium atoms was placed midway between them. Well-understood physics shows that each calcium atom, as it returns to its normal, less energetic state, will emit two photons, traveling back to back, whose spins are perfectly correlated, just as in the example of correlated electron spins we have been discussing. Indeed, in Aspect's experiment, whenever the detector settings are the same, the two photons are measured to have spins that are perfectly aligned. If lights were hooked up to Aspect's detectors to flash red in response to a clockwise spin and blue in response to a counterclockwise spin, the incoming photons would cause the detectors to flash the same color.

But, and this is the crucial point, when Aspect examined data from a large number of runs of the experiment—data in which the left and right detector settings were not always the same but, rather, were randomly and independently varied from run to run—he found that *the detectors did not agree more than 50 percent of the time.*

This is an earth-shattering result. This is the kind of result that should take your breath away. But just in case it hasn't, let me explain further. Aspect's results show that Einstein, Podolsky, and Rosen were proven by experiment—not by theory, not by pondering, but by nature—to be wrong. And that means there has to be something wrong with the reasoning EPR used to conclude that particles possess definite values for features—like spin values about distinct axes—for which definite values are forbidden by the uncertainty principle.

But where could they have gone wrong? Well, remember that the Einstein, Podolsky, and Rosen argument hangs on one central assumption: if at a given moment you can determine a feature of an object by an experiment done on another, spatially distant object, then the first object must have had this feature all along. Their rationale for this assumption was simple and thoroughly reasonable. Your measurement was done over *here* while the first object was way over *there*. The two objects were spatially separate, and hence your measurement could not possibly have had any effect on the first object. More precisely, since nothing goes faster than the speed of light, if your measurement on one object were somehow to cause a change in the other—for example, to cause the other to take on an identical spinning motion about a chosen axis—there would have to be a delay before this could happen, a delay at least as long as the time it would take light to traverse the distance between the two objects. But in both our abstract reasoning and in the actual experiments, the two particles are examined by the detectors at the *same* time. Therefore, whatever we learn about the first particle by measuring the second must be a feature that the first particle possessed, completely independent of whether we happened to undertake the measurement at all. In short, the core of the Einstein, Podolsky, Rosen argument is that *an object over there does not care about what you do to another object over here.*

But as we just saw, this reasoning leads to the prediction that the detectors should find the same result more than half the time, a prediction that is refuted by the experimental results. We are forced to conclude that the assumption made by Einstein, Podolsky, and Rosen, no matter how reasonable it seems, cannot be how our quantum universe works. Thus, through this indirect but carefully considered reasoning, the experiments lead us to conclude that *an object over there does care about what you do to another object over here.*

Even though quantum mechanics shows that particles randomly acquire this or that property when measured, we learn that the randomness can be linked across space. Pairs of appropriately prepared particles—they're called *entangled* particles—don't acquire their measured properties independently. They are like a pair of magical dice, one thrown in Atlantic City and the other in Las Vegas, each of which *randomly* comes up one number or another, yet the two of which somehow manage always to agree. Entangled particles act similarly, except they require no magic. *Entangled particles, even though spatially separate, do not operate autonomously.*

Einstein, Podolsky, and Rosen set out to show that quantum mechanics provides an incomplete description of the universe. Half a century later, theoretical insights and experimental results inspired by their work require us to turn their analysis on its head and conclude that the most basic, intuitively reasonable, classically sensible part of their reasoning is wrong: the universe is not local. The outcome of what you do at one place can be linked with what happens at another place, even if nothing travels between the two locations—even if there isn't enough time for anything to complete the journey between the two locations. Einstein's, Podolsky's, and Rosen's intuitively pleasing suggestion that such long-range correlations arise merely because particles have definite, preexisting, correlated properties is ruled out by the data. That's what makes this all so shocking.[14]

In 1997, Nicolas Gisin and his team at the University of Geneva carried out a version of the Aspect experiment in which the two detectors were placed 11 kilometers apart. The results were unchanged. On the microscopic scale of the photon's wavelengths, 11 kilometers is gargantuan. It might as well be 11 million kilometers—or 11 billion light-years, for that matter. There is every reason to believe that the correlation between the photons would persist no matter how far apart the detectors are placed.

This sounds totally bizarre. But there is now overwhelming evidence for this so-called *quantum entanglement.* If two photons are entangled, the successful measurement of either photon's spin about one axis "forces" the other, distant photon to have the same spin about the same axis; the act of measuring one photon "compels" the other, possibly distant photon to snap out of the haze of probability and take on a definitive spin value—a value that precisely matches the spin of its distant companion. And that boggles the mind.*

Entanglement and Special Relativity: The Standard View

I have put the words "forces" and "compels" in quotes because while they convey the sentiment our classical intuition longs for, their precise meaning in this context is critical to whether or not we are in for even more of an upheaval. With their everyday definitions, these words conjure up an

---

*Many researchers, including me, believe that Bell's argument and Aspect's experiment establish convincingly that the observed correlations between widely separated particles cannot be explained by Scully-type reasoning—reasoning that attributes the correlations to nothing more surprising than the particles' having acquired definite, correlated properties when they were (previously) together. Others have sought to evade or lessen the stunning nonlocality conclusion to which this has led us. I don't share their skepticism, but some works for general readers that discuss some of these alternatives are cited in the note section.[15]

image of volitional causality: we choose to do something here so as to *cause* or *force* a particular something to happen over there. If that were the right description of how the two photons are interrelated, *special relativity would be on the ropes.* The experiments show that from the viewpoint of an experimenter in the laboratory, at the precise moment one photon's spin is measured, the other photon immediately takes on the same spin property. If something were traveling from the left photon to the right photon, alerting the right photon that the left photon's spin had been determined through a measurement, it would have to travel between the photons instantaneously, conflicting with the speed limit set by special relativity.

The consensus among physicists is that any such apparent conflict with special relativity is illusory. The intuitive reason is that even though the two photons are spatially separate, their common origin establishes a fundamental link between them. Although they speed away from each other and become spatially separate, their history entwines them; even when distant, they are still part of one physical system. And so, it's really not that a measurement on one photon forces or compels another distant photon to take on identical properties. Rather, the two photons are so intimately bound up that it is justified to consider them—even though they are spatially separate—as parts of one physical entity. Then we can say that one measurement on this single entity—an entity containing two photons—affects the entity; that is, it affects both photons at once.

While this imagery may make the connection between the photons a little easier to swallow, as stated it's vague—what does it really mean to say two spatially separate things are one? A more precise argument is the following. When special relativity says that nothing can travel faster than the speed of light, the "nothing" refers to familiar matter or energy. But the case at hand is subtler, because it doesn't appear that any matter or energy is traveling between the two photons, and so there isn't anything whose speed we are led to measure. Nevertheless, there is a way to learn whether we've run headlong into a conflict with special relativity. A feature common to matter and energy is that when traveling from place to place they can transmit information. Photons traveling from a broadcast station to your radio carry information. Electrons traveling through Internet cables to your computer carry information. So, in any situation where something—even something unidentified—is purported to have traveled faster than light speed, a litmus test is to ask whether it has, or at least could have, transmitted information. If the answer is no, the standard reasoning goes, then nothing has exceeded light speed, and special relativity remains unchallenged. In practice, this is the test that physicists often employ in determining whether some subtle process has violated the laws of special relativity. (None has ever survived this test.) Let's apply it here.

Is there any way that, by measuring the spin of the left-moving and the right-moving photons about some given axis, we can send information from one to the other? The answer is no. Why? Well, the output found in either the left or the right detector is nothing but a *random* sequence of clockwise and counterclockwise results, since on any given run there is an equal probability of the particle to be found spinning one way or the other. In no way can we control or predict the outcome of any particular measurement. Thus, there is no message, there is no hidden code, there is no information whatsoever in either of these two random lists. The only interesting thing about the two lists is that they are identical—but that can't be discerned until the two lists are brought together and compared by some conventional, slower-than-light means (fax, e-mail, phone call, etc.). The standard argument thus concludes that although measuring the spin of one photon appears instantaneously to affect the other, no information is transmitted from one to the other, and the speed limit of special relativity remains in force. Physicists say that the spin results are correlated—since the lists are identical—but do not stand in a traditional cause-and-effect relationship because nothing travels between the two distant locations.

Entanglement and Special Relativity: The Contrarian View

Is that it? Is the potential conflict between the nonlocality of quantum mechanics and special relativity fully resolved? Well, probably. On the basis of the above considerations, the majority of physicists sum it up by saying there is a harmonious coexistence between special relativity and Aspect's results on entangled particles. In short, special relativity survives by the skin of its teeth. Many physicists find this convincing, but others have a nagging sense that there is more to the story.

At a gut level I've always shared the coexistence view, but there is no denying that the issue is delicate. At the end of the day, no matter what holistic words one uses or what lack of information one highlights, two widely separated particles, each of which is governed by the randomness of quantum mechanics, somehow stay sufficiently "in touch" so that whatever one does, the other instantly does too. And that seems to suggest that some kind of faster-than-light *something* is operating between them.

Where do we stand? There is no ironclad, universally accepted answer. Some physicists and philosophers have suggested that progress hinges on our recognizing that the focus of the discussion so far is somewhat misplaced: the real core of special relativity, they rightly point out, is not so much that light sets a speed limit, as that light's speed is something that all observers, regardless of their own motion, agree upon.[16] More generally, these researchers emphasize, the central principle of special relativity is that no observational vantage point is singled out over any other. Thus, they propose (and many agree) that if the egalitarian treatment of all constant-velocity observers could be squared with the experimental results on entangled particles, the tension with special relativity would be resolved.[17] But achieving this goal is not a trivial task. To see this concretely, let's think about how good old-fashioned textbook quantum mechanics explains the Aspect experiment.

According to standard quantum mechanics, when we perform a measurement and find a particle to be here, we cause its probability wave to change: the previous range of potential outcomes is reduced to the one actual result our measurement finds, as illustrated in Figure 4.7. Physicists say the measurement causes the probability wave to *collapse* and they envision that the larger the initial probability wave at some location, the larger the likelihood that the wave will collapse to that point—that is, the larger the likelihood that the particle will be found at that point. In the standard approach, the collapse happens instantaneously across the whole universe: once you find the particle here, the thinking goes, the probability of its being found *anywhere else* immediately drops to zero, and this is reflected in an immediate collapse of the probability wave.

In the Aspect experiment, when the left-moving photon's spin is measured and is found, say, to be clockwise about some axis, this collapses its probability wave throughout all of space, instantaneously setting the counterclockwise part to zero. Since this collapse happens everywhere, it happens also at the location of the right-moving photon. And, it turns out, this affects the counterclockwise part of the right-moving photon's probability wave, causing it to collapse to zero too. Thus, no matter how far away the right-moving photon is from the left-moving photon, its probability wave is instantaneously affected by the change in the left-moving photon's probability wave, ensuring that it has the same spin as the left-moving photon along the chosen axis. In standard quantum mechanics, then, it is this

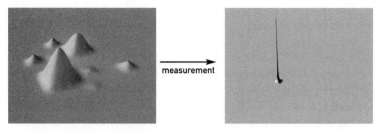

Figure 4.7 When a particle is observed at some location, the probability of finding it at any other location drops to zero, while its probability surges to 100 percent at the location where it is observed.

instantaneous change in probability waves that is responsible for the faster-than-light influence.

The mathematics of quantum mechanics makes this qualitative discussion precise. And, indeed, the long-range influences arising from collapsing probability waves change the prediction of how often Aspect's left and right detectors (when their axes are randomly and independently chosen) should find the same result. A mathematical calculation is required to get the exact answer (see notes section[18] if you're interested), but when the math is done, it predicts that the detectors should agree *precisely* 50 percent of the time (rather than predicting agreement more than 50 percent of the time—the result, as we've seen, found using EPR's hypothesis of a *local* universe). To impressive accuracy, *this is just what Aspect found in his experiments, 50 percent agreement.* Standard quantum mechanics matches the data impressively.

This is a spectacular success. Nevertheless, there is a hitch. *After more than seven decades, no one understands how or even whether the collapse of a probability wave really happens.* Over the years, the *assumption* that probability waves collapse has proven itself a powerful link between the probabilities that quantum theory predicts and the definite outcomes that experiments reveal. But it's an assumption fraught with conundrums. For one thing, the collapse does not emerge from the mathematics of quantum theory; it has to be put in by hand, and there is no agreed-upon or experimentally justified way to do this. For another, how is it possible that by finding an electron in your detector in New York City, you cause the electron's probability wave in the Andromeda galaxy to drop to zero instantaneously? To be sure, once you find the particle in New York City, it definitely won't be found in Andromeda, but what unknown mechanism enforces this with such spectacular efficiency? How, in looser language, does the part of the probability wave in Andromeda, and everywhere else, "know" to drop to zero simultaneously?[19]

We will take up this *quantum mechanical measurement problem* in Chapter 7 (and as we'll see, there are other proposals that avoid the idea of collapsing probability waves entirely), but suffice it here to note that, as we discussed in Chapter 3, something that is simultaneous from one perspective is not simultaneous from another moving perspective. (Remember Itchy and Scratchy setting their clocks on a moving train.) So if a probability wave were to undergo simultaneous collapse across space according to one observer, it will *not* undergo such simultaneous collapse according to another who is in motion. As a matter of fact, depending on their motion, some observers will report that the left photon was measured first, while other observers, equally trustworthy, will report that the right photon was measured first. Hence, even if the idea of collapsing probability waves were correct, there would fail to be an objective truth regarding which measurement—on the left or right photon—affected the other. Thus, the collapse of probability waves would seem to pick out one vantage point as special—the one according to which the collapse is simultaneous across space, the one according to which the left and right measurements occur at the same moment. But picking out a special perspective creates significant tension with the egalitarian core of special relativity. Proposals have been made to circumvent this problem, but debate continues regarding which, if any, are successful.[20]

Thus, although the majority view holds that there is a harmonious coexistence, some physicists and philosophers consider the exact relationship between quantum mechanics, entangled particles, and special relativity an open question. It's certainly possible, and in my view likely, that the majority view will ultimately prevail in some more definitive form. But history shows that subtle, foundational problems sometimes sow the seeds of future revolutions. On this one, only time will tell.

What Are We to Make of All This?

Bell's reasoning and Aspect's experiments show that the kind of universe Einstein envisioned may exist in the mind, but not in reality. Einstein's was a universe in which what you do right here has immediate relevance only for things that are also right here. Physics, in his view, was purely local. But we now see that the data rule out this kind of thinking; the data rule out this kind of universe.

Einstein's was also a universe in which objects possess definite values of all possible physical attributes. Attributes do not hang in limbo, waiting for an experimenter's measurement to bring them into existence. The majority of physicists would say that Einstein was wrong on this point, too. Particle properties, in this majority view, come into being when measurements force them to—an idea we will examine further in Chapter 7. When they are not being observed or interacting with the environment, particle properties have a nebulous, fuzzy existence characterized solely by a probability that one or another potentiality might be realized. The most extreme of those who hold this opinion would go as far as declaring that, indeed, when no one and no thing is "looking" at or interacting with the moon in any way, *it is not there.*

On this issue, the jury is still out. Einstein, Podolsky, and Rosen reasoned that the only sensible explanation for how measurements could reveal that widely separated particles had identical properties was that the particles possessed those definite properties all along (and, by virtue of their common past, their properties were correlated). Decades later, Bell's analysis and Aspect's data proved that this intuitively pleasing suggestion, based on the premise that particles always have definite properties, fails as an explanation of the experimentally observed nonlocal correlations. But the failure to explain away the mysteries of nonlocality does not mean that the notion of particles always possessing definite properties is itself ruled out. The data rule out a local universe, but they don't rule out particles having such hidden properties.

In fact, in the 1950s Bohm constructed his own version of quantum mechanics that incorporates both nonlocality *and* hidden variables. Particles, in this approach, always have both a definite position and a definite velocity, even though we can never measure both simultaneously. Bohm's approach made predictions that agreed fully with those of conventional quantum mechanics, but his formulation introduced an even more brazen element of nonlocality in which the *forces* acting on a particle at one location depend instantaneously on conditions at distant locations. In a sense, then, Bohm's version suggested how one might go partway toward Einstein's goal of restoring some of the intuitively sensible features of classical physics—particles having definite properties—that had been abandoned by the quantum revolution, but it also showed that doing so came at the price of accepting yet more blatant nonlocality. With this hefty cost, Einstein would have found little solace in this approach.

The need to abandon locality is the most astonishing lesson arising from the work of Einstein, Podolsky, Rosen, Bohm, Bell, and Aspect, as well as the many others who played important parts in this line of research. By virtue of their past, objects that at present are in vastly different regions of the universe can be part of a quantum mechanically entangled whole. Even though widely separated, such objects are committed to behaving in a random but coordinated manner.

We used to think that a basic property of space is that it separates and distinguishes one object from another. But we now see that quantum mechanics radically challenges this view. *Two things can be separated by an enormous amount of space and yet not have a fully independent existence.* A quantum connection can unite them, making the properties of each contingent on the properties of the other. Space does not distinguish such entangled objects. Space cannot overcome their interconnection. Space, even a huge amount of space, does not weaken their quantum mechanical interdependence.

Some people have interpreted this as telling us that "everything is connected to everything else" or that "quantum mechanics entangles us all in one universal whole." After all, the reasoning goes, at the big bang everything emerged from one place since, we believe, all places we now think of as different were the same place way back in the beginning. And since, like the two photons emerging from the same calcium atom, everything emerged from the same something in the beginning, everything should be quantum mechanically entangled with everything else.

While I like the sentiment, such gushy talk is loose and overstated. The quantum connections between the two photons emerging from the calcium atom are there, certainly, but they are extremely delicate. When Aspect and others carry out their experiments, it is crucial that the photons be allowed to travel absolutely unimpeded from their source to the detectors. Should they be jostled by stray particles or bump into pieces of equipment before reaching one of the detectors, the quantum connection between the photons will become monumentally more difficult to identify. Rather than looking for correlations in the properties of two photons, one would now need to look for a complex pattern of correlations involving the photons and everything else they may have bumped into. And as all these particles go their ways, bumping and jostling yet other particles, the quantum entanglement would become so spread out through these interactions with the environment that it would become virtually impossible to detect. For all intents and purposes, the original entanglement between the photons would have been erased.

Nevertheless, it is truly amazing that these connections do exist, and that in carefully arranged laboratory conditions they can be directly observed over significant distances. They show us, fundamentally, that space is not what we once thought it was.

What about time?

II

TIME AND EXPERIENCE

5

The Frozen River

Time is among the most familiar yet least understood concepts that humanity has ever encountered. We say that it flies, we say that it's money, we try to save it, we get annoyed when we waste it. But what *is* time? To paraphrase St. Augustine and Justice Potter Stewart, we know it when we see it, but surely, at the dawn of the third millennium our understanding of time must be deeper than that. In some ways, it is. In other ways, it's not. Through centuries of puzzling and pondering, we have gained insight into some of time's mysteries, but many remain. Where does time come from? What would it mean to have a universe without time? Could there be more than one time dimension, just as there is more than one space dimension? Can we "travel" to the past? If we did, could we change the subsequent unfolding of events? Is there an absolute, smallest amount of time? Is time a truly fundamental ingredient in the makeup of the cosmos, or simply a useful construct to organize our perceptions, but one not found in the lexicon with which the most fundamental laws of the universe are written? Could time be a derivative notion, emerging from some more basic concept that has yet to be discovered?

Finding complete and fully convincing answers to these questions ranks among the most ambitious goals of modern science. Yet the big questions are by no means the only ones. Even the everyday experience of time taps into some of the universe's thorniest conundrums.

Time and Experience

Special and general relativity shattered the universality, the oneness, of time. These theories showed that we each pick up a shard of Newton's old universal time and carry it with us. It becomes our own personal clock, our own personal lead relentlessly pulling us from one moment to the next. We are shocked by the theories of relativity, by the universe that is, because while our personal clock seems to tick away uniformly, in concert with our intuitive sense of time, comparison with other clocks reveals differences. Time for you need not be the same as time for me.

Let's accept that lesson as a given. But what *is* the true nature of time for me? What is the full character of time as experienced and conceived by the individual, without primary focus on comparisons with the experiences of others? Do these experiences accurately reflect the true nature of time? And what do they tell us about the nature of reality?

Our experiences teach us, overwhelmingly so, that the past is different from the future. The future seems to present a wealth of possibilities, while the past is bound to one thing, the fact of what actually happened. We feel able to influence, to affect, and to mold the future to one degree or another, while the past seems immutable. And in between *past* and *future* is the slippery concept of *now*, a temporal holding point that reinvents itself moment to moment, like the frames in a movie film as they sweep past the projector's intense light beam and become the momentary present. Time seems to march to an endless, perfectly uniform rhythm, reaching the fleeting destination of *now* with every beat of the drummer's stick.

Our experiences also teach us that there is an apparent lopsidedness to how things unfold in time. There is no use crying over spilled milk, because once spilled it can never be unspilled: we never see splattered milk gather itself together, rise off the floor, and coalesce in a glass that sets itself upright on a kitchen counter. Our world seems to adhere perfectly to a one-way temporal arrow, never deviating from the fixed stipulation that things can start like *this* and end like *that*, but they can never start like *that* and end like *this*.

Our experiences, therefore, teach us two overarching things about time. First, *time seems to flow*. It's as if we stand on the riverbank of time as the mighty current rushes by, sweeping the future toward us, becoming *now* at the moment it reaches us, and rushing onward as it recedes downstream into the past. Or, if that is too passive for your taste, invert the metaphor: we ride the river of time as it relentlessly rushes forward, sweeping us from one now to the next, as the past recedes with the passing scenery and the future forever awaits us downstream. (Our experiences have also taught us that time can inspire some of the mushiest metaphors.) Second, *time seems to have an arrow.* The flow of time seems to go one way and only one way, in the sense that things happen in one and only one temporal sequence. If someone hands you a box containing a short film of a glass of milk being spilled, but the film has been cut up into its individual frames, by examining the pile of images you can reassemble the frames in the right order without any help or instruction from the filmmaker. Time seems to have an intrinsic direction, pointing from what we call the past toward what we call the future, and things appear to change—milk spills, eggs break, candles burn, people age—in universal alignment with this direction.

These easily sensed features of time generate some of its most tantalizing puzzles. Does time really flow? If it does, what actually is flowing? And how fast does this time-stuff flow? Does time really have an arrow? Space, for example, does not appear to have an inherent arrow—to an astronaut in the dark recesses of the cosmos, left and right, back and forth, and up and down, would all be on equal footing—so where would an arrow of time come from? If there is an arrow of time, is it absolute? Or are there things that can evolve in a direction opposite to the way time's arrow seems to point?

Let's build up to our current understanding by first thinking about these questions in the context of classical physics. So, for the remainder of this and the next chapter (in which we'll discuss the flow of time and the arrow of time, respectively) we will ignore quantum probability and quantum uncertainty. A good deal of what we'll learn, nevertheless, translates directly to the quantum domain, and in Chapter 7 we will take up the quantum perspective.

Does Time Flow?

From the perspective of sentient beings, the answer seems obvious. As I type these words, I clearly *feel* time flowing. With every keystroke, each now gives way to the next. As you read these words, you no doubt feel time flowing, too, as your eyes scan from word to word across the page. Yet, as hard as physicists have tried, no one has found any convincing evidence within the laws of physics that supports this intuitive sense that time flows. In fact, a reframing of some of Einstein's insights from special relativity provides evidence that time does not flow.

To understand this, let's return to the loaf-of-bread depiction of space-time introduced in Chapter 3. Recall that the slices making up the loaf are the nows of a given observer; each slice represents space at one moment of time from his or her perspective. The union obtained by placing slice next to slice, in the order in which the observer experiences them, fills out a region of spacetime. If we take this perspective to a logical extreme and imagine that each slice depicts *all* of space at a given moment of time according to one observer's viewpoint, and if we include every possible slice, from the ancient past to the distant future, the loaf will encompass all of the universe throughout all time—the whole of spacetime. Every occurrence, regardless of when or where, is represented by some point in the loaf.

This is schematically illustrated in Figure 5.1, but the perspective should make you scratch your head. The "outside" perspective of the figure, in which we're looking at the whole universe, all of space at every moment of time, is a fictitious vantage point, one that none of us will ever

Figure 5.1 A schematic depiction of all space throughout all time (depicting, of course, only part of space through part of time) showing the formation of some early galaxies, the formation of the sun and the earth, and the earth's ultimate demise when the sun swells into a red giant, in what we now consider our distant future.

have. We are all *within* spacetime. Every experience you or I ever have occurs at some location in space at some moment of time. And since Figure 5.1 is meant to depict all of spacetime, it encompasses the totality of such experiences—yours, mine, and those of everyone and everything. If you could zoom in and closely examine all the comings and goings on planet earth, you'd be able to see Alexander the Great having a lesson with Aristotle, Leonardo da Vinci laying the final brushstroke on the Mona Lisa, and George Washington crossing the Delaware; as you continued scanning the image from left to right, you'd be able to see your grandmother playing as a little girl, your father celebrating his tenth birthday, and your own first day at school; looking yet farther to the right in the image, you could see yourself reading this book, the birth of your great-great-granddaughter, and, a little farther on, her inauguration as President. Given the coarse resolution of Figure 5.1, you can't actually see these moments, but you can see the (schematic) history of the sun and planet earth, from their birth out of a coalescing gas cloud to the earth's demise when the sun swells into a red giant. It's all there.

Unquestionably, Figure 5.1 is an imaginary perspective. It stands outside of space and time. It is the view from nowhere and nowhen. Even so—even though we can't actually step beyond the confines of spacetime and take in the full sweep of the universe—the schematic depiction of Figure 5.1 provides a powerful means of analyzing and clarifying basic properties of space and time. As a prime example, the intuitive sense of time's flow can be vividly portrayed in this framework by a variation on the movie-projector metaphor. We can envision a light that illuminates one time slice after another, momentarily making the slice come alive in the present—making it the momentary *now*—only to let it go instantly dark again as the light moves on to the next slice. Right now, in this intuitive way of thinking about time, the light is illuminating the slice in which you, sitting on planet earth, are reading *this* word, and now it is illuminating the slice in which you are reading *this* word. But, again, while this image seems to match experience, scientists have been unable to find anything in the laws of physics that embodies such a moving light. They have found no physical mechanism that singles out moment after moment to be momentarily real—to be the momentary *now*—as the mechanism flows ever onward toward the future.

Quite the contrary. While the *perspective* of Figure 5.1 is certainly imaginary, there is convincing evidence that the spacetime loaf—the totality of spacetime, not slice by single slice—is real. A less than widely appreciated implication of Einstein's work is that special relativistic reality treats all times equally. Although the notion of *now* plays a central role in our worldview, relativity subverts our intuition once again and declares ours an egalitarian universe in which every moment is as real as any other. We brushed up against this idea in Chapter 3 while thinking about the spinning bucket in the context of special relativity. There, through indirect reasoning analogous to Newton's, we concluded that spacetime is at least enough of a something to provide the benchmark for accelerated motion. Here we take up the issue from another viewpoint and go further. We argue that every part of the spacetime loaf in Figure 5.1 exists on the same footing as every other, suggesting, as Einstein believed, that reality embraces past, present, and future *equally* and that the flow we envision bringing one section to light as another goes dark is illusory.

The Persistent Illusion of Past, Present, and Future

To understand Einstein's perspective, we need a working definition of reality, an algorithm, if you will, for determining what things exist at a given moment. Here's one common approach. When I contemplate reality—what exists at *this* moment—I picture in my mind's eye a kind of snapshot, a mental freeze-frame image of the entire universe right *now*. As I type these words, my sense of what exists right *now*, my sense of reality, amounts to a list of all those things—the tick of midnight on my kitchen clock; my cat stretched out in flight between floor and windowsill; the first ray of morning sunshine illuminating Dublin; the hubbub on the floor of the Tokyo stock exchange; the fusion of two particular hydrogen atoms in the sun; the emission of a photon from the Orion nebula; the last moment of a dying star before it collapses into a black hole—that are, at this moment, in my freeze-frame mental image. These are the things happening right *now*, so they are the things that I declare exist right *now*. Does Charlemagne exist right now? No. Does Nero exist right now? No. Does Lincoln exist right now? No. Does Elvis exist right now? No. None of them are on my current now-list. Does anyone born in the year 2300 or 3500 or 57000 exist now? No. Again, none of them are in my mind's-eye freeze-frame image, none of them are on my current time slice, and so, none of them are on my current now-list. Therefore, I say without hesitation that they do not currently exist. That is how I define reality at any given moment; it's an intuitive approach that most of us use, often implicitly, when thinking about existence.

I will make use of this conception below, but be aware of one tricky point. A now-list—reality in this way of thinking—is a funny thing. Nothing you see right *now* belongs on your now-list, because it takes time for light to reach your eyes. Anything you see right *now* has already happened. You are not seeing the words on this page as they are now; instead, if you are holding the book a foot from your face, you are seeing them as they were a billionth of a second ago. If you look out across an average room, you are seeing things as they were some 10 billionths to 20 billionths of a second ago; if you look across the Grand Canyon, you are seeing the other side as it was about one ten-thousandth of a second ago; if you look at the moon, you are seeing it as it was a second and a half ago; for the sun, you see it as it was about eight minutes ago; for stars visible to the naked eye, you see them as they were from roughly a few years ago to 10,000 years ago. Curiously, then, although a mental freeze-frame image captures our sense of reality, our intuitive sense of "what's out there," it consists of events that we can't experience, or affect, or even record right now. Instead, an actual now-list can be compiled only after the fact. If you know how far away something is, you can determine when it emitted the light you see *now* and so you can determine on which of your time slices it belongs—on which already past now-list it should be recorded. Nevertheless, and this is the main point, as we use this information to compile the now-list for any given moment, continually updating it as we receive light from ever more distant sources, the things that are listed are the things that we intuitively believe existed at that moment.

Remarkably, this seemingly straightforward way of thinking leads to an unexpectedly expansive conception of reality. You see, according to Newton's absolute space and absolute time, everyone's freeze-frame picture of the universe at a given moment contains exactly the same events; everyone's *now* is the same *now*, and so everyone's now-list for a given moment is identical. If someone or something is on your now-list for a given moment, then it is necessarily also on my now-list for that moment. Most people's intuition is still bound up with this way of thinking, but special relativity tells a very different story. Look again at Figure 3.4. Two observers in relative motion have *nows*—single moments in time, from each one's perspective—that are different: their *nows* slice through spacetime at different angles. And different *nows* mean different now-lists. *Observers mov-*

ing relative to each other have different conceptions of what exists at a given moment, and hence they have different conceptions of reality.

At everyday speeds, the angle between two observers' now-slices is minuscule; that's why in day-to-day life we never notice a discrepancy between our definition of *now* and anybody else's. For this reason, most discussions of special relativity focus on what would happen if we traveled at enormous speeds—speeds near that of light—since such motion would tremendously magnify the effects. But there is another way to magnify the distinction between two observers' conceptions of *now*, and I find that it provides a particularly enlightening approach to the question of reality. It is based on the following simple fact: if you and I slice up an ordinary loaf at slightly different angles, it will have hardly any effect on the resulting pieces of bread. But if the loaf is *huge*, the conclusion is different. Just as a tiny opening between the blades of an enormously long pair of scissors translates into a large separation between the blade tips, cutting an enormous loaf of bread at slightly different angles yields slices that deviate by a huge amount at distances far from where the slices cross. You can see this in Figure 5.2.

The same is true for spacetime. At everyday speeds, the slices depicting *now* for two observers in relative motion will be oriented at only slightly different angles. If the two observers are nearby, this will have hardly any effect. But, just as in the loaf of bread, tiny angles generate large separations between slices when their impact is examined over large distances. And for slices of spacetime, a large deviation between slices means a significant disagreement on which events each observer considers to be happening now. This is illustrated in Figures 5.3 and 5.4, and it implies that individuals moving relative to each other, even at ordinary, everyday speeds, will have increasingly different conceptions of *now* if they are increasingly far apart in space.

To make this concrete, imagine that Chewie is on a planet in a galaxy far, far away—10 billion light-years from earth—idly sitting in his living room. Imagine further that you (sitting still, reading these words) and Chewie are not moving relative to each other (for simplicity, ignore the motion of the planets, the expansion of the universe, gravitational effects, and so on). Since you are at rest relative to each other, you and Chewie agree fully on issues of space and time: you would slice up spacetime in an identical manner, and so your now-lists would coincide exactly. After a little while, Chewie stands up and goes for a walk—a gentle, relaxing amble—in a direction that turns out to be directly away from you. This

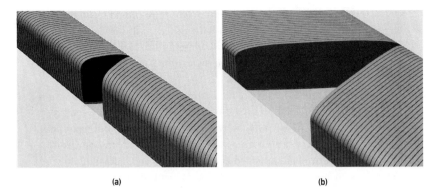

(a)   **(b)**

Figure 5.2 (a) In an ordinary loaf, slices cut at slightly different angles don't separate significantly. **(b)** But the larger the loaf, for the same angle, the greater the separation.

change in Chewie's state of motion means that his conception of *now*, his slicing up of spacetime, will rotate slightly (see Figure 5.3). This tiny angular change has no noticeable effect in Chewie's vicinity: the difference between his new *now* and that of anyone still sitting in his living room is minuscule. But over the enormous distance of 10 billion light-

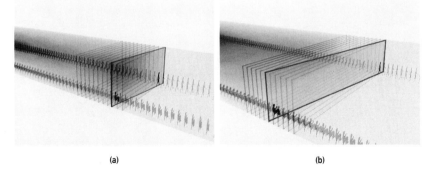

(a)   **(b)**

Figure 5.3 (a) Two individuals at rest relative to each other have identical conceptions of *now* and hence identical time slices. If one observer moves away from the other their time slices—what each observer considers *now*—rotate relative to each other; as illustrated, the darkened *now* slice for the moving observer rotates into the past of the stationary observer. **(b)** A greater separation between the observers yields a greater deviation between slices—a greater deviation in their conception of *now*.

years, this tiny shift in Chewie's notion of *now* is amplified (as in the passage from Figure 5.3a to 5.3b, but with the protagonists now being a huge distance apart, significantly accentuating the shift in their *nows*). *His now and your now, which were one and the same while he was sitting still, jump apart because of his modest motion.*

Figures 5.3 and 5.4 illustrate the key idea schematically, but by using the equations of special relativity we can calculate how different your *nows* become.[1] If Chewie walks away from you at about 10 miles per hour (Chewie has quite a stride) the events on earth that belong on his new now-list are events that happened about 150 years ago, according to you! According to his conception of *now*—a conception that is every bit as valid as yours and that up until a moment ago agreed fully with yours—you have not yet been born. If he moved toward you at the same speed, the angular shift would be opposite, as schematically illustrated in Figure 5.4, so that his *now* would coincide with what you would call 150 years in the future! Now, according to his *now*, you may no longer be a part of this world. And if, instead of just walking, Chewie hopped into the *Millennium Falcon* traveling at 1,000 miles per hour (less than the speed of a Concorde aircraft), his *now* would include events on earth that from your perspective took place 15,000 years ago or 15,000 years in the future,

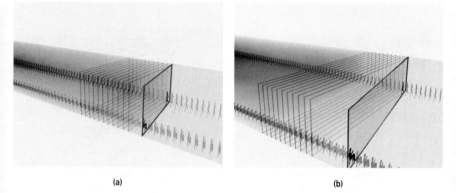

(a)   (b)

Figure 5.4 (a) Same as figure 5.3a, except when one observer moves toward the other, her *now* slice rotates into the future, not the past, of the other observer. **(b)** Same as 5.3b—a greater separation yields a greater deviation in conceptions of *now*, for the same relative velocity—with the rotation being toward the future instead of the past.

depending on whether he flew away or toward you. Given suitable choices of direction and speed of motion, Elvis or Nero or Charlemagne or Lincoln or someone born on earth way into what you call the future will belong on his new now-list.

While surprising, none of this generates any contradiction or paradox because, as we explained above, the farther away something is, the longer it takes to receive light it emits and hence to determine that it belongs on a particular now-list. For instance, even though John Wilkes Booth's approaching the State Box at Ford's Theatre will belong on Chewie's new now-list if he gets up and walks away from earth at about 9.3 miles per hour,[2] he can take no action to save President Lincoln. At such an enormous distance, it takes an enormous amount of time for messages to be received and exchanged, so only Chewie's descendants, billions of years later, will actually receive the light from that fateful night in Washington. The point, though, is that when his descendants use this information to update the vast collection of past now-lists, they will find that the Lincoln assassination belongs on the same now-list that contains Chewie's just getting up and walking away from earth. And yet, they will also find that a moment before Chewie got up, his now-list contained, among many other things, you, in earth's twenty-first century, sitting still, reading these words.[3]

Similarly, there are things about our future, such as who will win the U.S. presidential election in the year 2100, that seem completely open: more than likely, the candidates for that election haven't even been born, much less decided to run for office. But if Chewie gets up from his chair and walks toward earth at about 6.4 miles per hour, his now-slice—his conception of what exists, his conception of what has happened—*will* include the selection of the first president of the twenty-second century. Something that seems completely undecided to us is something that, for him, has already happened. Again, Chewie won't know the outcome of the election for billions of years, since that's how long it will take our television signals to reach him. But when word of the election results reaches Chewie's descendants and they use it to update Chewie's flip-card book of history, his collection of past now-lists, they will find that the election results belong on the same now-list in which Chewie got up and started walking toward earth—a now-list, Chewie's descendants note, that occurs just a moment after one that contains you, in the early years of earth's twenty-first century, finishing this paragraph.

This example highlights two important points. First, although we are used to the idea that relativistic effects become apparent at speeds near that of light, even at low velocities relativistic effects can be greatly amplified when considered over large distances in space. Second, the example gives insight into the issue of whether spacetime (the loaf) is really an entity or just an abstract concept, an abstract union of space right *now* together with its history and purported future.

You see, Chewie's conception of reality, his freeze-frame mental image, his conception of what exists *now*, is every bit as real for him as our conception of reality is for us. So, in assessing what constitutes reality, it would be stunningly narrow-minded if we didn't also include his perspective. For Newton, such an egalitarian approach wouldn't make the slightest difference, because, in a universe with absolute space and absolute time, everyone's now-slice coincides. But in a relativistic universe, our universe, it makes a big difference. Whereas our familiar conception of what exists right now amounts to a single now-slice—we usually view the past as gone and the future as yet to be—we must augment this image with Chewie's now-slice, a now-slice that, as the discussion revealed, can differ substantially from our own. Furthermore, since Chewie's initial location and the speed with which he moves are arbitrary, we should include the now-slices associated with all possibilities. These now-slices, as in our discussion above, would be centered on Chewie's—or some other real or hypothetical observer's—initial location in space and would be rotated at an angle that depends on the velocity chosen. (The only restriction comes from the speed limit set by light and, as explained in the notes, in the graphic depiction we are using this translates into a limit on the rotation angle of 45 degrees, either clockwise or counterclockwise.) As you can see, in Figure 5.5, the collection of all these now-slices fills out a substantial region of the spacetime loaf. In fact, if space is infinite—if now-slices extended infinitely far—then the rotated now-slices can be centered arbitrarily far away, and hence their union sweeps through *every* point in the spacetime loaf.*

So: *if you buy the notion that reality consists of the things in your freeze-frame mental image right now, and if you agree that your* now *is no more valid than the* now *of someone located far away in space who can*

---

*Pick any point in the loaf. Draw a slice that includes the point, and which intersects our current now-slice at an angle that is less than 45 degrees. This slice will represent the now-slice—*reality*—of a distant observer who was initially at rest relative to us, like Chewie, but is now moving relative to us at less than the speed of light. By design, this slice includes the (arbitrary) point in the loaf you happened to pick.[4]

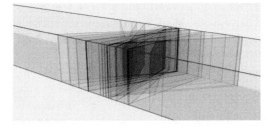

Figure 5.5 A sample of now-slices for a variety of observers (real or hypothetical) situated at a variety of distances from earth, moving with a variety of velocities.

move freely, then reality encompasses all of the events in spacetime. The total loaf exists. Just as we envision all of space as *really* being out there, as *really* existing, we should also envision all of time as *really* being out there, as *really* existing, too. Past, present, and future certainly appear to be distinct entities. But, as Einstein once said, "For we convinced physicists, the distinction between past, present, and future is only an illusion, however persistent."[5] The only thing that's real is the whole of spacetime.

Experience and the Flow of Time

In this way of thinking, events, regardless of when they happen from any particular perspective, just *are*. They all exist. They eternally occupy their particular point in spacetime. There is no flow. If you were having a great time at the stroke of midnight on New Year's Eve, 1999, you still are, since that is just one immutable location in spacetime. It is tough to accept this description, since our worldview so forcefully distinguishes between past, present, and future. But if we stare intently at this familiar temporal scheme and confront it with the cold hard facts of modern physics, its only place of refuge seems to lie within the human mind.

Undeniably, our conscious experience seems to sweep through the slices. It is as though our minds provide the projector light referred to earlier, so that moments of time come to life when they are illuminated by the power of consciousness. The flowing sensation from one moment to the next arises from our conscious recognition of change in our thoughts, feelings, and perceptions. And the sequence of change seems to have a continuous motion; it seems to unfold into a coherent story. But—without any pretense of psychological or neurobiological precision—we can envision how we might experience a flow of time even though, in actuality, there may be no such thing. To see what I mean, imagine playing *Gone with the Wind* through a faulty DVD player that randomly jumps forward and backward: one still frame flashes momentarily on the screen and is followed immediately by another from a completely different part of the film. When you watch this jumbled version, it will be hard for you to make sense of what's going on. But Scarlett and Rhett have no problem. In each frame, they do what they've always done in that frame. Were you able to stop the DVD on some particular frame and ask them about their thoughts and memories, they'd respond with the same answers they would have given had you played the DVD in a properly functioning player. If you asked them whether it was confusing to romp through the Civil War out of order, they'd look at you quizzically and figure you'd tossed back one too many mint juleps. In any given frame, they'd have the thoughts and memories they've always had in that frame—and, in particular, those thoughts and memories would give them the sensation that time is smoothly and coherently flowing forward, as usual.

Similarly, each moment in spacetime—each time slice—is like one of the still frames in a film. It exists whether or not some light illuminates it. As for Scarlett and Rhett, to the you who is in any such moment, it *is* the *now*, it *is* the moment you experience at *that* moment. And it always will be. Moreover, within each individual slice, your thoughts and memories are sufficiently rich to yield a sense that time has continuously flowed to that moment. This feeling, this sensation that time is flowing, doesn't require previous moments—previous frames—to be "sequentially illuminated."[6]

And if you think about it for one more moment, you'll realize that's a very good thing, because the notion of a projector light sequentially bringing moments to life is highly problematic for another, even more basic reason. If the projector light properly did its job and illuminated a given moment—say, the stroke of midnight, New Year's Eve, 1999—what would it mean for that moment to then go dark? If the moment were lit, then being illuminated would be a feature of the moment, a feature as everlasting and unchanging as everything else happening at that moment.

To experience illumination—to be "alive," to be the present, to be *the now*—and to then experience darkness—to be "dormant," to be the past, to be what was—is to experience change. *But the concept of change has no meaning with respect to a single moment in time.* The change would have to occur through time, the change would mark the passing of time, but what notion of time could that possibly be? By definition, moments *don't* include the passing of time—at least, not the time we're aware of—because moments just are, they are the raw material of time, they *don't* change. A particular moment can no more change in time than a particular location can move in space: if the location were to move, it would be a different location in space; if a moment in time were to change, it would be a different moment in time. The intuitive image of a projector light that brings each new *now* to life just doesn't hold up to careful examination. Instead, every moment is illuminated, and every moment remains illuminated. Every moment *is*. Under close scrutiny, the flowing river of time more closely resembles a giant block of ice with every moment forever frozen into place.[7]

This conception of time is significantly different from the one most of us have internalized. Even though it emerged from his own insights, Einstein was not hardened to the difficulty of fully absorbing such a profound change in perspective. Rudolf Carnap[8] recounts a wonderful conversation he had with Einstein on this subject: "Einstein said that the problem of the now worried him seriously. He explained that the experience of the now means something special for man, something essentially different from the past and the future, but that this important difference does not and cannot occur within physics. That this experience cannot be grasped by science seemed to him a matter of painful but inevitable resignation."

This resignation leaves open a pivotal question: Is science unable to grasp a fundamental quality of time that the human mind embraces as readily as the lungs take in air, or does the human mind impose on time a quality of its own making, one that is artificial and that hence does not show up in the laws of physics? If you were to ask me this question during the working day, I'd side with the latter perspective, but by nightfall, when critical thought eases into the ordinary routines of life, it's hard to maintain full resistance to the former viewpoint. Time is a subtle subject and we are far from understanding it fully. It is possible that some insightful person will one day devise a new way of looking at time and reveal a bona fide physical foundation for a time that flows. Then again, the discussion above, based on logic and relativity, may turn out to be the full story. Certainly, though, the feeling that time flows is deeply ingrained in our experience and thoroughly pervades our thinking and language. So much so, that we have lapsed, and will continue to lapse, into habitual, colloquial descriptions that refer to a flowing time. But don't confuse language with reality. Human language is far better at capturing human experience than at expressing deep physical laws.

6

Chance and the Arrow

DOES TIME HAVE A DIRECTION?

Even if time doesn't flow, it still makes sense to ask whether it has an arrow—whether there is a direction to the way things unfold *in* time that can be discerned in the laws of physics. It is the question of whether there is some intrinsic order in how events are sprinkled along spacetime and whether there is an essential scientific difference between one ordering of events and the reverse ordering. As everyone already knows, there certainly appears to be a huge distinction of this sort; it's what gives life promise and makes experience poignant. Yet, as we'll see, explaining the distinction between past and future is harder than you'd think. Rather remarkably, the answer we'll settle upon is intimately bound up with the precise conditions at the origin of the universe.

The Puzzle

A thousand times a day, our experiences reveal a distinction between things unfolding one way in time and the reverse. A piping hot pizza cools down en route from Domino's, but we never find a pizza arriving hotter than when it was removed from the oven. Cream stirred into coffee forms a uniformly tan liquid, but we never see a cup of light coffee unstir and separate into white cream and black coffee. Eggs fall, cracking and splattering, but we never see splattered eggs and eggshells gather together and coalesce into uncracked eggs. The compressed carbon dioxide gas in a bottle of Coke rushes outward when we twist off the cap, but we never find spread-out carbon dioxide gas gathering together and swooshing back into the bottle. Ice cubes put into a glass of room-temperature water melt, but we never see globules in a room-temperature glass of water coalesce into solid cubes of ice. These common sequences of events, as well as countless others, happen in only one temporal order. They never happen in reverse, and so they provide a notion of before and after—they give us a consistent and seemingly universal conception of past and future. These observations convince us that were we to examine all of spacetime from the outside (as in Figure 5.1), we would see significant asymmetry along the time axis. Splattered eggs the world over would lie to one side—the side we conventionally call the future—of their whole, unsplattered counterparts.

Perhaps the most pointed example of all is that our minds seem to have access to a collection of events that we call the past—our memories—but none of us seems able to remember the collection of events we call the future. So it seems obvious that there is a big difference between the past and the future. There seems to be a manifest orientation to how an enormous variety of things unfold in time. There seems to be a manifest distinction between the things we can remember (the past) and the things we cannot (the future). This is what we mean by time's having an orientation, a direction, or an arrow.[1]

Physics, and science more generally, is founded on regularities. Scientists study nature, find patterns, and codify these patterns in natural laws. You would think, therefore, that the enormous wealth of regularity leading us to perceive an apparent arrow of time would be evidence of a fundamental law of nature. A silly way to formulate such a law would be to introduce the Law of Spilled Milk, stating that glasses of milk spill but don't unspill, or the Law of Splattered Eggs, stating that eggs break and splatter but never unsplatter and unbreak. But that kind of law buys us nothing: it is merely descriptive, and offers no explanation beyond a simple observation of what happens. Yet we expect that somewhere in the depths of physics there must be a less silly law describing the motion and properties of the particles that make up pizza, milk, eggs, coffee, people, and stars—the fundamental ingredients of everything—that shows why things evolve through one sequence of steps but never the reverse. Such a law would give a fundamental explanation to the observed arrow of time.

The perplexing thing is that no one has discovered any such law. What's more, the laws of physics that have been articulated from Newton through Maxwell and Einstein, and up until today, show a *complete sym-*

metry between past and future.* Nowhere in any of these laws do we find a stipulation that they apply one way in time but not in the other. Nowhere is there any distinction between how the laws look or behave when applied in either direction in time. The laws treat what we call past and future on a completely equal footing. Even though experience reveals over and over again that there is an arrow of how events unfold in time, this arrow seems not to be found in the fundamental laws of physics.

Past, Future, and the Fundamental Laws of Physics

How can this be? Do the laws of physics provide no underpinning that distinguishes past from future? How can there be no law of physics explaining that events unfold in *this* order but never in reverse?

The situation is even more puzzling. The known laws of physics actually declare—contrary to our lifetime of experiences—that light coffee can separate into black coffee and white cream; a splattered yolk and a collection of smashed shell pieces can gather themselves together and form a perfectly smooth unbroken egg; the melted ice in a glass of room-temperature water can fuse back together into cubes of ice; the gas released when you open your soda can rush back into the bottle. All the physical laws that we hold dear fully support what is known as *time-reversal symmetry.* This is the statement that if some sequence of events can unfold in one temporal order (cream and coffee mix, eggs break, gas rushes outward) then these events can also unfold in reverse (cream and coffee unmix, eggs unbreak, gas rushes inward). I'll elaborate on this shortly, but the one-sentence summary is that not only do known laws fail to tell us why we see events unfold in only one order, they also tell us that, in theory, events can unfold in reverse order.†

The burning question is *Why don't we ever see such things?* I think it's a safe bet that no one has ever actually witnessed a splattered egg unsplat-

---

*There is an exception to this statement having to do with a certain class of exotic particles. As far as the questions discussed in this chapter are concerned, I consider this likely to be of little relevance and so won't mention this qualification further. If you are interested, it is briefly discussed in note 2.

†Note that time-reversal symmetry is not about time itself being reversed or "running" backward. Instead, as we've described, time-reversal symmetry is concerned with whether events that happen *in* time, in one particular temporal order, can also happen in the reverse order. A more appropriate phrase might be *event reversal* or *process reversal* or *event order reversal*, but we'll stick with the conventional term.

tering. But if the laws of physics allow it, and if, moreover, those laws treat splattering and unsplattering equally, why does one never happen while the other does?

Time-Reversal Symmetry

As a first step toward resolving this puzzle, we need to understand in more concrete terms what it means for the known laws of physics to be time-reversal symmetric. To this end, imagine it's the twenty-fifth century and you're playing tennis in the new interplanetary league with your partner, Coolstroke Williams. Somewhat unused to the reduced gravity on Venus, Coolstroke hits a gargantuan backhand that launches the ball into the deep, isolated darkness of space. A passing space shuttle films the ball as it goes by and sends the footage to CNN (Celestial News Network) for broadcast. Here's the question: If the technicians at CNN were to make a mistake and run the film of the tennis ball in reverse, would there be any way to tell? Well, if you knew the heading and orientation of the camera during the filming you might be able to recognize their error. But could you figure it out solely by looking at the footage itself, with no additional information? The answer is no. If in the correct (forward) time direction the footage showed the ball floating by from left to right, then in reverse it would show the ball floating by from right to left. And certainly, the laws of classical physics allow tennis balls to move either left or right. So the motion you see when the film is run in either the forward time direction or the reverse time direction is perfectly consistent with the laws of physics.

We've so far imagined that no forces were acting on the tennis ball, so that it moved with constant velocity. Let's now consider the more general situation by including forces. According to Newton, the effect of a force is to change the velocity of an object: forces impart accelerations. Imagine, then, that after floating awhile through space, the ball is captured by Jupiter's gravitational pull, causing it to move with increasing speed in a downward, rightward-sweeping arc toward Jupiter's surface, as in Figures 6.1a and 6.1b. If you play a film of this motion in reverse, the tennis ball will appear to move in an arc that sweeps upward and toward the left, away from Jupiter, as in Figure 6.1c. Here's the new question: is the motion depicted by the film when played backward—the time-reversed

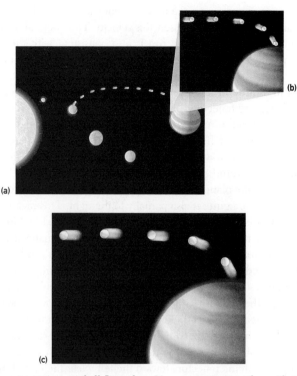

Figure 6.1 (a) A tennis ball flying from Venus to Jupiter together with **(b)** a close-up. **(c)** Tennis ball's motion if its velocity is reversed just before it hits Jupiter.

motion of what was actually filmed—allowed by the classical laws of physics? Is it motion that could happen in the real world? At first, the answer seems obviously to be yes: tennis balls can move in downward arcs to the right or upward arcs to the left, or, for that matter, in innumerable other trajectories. So what's the difficulty? Well, although the answer is indeed yes, this reasoning is too glib and misses the real intent of the question.

When you run the film in reverse, you see the tennis ball leap from Jupiter's surface, moving upward and toward the left, with exactly the same speed (but in exactly the opposite direction) from when it hit the planet. This initial part of the film is certainly consistent with the laws of physics: we can imagine, for example, someone launching the tennis ball from Jupiter's surface with precisely this velocity. The essential question is whether the *rest* of the reverse run is also consistent with the laws of physics. Would a ball launched with this initial velocity—and subject to Jupiter's downward-pulling gravity—actually move along the trajectory depicted in the rest of the reverse run film? Would it exactly retrace its original downward trajectory, but in reverse?

The answer to this more refined question *is* yes. To avoid any confusion, let's spell this out. In Figure 6.1a, before Jupiter's gravity had any significant effect, the ball was heading purely to the right. Then, in Figure 6.1b, Jupiter's powerful gravitational force caught hold of the ball and pulled it toward the planet's center—a pull that's mostly downward but, as you can see in the figure, is also partially to the right. This means that as the ball closed in on Jupiter's surface, its rightward speed had increased somewhat, but its downward speed had increased dramatically. In the reverse run film, therefore, the ball's launch from Jupiter's surface would be headed somewhat *leftward* but predominantly *upward*, as in Figure 6.1c. With this starting velocity, Jupiter's gravity would have had its greatest impact on the ball's upward speed, causing it to go slower and slower, while also decreasing the ball's leftward speed, but less dramatically. And with the ball's upward speed rapidly diminishing, its motion would become dominated by its speed in the leftward direction, causing it to follow an upward-arcing trajectory toward the left. Near the end of this arc, gravity would have sapped all the upward motion as well as the additional rightward velocity Jupiter's gravity imparted to the ball on its way down, leaving the ball moving purely to the left with exactly the same speed it had on its initial approach.

All this can be made quantitative, but the point to notice is that this trajectory is exactly the reverse of the ball's original motion. Simply by reversing the ball's velocity, as in Figure 6.1c—by setting it off with the same speed but in the opposite direction—one can make it fully retrace its original trajectory, but in reverse. Bringing the film back into the discussion, we see that the upward-arcing trajectory to the left—the trajectory we just figured out with reasoning based on Newton's laws of motion—is exactly what we would see upon running the film in reverse. So the ball's time-reversed motion, as depicted in the reverse-run film, conforms to the laws of physics just as surely as its forward-time motion. The motion we'd see upon running the film in reverse is motion that could *really* happen in the real world.

Although there are a few subtleties I've relegated to the endnotes, this conclusion is general.[2] All the known and accepted laws relating to motion—from Newton's mechanics just discussed, to Maxwell's electromagnetic theory, to Einstein's special and general theories of relativity (remember, we are putting off quantum mechanics until the next chapter)—embody time-reversal symmetry: motion that can occur in the usual forward-time direction can equally well occur in reverse. As the terminology can be a bit confusing, let me reemphasize that we are not reversing time. Time is doing what it always does. Instead, our conclusion is that *we can make an object trace its trajectory in reverse by the simple procedure of reversing its velocity at any point along its path*. Equivalently, the same procedure—reversing the object's velocity at some point along its path—would make the object execute the motion we'd see in a reverse-run film.

Tennis Balls and Splattering Eggs

Watching a tennis ball shoot between Venus and Jupiter—in either direction—is not particularly interesting. But as the conclusion we've reached is widely applicable, let's now go someplace more exciting: your kitchen. Place an egg on your kitchen counter, roll it toward the edge, and let it fall to the ground and splatter. To be sure, there is a lot of motion in this sequence of events. The egg falls. The shell cracks apart. Yolk splatters this way and that. The floorboards vibrate. Eddies form in the surrounding air. Friction generates heat, causing the atoms and molecules of the egg, floor, and air to jitter a little more quickly. But just as the laws of physics show us how we can make the tennis ball trace its precise path in reverse, the same laws show how we can make every piece of eggshell, every drop of yolk, every section of flooring, and every pocket of air exactly trace its motion in reverse, too. "All" we need do is reverse the velocity of each and every constituent of the splatter. More precisely, the reasoning used with the tennis ball implies that if, hypothetically, we were able to simultaneously reverse the velocity of *every* atom and molecule involved directly or indirectly with the splattering egg, *all* the splattering motion would proceed in reverse.

Again, just as with the tennis ball, if we succeeded in reversing all these velocities, what we'd see would look like a reverse-run film. But, unlike the tennis ball's, the egg-splattering's reversal of motion would be extremely impressive. A wave of jostling air molecules and tiny floor vibrations would converge on the collision site from all parts of the kitchen, causing every bit of shell and drop of yolk to head back toward the impact location. Each ingredient would move with exactly the same speed it had in the original splattering process, but each would now move in the opposite direction. The drops of yolk would fly back into a globule just as scores of little shell pieces arrived on the outskirts, perfectly aligned to fuse together into a smooth ovoid container. The air and floor vibrations would precisely conspire with the motion of the myriad coalescing yolk drops and shell pieces to give the newly re-formed egg just the right kick to jump off the floor in one piece, rise up to the kitchen counter, and land gently on the edge with just enough rotational motion to roll a few inches and gracefully come to rest. This *is* what would happen if we could perform the task of total and exact velocity reversal of everything involved.[3]

Thus, whether an event is simple, like a tennis ball arcing, or something more complex, like an egg splattering, the laws of physics show that what happens in one temporal direction can, at least in principle, also happen in reverse.

Principle and Practice

The stories of the tennis ball and the egg do more than illustrate the time-reversal symmetry of nature's laws. They also suggest why, in the real world of experience, we see many things happen one way but never in reverse. To get the tennis ball to retrace its path was not that hard. We grabbed it and sent it off with the same speed but in the opposite direction. That's it. But to get all the chaotic detritus of the egg to retrace its path would be monumentally more difficult. We'd need to grab every bit of splatter, and simultaneously send each off at the same speed but in the opposite direction. Clearly, that's beyond what we (or even all the King's horses and all the King's men) can really do.

Have we found the answer we've been looking for? Is the reason why eggs splatter but don't unsplatter, even though both actions are allowed by the laws of physics, a matter of what is and isn't practical? Is the answer simply that it's easy to make an egg splatter—roll it off a counter—but extraordinarily difficult to make it unsplatter?

Well, if it were the answer, trust me, I wouldn't have made it into such a big deal. The issue of ease versus difficulty *is* an essential part of the answer, but the full story within which it fits is far more subtle and surprising. We'll get there in due course, but we must first make the discussion of this section a touch more precise. And that takes us to the concept of entropy.

Entropy

Etched into a tombstone in the Zentralfriedhof in Vienna, near the graves of Beethoven, Brahms, Schubert, and Strauss, is a single equation, $S = k \log W$, which expresses the mathematical formulation of a powerful concept known as *entropy*. The tombstone bears the name of Ludwig Boltzmann, one of the most insightful physicists working at the turn of the last century. In 1906, in failing health and suffering from depression, Boltzmann committed suicide while vacationing with his wife and daughter in Italy. Ironically, just a few months later, experiments began to confirm that ideas Boltzmann had spent his life passionately defending were correct.

The notion of entropy was first developed during the industrial revolution by scientists concerned with the operation of furnaces and steam engines, who helped develop the field of thermodynamics. Through many years of research, the underlying ideas were sharply refined, culminating in Boltzmann's approach. His version of entropy, expressed concisely by the equation on his tombstone, uses statistical reasoning to provide a link between the huge number of individual ingredients that make up a physical system and the overall properties the system has.[4]

To get a feel for the ideas, imagine unbinding a copy of *War and Peace*, throwing its 693 double-sided pages high into the air, and then gathering the loose sheets into a neat pile.[5] When you examine the resulting stack, it is enormously more likely that the pages will be out of order than in order. The reason is obvious. There are many ways in which the order of the pages can be jumbled, but only one way for the order to be correct. To be in order, of course, the pages must be arranged precisely as 1, 2; 3, 4; 5, 6; and so on, up to 1,385, 1,386. Any other arrangement is out of order. A simple but essential observation is that, all else being equal, the more ways something can happen, the more likely it is that it will happen. And if something can happen in *enormously* more ways, like the pages landing in the wrong numerical order, it is *enormously* more likely that it will happen. We all know this intuitively. If you buy one lottery ticket, there is only one way you can win. If you buy a million tickets, each with different numbers, there are a million ways you can win, so your chances of striking it rich are a million times higher.

Entropy is a concept that makes this idea precise by counting the number of ways, consistent with the laws of physics, in which any given physical situation can be realized. *High entropy means that there are many ways; low entropy means there are few ways.* If the pages of *War and Peace* are stacked in proper numerical order, that is a low-entropy configuration, because there is one and only one ordering that meets the criterion. If the pages are out of numerical order, that is a high-entropy situation, because a little calculation shows that there are

124552198453778343366002935370498829163361101246389045136887691264686895591852984504377394069294743950794189338751876527656714059286627151367074739129571382353800016108126465301823420562057147320617202938290291250213170227821191347358265588154107136014311932215753415973385542846729869139815159925119085867260993481056143034134383056377136715110570478694133391293419244096105142887984779085360950895401401259328506329060341095131494663898390526767610427804166730154945522818861025024633866260360150888664701014297085458481514159839254687623129529334782951868123707745965224321488873516792844834030007871706366846238435362424516736228610919853939181503076046890466491297894062503326518685837322713637024739040189109406498813983380265451114876864895816491403426444110871911844164280902757137738090672587084302157950158991623204581301295083438653790819182377773852143753631225316415985892681059765281448013877486970265254626439371893927305921796747169166978155198569769269249467383642278227334577671807331624043363695277118367410428449347223477922340272256307211938539124728809290720342716923779362076501904571097887744535443586803319160959249877443194986997700333249463073243755353229067448176579539562184032951681442710422276081242890487164286648724030707364864934832509996672897344642531034930062662201460431205110109328239624925119689782833061921508282708143936599873268490479941668396577478902124562796195600187060805768778947870098610692265944872693410000872699876339900302559168582063973485103562967646116002251592001137227412733180748295472481928076532664070230832754286312646671501355905966429773337131834654748547607012423301287213532123732873272187482526403991104970017214756

70049929226458643522650111999999999999999999999999999999999
99999999999999999999999999999999999999999999999999999999999999
99999999999999999999999999999999999999999999999999999999999999
9999999999999999999999999—about 10^{1878}—different out-of-order page arrangements.[6] If you throw the pages in the air and then gather them in a neat stack, it is almost certain that they will wind up out of numerical order, because such configurations have enormously higher entropy—there are many more ways to achieve an out-of-order outcome—than the sole arrangement in which they are in correct numerical order.

In principle, we could use the laws of classical physics to figure out exactly where each page will land after the whole stack has been thrown in the air. So, again in principle, we could precisely predict the resulting arrangement of the pages[7] and hence (unlike in quantum mechanics, which we ignore until the next chapter) there would seem to be no need to rely on probabilistic notions such as which outcome is more or less likely than another. But statistical reasoning is both powerful and useful. If *War and Peace* were a pamphlet of only a couple of pages we just might be able to successfully complete the necessary calculations, but it would be impossible to do this for the real *War and Peace*.[8] Following the precise motion of 693 floppy pieces of paper as they get caught by gentle air currents and rub, slide, and flap against one another would be a monumental task, well beyond the capacity of even the most powerful supercomputer.

Moreover—and this is critical—having the exact answer wouldn't even be that useful. When you examine the resulting stack of pages, you are far less interested in the exact details of which page happens to be where than you are in the general question of whether the pages are in the correct order. If they are, great. You could sit down and continue reading about Anna Pavlovna and Nikolai Ilych Rostov, as usual. But if you found that the pages were not in their correct order, the precise details of the page arrangement are something you'd probably care little about. If you've seen one disordered page arrangement, you've pretty much seen them all. Unless for some strange reason you get mired in the minutiae of which pages happen to appear here or there in the stack, you'd hardly notice if someone further jumbled an out-of-order page arrangement you'd initially been given. The initial stack would look disordered and the further jumbled stack would also look disordered. So not only is the statistical reasoning enormously easier to carry out, but the answer it yields—ordered versus disordered—is more relevant to our real concern, to the kind of thing of which we would typically take note.

This sort of big-picture thinking is central to the statistical basis of entropic reasoning. Just as any lottery ticket has the same chance of winning as any other, after many tosses of *War and Peace* any particular ordering of the pages is just as likely to occur as any other. What makes the statistical reasoning fly is our declaration that there are two *interesting classes* of page configurations: ordered and disordered. The first class has one member (the correct page ordering 1, 2; 3, 4; and so on) while the second class has a huge number of members (every other possible page ordering). These two classes are a sensible set to use since, as above, they capture the overall, gross assessment you'd make on thumbing through any given page arrangement.

Even so, you might suggest making finer distinctions between these two classes, such as arrangements with just a handful of pages out of order, arrangements with only pages in the first chapter out of order, and so on. In fact, it can sometimes be useful to consider these intermediate classes. However, the number of possible page arrangements in each of these new subclasses is still extremely small compared with the number in the fully disordered class. For example, the total number of out-of-order arrangements that involve only the pages in Part One of *War and Peace* is 10^{-178} of 1 percent of the total number of out-of-order arrangements involving all pages. So, although on the initial tosses of the unbound book the resulting page arrangement will likely belong to one of the intermediate, not fully disordered classes, it is almost certain that if you repeat the tossing action many times over, the page order will ultimately exhibit no obvious pattern whatsoever. The page arrangement evolves toward the fully disordered class, since there are so many page arrangements that fit this bill.

The example of *War and Peace* highlights two essential features of entropy. First, *entropy is a measure of the amount of disorder in a physical system.* High entropy means that many rearrangements of the ingredients making up the system would go unnoticed, and this in turn means the system is highly disordered (when the pages of *War and Peace* are all mixed up, any further jumbling will hardly be noticed since it simply leaves the pages in a mixed-up state). Low entropy means that very few rearrangements would go unnoticed, and this in turn means the system is highly ordered (when the pages of *War and Peace* start in their proper order, you can easily detect almost any rearrangement). Second, in physical systems with many constituents (for instance, books with many pages being tossed in the air) there is a natural evolution toward greater disorder, since disorder can be achieved in so many more ways than order. In the language of entropy, this is the statement that *physical systems tend to evolve toward states of higher entropy.*

Of course, in making the concept of entropy precise and universal, the physics definition does not involve counting the number of page rearrangements of one book or another that leave it looking the same, either ordered or disordered. Instead, the physics definition counts the number of rearrangements of fundamental constituents—atoms, subatomic particles, and so on—that leave the gross, overall, "big-picture" properties of a given physical system unchanged. As in the example of *War and Peace,* low entropy means that very few rearrangements would go unnoticed, so the system is highly ordered, while high entropy means that many rearrangements would go unnoticed, and that means the system is very disordered.*

For a good physics example, and one that will shortly prove handy, let's think about the bottle of Coke referred to earlier. When gas, like the carbon dioxide that was initially confined in the bottle, spreads evenly throughout a room, there are *many* rearrangements of the individual molecules that will have no noticeable effect. For example, if you flail your arms, the carbon dioxide molecules will move to and fro, rapidly changing positions and velocities. But overall, there will be no qualitative effect on their arrangement. The molecules were spread uniformly before you flailed your arms, and they will be spread uniformly after you're done. The uniformly spread gas configuration is insensitive to an enormous number of rearrangements of its molecular constituents, and so is in a state of high entropy. By contrast, if the gas were spread in a smaller space, as when it was in the bottle, or confined by a barrier to a corner of the room, it has significantly lower entropy. The reason is simple. Just as thinner books have fewer page reorderings, smaller spaces provide fewer places for molecules to be located, and so allow for fewer rearrangements.

But when you twist off the bottle's cap or remove the barrier, you open up a whole new universe to the gas molecules, and through their bumping and jostling they quickly disperse to explore it. Why? It's the same statistical reasoning as with the pages of *War and Peace.* No doubt, some of the jostling will move a few gas molecules purely within the initial blob of gas or nudge a few that have left the blob back toward the ini-

---

*Entropy is another example in which terminology complicates ideas. Don't worry if you have to remind yourself repeatedly that *low* entropy means *high* order and that *high* entropy means *low* order (equivalently, high disorder). I often have to.

tial dense gas cloud. But since the volume of the room exceeds that of the initial cloud of gas, there are *many* more rearrangements available to the molecules if they disperse out of the cloud than there are if they remain within it. On average, then, the gas molecules will diffuse from the initial cloud and slowly approach the state of being spread uniformly throughout the room. Thus, the lower-entropy initial configuration, with the gas all bunched in a small region, naturally evolves toward the higher-entropy configuration, with the gas uniformly spread in the larger space. And once it has reached such uniformity, the gas will tend to maintain this state of high entropy: bumping and jostling still causes the molecules to move this way and that, giving rise to one rearrangement after another, but the overwhelming majority of these rearrangements do not affect the gross, overall appearance of the gas. That's what it means to have high entropy.[9]

In principle, as with the pages of *War and Peace*, we could use the laws of classical physics to determine precisely where each carbon dioxide molecule will be at a given moment of time. But because of the enormous number of CO_2 molecules—about 10^{24} in a bottle of Coke—actually carrying out such calculations is practically impossible. And even if, somehow, we were able to do so, having a list of a million billion billion particle positions and velocities would hardly give us a sense of how the molecules were distributed. Focusing on big-picture statistical features— is the gas spread out or bunched up, that is, does it have high or low entropy?—is far more illuminating.

Entropy, the Second Law, and the Arrow of Time

The tendency of physical systems to evolve toward states of higher entropy is known as the *second law of thermodynamics*. (The first law is the familiar conservation of energy.) As above, the basis of the law is simple statistical reasoning: there are more ways for a system to have higher entropy, and "more ways" means it is more likely that a system will evolve into one of these high-entropy configurations. Notice, though, that this is not a law in the conventional sense since, although such events are rare and unlikely, something *can* go from a state of high entropy to one of lower entropy. When you toss a jumbled stack of pages into the air and then gather them into a neat pile, they *can* turn out to be in perfect numerical order. You wouldn't want to place a high wager on its happening, but it *is* possible. It is also possible that the bumping and jostling will be just right to cause all the dispersed carbon dioxide molecules to move in concert and swoosh back into your open bottle of Coke. Don't hold your breath waiting for this outcome either, but it *can* happen.[10]

The large number of pages in *War and Peace* and the large number of gas molecules in the room are what makes the entropy difference between the disordered and ordered arrangements so huge, and what causes low-entropy outcomes to be so terribly unlikely. If you tossed only two double-sided pages in the air over and over again, you'd find that they landed in the correct order about 12.5 percent of the time. With three pages this would drop to about 2 percent of the tosses, with four pages it's about .3 percent, with five pages it's about .03 percent, with six pages it's about .002 percent, with ten pages it's .000000027 percent, and with 693 pages the percentage of tosses that would yield the correct order is so small—it involves so many zeros after the decimal point—that I've been convinced by the publisher not to use another page to write it out explicitly. Similarly, if you dropped only two gas molecules side by side into an empty Coke bottle, you'd find that at room temperature their random motion would bring them back together (within a millimeter of each other), on average, roughly every few seconds. But for a group of three molecules, you'd have to wait days, for four molecules you'd have to wait years, and for an initial dense blob of a million billion billion molecules it would take a length of time far greater than the current age of the universe for their random, dispersive motion to bring them back together into a small, ordered bunch. With more certainty than death and taxes, we can count on systems with many constituents evolving toward disorder.

Although it may not be immediately apparent, we have now come to an intriguing point. The second law of thermodynamics seems to have given us an arrow of time, *one that emerges when physical systems have a large number of constituents.* If you were to watch a film of a couple of carbon dioxide molecules that had been placed together in a small box (with a tracer showing the movements of each), you'd be hard pressed to say whether the film was running forward or in reverse. The two molecules would flit this way and that, sometimes coming together, sometimes moving apart, but they would not exhibit any gross, overall behavior distinguishing one direction in time from the reverse. However, if you were to watch a film of 10^{24} carbon dioxide molecules that had been placed together in the box (as a small, dense cloud of molecules, say), you could easily determine whether the film was being shown forward or in reverse: it is overwhelmingly likely that the forward time direction is the one in which the gas molecules become more and more uniformly spread out, *achieving higher and higher entropy*. If, instead, the film showed uniformly dispersed gas molecules swooshing together into a tight group, you'd immediately recognize that you were watching it in reverse.

The same reasoning holds for essentially all the things we encounter in daily life—things, that is, which have a large number of constituents: the forward-in-time arrow points in the direction of increasing entropy. If you watch a film of a glass of ice water placed on a bar, you can determine which direction is forward in time by checking that the ice melts—its H_2O molecules disperse throughout the glass, thereby achieving higher entropy. If you watch a film of a splattering egg, you can determine which direction is forward in time by checking that the egg's constituents become more and more disordered—that the egg splatters rather than unsplatters, thereby also achieving higher entropy.

As you can see, the concept of entropy provides a precise version of the "easy versus difficult" conclusion we found earlier. It's easy for the pages of *War and Peace* to fall out of order because there are *so many* out-of-order arrangements. It's difficult for the pages to fall in perfect order because hundreds of pages would need to move in just the right way to land in the unique sequence Tolstoy intended. It's easy for an egg to splatter because there are *so* many ways to splatter. It's difficult for an egg to unsplatter, because an enormous number of splattered constituents must move in perfect coordination to produce the single, unique result of a pristine egg resting on the counter. For things with many constituents, going from lower to higher entropy—from order to disorder—is easy, so it happens all the time. Going from higher to lower entropy—from disorder to order—is harder, so it happens rarely, at best.

Notice, too, that this entropic arrow is not completely rigid; there is no claim that this definition of time's direction is 100 percent foolproof. Instead, the approach has enough flexibility to allow these and other processes to happen in reverse as well. Since the second law proclaims that entropy increase is only a statistical likelihood, not an inviolable fact of nature, it allows for the rare possibility that pages can fall into perfect numerical order, that gas molecules can coalesce and reenter a bottle, and that eggs can unsplatter. By using the mathematics of entropy, the second law expresses precisely how statistically unlikely these events are (remember, the huge number on pages 152–53 reflects how much more likely it is that pages will land out of order), but it recognizes that they can happen.

This seems like a convincing story. Statistical and probabilistic reasoning has given us the second law of thermodynamics. In turn, the second law has provided us with an intuitive distinction between what we call past and what we call future. It has given us a practical explanation for why things in daily life, things that are typically composed of huge numbers of constituents, start like *this* and end like *that*, while we never see them start like *that* and end like *this*. But over the course of many years — and thanks to important contributions by physicists like Lord Kelvin, Josef Loschmidt, Henri Poincaré, S. H. Burbury, Ernst Zermelo, and Willard Gibbs — Ludwig Boltzmann came to appreciate that the full story of time's arrow is more surprising. Boltzmann realized that although entropy had illuminated important aspects of the puzzle, it had *not* answered the question of why the past and the future seem so different. Instead, entropy had redefined the question in an important way, one that leads to an unexpected conclusion.

Entropy: Past and Future

Earlier, we introduced the dilemma of past versus future by comparing our everyday observations with properties of Newton's laws of classical physics. We emphasized that we continually experience an obvious directionality to the way things unfold in time but the laws themselves treat what we call forward and backward in time on an exactly equal footing. As there is no arrow within the laws of physics that assigns a direction to time, no pointer that declares, "Use these laws in this temporal orientation but not in the reverse," we were led to ask: If the laws underlying experience treat both temporal orientations symmetrically, why are the experiences themselves so temporally lopsided, always happening in one direction but not the other? Where does the observed and experienced directionality of time come from?

In the last section we seemed to have made progress, through the second law of thermodynamics, which apparently singles out the future as the direction in which entropy increases. But on further thought it's not that simple. Notice that in our discussion of entropy and the second law, we did not modify the laws of classical physics in any way. Instead, all we did was use the laws in a "big picture" statistical framework: we ignored fine details (the precise order of *War and Peace*'s unbound pages, the precise locations and velocities of an egg's constituents, the precise locations and velocities of a bottle of Coke's CO_2 molecules) and instead focused our attention on gross, overall features (pages ordered vs. unordered, egg splattered vs. not splattered, gas molecules spread out vs. not spread out). We found that when physical systems are sufficiently complicated (books with many pages, fragile objects that can splatter into many fragments, gas with many molecules), there is a huge difference in entropy between their ordered and disordered configurations. And this means that there is a huge likelihood that the systems will evolve from lower to higher entropy, which is a rough statement of the second law of thermodynamics. But the key fact to notice is that the second law is *derivative:* it is merely a consequence of probabilistic reasoning applied to Newton's laws of motion.

This leads us to a simple but astounding point: *Since Newton's laws of physics have no built-in temporal orientation, all of the reasoning we have used to argue that systems will evolve from lower to higher entropy toward the future works equally well when applied toward the past.* Again, since the underlying laws of physics are time-reversal symmetric, there is no way for them even to distinguish between what we call the past and what we call the future. Just as there are no signposts in the deep darkness of empty space that declare this direction up and that direction down, there is nothing in the laws of classical physics that says this direction is time future and that direction is time past. The laws offer no temporal orientation; it's a distinction to which they are completely insensitive. And since the laws of motion are responsible for how things change—both toward what we call the future and toward what we call the past—the statistical/probabilistic reasoning behind the second law of thermodynamics applies equally well in both temporal directions. Thus, *not only is there an overwhelming probability that the entropy of a physical system will be higher in what we call the future, but there is the same overwhelming probability that it was higher in what we call the past.* We illustrate this in Figure 6.2.

This is *the* key point for all that follows, but it's also deceptively subtle. A common misconception is that if, according to the second law of thermodynamics, entropy increases toward the future, then entropy necessarily *decreases* toward the past. But that's where the subtlety comes in. The second law actually says that if at any given moment of interest, a physical system happens not to possess the maximum possible entropy, it is extraordinarily likely that the physical system will subsequently have *and* previously had more entropy. That's the content of Figure 6.2b. With laws that are blind to the past-versus-future distinction, such time symmetry is inevitable.

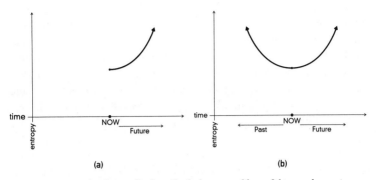

Figure 6.2 **(a)** As it's usually described, the second law of thermodynamics implies that entropy increases toward the future of any given moment. **(b)** Since the known laws of nature treat forward and backward in time identically, the second law actually implies that entropy increases both toward the future and toward the past from any given moment.

That's the essential lesson. It tells us that the entropic arrow of time is *double-headed.* From any specified moment, the arrow of entropy increase points toward the future *and* toward the past. And that makes it decidedly awkward to propose entropy as the explanation of the *one-way* arrow of experiential time.

Think about what the double-headed entropic arrow implies in concrete terms. If it's a warm day and you see partially melted ice cubes in a glass of water, you have full confidence that half an hour later the cubes will be more melted, since the more melted they are, the more entropy they have.[11] But you should have *exactly* the same confidence that half an hour earlier they were also more melted, since *exactly* the same statistical reasoning implies that entropy should increase toward the past. And the same conclusion applies to the countless other examples we encounter every day. Your assuredness that entropy increases toward the future — from partially dispersed gas molecules' further dispersing to partially jumbled page orders' getting more jumbled — should be matched by *exactly* the same assuredness that entropy was also higher in the past.

The troubling thing is that half of these conclusions seem to be flat-out wrong. Entropic reasoning yields accurate and sensible conclusions when applied in one time direction, toward what we call the future, but gives apparently inaccurate and seemingly ridiculous conclusions when applied toward what we call the past. Glasses of water with partially melted ice cubes do not usually start out as glasses of water with no ice cubes in which molecules of water coalesce and cool into chunks of ice, only to start melting once again. Unbound pages of *War and Peace* do not usually start thoroughly out of numerical order and through subsequent tosses get less jumbled, only to start getting more jumbled again. And going back to the kitchen, eggs do not generally start out splattered, and then coalesce into a pristine whole egg, only to splatter some time later.

Or do they?

Following the Math

Centuries of scientific investigations have shown that mathematics provides a powerful and incisive language for analyzing the universe. Indeed, the history of modern science is replete with examples in which the math made predictions that seemed counter to both intuition and experience (that the universe contains black holes, that the universe has anti-matter, that distant particles can be entangled, and so on) but which experiments and observations were ultimately able to confirm. Such developments have impressed themselves profoundly on the culture of theoretical physics. Physicists have come to realize that mathematics, when used with sufficient care, is a proven pathway to truth.

So, when a mathematical analysis of nature's laws shows that entropy should be higher toward the future *and* toward the past of any given moment, physicists don't dismiss it out of hand. Instead, something akin to a physicists' Hippocratic oath impels researchers to maintain a deep and healthy skepticism of the apparent truths of human experience and, with the same skeptical attitude, diligently follow the math and see where it leads. Only then can we properly assess and interpret any remaining mismatch between physical law and common sense.

Toward this end, imagine it's 10:30 p.m. and for the past half hour you've been staring at a glass of ice water (it's a slow night at the bar), watching the cubes slowly melt into small, misshapen forms. You have absolutely no doubt that a half hour earlier the bartender put fully formed ice cubes into the glass; you have no doubt because you trust your memory. And if, by some chance, your confidence regarding what happened during the last half hour should be shaken, you can ask the guy across the way, who was also watching the ice cubes melt (it's a *really* slow night at the bar), or perhaps check the video taken by the bar's surveillance camera, both of which would confirm that your memory is accurate. If you were then to ask yourself what you expect to happen to the ice cubes during the next half hour, you'd probably conclude that they'd continue to melt. And, if you'd gained sufficient familiarity with the concept of entropy, you'd explain your prediction by appealing to the overwhelming likelihood that entropy will increase from what you see, right now at 10:30 p.m., toward the future. All that makes good sense and jibes with our intuition and experience.

But as we've seen, such entropic reasoning—reasoning that simply says things are more likely to be disordered since there are more ways to be disordered, reasoning which is demonstrably powerful at explaining how things unfold toward the future—proclaims that entropy is just as likely to also have been higher in the past. This would mean that the partially melted cubes you see at 10:30 p.m. would actually have been *more* melted at earlier times; it would mean that at 10:00 p.m. they did not begin as solid ice cubes, but, instead, slowly coalesced out of room-temperature water on the way to 10:30 p.m., just as surely as they will slowly melt into room-temperature water on their way to 11:00 p.m.

No doubt, that sounds weird—or perhaps you'd say nutty. To be true, not only would H_2O molecules in a glass of room-temperature water have to coalesce spontaneously into partially formed cubes of ice, but the digital bits in the surveillance camera, as well as the neurons in your brain and those in the brain of the guy across the way, would all need to spontaneously arrange themselves by 10:30 p.m. to attest to there having been a collection of fully formed ice cubes that melted, even though there never was. Yet this bizarre-sounding conclusion is where a faithful application of entropic reasoning—the same reasoning that you embrace without hesitation to explain why the partially melted ice you see at 10:30 p.m. continues to melt toward 11:00 p.m.—leads when applied in the time-symmetric manner dictated by the laws of physics. This is the trouble with having fundamental laws of motion with no inbuilt distinction between past and future, laws whose mathematics treats the future and past of any given moment in exactly the same way.[12]

Rest assured that we will shortly find a way out of the strange place to which an egalitarian use of entropic reasoning has taken us; I'm not going to try to convince you that your memories and records are of a past that never happened (apologies to fans of *The Matrix*). But we will find it very useful to pinpoint precisely the disjuncture between intuition and the mathematical laws. So let's keep following the trail.

A Quagmire

Your intuition balks at a past with higher entropy because, when viewed in the usual forward-time unfolding of events, it would require a spontaneous rise in order: water molecules spontaneously cooling to 0 degrees Celsius and turning into ice, brains spontaneously acquiring memories of things that didn't happen, video cameras spontaneously producing images of things that never were, and so on, all of which seem extraordinarily unlikely—a proposed explanation of the past at which even Oliver Stone would scoff. On this point, the physical laws and the mathematics of entropy agree with your intuition completely. Such a sequence of events, when viewed in the forward time direction from 10 p.m. to 10:30 p.m., goes against the grain of the second law of thermodynamics—it results in a decrease in entropy—and so, although not impossible, it *is* very unlikely.

By contrast, your intuition and experience tell you that a far more likely sequence of events is that ice cubes that were fully formed at 10 p.m. partially melted into what you see in your glass, right now, at 10:30 p.m. But on this point, the physical laws and mathematics of entropy only partly agree with your expectation. Math and intuition concur that *if* there really were fully formed ice cubes at 10 p.m., then the most likely sequence of events would be for them to melt into the partial cubes you see at 10:30 p.m.: the resulting increase in entropy is in line both with the second law of thermodynamics and with experience. But where math and intuition deviate is that our intuition, unlike the math, fails to take account of the likelihood, or lack thereof, of actually having fully formed ice cubes at 10 p.m., *given the one observation we are taking as unassailable, as fully trustworthy, that right now, at 10:30 p.m., you see partially melted cubes.*

This is the pivotal point, so let me explain. The main lesson of the second law of thermodynamics is that physical systems have an overwhelming tendency to be in high-entropy configurations because there are so many ways such states can be realized. And once in such high-entropy states, physical systems have an overwhelming tendency to stay in them. High entropy is the natural state of being. You should never be surprised by or feel the need to explain why any physical system is in a high-entropy state. Such states are the norm. On the contrary, what does need explaining is why any given physical system is in a state of order, a state of low entropy. These states are not the norm. They can certainly happen. But from the viewpoint of entropy, such ordered states are rare aberrations that cry out for explanation. So the one fact in the episode we are taking as unquestionably true—your observation at 10:30 p.m. of low-entropy partially formed ice cubes—is a fact in need of an explanation.

And from the point of view of probability, it is absurd to explain this low-entropy state by invoking the even *lower*-entropy state, the *even less likely* state, that at 10 p.m. there were *even more ordered, more fully formed* ice cubes being observed in a *more* pristine, *more* ordered environment. Instead, it is enormously more likely that things began in an unsurprising, totally normal, high-entropy state: a glass of uniform liquid water with absolutely no ice. Then, through an unlikely but every-so-often-expectable statistical fluctuation, the glass of water went against the grain of the second law and evolved to a state of lower entropy in which partially formed ice cubes appeared. This evolution, although requiring rare and unfamiliar processes, completely avoids the even lower-entropy, the even less likely, the even more rare state of having *fully* formed ice cubes. At every moment between 10 p.m. and 10:30 p.m., this strange-sounding evolution has *higher* entropy than the normal ice-melting scenario, as you can see in Figure 6.3, and so it realizes the accepted observation at 10:30 p.m. in a way that is *more likely*—hugely more likely—than the scenario in which fully formed ice cubes melt.[13] That is the crux of the matter.*

---

*Remember, on pages 152–53 we showed the huge difference between the number of ordered and disordered configurations for a mere 693 double-sided sheets of paper. We are now discussing the behavior of roughly 10^{24} H_2O molecules, so the difference between the number of ordered and disordered configurations is breathtakingly monumental. Moreover, the same reasoning holds for all other atoms and molecules within you and within the environment (brains, security cameras, air molecules, and so on). Namely, in the standard explanation in which you can trust your memories, not only would the partially melted ice cubes have begun, at 10 p.m., in a more ordered—less likely—state, but so would everything else: when a video camera records a sequence of events, there is a net increase in entropy (from the heat and noise released by the recording process); similarly, when a brain records a memory, although we understand the microscopic details with less accuracy, there is a net increase in entropy (the brain may gain order but as with any order-producing process, if we take account of heat generated, there is a net increase in entropy). Thus, if we compare the total entropy in the bar between 10 p.m. and 10:30 p.m. in the two scenarios—one in which you trust your memories, and the other in which things spontaneously arrange themselves from an initial state of disorder to be consistent with what you see, now, at 10:30 p.m.—there is an enormous entropy difference. The latter scenario, every step of the way, has *hugely* more entropy than the former scenario, and so, from the standpoint of probability, is hugely more likely.

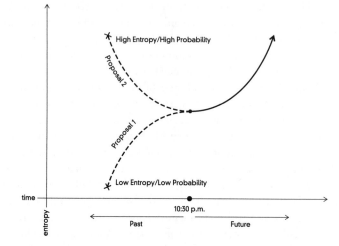

Figure 6.3 A comparison of two proposals for how the ice cubes got to their partially melted state, right now, at 10:30 p.m. Proposal 1 aligns with your memories of melting ice, but requires a comparatively low-entropy starting point at 10:00 p.m. Proposal 2 challenges your memories by describing the partially melted ice you see at 10:30 p.m. as having coalesced out of a glass of water, but starts off in a high-entropy, highly probable configuration of disorder at 10:00 p.m. Every step of the way toward 10:30 p.m., Proposal 2 involves states that are more likely than those in Proposal 1—because, as you can see in the graph, they have higher entropy—and so Proposal 2 is statistically favored.

It was a small step for Boltzmann to realize that the whole of the universe is subject to this same analysis. When you look around the universe right now, what you see reflects a great deal of biological organization, chemical structure, and physical order. Although the universe could be a totally disorganized mess, it's not. Why is this? Where did the order come from? Well, just as with the ice cubes, from the standpoint of probability it is extremely unlikely that the universe we see evolved from an even more ordered—an even less likely—state in the distant past that has slowly unwound to its current form. Rather, because the cosmos has so many constituents, the scales of ordered versus disordered are magnified intensely. And so what's true at the bar is true with a vengeance for the whole universe: it is *far* more likely—breathtakingly more likely—that the whole universe we now see arose as a statistically rare fluctuation from a normal, unsurprising, high-entropy, completely disordered configuration.

Think of it this way: if you toss a handful of pennies over and over again, sooner or later they will all land heads. If you have nearly the infinite patience needed to throw the jumbled pages of *War and Peace* in the air over and over again, sooner or later they will land in correct numerical order. If you wait with your open bottle of flat Coke, sooner or later the random jostling of the carbon dioxide molecules will cause them to reenter the bottle. And, for Boltzmann's kicker, if the universe waits long enough—for nearly an eternity, perhaps—its usual, high-entropy, highly probable, totally disordered state will, through its own bumping, jostling, and random streaming of particles and radiation, sooner or later just happen to coalesce into the configuration that we all see right now. Our bodies and brains would emerge fully formed from the chaos—stocked with memories, knowledge, and skills—even though the past they seem to reflect would never really have happened. Everything we know about, everything we value, would amount to nothing more than a rare but every-so-often-expectable statistical fluctuation momentarily interrupting a near eternity of disorder. This is schematically illustrated in Figure 6.4.

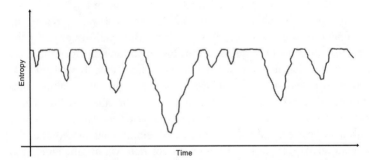

Figure 6.4 A schematic graph of the universe's total entropy through time. The graph shows the universe spending most of its time in a state of total disorder—a state of high entropy—and every so often experiencing fluctuations to states of varying degrees of order, varying states of lower entropy. The greater the entropy dip, the less likely the fluctuation. Significant dips in entropy, to the kind of order in the universe today, are extremely unlikely and would happen very rarely.

Taking a Step Back

When I first encountered this idea many years ago, it was a bit of a shock. Up until that point, I had thought I understood the concept of entropy fairly well, but the fact of the matter was that, following the approach of textbooks I'd studied, I'd only ever considered entropy's implications for the future. And, as we've just seen, while entropy applied toward the future confirms our intuition and experience, entropy applied toward the past just as thoroughly contradicts them. It wasn't quite as bad as suddenly learning that you've been betrayed by a longtime friend, but for me, it was pretty close.

Nevertheless, sometimes it's good not to pass judgment too quickly, and entropy's apparent failure to live up to expectations provides a case in point. As you're probably thinking, the idea that all we're familiar with just popped into existence is as tantalizing as it is hard to swallow. And it's not "merely" that this explanation of the universe challenges the veracity of everything we hold to be real and important. It also leaves critical questions unanswered. For instance, the more ordered the universe is today—the greater the dip in Figure 6.4—the more surprising and unlikely is the statistical aberration required to bring it into existence. So if the universe could have cut any corners, making things look more or less like what we see right now while skimping on the actual amount of order, probabilistic reasoning leads us to believe it would have. But when we examine the universe, there seem to be numerous lost opportunities, since there are many things that are more ordered than they have to be. If Michael Jackson never recorded *Thriller* and the millions of copies of this album now distributed worldwide all got there as part of an aberrant fluctuation toward lower entropy, the aberration would have been far less extreme if only a million or a half-million or just a few albums had formed. If evolution never happened and we humans got here via an aberrant jump toward lower entropy, the aberration would have been far less extreme if there weren't such a consistent and ordered evolutionary fossil record. If the big bang never happened and the more than 100 billion galaxies we now see arose as an aberrant jump toward lower entropy, the aberration would have been less extreme if there were 50 billion, or 5,000, or just a handful, or just one galaxy. And so if the idea that our universe is a statistical fluctuation—a happy fluke—has any validity, one would need to address how and why the universe went so far overboard and achieved a state of *such* low entropy.

Even more pressing, if you truly can't trust memories and records, then you also can't trust the laws of physics. Their validity rests on numerous experiments whose positive outcomes are attested to only by those very same memories and records. So all the reasoning based on the time-reversal symmetry of the accepted laws of physics would be totally thrown into question, thereby undermining our understanding of entropy and the whole basis for the current discussion. By embracing the conclusion that the universe we know is a rare but every-so-often-expectable statistical fluctuation from a configuration of total disorder, we're quickly led into a quagmire in which we lose all understanding, including the very chain of reasoning that led us to consider such an odd explanation in the first place.*

Thus, by suspending disbelief and diligently following the laws of physics and the mathematics of entropy—concepts which in combination tell us that it is overwhelmingly likely that disorder will increase both toward the future *and* toward the past from any given moment—we have gotten ourselves neck deep in quicksand. And while that might not sound pleasant, for two reasons it's a very good thing. First, it shows with precision why mistrust of memories and records—something at which we intuitively scoff—doesn't make sense. Second, by reaching a point where our whole analytical scaffolding is on the verge of collapse, we realize, forcefully, that we *must* have left something crucial out of our reasoning.

Therefore, to avoid the explanatory abyss, we ask ourselves: what new idea or concept, beyond entropy and the time symmetry of nature's laws, do we need in order to go back to trusting our memories and our records—our experience of room-temperature ice cubes melting and not unmelting, of cream and coffee mixing but not unmixing, of eggs splattering but not unsplattering? In short, where are we led if we try to explain

---

*A closely related point is that should we convince ourselves that the world we see right now just coalesced out of total disorder, the exact same reasoning—invoked anytime later—would require us to abandon our current belief and, instead, attribute the ordered world to a yet more recent fluctuation. Thus, in this way of thinking, every next moment invalidates the beliefs held in each previous moment, a distinctly unconvincing way of explaining the cosmos.

an asymmetric unfolding of events in spacetime, with entropy to our future higher, but entropy to our past *lower*? Is it possible?

It is. But only if things were very special early on.[14]

The Egg, the Chicken, and the Big Bang

To see what this means, let's take the example of a pristine, low-entropy, fully formed egg. How did this low-entropy physical system come into being? Well, putting our trust back in memories and records, we all know the answer. The egg came from a chicken. And that chicken came from an egg, which came from a chicken, which came from an egg, and so on. But, as emphasized most forcefully by the English mathematician Roger Penrose,[15] this chicken-and-egg story actually teaches us something deep and leads somewhere definite.

A chicken, or any living being for that matter, is a physical system of astonishingly high order. Where does this organization come from and how is it sustained? A chicken stays alive, and in particular, stays alive long enough to produce eggs, by eating and breathing. Food and oxygen provide the raw materials from which living beings extract the energy they require. But there is a critical feature of this energy that must be emphasized if we are to really understand what's going on. Over the course of its life, a chicken that stays fit takes in just about as much energy in the form of food as it gives back to the environment, mostly in the form of heat and other waste generated by its metabolic processes and daily activities. If there weren't such a balance of energy-in and energy-out, the chicken would get increasingly hefty.

The essential point, though, is that all forms of energy are not equal. The energy a chicken gives off to the environment in the form of heat is highly disordered—it often results in some air molecules here or there jostling around a touch more quickly than they otherwise would. Such energy has high entropy—it is diffuse and intermingled with the environment—and so cannot easily be harnessed for any useful purpose. To the contrary, the energy the chicken takes in from its feed has low entropy and is readily harnessed for important life-sustaining activities. So the chicken, and every life form in fact, is a conduit for taking in low-entropy energy and giving off high-entropy energy.

This realization pushes the question of where the low entropy of an egg originates one step further back. How is it that the chicken's energy source, the food, has such low entropy? How do we explain this aberrant source of order? If the food is of animal origin, we are led back to the initial question of how animals have such low entropy. But if we follow the food chain, we ultimately come upon animals (like me) that eat only plants. How do plants and their products of fruits and vegetables maintain low entropy? Through photosynthesis, plants use sunlight to separate ambient carbon dioxide into oxygen, which is given back to the environment, and carbon, which the plants use to grow and flourish. So we can trace the low-entropy, nonanimal sources of energy to the sun.

This pushes the question of explaining low entropy another step further back: where did our highly ordered sun come from? The sun formed about 5 billion years ago from an initially diffuse cloud of gas that began to swirl and clump under the mutual gravitational attraction of all its constituents. As the gas cloud got denser, the gravitational pull of one part on another got stronger, causing the cloud to collapse further in on itself. And as gravity squeezed the cloud tighter, it got hotter. Ultimately, it got hot enough to ignite nuclear processes that generated enough outward-flowing radiation to stem further gravitational contraction of the gas. A hot, stable, brightly burning star was born.

So where did the diffuse cloud of gas come from? It likely formed from the remains of older stars that reached the end of their lives, went supernova, and spewed their contents out into space. Where did the diffuse gas responsible for these early stars come from? We believe that the gas was formed in the aftermath of the big bang. Our most refined theories of the origin of the universe—our most refined *cosmological* theories—tell us that by the time the universe was a couple of minutes old, it was filled with a *nearly uniform hot gas* composed of roughly 75 percent hydrogen, 23 percent helium, and small amounts of deuterium and lithium. The essential point is that this gas filling the universe had extraordinarily *low* entropy. The big bang started the universe off in a state of low entropy, and that state appears to be the source of the order we currently see. In other words, *the current order is a cosmological relic.* Let's discuss this important realization in a little more detail.

Entropy and Gravity

Because theory and observation show that within a few minutes after the big bang, primordial gas was uniformly spread throughout the young universe, you might think, given our earlier discussion of the Coke and its carbon dioxide molecules, that the primordial gas was in a high-entropy, disordered state. But this turns out not to be true. Our earlier discussion of entropy completely ignored gravity, a sensible thing to do because gravity hardly plays a role in the behavior of the minimal amount of gas emerging from a bottle of Coke. And with that assumption, we found that uniformly dispersed gas has high entropy. But when gravity matters, the story is very different. Gravity is a universally attractive force; hence, if you have a large enough mass of gas, every region of gas will pull on every other and this will cause the gas to fragment into clumps, somewhat as surface tension causes water on a sheet of wax paper to fragment into droplets. When gravity matters, as it did in the high-density early universe, clumpiness—not uniformity—is the norm; it is the state toward which a gas tends to evolve, as illustrated in Figure 6.5.

Even though the clumps appear to be more ordered than the initially diffuse gas—much as a playroom with toys that are neatly grouped in trunks and bins is more ordered than one in which the toys are uniformly strewn around the floor—in calculating entropy you need to tally up the contributions from *all* sources. For the playroom, the entropy decrease in going from wildly strewn toys to their all being "clumped" in trunks and bins is more than compensated for by the entropy increase from the fat burned and heat generated by the parents who spent hours cleaning and arranging everything. Similarly, for the initially diffuse gas cloud, you find that the entropy decrease through the formation of orderly clumps is more than compensated by the heat generated as the gas compresses, and,

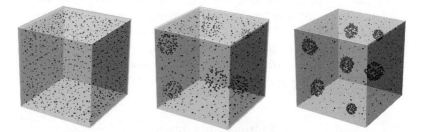

Figure 6.5 For huge volumes of gas, when gravity matters, atoms and molecules evolve from a smooth, evenly spread configuration, into one involving larger and denser clumps.

ultimately, by the enormous amount of heat and light released when nuclear processes begin to take place.

This is an important point that is sometimes overlooked. The overwhelming drive toward disorder does not mean that orderly structures like stars and planets, or orderly life forms like plants and animals, can't form. They can. And they obviously do. What the second law of thermodynamics entails is that in the formation of order there is generally a more-than-compensating generation of disorder. The entropy balance sheet is still in the black even though certain constituents have become more ordered. And of the fundamental forces of nature, gravity is the one that exploits this feature of the entropy tally to the hilt. Because gravity operates across vast distances and is universally attractive, it instigates the formation of the ordered clumps—stars—that give off the light we see in a clear night sky, all in keeping with the net balance of entropy increase.

The more squeezed, dense, and massive the clumps of gas are, the larger the overall entropy. Black holes, the most extreme form of gravitational clumping and squeezing in the universe, take this to the limit. The gravitational pull of a black hole is so strong that nothing, not even light, is able to escape, which explains why black holes are black. Thus, unlike ordinary stars, black holes stubbornly hold on to all the entropy they produce: none of it can escape the black hole's powerful gravitational grip.[16] In fact, as we will discuss in Chapter 16, nothing in the universe contains more disorder—more entropy—than a black hole.* This makes good intuitive sense: high entropy means that many rearrangements of the constituents of an object go unnoticed. Since we can't see inside a black hole, it is impossible for us to detect *any* rearrangement of its constituents—whatever those constituents may be—and hence black holes have maximum entropy. When gravity flexes its muscles to the limit, it becomes the most efficient generator of entropy in the known universe.

We have now come to the place where the buck finally stops. *The ultimate source of order, of low entropy, must be the big bang itself.* In its earliest moments, rather than being filled with gargantuan containers of entropy such as black holes, as we would expect from probabilistic considerations, for some reason the nascent universe was filled with a hot, uniform, gaseous mixture of hydrogen and helium. Although this configu-

---

*That is, a black hole of a given size contains more entropy than *anything* else of the same size.

ration has high entropy when densities are so low that we can ignore gravity, the situation is otherwise when gravity can't be ignored; then, such a uniform gas has extremely low entropy. In comparison with black holes, the diffuse, nearly uniform gas was in an extraordinarily low-entropy state. Ever since, in accordance with the second law of thermodynamics, the overall entropy of the universe has been gradually getting higher and higher; the overall, net amount of disorder has been gradually increasing. After about a billion years or so, gravity caused the primordial gas to clump, and the clumps ultimately formed stars, galaxies, and some lighter clumps that became planets. At least one such planet had a nearby star that provided a relatively low-entropy source of energy that allowed low-entropy life forms to evolve, and among such life forms there eventually was a chicken that laid an egg that found its way to your kitchen counter, and much to your chagrin that egg continued on the relentless trajectory to a higher entropic state by rolling off the counter and splattering on the floor. The egg splatters rather than unsplatters because it is carrying forward the drive toward higher entropy that was initiated by the extraordinarily low entropy state with which the universe began. Incredible order at the beginning is what started it all off, and we have been living through the gradual unfolding toward higher disorder ever since.

This is the stunning connection we've been leading up to for the entire chapter. *A splattering egg tells us something deep about the big bang*. It tells us that the big bang gave rise to an extraordinarily ordered nascent cosmos.

The same idea applies to all other examples. The reason why tossing the newly unbound pages of *War and Peace* into the air results in a state of higher entropy is that they *began* in such a highly ordered, low entropy form. Their initial ordered form made them ripe for entropy increase. By contrast, if the pages initially were totally out of numerical order, tossing them in the air would hardly make a difference, as far as entropy goes. So the question, once again, is: how did they become so ordered? Well, Tolstoy wrote them to be presented in that order and the printer and binder followed his instructions. And the highly ordered bodies and minds of Tolstoy and the book producers, which allowed them, in turn, to create a volume of such high order, can be explained by following the same chain of reasoning we just followed for an egg, once again leading us back to the big bang. How about the partially melted ice cubes you saw at 10:30 p.m.? Now that we are trusting memories and records, you remember that just before 10 p.m. the bartender put fully formed ice cubes in your glass. He got the ice cubes from a freezer, which was designed by a clever engineer and fabricated by talented machinists, all of whom are capable of creating something of such high order because they themselves are highly ordered life forms. And again, we can sequentially trace their order back to the highly ordered origin of the universe.

The Critical Input

The revelation we've come to is that we can trust our memories of a past with lower, not higher, entropy only if the big bang—the process, event, or happening that brought the universe into existence—started off the universe in an extraordinarily special, highly ordered state of low entropy. Without that critical input, our earlier realization that entropy should increase toward both the future and the past from any given moment would lead us to conclude that all the order we see arose from a chance fluctuation from an ordinary disordered state of high entropy, a conclusion, as we've seen, that undermines the very reasoning on which it's based. But by including the unlikely, low-entropy starting point of the universe in our analysis, we now see that the correct conclusion is that entropy increases toward the future, since probabilistic reasoning operates fully and without constraint in that direction; but entropy does not increase toward the past, since *that* use of probability would run afoul of our new proviso that the universe began in a state of low, not high, entropy.[17] Thus, conditions at the birth of the universe are critical to directing time's arrow. *The future is indeed the direction of increasing entropy. The arrow of time—the fact that things start like* this *and end like* that *but never start like* that *and end like* this *—began its flight in the highly ordered, low-entropy state of the universe at its inception.*[18]

The Remaining Puzzle

That the early universe set the direction of time's arrow is a wonderful and satisfying conclusion, but we are not done. A huge puzzle remains. How is it that the universe began in such a highly ordered configuration, setting things up so that for billions of years to follow everything could slowly evolve through steadily less ordered configurations toward higher and higher entropy? Don't lose sight of how remarkable this is. As we emphasized, from the standpoint of probability it is much more likely that the partially melted ice cubes you saw at 10:30 p.m. got there because a statistical fluke acted itself out in a glass of liquid water, than that they originated in the even less likely state of fully formed ice cubes. And what's true for ice cubes is true a gazillion times over for the whole universe. Probabilistically speaking, it is mind-bogglingly more likely that everything we now see in the universe arose from a rare but every-so-often-expectable statistical aberration away from total disorder, rather than having slowly evolved from the even more unlikely, the incredibly more ordered, the astoundingly low-entropy starting point required by the big bang.[19]

Yet, when we went with the odds and imagined that everything popped into existence by a statistical fluke, we found ourselves in a quagmire: that route called into question the laws of physics themselves. And so we are inclined to buck the bookies and go with a low-entropy big bang as the explanation for the arrow of time. The puzzle then is to explain how the universe began in such an unlikely, highly ordered configuration. *That* is the question to which the arrow of time points. It all comes down to cosmology.[20]

We will take up a detailed discussion of cosmology in Chapters 8 through 11, but notice first that our discussion of time suffers from a serious shortcoming: everything we've said has been based purely on classical physics. Let's now consider how quantum mechanics affects our understanding of time and our pursuit of its arrow.

7

Time and the Quantum

When we think about something like time, something we are within, something that is fully integrated into our day-to-day existence, something that is so pervasive, it is impossible to excise—even momentarily—from common language, our reasoning is shaped by the preponderance of our experiences. These day-to-day experiences are classical experiences; with a high degree of accuracy, they conform to the laws of physics set down by Newton more than three centuries ago. But of all the discoveries in physics during the last hundred years, quantum mechanics is far and away the most startling, since it undermines the whole conceptual schema of classical physics.

So it is worthwhile to expand upon our classical experiences by considering some experiments that reveal eyebrow-raising features of how quantum processes unfold in time. In this broadened context, we will then continue the discussion of the last chapter and ask whether there is a temporal arrow in the quantum mechanical description of nature. We will come to an answer, but one that is still controversial, even among physicists. And once again it will take us back to the origin of the universe.

The Past According to the Quantum

Probability played a central role in the last chapter, but as I stressed there a couple of times, it arose only because of its practical convenience and the utility of the information it provides. Following the exact motion of the 10^{24} H_2O molecules in a glass of water is well beyond our computational capacity, and even if it were possible, what would we do with the resulting mountain of data? To determine from a list of 10^{24} positions and velocities whether there were ice cubes in the glass would be a Herculean task. So we turned instead to probabilistic reasoning, which is computationally tractable and, moreover, deals with the macroscopic properties— order versus disorder; for example, ice versus water—we are generally interested in. But keep in mind that probability is by no means fundamentally stitched into the fabric of classical physics. In principle, if we knew precisely how things were now—knew the positions and velocities of every single particle making up the universe—classical physics says we could use that information to predict how things would be at any given moment in the future or how they were at any given moment in the past. Whether or not you actually follow its moment-to-moment development, according to classical physics you can talk about the past and the future, in principle, with a confidence that is controlled by the detail and the accuracy of your observations of the present.[1]

Probability will also play a central role in this chapter. But because probability *is* an inescapable element of quantum mechanics, it fundamentally alters our conceptualization of past and future. We've already seen that quantum uncertainty prevents simultaneous knowledge of exact positions and exact velocities. Correspondingly, we've also seen that quantum physics predicts only the probability that one or another future will be realized. We have confidence in these probabilities, to be sure, but since they are probabilities we learn that there is an *unavoidable* element of chance when it comes to predicting the future.

When it comes to describing the past, there is also a critical difference between classical and quantum physics. In classical physics, in keeping with its egalitarian treatment of all moments in time, the events leading up to something we observe are described using exactly the same language, employing exactly the same attributes, we use to describe the observation itself. If we see a fiery meteor in the night sky, we talk of its position and its velocity; if we reconstruct how it got there, we also talk of a unique succession of positions and velocities as the meteor hurtled through space toward earth. In quantum physics, though, once we observe something we enter the rarefied realm in which we know something with 100 percent certainty (ignoring issues associated with the accuracy of our equipment, and the like). But the past—by which we specifically mean the "unobserved" past, the time before we, or anyone, or anything has carried out a given observation—remains in the usual realm of quantum uncertainty, of probabilities. Even though we measure an electron's position as right here right now, a moment ago all it had were probabilities of being here, or there, or way over there.

And as we've seen, it is not that the electron (or any particle for that matter) really was located at only one of these possible positions, but we simply don't know which.[2] Rather, there is a sense in which the electron was at all of the locations, because each of the possibilities—each of the possible histories—contributes to what we now observe. Remember, we saw evidence of this in the experiment, described in Chapter 4, in which electrons were forced to pass through two slits. Classical physics, which relies on the commonly held belief that happenings have unique, conventional histories, would say that any electron that makes it to the detector screen went through either the left slit *or* the right slit. But this view of the past would lead us astray: it would predict the results illustrated in Figure 4.3a, which do not agree with what actually happens, as illustrated in Figure 4.3b. The observed interference pattern can be explained only by invoking an overlap between something that passes through *both* slits.

Quantum physics provides just such an explanation, but in doing so it drastically changes our stories of the past—our descriptions of how the particular things we observe came to be. According to quantum mechanics, each electron's probability wave *does* pass through both slits, and because the parts of the wave emerging from each slit commingle, the resulting probability profile manifests an interference pattern, and hence the electron landing positions do, too.

Compared with everyday experience, this description of the electron's past in terms of criss-crossing waves of probability is thoroughly unfamiliar. But, throwing caution to the wind, you might suggest taking this quantum mechanical description one step further, leading to a yet more bizarre-sounding possibility. Maybe each individual electron itself actually travels through both slits on its way to the screen, and the data result from an interference between these two classes of histories. That is, it's tempting to think of the waves emerging from the two slits as representing two possible histories for an individual electron—going through the left slit or going through the right slit—and since both waves contribute to what we observe on the screen, perhaps quantum mechanics is telling us that both potential histories of the electron contribute as well.

Surprisingly, this strange and wonderful idea—the brainchild of the Nobel laureate Richard Feynman, one of the twentieth century's most creative physicists—provides a perfectly viable way of thinking about quantum mechanics. According to Feynman, if there are alternative ways in which a given outcome can be achieved—for instance, an electron hits a point on the detector screen by traveling through the left slit, or hits the same point on the screen but by traveling through the right slit—then there is a sense in which the alternative histories all happen, and happen simultaneously. Feynman showed that each such history would contribute to the probability that their common outcome would be realized, and if these contributions were correctly added together, the result would agree with the total probability predicted by quantum mechanics.

Feynman called this the *sum over histories* approach to quantum mechanics; it shows that a probability wave embodies all possible pasts that could have preceded a given observation, and illustrates well that to succeed where classical physics failed, quantum mechanics had to substantially broaden the framework of history.[3]

To Oz

There is a variation on the double-slit experiment in which the interference between alternative histories is made even more evident because the two routes to the detector screen are more fully separated. It is a little easier to describe the experiment using photons rather than electrons, so we begin with a photon source—a laser—and we fire it toward what is known as a *beam splitter.* This device is made from a half-silvered mirror, like the kind used for surveillance, which reflects half of the light that hits it while allowing the other half to pass through. The initial single light beam is thus split in two, the left beam and the right beam, similar to what happens to a light beam that impinges on the two slits in the double-slit setup. Using judiciously placed fully reflecting mirrors, as in Figure 7.1, the two beams are brought back together further downstream at the location of the detector. Treating the light as a wave, as in the description by Maxwell, we expect—and, indeed, we find—an interference pattern on the screen. The length of the journey to all but the center point on the screen is slightly different for the left and right routes and so while the left beam might be reaching a peak at a given point on the detector screen, the right beam might be reaching a peak, a trough, or something in between. The detector records the combined height of the two waves and hence has the characteristic interference pattern.

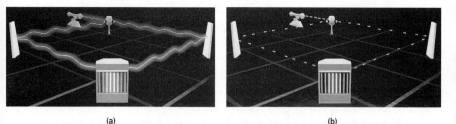

Figure 7.1 (a) In a beam-splitter experiment, laser light is split into two beams that travel two separate paths to the detector screen. **(b)** The laser can be turned down so that it fires individual photons; over time, the photon impact locations build up an interference pattern.

The classical/quantum distinction becomes apparent as we drastically lower the intensity of the laser so that it emits photons singly, say, one every few seconds. When a single photon hits the beam splitter, classical intuition says that it will either pass through or will be reflected. Classical reasoning doesn't even allow a hint of any kind of interference, since there is nothing to interfere: all we have are single, individual, particulate photons passing from source to detector, one by one, some going left, some going right. But when the experiment is done, the individual photons recorded over time, much as in Figure 4.4, *do* yield an interference pattern, as in Figure 7.1b. According to quantum physics, the reason is that each detected photon *could* have gotten to the detector by the left route or by going via the right route. Thus, we are obliged to combine these two possible histories in determining the probability that a photon will hit the screen at one particular point or another. When the left and right probability waves for each individual photon are merged in this way, they yield the undulating probability pattern of wave interference. And so, unlike Dorothy, who is perplexed when the Scarecrow points both left and right in giving her directions to Oz, the data can be explained perfectly by imagining that each photon takes both left and right routes toward the detector.

Prochoice

Although we have described the merging of possible histories in the context of only a couple of specific examples, this way of thinking about quantum mechanics is general. Whereas classical physics describes the present as having a unique past, the probability waves of quantum mechanics enlarge the arena of history: in Feynman's formulation, the observed present represents an amalgam—a particular kind of *average*—of all possible pasts compatible with what we now see.

In the case of the double-slit and beam-splitter experiments, there are two ways for an electron or photon to get from the source to the detector screen—going left or going right—and only by combining the possible histories do we get an explanation for what we observe. If the barrier had three slits, we'd have to take account of three kinds of histories; with 300 slits, we'd need to include the contributions of the whole slew of resulting possible histories. Taking this to the limit, if we imagine cutting an enormous number of slits—so many, in fact, that the barrier effectively disappears—quantum physics says that each electron would then traverse *every* possible path on its way to a particular point on the screen, and only by combining the probabilities associated with each such history could we explain the resulting data. That may sound strange. (It is strange.) But this bizarre treatment of times past explains the data of Figure 4.4, Figure 7.1b, and every other experiment dealing with the microworld.

You might wonder how literally you should take the sum over histories description. Does an electron that strikes the detector screen *really* get there by traveling along all possible routes, or is Feynman's prescription merely a clever mathematical contrivance that gets the right answer? This is among the key questions for assessing the true nature of quantum reality, so I wish I could give you a definitive answer. But I can't. Physicists often find it extremely useful to envision a vast assemblage of combining histories; I use this picture in my own research so frequently that it certainly feels real. But that's not the same thing as saying that it *is* real. The point is that quantum calculations unambiguously tell us the probability that an electron will land at one or another point on the screen, and these predictions agree with the data, spot on. As far as the theory's verification and predictive utility are concerned, the story we tell of how the electron got to that point on the screen is of little relevance.

But surely, you'd continue to press, we can settle the issue of what really happens by changing the experimental setup so that we can also watch the supposed fuzzy mélange of possible pasts melding into the observed present. It's a good suggestion, but we already know that there has to be a hitch. In Chapter 4, we learned that probability waves are not directly observable; since Feynman's coalescing histories are nothing but a particular way of thinking about probability waves, they, too, must evade direct observation. And they do. Observations cannot tease apart individual histories; rather, observations reflect *averages* of all possible histories. So, if you change the setup to observe the electrons in flight, you will see each electron pass by your additional detector in one location or another; you will never see any fuzzy multiple histories. When you use quantum mechanics to explain *why* you saw the electron in one place or another, the answer will involve averaging over all possible histories that could have led to that intermediate observation. But the observation itself has access only to histories that have already merged. By looking at the electron in flight, you have merely pushed back the notion of what you mean by a history. Quantum mechanics is starkly efficient: it explains what you see but prevents you from seeing the explanation.

You might further ask: Why, then, is classical physics—commonsense physics—which describes motion in terms of unique histories and trajectories, at all relevant to the universe? Why does it work so well in explaining and predicting the motion of everything from baseballs to planets to comets? How come there is no evidence in day-to-day life of the strange way in which the past apparently unfolds into the present? The reason, discussed briefly in Chapter 4 and to be elaborated shortly with greater precision, is that baseballs, planets, and comets are comparatively large, at least when compared with particles like electrons. And in quantum mechanics, the larger something is, the more skewed the averaging becomes: All possible trajectories *do* contribute to the motion of a baseball in flight, but the usual path—the one single path predicted by Newton's laws—contributes *much* more than do all other paths combined. For large objects, it turns out that classical paths are, by an enormous amount, the dominant contribution to the averaging process and so they are the ones we are familiar with. But when objects are small, like electrons, quarks, and photons, many different histories contribute at roughly the same level and hence all play important parts in the averaging process.

You might finally ask: What is so special about the act of observing or measuring that it can compel all the possible histories to ante up, merge together, and yield a single outcome? How does our act of observing somehow tell a particle it's time to tally up the histories, average them out, and commit to a definite result? Why do we humans and equipment of our making have this special power? Is it special? Or might the human act of observation fit into a broader framework of environmental influence that shows, quantum mechanically speaking, we aren't so special after all? We will take up these perplexing and controversial issues in the latter half of this chapter, since not only are they pivotal to the nature of quantum reality, but they provide an important framework for thinking about quantum mechanics and the arrow of time.

Calculating quantum mechanical averages requires significant technical training. And understanding fully how, when, and where the averages are tallied requires concepts that physicists are still working hard to formulate. But one key lesson can be stated simply: quantum mechanics is the ultimate prochoice arena: every possible "choice" something might make in going from here to there is included in the quantum mechanical probability associated with one possible outcome or another.

Classical and quantum physics treat the past in very different ways.

Pruning History

It is totally at odds with our classical upbringing to imagine one indivisible object—one electron or one photon—simultaneously moving along more than one path. Even those of us with the greatest of self-control would have a hard time resisting the temptation to sneak a peek: as the electron or photon passes through the doubly slit screen or the beam splitter, why not take a quick look to see what path it *really* follows on its way to the detector? In the double-slit experiment, why not put little detectors in front of each slit to tell you whether the electron went through one opening, the other, or both (while still allowing the electron to carry on toward the main detector)? In the beam-splitter experiment, why not put, on each pathway leading from the beam splitter, a little detector that will tell if the photon took the left route, the right route, or both routes (again, while allowing the photon to keep going onward toward the detector)?

The answer is that you *can* insert these additional detectors, but if you do, you will find two things. First, each electron and each photon will always be found to go through one and only one of the detectors; that is, you can determine which path each electron or photon follows, and you will find that it always goes one way or the other, not both. Second, you will also find that the resulting data recorded by the main detectors have changed. Instead of getting the interference patterns of Figure 4.3b and 7.1b, you get the results expected from classical physics, as in Figure 4.3a.

By introducing new elements—the new detectors—you have inadvertently changed the experiments. And the change is such that the paradox you were *just* about to reveal—that you now know which path each particle took, so how could there be any interference with another path that the particle demonstrably did not take?—is averted. The reason follows immediately from the last section. Your new observation singles out those histories that could have preceded whatever your new observation revealed. And since this observation determined which path the photon took, *we consider only those histories that traverse this path, thus eliminating the possibility of interference.*

Niels Bohr liked to summarize such things using his *principle of complementarity*. Every electron, every photon, every*thing*, in fact, has both wavelike and particlelike aspects. They are complementary features. Thinking purely in the conventional particle framework—in which particles move along single, unique trajectories—is incomplete, because it misses the wavelike aspects demonstrated by interference patterns.* Thinking purely in the wavelike framework is incomplete, because it misses the particlelike aspects demonstrated by measurements that find localized particles that can be, for example, recorded by a single dot on a screen. (See Figure 4.4.) A complete picture requires both complementary aspects to be taken into account. In any given situation you can force one feature to be more prominent by virtue of how you choose to interact. If you allow the electrons to travel from source to screen unobserved, their wavelike qualities can emerge, yielding interference. But if you observe the electron en route, you know which path it took, so you'd be at a loss to explain interference. Reality comes to the rescue. Your observation prunes the branches of quantum history. It forces the electron to behave as a particle; since particles go one way *or* the other, no interference pattern forms, so there's nothing to explain.

Nature does weird things. It lives on the edge. But it is careful to bob and weave from the fatal punch of logical paradox.

---

*Even though Feynman's sum over histories approach might seem to make the particle aspect prominent, it is just a particular interpretation of probability *waves* (since it involves many histories for a single particle, each making its own probabilistic contribution), and so is subsumed by the wavelike side of complementarity. When we speak of something behaving like a particle, we will always mean a conventional particle that travels along one and only one trajectory.

The Contingency of History

These experiments are remarkable. They provide simple but powerful proof that our world is governed by the quantum laws found by physicists in the twentieth century, and not by the classical laws found by Newton, Maxwell, and Einstein—laws we now recognize as powerful and insightful approximations for describing events at large enough scales. Already we have seen that the quantum laws challenge conventional notions of what happened in the past—those unobserved events that are responsible for what we now see. Some simple variations of these experiments take this challenge to our intuitive notion of how things unfold in time to an even greater, even more surprising level.

The first variation is called the *delayed-choice* experiment and was suggested in 1980 by the eminent physicist John Wheeler. The experiment brushes up against an eerily odd-sounding question: Does the past depend on the future? Note that this is not the same as asking whether we can go back and change the past (a subject we take up in Chapter 15). Instead, Wheeler's experiment, which has been carried out and analyzed in considerable detail, exposes a provocative interplay between events we imagine having taken place in the past, even the distant past, and those we see taking place right now.

To get a feel for the physics, imagine you are an art collector and Mr. Smithers, chairman of the new Springfield Art and Beautification Society, is coming to look at various works you have put up for sale. You know, however, that his real interest is in *The Full Monty*, a painting in your collection that you never felt quite fit, but one that was left to you by your beloved great-uncle Monty Burns, so that deciding whether to sell it is quite an emotional struggle. After Mr. Smithers arrives, you talk about your collection, recent auctions, the current show at the Metropolitan; surprisingly, you learn that, years back, Smithers was your great-uncle's top aide. By the end of the conversation you decide that you are willing to part with *The Full Monty*: There are so many other works you want, and you must exercise restraint or your collection will have no focus. In the world of art collecting, you have always told yourself, sometimes more is less.

As you reflect back upon this decision, in retrospect it seems that you had actually already decided to sell before Mr. Smithers arrived. Although you have always had a certain affection for *The Full Monty*, you have long been wary of amassing a sprawling collection and late-twentieth-century erotic-nuclear realism is an intimidating area for all but the most seasoned collector. Even though you remember that before your visitor's arrival you had been thinking that you didn't know what to do, from your current vantage point it seems as though you really did. It is not quite that future events have affected the past, but your enjoyable meeting with Mr. Smithers and your subsequent declaration of your willingness to sell have illuminated the past in a way that makes definite particular things that seemed undecided at the time. It is as though the meeting and your declaration helped you to accept a decision that was already made, one that was waiting to be ushered forth into the light of day. The future has helped you tell a more complete story of what was going on in the past.

Of course, in this example, future events are affecting only your perception or interpretation of the past, so the events are neither puzzling nor surprising. But the delayed-choice experiment of Wheeler transports this psychological interplay between the future and the past into the quantum realm, where it becomes both precise and startling. We begin with the experiment in Figure 7.1a, modified by turning the laser down so it fires one photon at a time, as in Figure 7.1b, and also by attaching a new photon detector next to the beam splitter. If the new detector is switched off (see Figure 7.2b), then we are back in the original experimental setup and the photons generate an interference pattern on the photographic screen. But if the new detector is switched on (Figure 7.2a), it tells us which path each photon traveled: if it detects a photon, then the photon took that path; if it fails to detect a photon, then the photon took the other path. Such "which-path" information, as it's called, compels the photon

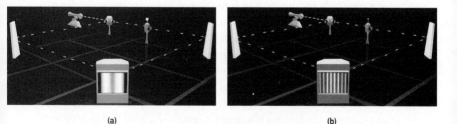

(a)   (b)

Figure 7.2 (a) By turning on "which-path" detectors, we spoil the interference pattern. **(b)** When the new detectors are switched off, we're back in the situation of Figure 7.1 and the interference pattern gets built up.

to act like a particle, so the wavelike interference pattern is no longer generated.

Now let's change things, à la Wheeler, by moving the new photon detector far downstream along one of the two pathways. In principle, the pathways can be as long as you like, so the new detector can be a considerable distance away from the beam splitter. Again, if this new photon detector is switched off, we are in the usual situation and the photons fill out an interference pattern on the screen. If it is switched on, it provides which-path information and thus precludes the existence of an interference pattern.

The new weirdness comes from the fact that the which-path measurement takes place long *after* the photon had to "decide" at the beam splitter whether to act as a wave and travel both paths or to act as a particle and travel only one. When the photon is passing through the beam splitter, it can't "know" whether the new detector is switched on or off—as a matter of fact, the experiment can be arranged so that the on/off switch on the detector is set *after* the photon has passed the splitter. To be prepared for the possibility that the detector is off, the photon's quantum wave had better split and travel both paths, so that an amalgam of the two can produce the observed interference pattern. But if the new detector turns out to have been on—or if it was switched on after the photon fully cleared the splitter—it would seem to present the photon with an identity crisis: on passing through the splitter, it had already committed itself to its wavelike character by traveling both paths, but now, sometime after making this choice, it "realizes" that it needs to come down squarely on the side of being a particle that travels one and only one path.

Somehow, though, the photons always get it right. Whenever the detector is on—again, even if the choice to turn it on is delayed until long after a given photon has passed through the beam splitter—the photon acts fully like a particle. It is found to be on one and only one route to the screen (if we were to put photon detectors way downstream along both routes, each photon emitted by the laser would be detected by one or the other detector, never both); the resulting data show no interference pattern. Whenever the new detector is off—again, even if this decision is made after each photon has passed the splitter—the photons act fully like a wave, yielding the famous interference pattern showing that they've traveled both paths. It's as if the photons adjust their behavior in the past according to the future choice of whether the new detector is switched on; it's as though the photons have a "premonition" of the experimental situation they will encounter farther downstream, and act accordingly. It's as if a consistent and definite history becomes manifest only after the future to which it leads has been fully settled.[4]

There is a similarity to your experience of deciding to sell *The Full Monty*. Before meeting with Mr. Smithers, you were in an ambiguous, undecided, fuzzy, mixed state of being both willing and unwilling to sell the painting. But talking together about the art world and learning of Smithers's affection for your great-uncle made you increasingly comfortable with the idea of selling. The conversation led to a firm decision, which in turn allowed a history of the decision to crystallize out of the previous uncertainty. In retrospect it felt as if the decision had really been made all along. But if you hadn't gotten on so well with Mr. Smithers, if he hadn't given you confidence that *The Full Monty* would be in trustworthy hands, you might very well have decided not to sell. And the story of the past that you might tell in this situation could easily involve a recognition that you'd actually decided long ago *not* to sell—that no matter how sensible it might be to sell the painting, deep down you've always known that the sentimental connection was just too strong to let it go. The actual past, of course, did not change one bit. Yet a different experience now would lead you to describe a different history.

In the psychological arena, rewriting or reinterpreting the past is commonplace; our story of the past is often informed by our experiences in the present. But in the arena of physics—an arena we normally consider to be objective and set in stone—a future contingency of history makes one's head spin. To make the spinning even more severe, Wheeler imagines a cosmic version of the delayed choice experiment in which the light source is not a laboratory laser but, instead, a powerful quasar in deep space. The beam splitter is not a laboratory variety, either, but is an intervening galaxy whose gravitational pull can act like a lens that focuses passing photons and directs them toward earth, as in Figure 7.3. Although no one has as yet carried out this experiment, in principle, if enough photons from the quasar are collected, they should fill out an interference pattern on a long-exposure photographic plate, just as in the laboratory beam-splitter experiment. But if we were to put another photon detector right near the end of one route or the other, it would provide which-path information for the photons, thereby destroying the interference pattern.

What's striking about this version is that, from our perspective, the photons could have been traveling for many billions of years. Their decision to go one way around the galaxy, like a particle, or both ways, like a

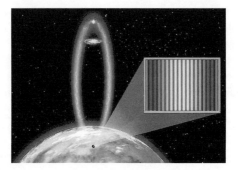

Figure 7.3 Light from a distant quasar, split and focused by an intervening galaxy, will, in principle, yield an interference pattern. If an additional detector, which allows the determination of the path taken by each photon, were switched on, the ensuing photons would no longer fill out an interference pattern.

wave, would seem to have been made long before the detector, any of us, or even the earth existed. Yet, billions of years later, the detector was built, installed along one of the paths the photons take to reach earth, and switched on. And these recent acts somehow ensure that the photons under consideration act like particles. They act as though they have been traveling along precisely one path or the other on their long journey to earth. But if, after a few minutes, we turn off the detector, the photons that subsequently reach the photographic plate start to build up an interference pattern, indicating that for billions of years they have been traveling in tandem with their ghostly partners, taking opposite paths around the galaxy.

Has our turning the detector on or off in the twenty-first century had an effect on the motion of photons some billions of years earlier? Certainly not. Quantum mechanics does not deny that the past has happened, and happened fully. Tension arises simply because the concept of *past* according to the quantum is different from the concept of *past* according to classical intuition. Our classical upbringing makes us long to say that a given photon *did* this or *did* that. But in a quantum world, our world, this reasoning imposes upon the photon a reality that is too restrictive. As we have seen, in quantum mechanics the norm is an indeterminate, fuzzy, hybrid reality consisting of many strands, which only crystallizes into a more familiar, definite reality when a suitable observation is carried out. It is not that the photon, billions of years ago, decided to go one way around the galaxy or the other, or both. Instead, for billions of years it has been in the quantum norm—a hybrid of the possibilities.

The act of observation links this unfamiliar quantum reality with everyday classical experience. Observations we make today cause one of the strands of quantum history to gain prominence in our recounting of the past. In this sense, then, although the quantum evolution from the past until now is unaffected by anything we do now, the story we tell of the past can bear the imprint of today's actions. If we insert photon detectors along the two pathways light takes to a screen, then our story of the past will include a description of which pathway each photon took; by inserting the photon detectors, we ensure that which-path information is an essential and definitive detail of our story. But, if we don't insert the photon detectors, our story of the past will, of necessity, be different. Without the photon detectors, we can't recount anything about which path the photons took; without the photon detectors, which-path details are fundamentally unavailable. Both stories are valid. Both stories are interesting. They just describe different situations.

An observation today can therefore help complete the story we tell of a process that began yesterday, or the day before, or perhaps a billion years earlier. An observation today can delineate the kinds of details we can and must include in today's recounting of the past.

Erasing the Past

It is essential to note that in these experiments the past is not in any way altered by today's actions, and that no clever modification of the experiments will accomplish that slippery goal. This raises the question: If you can't change something that has already happened, can you do the next best thing and erase its *impact* on the present? To one degree or another, sometimes this fantasy can be realized. A baseball player who, with two outs in the bottom of the ninth inning, drops a routine fly ball, allowing the opposing team to close within one run, can undo the impact of his error by a spectacular diving catch on the ball hit by the next batter. And, of course, such an example is not the slightest bit mysterious. Only when an event in the past seems definitively to preclude another event's happening in the future (as the dropped fly ball definitively precluded a perfect game) would we think there was something awry if we were subsequently told that the precluded event had actually happened. The *quantum eraser*, first suggested in 1982 by Marlan Scully and Kai Drühl, hints at this kind of strangeness in quantum mechanics.

A simple version of the quantum eraser experiment makes use of the double-slit setup, modified in the following way. A tagging device is placed in front of each slit; it marks any passing photon so that when the photon is examined later, you can tell through which slit it passed. The question of how you can place a mark on a photon—how you can do the equivalent of placing an "L" on a photon that passes through the left slit and an "R" on a photon that passes through the right slit—is a good one, but the details are not particularly important. Roughly, the process relies on using a device that allows a photon to pass freely through a slit but forces its spin axis to point in a particular direction. If the devices in front of the left and right slits manipulate the photon spins in specific but distinct ways, then a more refined detector screen that not only registers a dot at the photon's impact location, but also keeps a record of the photon's spin orientation, will reveal through which slit a given photon passed on its way to the detector.

When this double-slit-with-tagging experiment is run, the photons do not build up an interference pattern, as in Figure 7.4a. By now the explanation should be familiar: the new tagging devices allow which-path information to be gleaned, and which-path information singles out one history or another; the data show that any given photon passed through either the left slit or the right slit. And without the combination of left-slit and right-slit trajectories, there are no overlapping probability waves, so no interference pattern is generated.

Now, here is Scully and Drühl's idea. What if, just before the photon hits the detection screen, you eliminate the possibility of determining through which slit it passed by erasing the mark imprinted by the tagging device? Without the means, even in principle, to extract the which-path information from the detected photon, will both classes of histories come back into play, causing the interference pattern to reemerge? Notice that this kind of "undoing" the past would fall much further into the shocking category than the ballplayer's diving catch in the ninth inning. When the tagging devices are turned on, we imagine that the photon obediently acts as a particle, passing through the left slit *or* the right slit. If somehow, just before it hits the screen, we erase the which-slit mark it is carrying, it seems too late to allow an interference pattern to form. For interference, we need the photon to act like a wave. It must pass through both slits so

Figure 7.4 In the quantum eraser experiment, equipment placed in front of the two slits marks the photons so that subsequent examination can reveal through which slit each photon passed. In **(a)** we see that this which-path information spoils the interference pattern. In **(b)** a device that erases the mark on the photons is inserted just in front of the detector screen. Because the which-path information is eliminated, the interference pattern reappears.

that it can cross-mingle with itself on the way to the detector screen. But our initial tagging of the photon seems to ensure that it acts like a particle and travels either through the left or through the right slit, preventing interference from happening.

In an experiment carried out by Raymond Chiao, Paul Kwiat, and Aephraim Steinberg, the setup was, schematically, as in Figure 7.4, with a new erasure device inserted just in front of the detection screen. Again, the details are not of the essence, but briefly put, the eraser works by ensuring that regardless of whether a photon from the left slit or the right slit enters, its spin is manipulated to point in one and the same fixed direction. Subsequent examination of its spin therefore yields no information about which slit it passed through, and so the which-path mark has been erased. Remarkably, the photons detected by the screen after this erasure *do* produce an interference pattern. When the eraser is inserted just in front of the detector screen, it undoes—it erases—the effect of tagging the photons way back when they approached the slits. As in the delayed-choice experiment, in principle this kind of erasure could occur billions of years after the influence it is thwarting, in effect undoing the past, even undoing the ancient past.

How are we to make sense of this? Well, keep in mind that the data conform perfectly to the theoretical prediction of quantum mechanics. Scully and Drühl proposed this experiment because their quantum mechanical calculations convinced them it would work. And it does. So, as is usual with quantum mechanics, the puzzle doesn't pit theory against experiment. It pits theory, confirmed by experiment, against our intuitive sense of time and reality. To ease the tension, notice that were you to place a photon *detector* in front of each slit, the detector's readout would establish with certainty whether the photon went through the left slit or through the right slit, and there'd be no way to erase such definitive information—there'd be no way to recover an interference pattern. But the tagging devices are different because they provide only the potential for which-path information to be determined—and potentialities are just the kinds of things that can be erased. A tagging device modifies a passing photon in such a way, roughly speaking, that it still travels both paths, but the left part of its probability wave is blurred out relative to the right, or the right part of its probability wave is blurred out relative to the left. In turn, the orderly sequence of peaks and troughs that would normally emerge from each slit—as in Figure 4.2b—is also blurred out, so no interference pattern forms on the detector screen. The crucial realization, though, is that both the left and the right waves are still present. The eraser works because it refocuses the waves. Like a pair of glasses, it compensates for the blurring, brings both waves back into sharp focus, and allows them once again to combine into an interference pattern. It's as if after the tagging devices accomplish their task, the interference pattern disappears from view but patiently lies in wait for someone or something to resuscitate it.

That explanation may make the quantum eraser a little less mysterious, but here is the finale—a stunning variation on the quantum-eraser experiment that challenges conventional notions of space and time even further.

Shaping the Past*

This experiment, the *delayed-choice quantum eraser,* was also proposed by Scully and Drühl. It begins with the beam-splitter experiment of Figure 7.1, modified by inserting two so-called down-converters, one on each pathway. Down-converters are devices that take one photon as input and produce two photons as output, each with half the energy ("down-converted") of the original. One of the two photons (called the *signal* photon) is directed along the path that the original would have followed toward

---

*If you find this section tough going, you can safely move on to the next section without loss of continuity. But I encourage you to try to get through it, as the results are truly stupendous.

the detector screen. The other photon produced by the down-converter (called the *idler* photon) is sent in a different direction altogether, as in Figure 7.5a. On each run of the experiment, we can determine which path a signal photon takes to the screen by observing which down-converter spits out the idler-photon partner. And once again, the ability to glean which-path information about the signal photons—even though it is totally indirect, since we are not interacting with any signal photons at all—has the effect of preventing an interference pattern from forming.

Now for the weirder part. What if we manipulate the experiment so as to make it impossible to determine from which down-converter a given idler photon emerged? What if, that is, we erase the which-path information embodied by the idler photons? Well, something amazing happens: even though we've done nothing directly to the signal photons, by erasing the which-path information carried by their idler partners we can recover an interference pattern from the signal photons. Let me show you how this goes because it is truly remarkable.

Take a look at Figure 7.5b, which embodies all the essential ideas. But don't be intimidated. It's simpler than it appears, and we'll now go through it in manageable steps. The setup in Figure 7.5b differs from that of Figure 7.5a with regard to how we detect the idler photons after they've been emitted. In Figure 7.5a, we detected them straight out, and so we could immediately determine from which down-converter each was produced—that is, which path a given signal photon took. In the new experiment, each idler photon is sent through a maze, which compromises our ability to make such a determination. For example, imagine that an idler photon is emitted from the down-converter labeled "L." Rather than immediately entering a detector (as in Figure 7.5a), this photon is sent to a beam splitter (labeled "a"), and so has a 50 percent chance of heading onward along the path labeled "A," and a 50 percent chance of heading onward along the path labeled "B." Should it head along path A, it will enter a photon detector (labeled "1"), and its arrival will be duly recorded. But should the idler photon head along path B, it will be subject to yet further shenanigans. It will be directed to another beam splitter (labeled "c") and so will have a 50 percent chance of heading onward along path E to the detector labeled "2," and a 50 percent chance of heading onward along path F to the detector labeled "3." Now—stay with me, as there is a point to all this—the exact same reasoning, when applied to an idler photon emitted from the other down-converter, labeled "R," tells us that if the idler heads along path D it will be recorded by detector 4, but if it heads

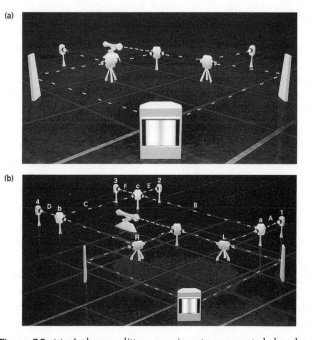

Figure 7.5 **(a)** A beam-splitter experiment, augmented by down-converters, does not yield an interference pattern, since the idler photons yield which-path information. **(b)** If the idler photons are not detected directly, but instead are sent through the maze depicted, then an interference pattern can be extracted from the data. Idler photons that are detected by detectors 2 or 3 do not yield which-path information and hence their signal photons fill out an interference pattern.

along path C it will be detected by either detector 3 or detector 2, depending on the path it follows after passing through beam splitter c.

Now for why we've added all this complication. Notice that if an idler photon is detected by detector 1, we learn that the corresponding signal photon took the left path, since there is no way for an idler that was emitted from down-converter R to find its way to this detector. Similarly, if an idler photon is detected by detector 4, we learn that its signal photon partner took the right path. But if an idler photon winds up in detector 2, we have no idea which path its signal photon partner took, since there is an equal chance that it was emitted by down-converter L and followed path

B–E, or that it was emitted by down-converter R and followed path C–E. Similarly, if an idler is detected by detector 3, it could have been emitted by down-converter L and have traveled path B–F, or by down-converter R and traveled path C–F. Thus, *for signal photons whose idlers are detected by detector 1 or 4, we have which-path information, but for those whose idlers are detected by detector 2 or 3, the which-path information is erased.*

Does this erasure of some of the which-path information—even though we've done *nothing directly* to the signal photons—mean interference effects are recovered? Indeed it does—but only for those signal photons whose idlers wind up in either detector 2 or detector 3. Namely, the totality of impact positions of the signal photons on the screen will look like the data in Figure 7.5a, *showing not the slightest hint of an interference pattern,* as is characteristic of photons that have traveled one path *or* the other. But if we focus on a *subset* of the data points—for example, those signal photons whose idlers entered detector 2—then that subset of points *will* fill out an interference pattern! These signal photons—whose idlers happened, by chance, not to provide any which-path information—act as though they've traveled both paths! If we were to hook up the equipment so that the screen displays a red dot for the position of each signal photon whose idler was detected by detector 2, and a green dot for all others, someone who is color-blind would see no interference pattern, but everyone else would see that the red dots were arranged with bright and dark bands—an interference pattern. The same holds true with detector 3 in place of detector 2. But there would be no such interference pattern if we single out signal photons whose idlers wind up in detector 1 or detector 4, since these are the idlers that yield which-path information about their partners.

These results—which have been confirmed by experiment[5]—are dazzling: by including down-converters that have the potential to provide which-path information, we lose the interference pattern, as in Figure 7.5a. And without interference, we would naturally conclude that each photon went along either the left path or the right path. But we now learn that this would be a hasty conclusion. By carefully eliminating the potential which-path information carried by some of the idlers, we can coax the data to yield up an interference pattern, indicating that some of the photons actually took both paths.

Notice, too, perhaps the most dazzling result of all: the three additional beam splitters and the four idler-photon detectors can be on the other side of the laboratory or even on the other side of the universe, since nothing in our discussion depended at all on whether they receive a given idler photon before or after its signal photon partner has hit the screen. Imagine, then, that these devices are all far away, say ten light-years away, to be definite, and think about what this entails. You perform the experiment in Figure 7.5b today, recording—one after another—the impact locations of a huge number of signal photons, and you observe that they show no sign of interference. If someone asks you to explain the data, you might be tempted to say that because of the idler photons, which-path information is available and hence each signal photon definitely went along either the left or the right path, eliminating any possibility of interference. But, as above, this would be a hasty conclusion about what happened; it would be a thoroughly premature description of the past.

You see, ten years later, the four photon detectors will receive—one after another—the idler photons. If you are subsequently informed about which idlers wound up, say, in detector 2 (e.g., the first, seventh, eighth, twelfth . . . idlers to arrive), and if you then go back to data you collected years earlier and highlight the corresponding signal photon locations on the screen (e.g., the first, seventh, eighth, twelfth . . . signal photons that arrived), you will find that the highlighted data points fill out an interference pattern, thus revealing that those signal photons should be described as having traveled both paths. Alternatively, if 9 years, 364 days after you collected the signal photon data, a practical joker should sabotage the experiment by removing beam splitters a and b—ensuring that when the idler photons arrive the next day, they all go to either detector 1 or detector 4, thus preserving *all* which-path information—then, when you receive this information, you will conclude that *every* signal photon went along either the left path or the right path, and there will be no interference pattern to extract from the signal photon data. Thus, as this discussion forcefully highlights, the story you'd tell to explain the signal photon data depends significantly on measurements conducted ten years after those data were collected.

Again, let me emphasize that the future measurements do not change anything at all about things that took place in your experiment today; the future measurements do not in any way change the data you collected today. But the future measurements *do* influence the kinds of details you can invoke when you subsequently describe what happened today. Before you have the results of the idler photon measurements, you really can't say anything at all about the which-path history of any given signal photon. However, once you have the results, you conclude that signal photons whose idler partners were successfully used to ascertain which-path information *can* be described as having—years earlier—traveled either left or right. You also conclude that signal photons whose idler partners had their which-path information erased *cannot* be described as having— years earlier—definitely gone one way or the other (a conclusion you can convincingly confirm by using the newly acquired idler photon data to expose the previously hidden interference pattern among this latter class of signal photons). We thus see that the future helps shape the story you tell of the past.

These experiments are a magnificent affront to our conventional notions of space and time. Something that takes place long after and far away from something else nevertheless is vital to our description of that something else. By any classical—commonsense—reckoning, that's, well, crazy. Of course, that's the point: classical reckoning is the wrong kind of reckoning to use in a quantum universe. We have learned from the Einstein-Podolsky-Rosen discussion that quantum physics is not local in space. If you have fully absorbed that lesson—a tough one to accept in its own right—these experiments, which involve a kind of entanglement across space and through time, may not seem thoroughly outlandish. But by the standards of daily experience, they certainly are.

Quantum Mechanics and Experience

For a few days after I first learned about these experiments, I remember feeling elated. I felt I'd been given a glimpse into a veiled side of reality. Common experience—mundane, ordinary, day-to-day activities—suddenly seemed part of a classical charade, hiding the true nature of our quantum world. The world of the everyday suddenly seemed nothing but an inverted magic act, lulling its audience into believing in the usual, familiar conceptions of space and time, while the astonishing truth of quantum reality lay carefully guarded by nature's sleights of hand.

In recent years, physicists have expended much effort in trying to explain nature's ruse—to figure out precisely how the fundamental laws of quantum physics morph into the classical laws that are so successful at explaining common experience—in essence, to figure out how the atomic and subatomic shed their magical weirdness when they combine to form macroscopic objects. Research continues, but much has already been learned. Let's look at some aspects of particular relevance to the question of time's arrow, but now from the standpoint of quantum mechanics.

Classical mechanics is based on equations that Newton discovered in the late 1600s. Electromagnetism is based on equations Maxwell discovered in the late 1800s. Special relativity is based on equations Einstein discovered in 1905, and general relativity is based on equations he discovered in 1915. What all these equations have in common, and what is central to the dilemma of time's arrow (as explained in the last chapter), is their completely symmetric treatment of past and future. Nowhere in any of these equations is there anything that distinguishes "forward" time from "backward" time. Past and future are on an equal footing.

Quantum mechanics is based on an equation that Erwin Schrödinger discovered in 1926.[6] You don't need to know anything about this equation beyond the fact that it takes as input the shape of a quantum mechanical probability wave at one moment of time, such as that in Figure 4.5, and allows one to determine what the probability wave looks like at any other time, earlier or later. If the probability wave is associated with a particle, such as an electron, you can use it to predict the probability that, at any specified time, an experiment will find the electron at any specified location. Like the classical laws of Newton, Maxwell, and Einstein, the quantum law of Schrödinger embraces an egalitarian treatment of time-future and time-past. A "movie" showing a probability wave starting like *this* and ending like *that* could be run in reverse—showing a probability wave starting like *that* and ending like *this*—and there would be no way to say that one evolution was right and the other wrong. Both would be equally valid solutions of Schrödinger's equation. Both would represent equally sensible ways in which things could evolve.[7]

Of course, the "movie" now referred to is quite different from the ones used in analyzing the motion of a tennis ball or a splattering egg in the last chapter. Probability waves are not things we can see directly; there are no cameras that can capture probability waves on film. Instead, we can describe probability waves using mathematical equations and, in our mind's eye, we can imagine the simplest of them having shapes such as those in Figures 4.5 and 4.6. But the only access we have to the probability waves themselves is indirect, through the process of measurement.

That is, as outlined in Chapter 4 and seen repeatedly in the experiments above, the standard formulation of quantum mechanics describes the unfolding of phenomena using *two* quite distinct stages. In stage one, the probability wave—or, in the more precise language of the field, the *wavefunction*—of an object such as an electron evolves according to the equation discovered by Schrödinger. This equation ensures that the shape of the wavefunction changes smoothly and gradually, much as a water wave changes shape as it travels from one side of a lake toward the other.* In the standard description of the second stage, we make contact with observable reality by measuring the electron's position, and when we do so, the shape of its wavefunction sharply and abruptly changes. The electron's wavefunction is unlike more familiar examples like water waves and sound waves: when we measure the electron's position, its wavefunction spikes or, as illustrated in Figure 4.7, it collapses, dropping to the value 0 everywhere the particle is not found and surging to 100 percent probability at the single location where the particle is found by the measurement.

Stage one—the evolution of wavefunctions according to Schrödinger's equation—is mathematically rigorous, totally unambiguous, and fully accepted by the physics community. Stage two—the collapse of a wavefunction upon measurement—is, to the contrary, something that during the last eight decades has, at best, kept physicists mildly bemused, and at worst, posed problems, puzzles, and potential paradoxes that have devoured careers. The difficulty, as mentioned at the end of Chapter 4, is that according to Schrödinger's equation, wavefunctions do *not* collapse. Wavefunction collapse is an add-on. It was introduced after Schrödinger discovered his equation, in an attempt to account for what experimenters actually see. Whereas a raw, uncollapsed wavefunction embodies the strange idea that a particle is here and there, experimenters never see this. They always find a particle definitely at one location or another; they never see it partially here and partially there; the needle on their measuring devices never hovers in some ghostly mixture of pointing at this value and also at that value.

The same goes, of course, for our own casual observations of the world around us. We never observe a chair to be both here and there; we never observe the moon to be in one part of the night sky as well as another; we never see a cat that is both dead and alive. The notion of wavefunction collapse aligns with our experience by postulating that the

---

*Quantum mechanics, rightly, has a reputation as being anything but smooth and gradual; rather, as we will see explicitly in later chapters, it reveals a turbulent and jittery microcosmos. The origin of this jitteriness is the probabilistic nature of the wavefunction—even though things can be one way at one moment, there is a probability that they will be significantly different a moment later—*not* an ever-present jittery quality of the wavefunction itself.

act of measurement induces the wavefunction to relinquish quantum limbo and usher one of the many potentialities (particle here, or particle there) into reality.

The Quantum Measurement Puzzle

But how does an experimenter's making a measurement cause a wavefunction to collapse? In fact, does wavefunction collapse really happen, and if it does, what really goes on at the microscopic level? Do any and all measurements cause collapse? When does the collapse happen and how long does it take? Since, according to the Schrödinger equation, wavefunctions do not collapse, what equation takes over in the second stage of quantum evolution, and how does the new equation dethrone Schrödinger's, usurping its usual ironclad power over quantum processes? And, of importance to our current concern with time's arrow, while Schrödinger's equation, the equation that governs the first stage, makes no distinction between forward and backward in time, does the equation for stage two introduce a fundamental asymmetry between time before and time after a measurement is carried out? That is, does quantum mechanics, *including its interface with the world of the everyday via measurements and observations,* introduce an arrow of time into the basic laws of physics? After all, we discussed earlier how the quantum treatment of the past differs from that of classical physics, and by *past* we meant before a particular observation or measurement had taken place. So do measurements, as embodied by stage-two wavefunction collapse, establish an asymmetry between past and future, between before and after a measurement is made?

These questions have stubbornly resisted complete solution and they remain controversial. Yet, through the decades, the predictive power of quantum theory has hardly been compromised. The stage one / stage two formulation of quantum theory, even though stage two has remained mysterious, predicts probabilities for measuring one outcome or another. And these predictions have been confirmed by repeating a given experiment over and over again and examining the frequency with which one or another outcome is found. The fantastic experimental success of this approach has far outweighed the discomfort of not having a precise articulation of what actually happens in stage two.

But the discomfort has always been there. And it is not simply that some details of wavefunction collapse have not quite been worked out. The *quantum measurement problem*, as it is called, is an issue that speaks to the limits and the universality of quantum mechanics. It's simple to see this. The stage one / stage two approach introduces a split between what's being observed (an electron, or a proton, or an atom, for example) and the experimenter who does the observing. Before the experimenter gets into the picture, wavefunctions happily and gently evolve according to Schrödinger's equation. But then, when the experimenter meddles with things to perform a measurement, the rules of the game suddenly change. Schrödinger's equation is cast aside and stage-two collapse takes over. Yet, since there is no difference between the atoms, protons, and electrons that make up the experimenter and the equipment he or she uses, and the atoms, protons, and electrons that he or she studies, why in the world is there a split in how quantum mechanics treats them? If quantum mechanics is a universal theory that applies without limitations to *everything*, the observed and the observer should be treated in exactly the same way.

Niels Bohr disagreed. He claimed that experimenters and their equipment *are* different from elementary particles. Even though they are made from the same particles, they are "big" collections of elementary particles and hence governed by the laws of classical physics. Somewhere between the tiny world of individual atoms and subatomic particles and the familiar world of people and their equipment, the rules change because the sizes change. The motivation for asserting this division is clear: a tiny particle, according to quantum mechanics, can be located in a fuzzy mixture of here and there, yet we don't see such behavior in the big, everyday world. But exactly where is the border? And, of vital importance, how do the two sets of rules interface when the big world of the everyday confronts the minuscule world of the atomic, as in the case of a measurement? Bohr forcefully declared these questions to be out of bounds, by which he meant, truth be told, that they were beyond the bounds of what he or anyone else could answer. And since even without addressing them the theory makes astonishingly accurate predictions, for a long time such issues were far down on the list of critical questions that physicists were driven to settle.

But to understand quantum mechanics completely, to determine fully what it says about reality, and to establish what role it might play in setting a direction to time's arrow, we must come to grips with the quantum measurement problem.

In the next two sections, we'll describe some of the most prominent and promising attempts to do so. The upshot, should you at any point want to rush ahead to the last section focusing on quantum mechanics and the arrow of time, is that much ingenious work on the quantum measurement problem has yielded significant progress, but a broadly accepted solution still seems just beyond our reach. Many view this as the single most important gap in our formulation of quantum law.

Reality and the Quantum Measurement Problem

Over the years, there have been many proposals for solving the quantum measurement problem. Ironically, although they entail differing conceptions of reality—some drastically different—when it comes to predictions for what a researcher will measure in most every experiment, they all agree and each one works like a charm. Each proposal puts on the same show, even though, were you to peek backstage, you'd see that their modi operandi differ substantially.

When it comes to entertainment, you generally don't want to know what's happening off in the wings; you are perfectly content to focus solely on the production. But when it comes to understanding the universe, there is an insatiable urge to pull back all curtains, open all doors, and expose completely the deep inner workings of reality. Bohr considered this urge baseless and misguided. To him, reality *was* the performance. Like a Spalding Gray soliloquy, an experimenter's bare-bones measurements *are* the whole show. There isn't anything else. According to Bohr, there is no backstage. Trying to analyze how, and when, and why a quantum wavefunction relinquishes all but one possibility and produces a single definite number on a measuring device is missing the point. The measured number itself is all that's worthy of attention.

For decades, this perspective held sway. However, its calmative effect on the mind struggling with quantum theory notwithstanding, one can't help feeling that the fantastic predictive power of quantum mechanics means that it *is* tapping into a hidden reality that underlies the workings of the universe. One can't help wanting to go further and understand how quantum mechanics interfaces with common experience—how it bridges the gap between wavefunction and observation, and what hidden reality underlies the observations. Over the years, a number of researchers have taken up this challenge; here are some proposals they've developed.

One approach, with historical roots that go back to Heisenberg, is to abandon the view that wavefunctions are objective features of quantum reality and, instead, view them merely as an embodiment of what we know about reality. Before we perform a measurement, we don't know where the electron is and, this view proposes, our ignorance of its location is reflected by the electron's wavefunction describing it as possibly being at a variety of different positions. At the moment we measure its position, though, our knowledge of its whereabouts suddenly changes: we now know its position, in principle, with total precision. (By the uncertainty principle, if we know its location we will necessarily be completely ignorant of its velocity, but that's not an issue for the current discussion.) This sudden change in our knowledge, according to this perspective, is reflected in a sudden change in the electron's wavefunction: it suddenly collapses and takes on the spiked shape of Figure 4.7, indicating our definite knowledge of the electron's position. In this approach, then, the abrupt collapse of a wavefunction is completely unsurprising: it is nothing more than the abrupt change in knowledge that we all experience when we learn something new.

A second approach, initiated in 1957 by Wheeler's student Hugh Everett, denies that wavefunctions ever collapse. Instead, each and every potential outcome embodied in a wavefunction sees the light of day; the daylight each sees, however, streams through its own separate universe. In this approach, the *Many Worlds interpretation*, the concept of "the universe" is enlarged to include innumerable "parallel universes" — innumerable versions of our universe — so that anything that quantum mechanics predicts *could* happen, even if only with minuscule probability, *does* happen in at least one of the copies. If a wavefunction says that an electron can be here, there, and way over there, then in one universe a version of you will find it here; in another universe, another copy of you will find it there; and in a third universe, yet another you will find the electron way over there. The sequence of observations that we each make from one second to the next thus reflects the reality taking place in but one part of this gargantuan, infinite network of universes, each one populated by copies of you and me and everyone else who is still alive in a universe in which certain observations have yielded certain outcomes. In one such universe you are now reading these words, in another you've taken a break to surf the Web, in yet another you're anxiously awaiting the curtain to rise for your Broadway debut. It's as though there isn't a single spacetime block as depicted in Figure 5.1, but an infinite number, with each realizing one possible course of events. In the Many Worlds approach, then, no potential outcome remains merely a potential. Wavefunctions don't collapse. Every potential outcome comes out in one of the parallel universes.

A third proposal, developed in the 1950s by David Bohm—the same physicist we encountered in Chapter 4 when discussing the Einstein-Podolsky-Rosen paradox—takes a completely different approach.[8] Bohm argued that particles such as electrons *do* possess definite positions and definite velocities, just as in classical physics, and just as Einstein had hoped. But, in keeping with the uncertainty principle, these features are hidden from view; they are examples of the *hidden variables* mentioned in Chapter 4. You can't determine both simultaneously. For Bohm, such uncertainty represented a limit on what we can know, but implied nothing about the actual attributes of the particles themselves. His approach does not fall afoul of Bell's results because, as we discussed toward the end of Chapter 4, possessing definite properties forbidden by quantum uncertainty is *not* ruled out; only locality is ruled out, and Bohm's approach is *not* local.[9] Instead, Bohm imagined that the wavefunction of a particle is another, *separate element of reality*, one that exists *in addition to the particle itself*. It's not particles *or* waves, as in Bohr's complementarity philosophy; according to Bohm, it's particles *and* waves. Moreover, Bohm posited that a particle's wavefunction interacts with the particle itself—it "guides" or "pushes" the particle around—in a way that determines its subsequent motion. While this approach agrees fully with the successful predictions of standard quantum mechanics, Bohm found that changes to the wavefunction in one location are able to immediately push a particle at a distant location, a finding that explicitly reveals the nonlocality of his approach. In the double-slit experiment, for example, each particle goes through one slit or the other, while its wavefunction goes through both and suffers interference. Since the wavefunction guides the particle's motion, it should not be terribly surprising that the equations show the particle is likely to land where the wavefunction value is large and it is unlikely to land where it is small, explaining the data in Figure 4.4. In Bohm's approach, there is no separate stage of wavefunction collapse since, if you measure a particle's position and find it *here*, that is truly where it was a moment before the measurement took place.

A fourth approach, developed by the Italian physicists Giancarlo Ghirardi, Alberto Rimini, and Tullio Weber, makes the bold move of modifying Schrödinger's equation in a clever way that results in hardly any effect on the evolution of wavefunctions for individual particles, but has a dramatic impact on quantum evolution when applied to "big" everyday objects. The proposed modification envisions that wavefunctions are inherently unstable; even without any meddling, these researchers suggest, sooner or later every wavefunction collapses, of its own accord, to a spiked shape. For an individual particle, Ghirardi, Rimini, and Weber postulate that wavefunction collapse happens spontaneously and randomly, kicking in, on average, only once every billion years or so.[10] This is so infrequent that it entails only the slightest change to the usual quantum mechanical description of individual particles, and that's good, since quantum mechanics describes the microworld with unprecedented accuracy. But for large objects such as experimenters and their equipment, which have billions and billions of particles, the odds are high that in a tiny fraction of any given second the posited spontaneous collapse will kick in for at least one constituent particle, causing its wavefunction to collapse. And, as argued by Ghirardi, Rimini, Weber, and others, the entangled nature of all the individual wavefunctions in a large object ensures that this collapse initiates a kind of quantum domino effect in which the wavefunctions of all the constituent particles collapse as well. As this happens in a brief fraction of a second, the proposed modification ensures that large objects are essentially always in one definite configuration: pointers on measuring equipment always point to one definite value; the moon is always at one definite location in the sky; brains inside experimenters always have one definite experience; cats are always either dead or alive.

Each of these approaches, as well as a number of others I won't discuss, has its supporters and detractors. The "wavefunction as knowledge" approach finesses the issue of wavefunction collapse by denying any reality for wavefunctions, turning them instead into mere descriptors of what we know. But why, a detractor asks, should fundamental physics be so closely tied to human awareness? If we were not here to observe the world, would wavefunctions never collapse, or, perhaps, would the very concept of a wavefunction not exist? Was the universe a vastly different place before human consciousness evolved on planet earth? What if, instead of human experimenters, mice or ants or amoebas or computers are the only observers? Is the change in their "knowledge" adequate to be associated with the collapse of a wavefunction?[11]

By contrast, the Many Worlds interpretation avoids the whole matter of wavefunction collapse, since in this approach wavefunctions don't collapse. But the price to pay is an enormous proliferation of universes, something that many a detractor has found intolerably exorbitant.[12] Bohm's approach also avoids wavefunction collapse; but, its detractors claim, in granting independent reality to both particles and waves, the theory lacks economy. Moreover, the detractors correctly argue, in Bohm's formulation the wavefunction can exert faster-than-light influences on the particles it pushes. Supporters note that the former complaint is subjective at best, and the latter conforms to the nonlocality Bell proved unavoidable, so neither criticism is convincing. Nevertheless, perhaps unjustifiably, Bohm's approach has never caught on.[13] The Ghirardi-Rimini-Weber approach deals with wavefunction collapse directly, by changing the equations to incorporate a new spontaneous collapse mechanism. But, detractors point out, there is as yet not a shred of experimental evidence supporting the proposed modification to Schrödinger's equation.

Research seeking a solid and fully transparent connection between the formalism of quantum mechanics and the experience of everyday life will no doubt go on for some time to come, and it's hard to say which, if any, of the known approaches will ultimately achieve a majority consensus. Were physicists to be polled today, I don't think there would be an overwhelming favorite. Unfortunately, experimental input is of limited help. While the Ghirardi-Rimini-Weber proposal does make predictions that can, in certain situations, differ from standard stage one / stage two quantum mechanics, the deviations are too small to be tested with today's technology. The situation with the other three proposals is worse because they stymie experimental adjudication even more definitively. They agree fully with the standard approach, and so each yields the same predictions for things that can be observed and measured. They differ only regarding what happens backstage, as it were. They only differ, that is, regarding what quantum mechanics implies for the underlying nature of reality.

Even though the quantum measurement problem remains unsolved, during the last few decades a framework has been under development that, while still incomplete, has widespread support as a likely ingredient of any viable solution. It's called *decoherence*.

Decoherence and Quantum Reality

When you first encounter the probabilistic aspect of quantum mechanics, a natural reaction is to think that it is no more exotic than the probabilities that arise in coin tosses or roulette wheels. But when you learn about quantum interference, you realize that probability enters quantum mechanics in a far more fundamental way. In everyday examples, various outcomes—heads versus tails, red versus black, one lottery number versus another—are assigned probabilities with the understanding that one or another result will definitely happen and that each result is the end product of an independent, definite history. When a coin is tossed, sometimes the spinning motion is just right for the toss to come out heads and sometimes it's just right for the toss to come out tails. The 50-50 probability we assign to each outcome refers not just to the final result—heads or tails—but also to the histories that lead to each outcome. Half of the possible ways you can toss a coin result in heads, and half result in tails. The histories themselves, though, are totally separate, isolated alternatives. There is no sense in which different motions of the coin reinforce each other or cancel each other out. They're all independent.

But in quantum mechanics, things are different. The alternate paths an electron can follow from the two slits to the detector are not separate, isolated histories. The possible histories commingle to produce the observed outcome. Some paths reinforce each other, while others cancel each other out. Such quantum interference between the various possible histories is responsible for the pattern of light and dark bands on the detector screen. Thus, *the telltale difference between the quantum and the classical notions of probability is that the former is subject to interference and the latter is not.*

Decoherence is a widespread phenomenon that forms a bridge between the quantum physics of the small and the classical physics of the not-so-small by suppressing quantum interference—that is, by diminishing sharply the core difference between quantum and classical probabilities. The importance of decoherence was realized way back in the early days of quantum theory, but its modern incarnation dates from a seminal paper by the German physicist Dieter Zeh in 1970,[14] and has since been developed by many researchers, including Erich Joos, also from Germany, and Wojciech Zurek, of the Los Alamos National Laboratory in New Mexico.

Here's the idea. When Schrödinger's equation is applied in a simple situation such as single, isolated photons passing through a screen with two slits, it gives rise to the famous interference pattern. But there are two very special features of this laboratory example that are not characteristic of real-world happenings. First, the things we encounter in day-to-day life are larger and more complicated than a single photon. Second, the things we encounter in day-to-day life are not isolated: they interact with us and with the environment. The book now in your hands is subject to human contact and, more generally, is continually struck by photons and air molecules. Moreover, since the book itself is made of many molecules and atoms, these constantly jittering constituents are continually bouncing off each other as well. The same is true for pointers on measuring devices, for cats, for human brains, and for just about everything you encounter in daily life. On astrophysical scales, the earth, the moon, asteroids, and the other planets are continually bombarded by photons from the sun. Even a grain of dust floating in the darkness of outer space is subject to continual hits from low-energy microwave photons that have been streaming through space since a short time after the big bang. And so, to understand what quantum mechanics says about real-world happenings—as opposed to pristine laboratory experiments—we should apply Schrödinger's equation to these more complex, messier situations.

In essence, this is what Zeh emphasized, and his work, together with that of many others who have followed, has revealed something quite wonderful. Although photons and air molecules are too small to have any significant effect on the motion of a big object like this book or a cat, they are able to do something else. They continually "nudge" the big object's wavefunction, or, in physics-speak, they disturb its *coherence*: they blur its orderly sequence of crest followed by trough followed by crest. This is critical, because a wavefunction's orderliness is central to generating interference effects (see Figure 4.2). And so, much as adding tagging devices to the double-slit experiment blurs the resulting wavefunction and thereby washes out interference effects, the constant bombardment of objects by constituents of their environment also washes out the possibility of intereference phenomena. In turn, once quantum interference is no longer possible, the probabilities inherent to quantum mechanics are, for all practical purposes, just like the probabilities inherent to coin tosses and roulette wheels. Once environmental decoherence blurs a wavefunction, the exotic nature of quantum probabilities melts into the more familiar probabilities of day-to-day life.[15] This suggests a resolution of the quantum measurement puzzle, one that, if realized, would be just about the best thing we could hope for. I'll describe it first in the most optimistic light, and then stress what still needs to be done.

If a wavefunction for an isolated electron shows that it has, say, a 50 percent chance of being here and a 50 percent chance of being there, we must interpret these probabilities using the full-fledged weirdness of quantum mechanics. Since both of the alternatives can reveal themselves by commingling and generating an interference pattern, we must think of them as equally real. In loose language, there's a sense in which the electron *is* at both locations. What happens now if we measure the electron's position with a nonisolated, everyday-sized laboratory instrument? Well, corresponding to the electron's ambiguous whereabouts, the pointer on the instrument has a 50 percent chance of pointing to this value and a 50 percent chance of pointing to that value. But because of decoherence, the pointer will *not* be in a ghostly mixture of pointing at both values; because of decoherence, we can interpret *these* probabilities in the usual, classical, everyday sense. Just as a coin has a 50 percent chance of landing heads and a 50 percent chance of landing tails, but lands *either* heads *or* tails, the pointer has a 50 percent chance of pointing to this value and a 50 percent chance of pointing to that value, but it will definitely point to one *or* the other.

Similar reasoning applies for all other complex, nonisolated objects. If a quantum calculation reveals that a cat, sitting in a closed box, has a 50 percent chance of being dead and a 50 percent chance of being alive— because there is a 50 percent chance that an electron will hit a booby-trap mechanism that subjects the cat to poison gas and a 50 percent chance that the electron misses the booby trap—decoherence suggests that the cat will *not* be in some absurd mixed state of being both dead and alive. Although decades of heated debate have been devoted to issues like What does it mean for a cat to be both dead and alive? How does the act of opening the box and observing the cat force it to choose a definite status, dead or alive?, decoherence suggests that long before you open the box, the environment has already completed billions of observations that, in almost no time at all, turned all mysterious quantum probabilities into their less mysterious classical counterparts. Long before you look at it, the environment has compelled the cat to take on one, single, definite condition. Decoherence forces much of the weirdness of quantum physics to "leak" from large objects since, bit by bit, the quantum weirdness is carried away by the innumerable impinging particles from the environment.

It's hard to imagine a more satisfying solution to the quantum measurement problem. By being more realistic and abandoning the simplifying assumption that ignores the environment—a simplification that was crucial to making progress during the early development of the field—we would find that quantum mechanics has a built-in solution. Human consciousness, human experimenters, and human observations would no longer play a special role since they (we!) would simply be elements of the environment, like air molecules and photons, which can interact with a given physical system. There would also no longer be a stage one / stage two split between the evolution of the objects and the experimenter who measures them. Everything—observed and observer—would be on an equal footing. Everything—observed and observer—would be subject to precisely the same quantum mechanical law as is set down in Schrödinger's equation. The act of measurement would no longer be special; it would merely be one specific example of contact with the environment.

Is that it? Does decoherence resolve the quantum measurement problem? Is decoherence responsible for wavefunctions' closing the door on all but one of the potential outcomes to which they can lead? Some think so. Researchers like Robert Griffiths, of Carnegie Mellon; Roland Omnès, of Orsay; the Nobel laureate Murray Gell-Mann, of the Santa Fe Institute; and Jim Hartle, of the University of California at Santa Barbara, have made great progress and claim that they have developed decoherence into a complete framework (called *decoherent histories*) that solves the measurement problem. Others, like myself, are intrigued but not yet fully convinced. You see, the power of decoherence is that it successfully removes the artificial barrier Bohr erected between large and small physical systems, making everything subject to the same quantum mechanical formulas. This is important progress and I think Bohr would have found it gratifying. Although the unresolved quantum measurement problem never diminished physicists' ability to reconcile theoretical calculations with experimental data, it did lead Bohr and his colleagues to articulate a quantum mechanical framework with some distinctly awkward features. Many found the framework's need for fuzzy words about wavefunction collapse or the imprecise notion of "large" systems belonging to the dominion of classical physics, unnerving. To a significant extent, by taking account of decoherence, researchers have rendered these vague ideas unnecessary.

However, a key issue that I skirted in the description above is that even though decoherence suppresses quantum interference and thereby coaxes weird quantum probabilities to be like their familiar classical counterparts, *each of the potential outcomes embodied in a wavefunction still vies for realization.* And so we are still left wondering how one outcome "wins" and where the many other possibilities "go" when that actually happens. When a coin is tossed, classical physics gives an answer to the analogous question. It says that if you examine the way the coin is set spinning with adequate precision, you can, in principle, *predict* whether it will land heads or tails. On closer inspection, then, precisely one outcome is determined by details you initially overlooked. The same cannot be said in quantum physics. Decoherence allows quantum probabilities to be interpreted much like classical ones, but does not provide any finer details that select one of the many possible outcomes to actually happen.

Much in the spirit of Bohr, some physicists believe that searching for such an explanation of how a single, definite outcome arises is misguided. These physicists argue that quantum mechanics, with its updating to include decoherence, is a sharply formulated theory whose predictions account for the behavior of laboratory measuring devices. And according to this view, *that* is the goal of science. To seek an explanation of *what's really going on*, to strive for an understanding of *how a particular outcome came to be*, to hunt for a level of *reality beyond detector readings and computer printouts* betrays an unreasonable intellectual greediness.

Many others, including me, have a different perspective. Explaining data *is* what science is about. But many physicists believe that science is also about embracing the theories data confirms and going further by using them to gain maximal insight into the nature of reality. I strongly suspect that there is much insight to be gained by pushing onward toward a complete solution of the measurement problem.

Thus, although there is wide agreement that environment-induced decoherence is a crucial part of the structure spanning the quantum-to-classical divide, and while many are hopeful that these considerations will one day coalesce into a complete and cogent connection between the two, far from everyone is convinced that the bridge has yet been fully built.

Quantum Mechanics and the Arrow of Time

So where do we stand on the measurement problem, and what does it mean for the arrow of time? Broadly speaking, there are two classes of proposals for linking common experience with quantum reality. In the first class (for example, wavefunction as knowledge; Many Worlds; decoherence), Schrödinger's equation is the be-all and end-all of the story; the proposals simply provide different ways of interpreting what the equation means for physical reality. In the second class (for example, Bohm;

Ghirardi-Rimini-Weber), Schrödinger's equation must be supplemented with other equations (in Bohm's case, an equation that shows how a wavefunction pushes a particle around) or it must be modified (in the Ghirardi-Rimini-Weber case, to incorporate a new, explicit collapse mechanism). A key question for determining the impact on time's arrow is whether these proposals introduce a fundamental asymmetry between one direction in time and the other. Remember, Schrödinger's equation, just like those of Newton, Maxwell, and Einstein, treats forward and backward in time on a completely equal footing. It provides no arrow to temporal evolution. Do any of the proposals change this?

In the first class of proposals, the Schrödinger framework is not at all modified, so temporal symmetry is maintained. In the second class, temporal symmetry may or may not survive, depending on the details. For example, in Bohm's approach, the new equation proposed does treat time future and time past on an equal footing and so no asymmetry is introduced. However, the proposal of Ghirardi, Rimini, and Weber introduces a collapse mechanism that *does* have a temporal arrow—an "uncollapsing" wavefunction, one that goes from a spiked to a spread-out shape, would not conform to the modified equations. Thus, depending on the proposal, quantum mechanics, together with a resolution to the quantum measurement puzzle, may or may not continue to treat each direction in time equally. Let's consider the implications of each possibility.

If time symmetry persists (as I suspect it will), all of the reasoning and all of the conclusions of the last chapter can be carried over with little change to the quantum realm. The core physics that entered our discussion of time's arrow was the time-reversal symmetry of classical physics. While the basic language and framework of quantum physics differ from those of classical physics—wavefunctions instead of positions and velocities; Schrödinger's equation instead of Newton's laws—time-reversal symmetry of all quantum equations would ensure that the treatment of time's arrow would be unchanged. Entropy in the quantum world can be defined much as in classical physics so long as we describe particles in terms of their wavefunctions. And the conclusion that entropy should always be on the rise—increasing both toward what we call the future and toward what we call the past—would still hold.

We would thus come to the same puzzle encountered in Chapter 6. If we take our observations of the world right now as given, as undeniably real, and if entropy should increase both toward the future and toward the past, how do we explain how the world got to be the way it is and how it will subsequently unfold? And the same two possibilities would present themselves: either all that we see popped into existence by a statistical fluke that you would expect to happen every so often in an eternal universe that spends the vast majority of its time being totally disordered, or, for some reason, entropy was astoundingly low just following the big bang and for the last 14 billion years things have been slowly unwinding and will continue to do so toward the future. As in Chapter 6, to avoid the quagmire of not trusting memories, records, and the laws of physics, we focus on the second option—a low-entropy bang—and seek an explanation for how and why things began in such a special state.

If, on the other hand, time symmetry is lost—if the resolution of the measurement problem that is one day accepted reveals a fundamental asymmetric treatment of future versus past within quantum mechanics— it could very well provide the most straightforward explanation of time's arrow. It might show, for instance, that eggs splatter but don't unsplatter because, unlike what we found using the laws of classical physics, splattering solves the full quantum equations but unsplattering doesn't. A reverse-run movie of a splattering egg would then depict motion that couldn't happen in the real world, which would explain why we've never seen it. And that would be that.

Possibly. But even though this would seem to provide a very different explanation of time's arrow, in reality it may not be as different as it appears. As we emphasized in Chapter 6, for the pages of *War and Peace* to become increasingly disordered they must begin ordered; for an egg to become disordered through splattering, it must begin as an ordered, pristine egg; for entropy to increase toward the future, entropy must be low in the past so things have the potential to become disordered. However, just because a law treats past and future differently does not ensure that the law dictates a past with lower entropy. The law might still imply higher entropy toward the past (perhaps entropy would increase asymmetrically toward past and future), and it's even possible that a time-asymmetric law would be unable to say anything about the past at all. The latter is true of the Ghirardi-Rimini-Weber proposal, one of the only substantive time-asymmetric proposals on the market. Once their collapse mechanism does its trick, there is no way to undo it—there is no way to start from the collapsed wavefunction and evolve it back to its previous spread-out form. The detailed form of the wavefunction is lost in the collapse—it turns into a spike—and so it's impossible to "retrodict" what things were like at any time before the collapse occurred.

Thus, even though a time-asymmetric law would provide a partial explanation for why things unfold in one temporal order but never in the reverse order, it could very well call for the same key supplement required by time-symmetric laws: an explanation for why entropy was low in the distant past. Certainly, this is true of the time-asymmetric modifications to quantum mechanics that have so far been proposed. And so, unless some future discovery reveals two features, both of which I consider unlikely—a time-asymmetric solution to the quantum measurement problem that, additionally, ensures that entropy decreases toward the past—our effort to explain the arrow of time leads us, once again, back to the origin of the universe, the subject of the next part of the book.

As these chapters will make clear, cosmological considerations wend their way through many mysteries at the heart of space, time, and matter. So on the journey toward modern cosmology's insights into time's arrow, it's worth our while not to rush through the landscape, but rather, to take a well-considered stroll through cosmic history.

III

SPACETIME
AND COSMOLOGY

8

Of Snowflakes
and Spacetime

SYMMETRY AND THE EVOLUTION OF THE COSMOS

Richard Feynman once said that if he had to summarize the most important finding of modern science in one sentence he would choose "The world is made of atoms." When we recognize that so much of our understanding of the universe relies on the properties and interactions of atoms—from the reason that stars shine and the sky is blue to the explanation for why you feel this book in your hand and see these words with your eyes—we can well appreciate Feynman's choice for encapsulating our scientific legacy. Many of today's leading scientists agree that if they were offered a second sentence, they'd choose "Symmetry underlies the laws of the universe." During the last few hundred years there have been many upheavals in science, but the most lasting discoveries have a common characteristic: they've identified features of the natural world that remain unchanged even when subjected to a wide range of manipulations. These unchanging attributes reflect what physicists call symmetries, and they have played an increasingly vital role in many major advances. This has provided ample evidence that symmetry—in all its mysterious and subtle guises—shines a powerful light into the darkness where truth awaits discovery.

In fact, we will see that the history of the universe is, to a large extent, the history of symmetry. The most pivotal moments in the evolution of the universe are those in which balance and order suddenly change, yielding cosmic arenas qualitatively different from those of preceding eras. Current theory holds that the universe went through a number of these transitions during its earliest moments and that *everything* we've ever encountered is a tangible remnant of an earlier, more symmetric cosmic epoch. But there is an even grander sense, a metasense, in which symmetry lies at the core of an evolving cosmos. Time itself is intimately entwined with symmetry. As will become clear, the practical connotation of time as a measure of change, as well as the very existence of a kind of cosmic time that allows us to speak sensibly of things like "the age and evolution of the universe as a whole," rely sensitively on aspects of symmetry. And as scientists have examined that evolution, looking back toward the beginning in search of the true nature of space and time, symmetry has established itself as the most sure-footed of guides, providing insights and answers that would otherwise have been completely out of reach.

Symmetry and the Laws of Physics

Symmetry abounds. Hold a cue ball in your hand and rotate it this way or that—spin it around any axis—and it looks exactly the same. Put a plain, round dinner plate on a placemat and rotate it about its center: it looks completely unchanged. Gently catch a newly formed snowflake and rotate it so that each tip is moved into the position previously held by its neighbor, and you'd be hard pressed to notice that you'd done anything at all. Take the letter "A," flip it about a vertical axis passing through its apex, and it will provide you with a perfect replica of the original.

As these examples make clear, the symmetries of an object are the manipulations, real or imagined, to which it can be subjected with no effect on its appearance. The more kinds of manipulations an object can sustain with no discernible effect, the more symmetric it is. A perfect sphere is highly symmetric, since any rotation about its center—using an up-down axis, a left-right axis, or any axis in fact—leaves it looking exactly the same. A cube is less symmetric, since only rotations in units of 90 degrees about axes that pass through the center of its faces (and combinations thereof) leave it looking unchanged. Of course, should someone perform any other rotation, such as in Figure 8.1c, you obviously can still recognize the cube, but you also can see clearly that someone has tampered with it. By contrast, symmetries are like the deftest of prowlers; they are manipulations that leave no evidence whatsoever.

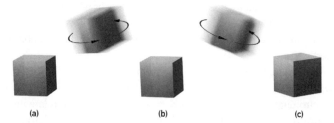

Figure 8.1 If a cube, as in (**a**), is rotated by 90 degrees, or multiples thereof, around axes passing through any of its faces, it looks unchanged, as in (**b**). But any other rotations can be detected, as in (**c**).

All these are examples of symmetries of objects *in* space. The symmetries underlying the known laws of physics are closely related to these, but zero in on a more abstract question: what manipulations—once again, real or imagined—can be performed on you or on the environment that will have absolutely no effect on the *laws* that explain the physical phenomena you observe? Notice that to be a symmetry, manipulations of this sort are not required to leave your observations unchanged. Instead, we are concerned with whether the laws governing those observations—the laws that explain what you see before, and then what you see after, some manipulation—are unchanged. As this is a central idea, let's see it at work in some examples.

Imagine that you're an Olympic gymnast and for the last four years you've been training diligently in your Connecticut gymnastics center. Through seemingly endless repetition, you've got every move in your various routines down perfectly—you know just how hard to push off the balance beam to execute an aerial walkover, how high to jump in the floor exercise for a double-twisting layout, how fast to swing on the parallel bars to launch your body on a perfect double-somersault dismount. In effect, your body has taken on an innate sense of Newton's laws, since it is these very laws that govern your body's motion. Now, when you finally do your routines in front of a packed audience in New York City, the site of the Olympic competition itself, you're banking on the same laws holding, since you intend to perform your routines exactly as you have in practice. Everything we know about Newton's laws lends credence to your strategy. Newton's laws are not specific to one location or another. They don't work one way in Connecticut and another way in New York. Rather, we believe his laws work in exactly the same way regardless of where you are. Even though you have changed location, the laws that govern your body's motion remain as unaffected as the appearance of a cue ball that has been rotated.

This symmetry is known as *translational symmetry* or *translational invariance*. It applies not only to Newton's laws but also to Maxwell's laws of electromagnetism, to Einstein's special and general relativities, to quantum mechanics, and to just about any proposal in modern physics that anyone has taken seriously.

Notice one important thing, though. The details of your observations and experiences can and sometimes will vary from place to place. Were you to perform your gymnastics routines on the moon, you'd find that the path your body took in response to the same upward jumping force of your legs would be very different. But we fully understand this particular difference and it is already integrated into the laws themselves. The moon is less massive than the earth, so it exerts less gravitational pull; as a result, your body travels along different trajectories. And this fact—that the gravitational pull of a body depends on its mass—is an *integral* part of Newton's law of gravity (as well as of Einstein's more refined general relativity). The difference between your earth and moon experiences doesn't imply that the law of gravity has changed from place to place. Instead, it merely reflects an environmental difference that the law of gravity already accommodates. So when we said that the known laws of physics apply equally well in Connecticut or New York—or, let's now add, on the moon—that was true, but bear in mind that you may need to specify environmental differences on which the laws depend. Nevertheless, and this is the key conclusion, the explanatory framework the laws provide is not at all changed by a change in location. A change in location does not require physicists to go back to the drawing board and come up with new laws.

The laws of physics didn't have to operate this way. We can imagine a universe in which physical laws are as variable as those of local and national governments; we can imagine a universe in which the laws of physics with which we are familiar tell us nothing about the laws of physics on the moon, in the Andromeda galaxy, in the Crab nebula, or on the other side of the universe. In fact, we don't know with absolute certainty that the laws that work here are the same ones that work in far-flung corners of the cosmos. But we do know that should the laws somehow change way out there, it must be *way* out there, because ever more precise astronomical observations have provided ever more convincing evidence that the laws are uniform throughout space, at least the space we can see. This highlights the amazing power of symmetry. We are bound to planet earth and its vicinity. And yet, because of translational symmetry, we can learn about fundamental laws at work in the entire universe without straying from home, since the laws we discover here *are* those laws.

Rotational symmetry or *rotational invariance* is a close cousin of translational invariance. It is based on the idea that every spatial direction is on an equal footing with every other. The view from earth certainly doesn't lead you to this conclusion. When you look up, you see very different things than you do when you look down. But, again, this reflects details of the environment; it is not a characteristic of the underlying laws themselves. If you leave earth and float in deep space, far from any stars, galaxies, or other heavenly bodies, the symmetry becomes evident: there is nothing that distinguishes one particular direction in the black void from another. They are all on a par. You wouldn't have to give a moment's thought to whether a deep-space laboratory you're setting up to investigate properties of matter or forces should be oriented this way or that, since the underlying laws are insensitive to this choice. If one night a prankster were to change the laboratory's gyroscopic settings, causing it to rotate some number of degrees about some particular axis, you'd expect this to have no consequences whatsoever for the laws of physics probed by your experiments. Every measurement ever done fully confirms this expectation. Thus, we believe that the laws that govern the experiments you carry out and explain the results you find are insensitive both to where you are—this is translational symmetry—and to how you happen to be oriented in space—this is rotational symmetry.[1]

As we discussed in Chapter 3, Galileo and others were well aware of another symmetry that the laws of physics should respect. If your deep-space laboratory is moving with constant velocity—regardless of whether you're moving 5 miles per hour this way or 100,000 miles per hour that way—the motion should have absolutely no effect on the laws that explain your observations, because you are as justified as the next guy in claiming that you are at rest and it's everything else that is moving. Einstein, as we have seen, extended this symmetry in a thoroughly unanticipated way by including the speed of light among the observations that would be unaffected by either your motion or the motion of the light's source. This was a stunning move because we ordinarily throw the particulars of an object's speed into the environmental details bin, recognizing that the speed observed generally depends upon the motion of the observer. But Einstein, seeing light's symmetry stream through the cracks in nature's Newtonian façade, elevated light's speed to an inviolable law of nature, declaring it to be as unaffected by motion as the cue ball is unaffected by rotations.

General relativity, Einstein's next major discovery, fits squarely within this march toward theories with ever greater symmetry. Just as you can think of special relativity as establishing symmetry among all observers moving relative to one another with constant velocity, you can think of general relativity as going one step farther and establishing symmetry among all accelerated vantage points as well. This is extraordinary because, as we've emphasized, although you can't feel constant velocity motion, you *can* feel accelerated motion. So it would seem that the laws of physics describing your observations must surely be different when you are accelerating, to account for the additional force you feel. Such *is* the case with Newton's approach; his laws, the ones that appear in all first-year physics textbooks, must be modified if utilized by an accelerating observer. But through the principle of equivalence, discussed in Chapter 3, Einstein realized that the force you feel from accelerating is indistinguishable from the force you feel in a gravitational field of suitable strength (the greater the acceleration, the greater the gravitational field). Thus, according to Einstein's more refined perspective, the laws of physics do *not* change when you accelerate, as long as you include an appropriate gravitational field in your description of the environment. General relativity treats all observers, even those moving at arbitrary nonconstant velocities, equally—they are completely symmetric—since each can claim to be at rest by attributing the different forces felt to the effect of different gravitational fields. The differences in the observations between one accelerating observer and another are therefore no more surprising and provide no greater evidence of a change in nature's laws than do the differences you find when performing your gymnastics routine on earth or the moon.[2]

These examples give some sense of why many consider, and I suspect Feynman would have agreed, that the copious symmetries underlying natural law present a close runner-up to the atomic hypothesis as a summary of our deepest scientific insights. But there is more to the story. Over the last few decades, physicists have elevated symmetry principles to the highest rung on the explanatory ladder. When you encounter a proposed law of nature, a natural question to ask is: Why this law? Why special relativity? Why general relativity? Why Maxwell's theory of electromagnetism? Why the Yang-Mills theories of the strong and weak nuclear forces (which we'll look at shortly)? One important answer is that these theories make predictions that have been repeatedly confirmed by precision experiments. This is essential to the confidence physicists have in the theories, certainly, but it leaves out something important.

Physicists also believe these theories are on the right track because, in some hard-to-describe way, they *feel* right, and ideas of symmetry are essential to this feeling. It feels right that no location in the universe is somehow special compared with any other, so physicists have confidence that translational symmetry should be among the symmetries of nature's laws. It feels right that no particular constant-velocity motion is somehow special compared with any other, so physicists have confidence that special relativity, by fully embracing symmetry among all constant-velocity observers, is an essential part of nature's laws. It feels right, moreover, that *any* observational vantage point—regardless of the possibly accelerated motion involved—should be as valid as any other, and so physicists believe that general relativity, the simplest theory incorporating this symmetry, is among the deep truths governing natural phenomena. And, as we shall shortly see, the theories of the three forces other than gravity—electromagnetism and the strong and weak nuclear forces—are founded on other, somewhat more abstract but equally compelling principles of symmetry. So the symmetries of nature are not merely consequences of nature's laws. From our modern perspective, symmetries are the foundation from which laws spring.

Symmetry and Time

Beyond their role in fashioning the laws governing nature's forces, ideas of symmetry are vital to the concept of time itself. No one has as yet found the definitive, fundamental definition of time, but, undoubtedly, part of time's role in the makeup of the cosmos is that it is the bookkeeper of change. We recognize that time has elapsed by noticing that things now are different from how they were then. The hour hand on your watch points to a different number, the sun is in a different position in the sky, the pages in your unbound copy of *War and Peace* are more disordered, the carbon dioxide gas that rushed from your bottle of Coke is more spread out—all this makes plain that things have changed, and time is what provides the potential for such change to be realized. To paraphrase

John Wheeler, time is nature's way of keeping everything—all change, that is—from happening all at once.

The existence of time thus relies on the *absence* of a particular symmetry: things in the universe must *change* from moment to moment for us even to define a notion of *moment to moment* that bears any resemblance to our intuitive conception. If there were perfect symmetry between how things are now and how they were then, if the change from moment to moment were of no more consequence than the change from rotating a cue ball, time as we normally conceive it wouldn't exist.[3] That's not to say the spacetime expanse, schematically illustrated in Figure 5.1, wouldn't exist; it could. But since everything would be completely uniform along the time axis, there'd be no sense in which the universe evolves or changes. Time would be an abstract feature of this reality's arena—the fourth dimension of the spacetime continuum—but otherwise, it would be unrecognizable.

Nevertheless, even though the existence of time coincides with the lack of one particular symmetry, its application on a cosmic scale requires the universe to be highly respectful of a different symmetry. The idea is simple and answers a question that may have occurred to you while reading Chapter 3. If relativity teaches us that the passage of time depends on how fast you move and on the gravitational field in which you happen to be immersed, what does it mean when astronomers and physicists speak of the entire universe's being a particular definite age—an age which these days is taken to be about 14 billion years? Fourteen billion years according to whom? Fourteen billion years on which clock? Would beings living in the distant Tadpole galaxy also conclude that the universe is 14 billion years old, and if so, what would have ensured that their clocks have been ticking away in synch with ours? The answer relies on symmetry—symmetry in space.

If your eyes could see light whose wavelength is much longer than that of orange or red, you would not only be able to see the interior of your microwave oven burst into activity when you push the start button, but you would also see a faint and nearly uniform glow spread throughout what the rest of us perceive as a dark night sky. More than four decades ago, scientists discovered that the universe is suffused with microwave radiation—long-wavelength light—that is a cool relic of the sweltering conditions just after the big bang.[4] This *cosmic microwave background radiation* is perfectly harmless. Early on, it was stupendously hot, but as the universe evolved and expanded, the radiation steadily diluted and cooled. Today it is just about 2.7 degrees above absolute zero, and its greatest claim to mischief is its contribution of a small fraction of the snow you see on your television set when you disconnect the cable and turn to a station that isn't broadcasting.

But this faint static gives astronomers what tyrannosaurus bones give paleontologists: a window onto earlier epochs that is crucial to reconstructing what happened in the distant past. An essential property of the radiation, revealed by precision satellite measurements over the last decade, is that it is extremely uniform. The temperature of the radiation in one part of the sky differs from that in another part by less than a thousandth of a degree. On earth, such symmetry would make the Weather Channel of little interest. If it were 85 degrees in Jakarta, you would immediately know that it was between 84.999 degrees and 85.001 degrees in Adelaide, Shanghai, Cleveland, Anchorage, and everywhere else for that matter. On a cosmic scale, by contrast, the uniformity of the radiation's temperature is *fantastically* interesting, as it supplies two critical insights.

First, it provides observational evidence that in its earliest stages the universe was not populated by large, clumpy, high-entropy agglomerations of matter, such as black holes, since such a heterogeneous environment would have left a heterogeneous imprint on the radiation. Instead, the uniformity of the radiation's temperature attests to the young universe being homogeneous; and, as we saw in Chapter 6, when gravity matters—as it did in the dense early universe—homogeneity implies low entropy. That's a good thing, because our discussion of time's arrow relied heavily on the universe's starting out with low entropy. One of our goals in this part of the book is to go as far as we can toward explaining this observation—we want to understand how the homogeneous, low-entropy, highly unlikely environment of the early universe came to be. This would take us a big step closer to grasping the origin of time's arrow.

Second, although the universe has been evolving since the big bang, on average the evolution must have been nearly identical across the cosmos. For the temperature here and in the Whirlpool galaxy, and in the Coma cluster, and everywhere else to agree to four decimal places, the physical conditions in every region of space must have evolved in essentially the same way since the big bang. This is an important deduction, but you must interpret it properly. A glance at the night sky certainly reveals a varied cosmos: planets and stars of various sorts sprinkled here and there throughout space. The point, though, is that when we analyze the evolution of the entire universe we take a macro perspective that averages over these "small"-scale variations, and large-scale averages *do* appear to be almost completely uniform. Think of a glass of water. On the scale of molecules, the water is extremely heterogeneous: there is an H_2O molecule over here, an expanse of empty space, another H_2O molecule over there, and so on. But if we average over the small-scale molecular lumpiness and examine the water on the "large" everyday scales we can see with the naked eye, the water in the glass looks perfectly uniform. The nonuniformity we see when gazing skyward is like the microscopic view from a single H_2O molecule. But as with the glass of water, when the universe is examined on large enough scales—scales on the order of hundreds of millions of light-years—it appears extraordinarily homogeneous. The uniformity of the radiation is thus a fossilized testament to the uniformity of both the laws of physics and the details of the environment across the cosmos.

This conclusion is of great consequence because the universe's uniformity is what allows us to define a concept of time applicable to the universe as a whole. If we take the measure of change to be a working definition of elapsed time, the uniformity of conditions throughout space is evidence of the uniformity of change throughout the cosmos, and thus implies the uniformity of elapsed time as well. Just as the uniformity of earth's geological structure allows a geologist in America, and one in Africa, and another in Asia to agree on earth's history and age, the uniformity of cosmic evolution throughout all of space allows a physicist in the Milky Way galaxy, and one in the Andromeda galaxy, and another in the Tadpole galaxy to all agree on the *universe's* history and age. Concretely, the homogeneous evolution of the universe means that a clock here, a clock in the Andromeda galaxy, and a clock in the Tadpole galaxy will, on average, have been subject to nearly identical physical conditions and hence will have ticked off time in nearly the same way. The homogeneity of space thus provides a universal synchrony.

While I have so far left out important details (such as the expansion of space, covered in the next section) the discussion highlights the core of the issue: time stands at the crossroads of symmetry. If the universe had perfect temporal symmetry—if it were completely unchanging—it would be hard to define what time even means. On the other hand, if the universe did not have symmetry in space—if, for example, the background radiation were thoroughly haphazard, having wildly different temperatures in different regions—time in a cosmological sense would have little meaning. Clocks in different locations would tick off time at different rates, and so if you asked what things were like when the universe was 3 billion years old, the answer would depend on whose clock you were looking at to see that those 3 billion years had elapsed. *That* would be complicated. Fortunately, our universe does not have so much symmetry as to render time meaningless, but does have enough symmetry that we can avoid such complexities, allowing us to speak of its overall age and its overall evolution through time.

So, let's now turn our attention to that evolution and consider the history of the universe.

Stretching the Fabric

The history of the universe sounds like a big subject, but in broad-brush outline it is surprisingly simple and relies in large part on one essential fact: The universe is expanding. As this is *the* central element in the unfolding of cosmic history, and, surely, is one of humanity's most profound discoveries, let's briefly examine how we know it is so.

In 1929, Edwin Hubble, using the 100-inch telescope at the Mount Wilson observatory in Pasadena, California, found that the couple of dozen galaxies he could detect were all rushing away.[5] In fact, Hubble found that the more distant a galaxy is, the faster its recession. To give a sense of scale, more refined versions of Hubble's original observations (that have studied thousands of galaxies using, among other equipment, the Hubble Space Telescope) show that galaxies that are 100 million light-years from us are moving away at about 5.5 million miles per hour, those at 200 million light-years are moving away twice as fast, at about 11 million miles per hour, those at 300 million light-years' distance are moving away three times as fast, at about 16.5 million miles per hour, and so on. Hubble's was a shocking discovery because the prevailing scientific and philosophical prejudice held that the universe was, on its largest scales, static, eternal, fixed, and unchanging. But in one stroke, Hubble shattered that view. And in a wonderful confluence of experiment and theory, Einstein's general relativity was able to provide a beautiful explanation for Hubble's discovery.

Actually, you might not think that coming up with an explanation would be particularly difficult. After all, if you were to pass by a factory and see all sorts of material violently flying outward in all directions, you would likely think that there had been an explosion. And if you traveled backward along the paths taken by the scraps of metal and chunks of concrete, you'd find them all converging on a location that would be a likely contender for where the explosion occurred. By the same reasoning, since the view from earth—as attested to by Hubble's and subsequent observations—shows that galaxies are rushing outward, you might think our position in space was the location of an ancient explosion that uniformly spewed out the raw material of stars and galaxies. The problem with this theory, though, is that it singles out one region of space—our region—as unique by making it the universe's birthplace. And were that the case, it would entail a deep-seated asymmetry: the physical conditions in regions far from the primordial explosion—far from us—would be very different from those here. As there is no evidence for such asymmetry in astronomical data, and furthermore, as we are highly suspect of anthropocentric explanations laced with pre-Copernican thinking, a more sophisticated interpretation of Hubble's discovery is called for, one in which our location does not occupy some special place in the cosmic order.

General relativity provides such an interpretation. With general relativity, Einstein found that space and time are flexible, not fixed, rubbery, not rigid; and he provided equations that tell us precisely how space and time respond to the presence of matter and energy. In the 1920s, the Russian mathematician and meteorologist Alexander Friedmann and the Belgian priest and astronomer Georges Lemaître independently analyzed Einstein's equations as they apply to the entire universe, and the two found something striking. Just as the gravitational pull of the earth implies that a baseball popped high above the catcher must either be heading farther upward or must be heading downward but certainly cannot be staying put (except for the single moment when it reaches its highest point), Friedmann and Lemaître realized that the gravitational pull of the matter and radiation spread throughout the entire cosmos implies that the fabric of space must either be stretching or contracting, but that it could not be staying fixed in size. In fact, this is one of the rare examples in which the metaphor not only captures the essence of the physics but also its mathematical content since, it turns out, the equations governing the baseball's height above the ground are nearly identical to Einstein's equations governing the size of the universe.[6]

The flexibility of space in general relativity provides a profound way to interpret Hubble's discovery. Rather than explaining the outward motion of galaxies by a cosmic version of the factory explosion, general relativity says that for billions of years space has been stretching. And as it has swelled, space has dragged the galaxies away from each other much as the black specks in a poppy seed muffin are dragged apart as the dough rises in baking. Thus, the origin of the outward motion is *not* an explosion that took place within space. Instead, the outward motion arises from the relentless outward swelling of space itself.

To grasp this key idea more fully, think also of the superbly useful balloon model of the expanding universe that physicists often invoke (an analogy that can be traced at least as far back as a playful cartoon, which you can see in the endnotes, that appeared in a Dutch newspaper in 1930 following an interview with Willem de Sitter, a scientist who made substantial contributions to cosmology[7]). This analogy likens our three-dimensional space to the easier-to-visualize two-dimensional surface of a spherical balloon, as in Figure 8.2a, that is being blown up to larger and larger size. The galaxies are represented by numerous evenly spaced pennies glued to the balloon's surface. Notice that as the balloon expands, the pennies all move away from one another, providing a simple analogy for how expanding space drives all galaxies to separate.

An important feature of this model is that there is complete symmetry among the pennies, since the view any particular Lincoln sees is the same as the view any other Lincoln sees. To picture it, imagine shrinking yourself, lying down on a penny, and looking out in all directions across the balloon's surface (remember, in this analogy the balloon's surface represents all of space, so looking off the balloon's surface has no meaning). What will you observe? Well, you will see pennies rushing away from you in all directions as the balloon expands. And if you lie down on a different penny what will you observe? The symmetry ensures you'll see the same thing: pennies rushing away in all directions. This tangible image captures well our belief—supported by increasingly precise astronomical surveys—that an observer in any one of the universe's more than 100 billion galaxies, gazing across his or her night sky with a powerful telescope, would, on average, see an image similar to the one we see: surrounding galaxies rushing away in all directions.

And so, unlike a factory explosion within a fixed, preexisting space, if outward motion arises because space itself is stretching, there need be no special point—no special penny, no special galaxy—that is the center of the outward motion. Every point—every penny, every galaxy—is completely on a par with every other. The view from any location *seems* like the view from the center of an explosion: each Lincoln sees all other Lincolns rushing away; an observer, like us, in any galaxy sees all other galaxies rushing away. But since this is true for all locations, there is no special or unique location that is *the* center from which the outward motion is emanating.

Moreover, not only does this explanation account qualitatively for the outward motion of galaxies in a manner that is spatially homogeneous, it also explains the quantitative details found by Hubble and confirmed with greater precision by subsequent observations. As illustrated in Figure 8.2b, if the balloon swells during some time interval, doubling in size for example, all spatial separations will double in size as well: pennies that were 1 inch apart will now be 2 inches apart, pennies that were 2 inches apart will now be 4 inches apart, pennies that were 3 inches apart will now be 6 inches apart, and so on. Thus, in any given time interval, the increase in separation between two pennies is propor-

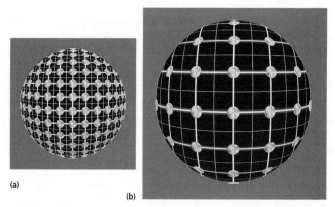

(a)

(b)

Figure 8.2 (a) If evenly spaced pennies are glued to the surface of a sphere, the view seen by any Lincoln is the same as that seen by any other. This aligns with the belief that the view from any galaxy in the universe, on average, is the same as that seen from any other. (b) If the sphere expands, the distances between all pennies increase. Moreover, the farther apart two pennies are in 8.2a, the greater the separation they experience from the expansion in 8.2b. This aligns well with measurements showing that the farther away from a given vantage point a galaxy is, the faster it moves away from that point. Note that no one penny is singled out as special, also in keeping with our belief that no galaxy in the universe is special or the center of the expansion of space.

tional to the initial distance between them. And since a greater increase in separation during a given time interval means a greater speed, pennies that are farther away from one another separate more quickly. In essence, the farther away from each other two pennies are, the more of the balloon's surface there is between them, and so the faster they're pushed apart when it swells. Applying exactly the same reasoning to expanding space and the galaxies it contains, we get an explanation for Hubble's observations. The farther away two galaxies are, the more space there is between them, so the faster they're pushed away from one another as space swells.

By attributing the observed motion of galaxies to the swelling of space, general relativity provides an explanation that not only treats all locations in space symmetrically, but also accounts for all of Hubble's data in one fell swoop. It is this kind of explanation, one that elegantly steps outside the box (in this case, one that actually *uses* the "box"—space, that is) to explain observations with quantitative precision and artful symmetry, that physicists describe as almost being too beautiful to be wrong. There is essentially universal agreement that the fabric of the space is stretching.

Time in an Expanding Universe

Using a slight variation on the balloon model, we can now understand more precisely how symmetry in space, even though space is expanding, yields a notion of time that applies uniformly across the cosmos. Imagine replacing each penny by an identical clock, as in Figure 8.3. We know from relativity that identical clocks will tick off time at different rates if they are subject to different physical influences—different motions, or different gravitational fields. But the simple yet key observation is that the complete symmetry among all Lincolns on the inflating balloon translates to complete symmetry among all the clocks. All the clocks experience identical physical conditions, so all tick at exactly the same rate and record identical amounts of elapsed time. Similarly, in an expanding universe in which there is a high degree of symmetry among all the galaxies, *clocks that move along with one or another galaxy must also tick at the same rate and hence record an identical amount of elapsed time.* How could it be otherwise? Each clock is on a par with every other, having experienced, on average, nearly identical physical conditions. This again shows the stunning power of symmetry. Without any calculation or detailed analysis, we realize that the uniformity of the physical environment, as evidenced by the uniformity of the microwave background radiation and the uniform distribution of galaxies throughout space,[8] allows us to infer uniformity of time.

Although the reasoning here is straightforward, the conclusion may nevertheless be confusing. Since the galaxies are all rushing apart as space expands, clocks that move along with one or another galaxy are also rushing apart. What's more, they're moving relative to each other at an enormous variety of speeds determined by the enormous variety of distances between them. Won't this motion cause the clocks to fall out of synchronization, as Einstein taught us with special relativity? For a number of reasons, the answer is no; here is one particularly useful way to think about it.

Recall from Chapter 3 that Einstein discovered that clocks that move *through* space in different ways tick off time at different rates (because they divert different amounts of their motion through time into motion through space; remember the analogy with Bart on his skateboard, first heading north and then diverting some of his motion to the east). But the

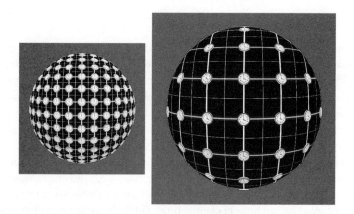

Figure 8.3 Clocks that move along with galaxies—whose motion, on average, arises only from the expansion of space—provide universal cosmic timepieces. They stay synchronized even though they separate from one another, since they move *with* space but not *through* space.

clocks we are now discussing are *not* moving through space at all. Just as each penny is glued to one point on the balloon and only moves relative to other pennies because of the swelling of the balloon's surface, each galaxy occupies one region of space and, for the most part, only moves relative to other galaxies because of the expansion of space. And this means that, with respect to space itself, all the clocks are actually stationary, so they tick off time identically. It is precisely these clocks—*clocks whose only motion comes from the expansion of space*—that provide the synchronized cosmic clocks used to measure the age of the universe.

Notice, of course, that you are free to take your clock, hop aboard a rocket, and zip this way and that across space at enormous speeds, undergoing motion significantly in excess of the cosmic flow from spatial expansion. If you do this, your clock *will* tick at a different rate and you *will* find a different length of elapsed time since the bang. This is a perfectly valid point of view, but it is completely individualistic: the elapsed time measured is tied to the history of your particular whereabouts and states of motion. When astronomers speak of the universe's age, though, they are seeking something universal—they are seeking a measure that has the same meaning everywhere. The uniformity of change throughout space provides a way of doing that.[9]

In fact, the uniformity of the microwave background radiation provides a ready-made test of whether you actually are moving with the cosmic flow of space. You see, although the microwave radiation is homogeneous across space, if you undertake additional motion beyond that from the cosmic flow of spatial expansion, you will not observe the radiation to be homogeneous. Just as the horn on a speeding car has a higher pitch when approaching and a lower pitch when receding, if you are zipping around in a spaceship, the crests and troughs of the microwaves heading toward the front of your ship will hit at a higher frequency than those traveling toward the back of your ship. Higher-frequency microwaves translate into higher temperatures, so you'd find the radiation in the direction you are heading to be a bit warmer than the radiation reaching you from behind. As it turns out, here on "spaceship" earth, astronomers *do* find the microwave background to be a little warmer in one direction in space and a little colder in the opposite direction. The reason is that not only does the earth move around the sun, and the sun move around the galactic center, but the entire Milky Way galaxy has a small velocity, in excess of cosmic expansion, toward the constellation Hydra. Only when astronomers correct for the effect these relatively slight additional motions have on the microwaves we receive does the radiation exhibit the exquisite uniformity of temperature between one part of the sky and another. It is this uniformity, this overall symmetry between one location and another, that allows us to speak sensibly of time when describing the entire universe.

Subtle Features of an Expanding Universe

A few subtle points in our explanation of cosmic expansion are worthy of emphasis. First, remember that in the balloon metaphor, it is only the balloon's *surface* that plays any role—a surface that is only two-dimensional (each location can be specified by giving two numbers analogous to latitude and longitude on earth), whereas the space we see when we look around has three dimensions. We make use of this lower-dimensional model because it retains the concepts essential to the true, three-dimensional story but is far easier to visualize. It's important to bear this in mind, especially if you have been tempted to say that there *is* a special point in the balloon model: the center point in the interior of the balloon away from which the whole rubber surface is moving. While this observation is true, it is meaningless in the balloon analogy because any point not on the balloon's surface plays no role. The surface of the balloon represents *all* of space; points that do not lie on the surface of the balloon are merely irrelevant by-products of the analogy and do not correspond to any location in the universe.*

---

*To go beyond the two-dimensional metaphor of a balloon's surface and have a spherical three-dimensional model is easy mathematically but difficult to picture, even for professional mathematicians and physicists. You might be tempted to think of a solid, three-dimensional ball, like a bowling ball without the finger holes. This, however, isn't an acceptable shape. We want all points in the model to be on a completely equal footing, since we believe that every place in the universe is (on average) just like any other. But the bowling ball has all sorts of different points: some are on the outside surface, others are embedded in the interior, one is right in the center. Instead, just as the two-dimensional surface of a balloon surrounds a *three*-dimensional spherical region (containing the balloon's air), an acceptable round three-dimensional shape would need to surround a *four*-dimensional spherical region. So the three-dimensional spherical surface of a balloon in a four-dimensional space is an acceptable shape. But if that still leaves you groping for an image, do what just about all professionals do: stick to the easy-to-visualize lower-dimensional analogies. They capture almost all of the essential features. A bit further on, we consider three-dimensional flat space, as opposed to the round shape of a sphere, and that flat space can be visualized.

Second, if the speed of recession is larger and larger for galaxies that are farther and farther away, doesn't that mean that galaxies that are sufficiently distant will rush away from us at a speed greater than the speed of light? The answer is a resounding, definite yes. Yet there is no conflict with special relativity. Why? Well, it's closely related to the reason clocks moving apart due to the cosmic flow of space stay synchronized. As we emphasized in Chapter 3, Einstein showed that nothing can move *through* space faster than light. But galaxies, on average, hardly move through space at all. Their motion is due almost completely to the *stretching of space itself*. And Einstein's theory does not prohibit space from expanding in a way that drives two points—two galaxies—away from each other at greater than light speed. His results only constrain speeds for which motion from spatial expansion has been subtracted out, motion in excess of that arising from spatial expansion. Observations confirm that for typical galaxies zipping along with the cosmic flow, such excess motion is minimal, fully in keeping with special relativity, even though their motion relative to each other, arising from the swelling of space itself, may exceed the speed of light.*

Third, if space is expanding, wouldn't that mean that in addition to galaxies being driven away from each other, the swelling space within each galaxy would drive all its stars to move farther apart, and the swelling space within each star, and within each planet, and within you and me and everything else, would drive all the constituent atoms to move farther apart, and the swelling of space within each atom would drive all the subatomic constituents to move farther apart? In short, wouldn't swelling space cause *everything* to grow in size, including our meter sticks, and in that way make it impossible to discern that any expansion had actually happened? The answer: no. Think again about the balloon-and-penny model. As the surface of the balloon swells, all the pennies are driven apart, but the pennies themselves surely do not expand. Of course, had we represented the galaxies by little circles drawn on the balloon with a black marker, then indeed, as the balloon grew in size the little circles would grow as well. But pennies, not blackened circles, capture what really happens. Each penny stays fixed in size because the forces holding its zinc

---

*Depending on whether the rate of the universe's expansion is speeding up or slowing down over time, the light emitted from such galaxies may fight a battle that would have made Zeno proud: the light may stream toward us at light speed while the expansion of space makes the distance the light has yet to cover ever larger, preventing the light from ever reaching us. See notes section for details.[10]

and copper atoms together are far stronger than the outward pull of the expanding balloon to which it is glued. Similarly, the nuclear force holding individual atoms together, and the electromagnetic force holding your bones and skin together, and the gravitational force holding planets and stars intact and bound together in galaxies, are stronger than the outward swelling of space, and so none of these objects expands. Only on the largest of scales, on scales much larger than individual galaxies, does the swelling of space meet little or no resistance (the gravitational pull between widely separated galaxies is comparatively small, because of the large separations involved) and so only on such supergalactic scales does the swelling of space drive objects apart.

Cosmology, Symmetry, and the Shape of Space

If someone were to wake you in the middle of the night from a deep sleep and demand you tell them the shape of the universe—the overall shape of space—you might be hard pressed to answer. Even in your groggy state, you know that Einstein showed space to be kind of like Silly Putty and so, in principle, it can take on practically any shape. How, then, can you possibly answer your interrogator's question? We live on a small planet orbiting an average star on the outskirts of a galaxy that is but one of hundreds of billions dispersed throughout space, so how in the world can you be expected to know anything at all about the shape of the entire universe? Well, as the fog of sleep begins to lift, you gradually realize that the power of symmetry once again comes to the rescue.

If you take account of scientists' widely held belief that, over large-scale averages, all locations and all directions in the universe are symmetrically related to one another, then you're well on your way to answering the interrogator's question. The reason is that almost all shapes *fail* to meet this symmetry criterion, because one part or region of the shape fundamentally differs from another. A pear bulges significantly at the bottom but less so at the top; an egg is flatter in the middle but pointier at its ends. These shapes, although exhibiting some degree of symmetry, do not possess complete symmetry. By ruling out such shapes, and limiting yourself only to those in which every region and direction is like every other, you are able to narrow down the possibilities fantastically.

We've already encountered one shape that fits the bill. The balloon's spherical shape was the key ingredient in establishing the symmetry between all the Lincolns on its swelling surface, and so the three-dimensional version of this shape, the so-called *three-sphere*, is one candidate for the shape of space. But this is not the only shape that yields complete symmetry. Continuing to reason with the more easily visualized two-dimensional models, imagine an *infinitely* wide and *infinitely* long rubber sheet—one that is completely uncurved—with evenly spaced pennies glued to its surface. As the entire sheet expands, there once again is complete spatial symmetry and complete consistency with Hubble's discovery: every Lincoln sees every other Lincoln rush away with a speed proportional to its distance, as in Figure 8.4. Hence, a three-dimensional version of this shape, like an infinite expanding cube of transparent rubber with galaxies evenly sprinkled throughout its interior, is another possible shape for space. (If you prefer culinary metaphors, think of an infinitely large version of the poppy seed muffin mentioned earlier, one that is shaped like a cube but goes on forever, with poppy seeds playing the role of galaxies. As the muffin bakes, the dough expands, causing each poppy seed to rush away from the others.) This shape is called *flat space* because, unlike the spherical example, it has no curvature (a meaning of "flat" that mathematicians and physicists use, but that differs from the colloquial meaning of "pancake-shaped.")[11]

One nice thing about both the spherical and the infinite flat shapes is that you can walk endlessly and never reach an edge or a boundary. This is appealing because it allows us to avoid thorny questions: What is beyond

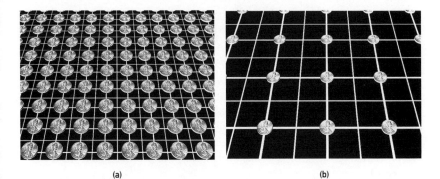

(a)   (b)

Figure 8.4 **(a)** The view from any penny on an infinite flat plane is the same as the view from any other. **(b)** The farther apart two pennies are in Figure 8.4a, the greater the increase in their separation when the plane expands.

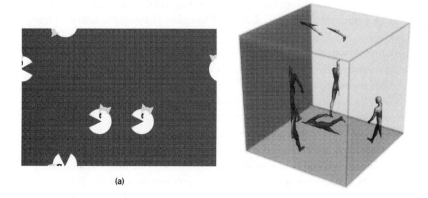

Figure 8.5 (a) A video game screen is flat (in the sense of "uncurved") and has a finite size, but contains no edges or boundaries since it "wraps around." Mathematically, such a shape is called a *two-dimensional torus*. **(b)** A three-dimensional version of the same shape, called a *three-dimensional torus*, is also flat (in the sense of uncurved) and has a finite volume, and also has no edges or boundaries, because it wraps around. If you pass through one face, you enter the opposite face.

the edge of space? What happens if you walk into a boundary of space? If space has no edges or boundaries, the question has no meaning. But notice that the two shapes realize this attractive feature in different ways. If you walk straight ahead in a spherically shaped space, you'll find, like Magellan, that sooner or later you return to your starting point, never having encountered an edge. By contrast, if you walk straight ahead in infinite flat space, you'll find that, like the Energizer Bunny, you can keep going and going, again never encountering an edge, but also never returning to where your journey began. While this might seem like a fundamental difference between the geometry of a curved and a flat shape, there is a simple variation on flat space that strikingly resembles the sphere in this regard.

To picture it, think of one of those video games in which the screen appears to have edges but in reality doesn't, since you can't actually fall off: if you move off the right edge, you reappear on the left; if you move off the top edge, you reappear on the bottom. The screen "wraps around," identifying top with bottom and left with right, and in that way the shape is flat (uncurved) and has *finite* size, but has no edges. Mathematically, this shape is called a *two-dimensional torus*; it is illustrated in Figure 8.5a.[12] The three-dimensional version of this shape—a three-dimensional torus—provides another possible shape for the fabric of space. You can think of this shape as an enormous cube that wraps around along all three axes: when you walk through the top you reappear at the bottom, when you walk through the back, you reappear at the front, when you walk through the left side, you reappear at the right, as in Figure 8.5b. Such a shape is flat—again, in the sense of being uncurved, not in the sense of being like a pancake—three-dimensional, finite in all directions, and yet has no edges or boundaries.

Beyond these possibilities, there is still another shape consistent with the symmetric expanding space explanation for Hubble's discovery. Although it's hard to picture in three dimensions, as with the spherical example there is a good two-dimensional stand-in: an infinite version of a Pringle's potato chip. This shape, often referred to as a *saddle*, is a kind of inverse of the sphere: Whereas a sphere is symmetrically bloated outward, the saddle is symmetrically shrunken inward, as illustrated in Figure 8.6. Using a bit of mathematical terminology, we say that the sphere is *positively curved* (bloats outward), the saddle is *negatively curved* (shrinks inward), and flat space—whether infinite or finite—has *no curvature* (no bloating or shrinking).*

Researchers have proven that this list—uniformly positive, negative, or zero—exhausts the possible curvatures for space that are consistent with the requirement of symmetry between all locations and in all directions. And that is truly stunning. We are talking about the shape of the *entire universe*, something for which there are endless possibilities. Yet, by invoking the immense power of symmetry, researchers have been able to narrow the possibilities sharply. And so, if you allow symmetry to guide your answer, and your late-night interrogator grants you a mere handful of guesses, you'll be able to meet his challenge.[13]

All the same, you might wonder why we've come upon a variety of

---

*Just as the video game screen gives a finite-sized version of flat space that has no edges or boundaries, there are finite-sized versions of the saddle shape that also have no edges or boundaries. I won't discuss this further, save to note that it implies that all three possible curvatures (positive, zero, negative) can be realized by finite-sized shapes without edges or boundaries. (In principle, then, a space-faring Magellan could carry out a cosmic version of his voyage in a universe whose curvature is given by any of the three possibilities.)

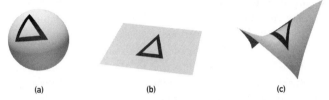

Figure 8.6 Using the two-dimensional analogy for space, there are three types of curvature that are completely symmetric—that is, curvatures in which the view from any location is the same as that from any other. They are (a) *positive* curvature, which uniformly bloats outward, as on a sphere; (b) *zero* curvature, which does not bloat at all, as on an infinite plane or finite video game screen; (c) *negative* curvature, which uniformly shrinks inward, as on a saddle.

possible shapes for the fabric of space. We inhabit a single universe, so why can't we specify a unique shape? Well, the shapes we've listed are the only ones consistent with our belief that every observer, regardless of where in the universe they're located, should see on the largest of scales an identical cosmos. But such considerations of symmetry, while highly selective, are not able to go all the way and pick out a unique answer. For that we need Einstein's equations from general relativity.

As input, Einstein's equations take the amount of matter and energy in the universe (assumed, again by consideration of symmetry, to be distributed uniformly) and as output, they give the curvature of space. The difficulty is that for many decades astronomers have been unable to agree on how much matter and energy there actually is. If all the matter and energy in the universe were to be smeared uniformly throughout space, and if, after this was done, there turned out to be more than the so-called *critical density* of about .00000000000000000000001 (10^{-23}) grams in every cubic meter*—about five hydrogen atoms per cubic meter—Einstein's equations would yield a positive curvature for space; if there were less than the critical density, the equations would imply negative curva-

---

*Today, matter in the universe is more abundant than radiation, so it's convenient to express the critical density in units most relevant for mass—grams per cubic meter. Note too that while 10^{-23} grams per cubic meter might not sound like a lot, there are *many* cubic meters of space out there in the cosmos. Moreover, the farther back in time you look, the smaller the space into which the mass/energy is squeezed, so the denser the universe becomes.

ture; if there were exactly the critical density, the equations would tell us that space has no overall curvature. Although this observational issue is yet to be settled definitively, the most refined data are tipping the scales on the side of no curvature—the flat shape. But the question of whether the Energizer Bunny could move forever in one direction and vanish into the darkness, or would one day circle around and catch you from behind—whether space goes on forever or wraps back like a video screen—is still completely open.[14]

Even so, even without a final answer to the shape of the cosmic fabric, what's abundantly clear is that symmetry is *the* essential consideration allowing us to comprehend space and time when applied to the universe as a whole. Without invoking the power of symmetry, we'd be stuck at square one.

Cosmology and Spacetime

We can now illustrate cosmic history by combining the concept of expanding space with the loaf-of-bread description of spacetime from Chapter 3. Remember, in the loaf-of-bread portrayal, each slice—even though two-dimensional—represents all of three-dimensional space at a single moment of time from the perspective of one particular observer. Different observers slice up the loaf at different angles, depending on details of their relative motion. In the examples encountered previously, we did not take account of expanding space and, instead, imagined that the fabric of the cosmos was fixed and unchanging over time. We can now refine those examples by including cosmological evolution.

To do so, we will take the perspective of observers who are at rest with respect to space—that is, observers whose only motion arises from cosmic expansion, just like the Lincolns glued to the balloon. Again, even though they are moving relative to one another, there is symmetry among all such observers—their watches all agree—and so they slice up the spacetime loaf in exactly the same way. Only relative motion in excess of that coming from spatial expansion, only relative motion *through* space as opposed to motion *from* swelling space, would result in their watches falling out of synch and their slices of the spacetime loaf being at different angles. We also need to specify the shape of space, and for purposes of comparison we will consider some of the possibilities discussed above.

The easiest example to draw is the flat and finite shape, the video game shape. In Figure 8.7a, we show one slice in such a universe, a schematic image you should think of as representing all of space right *now*. For simplicity, imagine that our galaxy, the Milky Way, is in the middle of the figure, but bear in mind that no location is in any way special compared with any other. Even the edges are illusory. The top side is not a place where space ends, since you can pass through and reappear at the bottom; similarly, the left side is not a place where space ends, since you can pass through and reappear on the right side. To accommodate astronomical observations, each side should extend at least 14 billion light-years (about 85 billion trillion miles) from its midpoint, but each could be much longer.

Note that right now we can't literally see the stars and galaxies as drawn on this *now* slice since, as we discussed in Chapter 5, it takes time for the light emitted by any object right now to reach us. Instead, the light we see when we look up on a clear, dark night was emitted long ago—millions and even billions of years ago—and only now has completed the long journey to earth, entered our telescopes, and allowed us to marvel at the wonders of deep space. Since space is expanding, eons ago, when this light was emitted, the universe was a lot smaller. We illustrate this in Fig-

(a)   (b)

Figure 8.7 (a) A schematic image depicting all of space right now, assuming space is flat and finite in extent, i.e. shaped like a video game screen. Note that the galaxy on the upper right wraps around on the left. (b) A schematic image depicting all of space as it evolves through time, with a few time slices emphasized for clarity. Note that the overall size of space and the separation between galaxies decrease as we look farther back in time.

ure 8.7b in which we have put our current *now* slice on the right-hand side of the loaf and included a sequence of slices to the left that depict our universe at ever earlier moments of time. As you can see, the overall size of space and the separations between individual galaxies both decrease as we look at the universe at ever earlier moments.

In Figure 8.8, you can also see the history of light, emitted by a distant galaxy perhaps a billion years ago, as it has traveled toward us here in the Milky Way. On the initial slice in Figure 8.8a, the light is first emitted, and on subsequent slices you can see the light getting closer and closer even as the universe gets larger and larger, and finally you can see it reach-

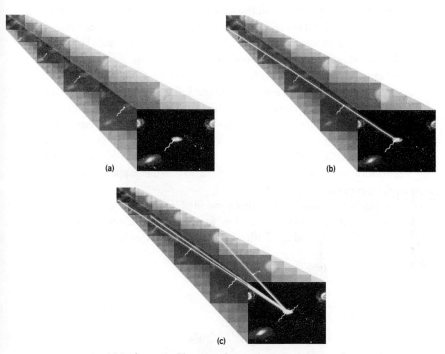

Figure 8.8 (a) Light emitted long ago from a distant galaxy gets closer and closer to the Milky Way on subsequent time slices. **(b)** When we finally see the distant galaxy, we are looking at it across both space and time, since the light we see was emitted long ago. The path through spacetime followed by the light is highlighted. **(c)** The paths through spacetime taken by light emitted from various astronomical bodies that we see today.

ing us on the rightmost time slice. In Figure 8.8b, by connecting the locations on each slice that the light's leading edge passed through during its journey, we show the light's path through spacetime. Since we receive light from many directions, Figure 8.8c shows a sample of trajectories through space and time that various light beams take to reach us now.

The figures dramatically show how light from space can be used as a cosmic time capsule. When we look at the Andromeda galaxy, the light we receive was emitted some 3 million years ago, so we are seeing Andromeda as it was in the distant past. When we look at the Coma cluster, the light we receive was emitted some 300 million years ago and hence we are seeing the Coma cluster as it was in an even earlier epoch. If right now all the stars in all the galaxies in this cluster were to go supernova, we would still see the same undisturbed image of the Coma cluster and would do so for another 300 million years; only then would light from the exploding stars have had enough time to reach us. Similarly, should an astronomer in the Coma cluster who is on our current now-slice turn a superpowerful telescope toward earth, she will see an abundance of ferns, arthropods, and early reptiles; she won't see the Great Wall of China or the Eiffel Tower for almost another 300 million years. Of course, this astronomer, well trained in basic cosmology, realizes that she is seeing light emitted in earth's distant past, and in laying out her own cosmic spacetime loaf will assign earth's early bacteria to their appropriate epoch, their appropriate set of time slices.

All of this assumes that both we and the Coma cluster astronomer are moving only with the cosmic flow from spatial expansion, since this ensures that her slicing of the spacetime loaf coincides with ours—it

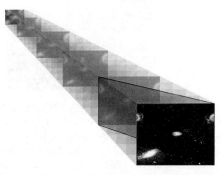

Figure 8.9 The time slice of an observer moving significantly in excess of the cosmic flow from spatial expansion.

ensures that her *now*-lists agree with ours. However, should she break ranks and move through space substantially in excess of the cosmic flow, her slices will tilt relative to ours, as in Figure 8.9. In this case, as we found with Chewie in Chapter 5, this astronomer's *now* will coincide with what we consider to be our future or our past (depending on whether the additional motion is toward or away from us). Notice, though, that her slices will no longer be spatially homogeneous. Each angled slice in Figure 8.9 intersects the universe in a range of different epochs and so the slices are far from uniform. This significantly complicates the description of cosmic history, which is why physicists and astronomers generally don't contemplate such perspectives. Instead, they usually consider only the perspective of observers moving solely with the cosmic flow, since this yields slices that are homogeneous—but fundamentally speaking, each viewpoint is as valid as any other.

As we look farther to the left on the cosmic spacetime loaf, the universe gets ever smaller and ever denser. And just as a bicycle tire gets hotter and hotter as you squeeze more and more air into it, the universe gets hotter and hotter as matter and radiation are compressed together more and more tightly by the shrinking of space. If we head back to a mere ten millionths of a second after the beginning, the universe gets so dense and so hot that ordinary matter disintegrates into a primordial plasma of nature's elementary constituents. And if we continue our journey, right back to nearly time zero itself—the time of the *big bang*—the entire known universe is compressed to a size that makes the dot at the end of

Figure 8.10 Cosmic history—the spacetime "loaf"—for a universe that is flat and of finite spatial extent. The fuzziness at the top denotes our lack of understanding near the beginning of the universe.

this sentence look gargantuan. The densities at such an early epoch were so great, and the conditions were so extreme, that the most refined physical theories we currently have are unable to give us insight into what happened. For reasons that will become increasingly clear, the highly successful laws of physics developed in the twentieth century break down under such intense conditions, leaving us rudderless in our quest to understand the beginning of time. We will see shortly that recent developments are providing a hopeful beacon, but for now we acknowledge our incomplete understanding of what happened at the beginning by putting a fuzzy patch on the far left of the cosmic spacetime loaf—our version of the terra incognita on maps of old. With this finishing touch, we present Figure 8.10 as a broad-brush illustration of cosmic history.

Alternative Shapes

We've so far assumed that space is shaped like a video game screen, but the story has many of the same features for the other possibilities. For example, if the data ultimately show that the shape of space is spherical, then, as we go ever farther back in time, the size of the sphere gets ever smaller, the universe gets ever hotter and denser, and at time zero we encounter some kind of big bang beginning. Drawing an illustration analogous to Figure 8.10 is challenging since spheres don't neatly stack one next to the other (you can, for example, imagine a "spherical loaf" with each slice being a sphere that surrounds the previous), but aside from the graphic complications, the physics is largely the same.

The cases of infinite flat space and of infinite saddle-shaped space also share many features with the two shapes already discussed, but they do differ in one essential way. Take a look at Figure 8.11, in which the slices represent flat space that goes on forever (of which we can show only a portion, of course). As you look at ever earlier times, space shrinks; galaxies get closer and closer together the farther back you look in Figure 8.11b. However, the overall size of space stays the same. Why? Well, infinity is a funny thing. If space is infinite and you shrink all distances by a factor of two, the size of space becomes half of infinity, and that is still infinite. So although everything gets closer together and the densities get ever higher as you head further back in time, the overall size of the universe stays infinite; things get dense everywhere on an infinite spatial expanse. This yields a rather different image of the big bang.

Normally, we imagine the universe began as a dot, roughly as in Figure 8.10, in which there is no exterior space or time. Then, from some kind of eruption, space and time unfurled from their compressed form and the expanding universe took flight. But if the universe is spatially infinite, *there was already an infinite spatial expanse at the moment of the big bang.* At this initial moment, the energy density soared and an incomparably large temperature was reached, but these extreme conditions existed everywhere, not just at one single point. In this setting, the big bang did not take place at one point; instead, the big bang eruption took place *everywhere* on the infinite expanse. Comparing this to the conventional single-dot beginning, it is as though there were many big bangs, one at each point on the infinite spatial expanse. After the bang, space swelled, but its overall size didn't increase since something already infinite can't get any bigger. What did increase are the separations between objects like galaxies (once they formed), as you can see by looking from left to right in Figure 8.11b. An observer like you or me, looking out from one galaxy or another, would see surrounding galaxies all rushing away, just as Hubble discovered.

Bear in mind that this example of infinite flat space is far more than academic. We will see that there is mounting evidence that the overall

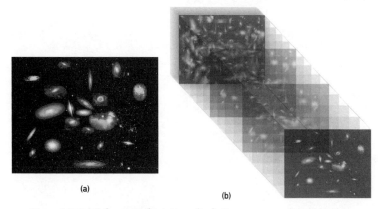

(a)

(b)

Figure 8.11 (a) Schematic depiction of infinite space, populated by galaxies. **(b)** Space shrinks at ever earlier times—so galaxies are closer and more densely packed at earlier times—but the overall size of infinite space stays infinite. Our ignorance of what happens at the earliest times is again denoted by a fuzzy patch, but here the patch extends through the infinite spatial expanse.

shape of space is not curved, and since there is no evidence as yet that space has a video game shape, the flat, infinitely large spatial shape is the front-running contender for the large-scale structure of spacetime.

Cosmology and Symmetry

Considerations of symmetry have clearly been indispensable in the development of modern cosmological theory. The meaning of time, its applicability to the universe as a whole, the overall shape of space, and even the underlying framework of general relativity, all rest on foundations of symmetry. Even so, there is yet another way in which ideas of symmetry have informed the evolving cosmos. Through the course of its history, the temperature of the universe has swept across an enormous range, from the ferociously hot moments just after the bang to the few degrees above absolute zero you'd find today if you took a thermometer into deep space. And, as I will explain in the next chapter, because of a critical interdependence between heat and symmetry, what we see today is likely but a cool remnant of the far richer symmetry that molded the early universe and determined some of the most familiar and essential features of the cosmos.

9

Vaporizing the Vacuum

HEAT, NOTHINGNESS, AND UNIFICATION

For as much as 95 percent of the universe's history, a cosmic correspondent concerned with the broad-brush, overall form of the universe would have reported more or less the same story: *Universe continues to expand. Matter continues to spread due to expansion. Density of universe continues to diminish. Temperature continues to drop. On largest of scales, universe maintains symmetric, homogeneous appearance.* But it wouldn't always have been so easy to cover the cosmos. The earliest stages would have required furiously hectic reporting, because in those initial moments the universe underwent rapid change. And we now know that what happened way back then has played a dominant role in what we experience today.

In this chapter, we will focus on critical moments in the first fraction of a second after the big bang, when the amount of symmetry embodied by the universe is believed to have changed abruptly, with each change launching a profoundly different epoch in cosmic history. While the correspondent can now leisurely fax in the same few lines every few billion years, in those early moments of briskly changing symmetry the job would have been considerably more challenging, because the basic structure of matter and the forces responsible for its behavior would have been completely unfamiliar. The reason is tied up with an interplay between *heat* and *symmetry*, and requires a complete rethinking of what we mean by the notions of empty space and of nothingness. As we will see, such rethinking not only enriches substantially our understanding of the universe's first moments, but also takes us a step closer to realizing a dream that harks back to Newton, Maxwell, and, in particular, Einstein—the dream of *unification*. Of equal importance, these developments set the stage for the most modern cosmological framework, *inflationary cosmology*, an approach that announces answers to some of the most pressing questions and thorniest puzzles on which the standard big bang model is mute.

Heat and Symmetry

When things get very hot or very cold, they sometimes change. And sometimes the change is so pronounced that you can't even recognize the things with which you began. Because of the torrid conditions just after the bang, and the subsequent rapid drop in temperature as space expanded and cooled, understanding the effects of temperature change is crucial in grappling with the early history of the universe. But let's start simpler. Let's start with ice.

If you heat a very cold piece of ice, at first not much happens. Although its temperature rises, its appearance remains pretty much unchanged. But if you raise its temperature all the way to 0 degrees Celsius and you keep the heat on, suddenly something dramatic does happen. The solid ice starts to melt and turns into liquid water. Don't let the familiarity of this transformation dull the spectacle. Without previous experiences involving ice and water, it would be a challenge to realize the intimate connection between them. One is a rock-hard solid while the other is a viscous liquid. Simple observation reveals no direct evidence that their molecular makeup, H_2O, is identical. If you'd never before seen ice or water and were presented with a vat of each, at first you would likely think they were unrelated. And yet, as either crosses through 0 degrees Celsius, you'd witness a wondrous alchemy as each transmutes into the other.

If you continue to heat liquid water, you again find that for a while not much happens beyond a steady rise in temperature. But then, when you reach 100 degrees Celsius, there is another sharp change: the liquid water starts to boil and transmute into steam, a hot gas that again is not obviously connected to liquid water or to solid ice. Yet, of course, all three share the same molecular composition. The changes from solid to liquid and liquid to gas are known as *phase transitions*. Most substances go through a similar sequence of changes if their temperatures are varied through a wide enough range.[1]

Symmetry plays a central role in phase transitions. In almost all cases, if we compare a suitable measure of something's symmetry before and after it goes through a phase transition, we find a significant change. On a molecular scale, for instance, ice has a crystalline form with H_2O molecules arranged in an ordered, hexagonal lattice. Like the symmetries of the box in Figure 8.1, the overall pattern of the ice molecules is left unchanged only by certain special manipulations, such as rotations in units of 60 degrees about particular axes of the hexagonal arrangement. By contrast, when we heat ice, the crystalline arrangement melts into a jumbled, uniform clump of molecules—liquid water—that remains unchanged under rotations by any angle, about any axis. So, by heating ice and causing it to go through a solid-to-liquid phase transition, we have made it more symmetric. (Remember, although you might intuitively think that something more ordered, like ice, is more symmetric, quite the opposite is true; something is more symmetric if it can be subjected to more transformations, such as rotations, while its appearance remains unchanged.)

Similarly, if we heat liquid water and it turns into gaseous steam, the phase transition also results in an increase of symmetry. In a clump of water, the individual H_2O molecules are, on average, packed together with the hydrogen side of one molecule next to the oxygen side of its neighbor. If you were to rotate one or another molecule in a clump it would noticeably disrupt the molecular pattern. But when the water boils and turns into steam, the molecules flit here and there freely; there is no longer any pattern to the orientations of the H_2O molecules and hence, were you to rotate a molecule or group of molecules, the gas would look the same. Thus, just as the ice-to-water transition results in an increase in symmetry, the water-to-steam transition does so as well. Most (but not all[2]) substances behave in a similar way, experiencing an increase of symmetry when they undergo solid-to-liquid and liquid-to-gas phase transitions.

The story is much the same when you cool water or almost any other substance; it just takes place in reverse. For example, when you cool gaseous steam, at first not much happens, but as its temperature drops to 100 degrees Celsius, it suddenly starts to condense into liquid water; when you cool liquid water, not much happens until you reach 0 degrees

Celsius, at which point it suddenly starts to freeze into solid ice. And, following the same reasoning regarding symmetries—but in reverse—we conclude that both of these phase transitions are accompanied by a *decrease* in symmetry.*

So much for ice, water, steam, and their symmetries. What does all this have to do with cosmology? Well, in the 1970s, physicists realized that not only can objects *in* the universe undergo phase transitions, *but the cosmos as a whole can do so as well.* Over the last 14 billion years, the universe has steadily expanded and decompressed. And just as a decompressing bicycle tire cools off, the temperature of the expanding universe has steadily dropped. During much of this decrease in temperature, not much happened. But there is reason to believe that when the universe passed through particular critical temperatures—the analogs of 100 degrees Celsius for steam and 0 degrees Celsius for water—it underwent radical change and experienced a drastic reduction in symmetry. Many physicists believe that we are now living in a "condensed" or "frozen" phase of the universe, one that is very different from earlier epochs. The cosmological phase transitions did not literally involve a gas condensing into a liquid, or a liquid freezing into a solid, although there are many qualitative similarities with these more familiar examples. Rather, the "substance" that condensed or froze when the universe cooled through particular temperatures is a field—more precisely, a *Higgs field*. Let's see what this means.

Force, Matter, and Higgs Fields

Fields provide the framework for much of modern physics. The electromagnetic field, discussed in Chapter 3, is perhaps the simplest and most widely appreciated of nature's fields. Living among radio and television broadcasts, cell phone communications, the sun's heat and light, we are all constantly awash in a sea of electromagnetic fields. Photons are the elementary constituents of electromagnetic fields and can be thought of as the microscopic transmitters of the electromagnetic force. When you see something, you can think of it in terms of a waving electromagnetic

---

*Even though a decrease in symmetry means that fewer manipulations go unnoticed, the heat released to the environment during these transformations ensures that overall entropy—including that of the environment—still increases.

field entering your eye and stimulating your retina, or in terms of photon particles entering your eye and doing the same thing. For this reason, the photon is sometimes described as the *messenger particle* of the electromagnetic force.

The gravitational field is also familiar since it constantly and consistently anchors us, and everything around us, to the earth's surface. As with electromagnetic fields, we are all immersed in a sea of gravitational fields; the earth's is dominant, but we also feel the gravitational fields of the sun, the moon, and the other planets. Just as photons are particles that constitute an electromagnetic field, physicists believe that *gravitons* are particles that constitute a gravitational field. Graviton particles have yet to be discovered experimentally, but that's not surprising. Gravity is by far the weakest of all forces (for example, an ordinary refrigerator magnet can pick up a paper clip, thereby overcoming the pull of the *entire* earth's gravity) and so it's understandable that experimenters have yet to detect the smallest constituents of the feeblest force. Even without experimental confirmation, though, most physicists believe that just as photons transmit the electromagnetic force (they are the electromagnetic force's messenger particles), gravitons transmit the gravitational force (they are the gravitational force's messenger particles). When you drop a glass, you can think of the event in terms of the earth's gravitational field pulling on the glass, or, using Einstein's more refined geometrical description, you can think of it in terms of the glass's sliding along an indentation in the spacetime fabric caused by the earth's presence, or—if gravitons do indeed exist—you can also think of it in terms of graviton particles firing back and forth between the earth and the glass, communicating a gravitational "message" that "tells" the glass to fall toward the earth.

Beyond these well-known force fields, there are two other forces of nature, the *strong nuclear force* and the *weak nuclear force*, and they also exert their influence via fields. The nuclear forces are less familiar than electromagnetism and gravity because they operate only on atomic and subatomic scales. Even so, their impact on daily life, through nuclear fusion that causes the sun to shine, nuclear fission at work in atomic reactors, and radioactive decay of elements like uranium and plutonium, is no less significant. The strong and weak nuclear force fields are called *Yang-Mills fields* after C. N. Yang and Robert Mills, who worked out their theoretical underpinnings in the 1950s. And just as electromagnetic fields are composed of photons, and gravitational fields are believed to be composed of gravitons, the strong and weak fields also have particulate constituents. The particles of the strong force are called *gluons* and those of the weak force are called W and Z particles. The existence of these force particles was confirmed by accelerator experiments carried out in Germany and Switzerland in the late 1970s and early 1980s.

The field framework also applies to matter. Roughly speaking, the probability waves of quantum mechanics may themselves be thought of as space-filling fields that provide the probability that some or other particle of matter is at some or other location. An electron, for instance, can be thought of as a particle—one that can leave a dot on a phosphor screen, as in Figure 4.4—but it can (and must) also be thought of in terms of a waving field, one that can contribute to an interference pattern on a phosphor screen as in Figure 4.3b.[3] In fact, although I won't go into it in greater detail here,[4] an electron's probability wave is closely associated with something called an *electron field*—a field that in many ways is similar to an electromagnetic field but in which the electron plays a role analogous to the photon's, being the electron field's smallest constituent. The same kind of field description holds true for all other species of matter particles as well.

Having discussed both matter fields and force fields, you might think we've covered everything. But there is general agreement that the story told thus far is not quite complete. Many physicists strongly believe that there is yet a third kind of field, one that has never been experimentally detected but that over the last couple of decades has played a pivotal role both in modern cosmological thought and in elementary particle physics. It is called a Higgs field, after the Scottish physicist Peter Higgs.[5] And if the ideas in the next section are right, the entire universe is permeated by an ocean of Higgs field—a cold relic of the big bang—that is responsible for many of the properties of the particles that make up you and me and everything else we've ever encountered.

Fields in a Cooling Universe

Fields respond to temperature much as ordinary matter does. The higher the temperature, the more ferociously the value of a field will—like the surface of a rapidly boiling pot of water—undulate up and down. At the chilling temperature characteristic of deep space today (2.7 degrees above absolute zero, or 2.7 Kelvin, as it is usually denoted), or even at the warmer temperatures here on earth, field undulations are minuscule. But the temperature just after the big bang was so enormous—at 10^{-43} seconds after the bang, the temperature is believed to have been about 10^{32} Kelvin—that all fields violently heaved to and fro.

As the universe expanded and cooled, the initially huge density of matter and radiation steadily dropped, the vast expanse of the universe became ever emptier, and field undulations became ever more subdued. For most fields this meant that their values, on average, got closer to zero. At some moment, the value of a particular field might jitter slightly above zero (a peak) and a moment later it might dip slightly below zero (a trough), but on average the value of most fields closed in on zero—the value we intuitively associate with absence or emptiness.

Here's where the Higgs field comes in. It's a variety of field, researchers have come to realize, that had properties similar to other fields' at the scorchingly high temperatures just after the big bang: it fluctuated wildly up and down. But researchers believe that (just as steam condenses into liquid water when its temperature drops sufficiently) when the temperature of the universe dropped sufficiently, the Higgs field condensed into a particular *nonzero* value throughout all of space. Physicists refer to this as the formation of a *nonzero Higgs field vacuum expectation value*—but to ease the technical jargon, I'll refer to this as the formation of a *Higgs ocean*.

It's kind of like what would happen if you were to drop a frog into a hot metal bowl, as in Figure 9.1a, with a pile of worms lying in the center. At first, the frog would jump this way and that—high up, low down, left, right—in a desperate attempt to avoid burning its legs, and on average would stay so far from the worms that it wouldn't even know they were there. But as the bowl cooled, the frog would calm itself, would hardly jump at all, and, instead, would gently slide down to the most restful spot at the bowl's bottom. There, having closed in on the bowl's center, it would finally rendezvous with its dinner, as in Figure 9.1b.

But if the bowl were shaped differently, as in Figure 9.1c, things would turn out differently. Imagine again that the bowl starts out very hot and that the worm pile still lies at the bowl's center, now high up on the central bump. Were you to drop the frog in, it would again wildly jump this way and that, remaining oblivious to the prize perched on the central plateau. Then, as the bowl cooled, the frog would again settle itself, reduce its jumping, and slide down the bowl's smooth sides. But because

(a)   (b)

Figure 9.1 (a) A frog dropped into a hot metal bowl incessantly jumps around. **(b)** When the bowl cools, the frog calms down, jumps much less, and slides down to the bowl's middle.

of the new shape, the frog would never make it to the bowl's center. Instead, it would slide down into the bowl's valley and remain at a distance from the worm pile, as in Figure 9.1d.

If we imagine that the distance between the frog and the worm pile represents the value of a field—the farther the frog is from the worms, the larger value of the field—and the height of the frog represents the energy contained in that field value—the higher up on the bowl the frog happens to be, the more energy the field contains—then these examples

(c)   (d)

Figure 9.1 (c) As in **(a)**, but with a hot bowl of a different shape. **(d)** As in **(b)**, but now when the bowl cools, the frog slides down to the valley, which is some distance from the bowl's center (where the worms are located).

convey well the behavior of fields as the universe cools. When the universe is hot, fields jump wildly from value to value, much as the frog jumps from place to place in the bowl. As the universe cools, fields "calm down," they jump less often and less frantically, and their values slide downward to lower energy.

But here's the thing. As with the frog example, there's a possibility of two qualitatively different outcomes. If the shape of the field's energy bowl—its so-called *potential energy*—is similar to that in Figure 9.1a, the field's value throughout space will slide all the way down to zero, the bowl's center, just as the frog slides all the way down to the worm pile. However, if the field's potential energy looks like that in Figure 9.1c, the field's value will not make it all the way to zero, to the energy bowl's center. Instead, just as the frog will slide down to the valley, which is a *nonzero* distance from the worm pile, the field's value will also slide down to the valley—a nonzero distance from the bowl's center—and that means the field will have a *nonzero* value.[6] The latter behavior is characteristic of Higgs fields. As the universe cools, a Higgs field's value gets caught in the valley and never makes it to zero. And since what we're describing would happen uniformly throughout space, the universe would be permeated by a uniform and nonzero Higgs field—a Higgs ocean.

The reason this happens sheds light on the fundamental peculiarity of Higgs fields. As a region of space becomes ever cooler and emptier—as matter and radiation get ever more sparse—the energy in the region gets ever lower. Taking this to the limit, you know you've reached the emptiest a region of space can be when you've lowered its energy as far as possible. For ordinary fields suffusing a region of space, their energy contribution is lowest when their value has slid all the way down to the center of the bowl as in Figure 9.1b; they have zero energy when their value is zero. That makes good, intuitive sense since we associate emptying a region of space with setting everything, including field values, to zero.

But for a Higgs field, things work differently. Just as a frog can reach the central plateau in Figure 9.1c and be *zero* distance from the worm pile only if it has enough energy to jump up from the surrounding valley, a Higgs field can reach the bowl's center, and have *value zero*, only if it too embodies enough energy to surmount the bowl's central bump. If, to the contrary, the frog has little or no energy, it will slide to the valley in Figure 9.1d—a *nonzero* distance from the worm pile. Similarly, a Higgs field with little or no energy will also slide to the bowl's valley—a nonzero distance from the bowl's center—and hence it will have a *nonzero* value.

To force a Higgs field to have a value of zero—the value that would seem to be the closest you can come to completely removing the field from the region, the value that would seem to be the closest you can come to a state of nothingness—you would have to *raise* its energy and, energetically speaking, the region of space would not be as empty as it possibly could. Even though it sounds contradictory, removing the Higgs field—reducing its value to zero, that is—is tantamount to adding energy to the region. As a rough analogy, think of one of those fancy noise reduction headphones that produce sound waves to cancel those coming from the environment that would otherwise impinge on your eardrums. If the headphones work perfectly, you hear silence when they produce their sounds, but you hear the ambient noise if you shut them off. Researchers have come to believe that just as you hear *less* when the headphones are suffused with the sounds they are programmed to produce, so cold, empty space harbors as little energy as it possibly can—it is as empty as it can be—when it is suffused with an ocean of Higgs field. Researchers refer to the emptiest space can be as the *vacuum*, and so we learn that the vacuum may actually be permeated by a uniform Higgs field.

The process of a Higgs field's assuming a nonzero value throughout space—forming a Higgs ocean—is called *spontaneous symmetry breaking**
and is one of the most important ideas to emerge in the later decades of twentieth-century theoretical physics. Let's see why.

The Higgs Ocean and the Origin of Mass

If a Higgs field has a nonzero value—if we are all immersed in an ocean of Higgs field—then shouldn't we feel it or see it or otherwise be aware of it in some way? Absolutely. And modern theory claims we do. Take your

---

*The terminology isn't particularly important, but briefly, here's where it comes from. The valley in Figure 9.1c and 9.1d has a symmetric shape—it's circular—with every point being on a par with every other (each point denotes a Higgs field value of lowest possible energy). Yet, when the Higgs field's value slides down the bowl, it lands on *one* particular point on the circular valley, and in so doing "spontaneously" selects one location on the valley as special. In turn, the points on the valley are no longer all on an equal footing, since one has been picked out, and so the Higgs field disrupts or "breaks" the previous symmetry between them. Thus, putting the words together, the process in which the Higgs slides down to one particular nonzero value in the valley is called *spontaneous symmetry breaking*. Later in the text, we will describe more tangible aspects of the reduction of symmetry associated with such a formation of a Higgs ocean.[7]

arm and swing it back and forth. You can feel your muscles at work driving the mass of your arm left and right and back again. If you take hold of a bowling ball, your muscles will have to work harder, since the greater the mass to be moved the greater the force they must exert. In this sense, the mass of an object represents the resistance it has to being moved; more precisely, the mass represents the resistance an object has to changes in its motion—to accelerations—such as first going left and then right and then left again. But where does this resistance to being accelerated come from? Or, in physics-speak, what gives an object its inertia?

In Chapters 2 and 3 we encountered various proposals Newton, Mach, and Einstein advanced as *partial* answers to this question. These scientists sought to specify a standard of rest with respect to which accelerations, such as those arising in the spinning-bucket experiment, could be defined. For Newton, the standard was absolute space; for Mach, it was the distant stars; and for Einstein, it was initially absolute spacetime (in special relativity) and then the gravitational field (in general relativity). But once delineating a standard of rest, and, in particular, specifying a benchmark for defining accelerations, none of these scientists took the next step to explain *why* objects resist accelerations. That is, none of them specified a mechanism whereby an object acquires its mass—its inertia— the attribute that fights accelerations. With the Higgs field, physicists have now suggested an answer.

The atoms that make up your arm, and the bowling ball you may have picked up, are all made of protons, neutrons, and electrons. The protons and neutrons, experimenters revealed in the late 1960s, are each composed of three smaller particles known as quarks. So, when you swing your arm back and forth, you are actually swinging all the constituent quarks and electrons back and forth, which brings us to the point. The Higgs ocean in which modern theory claims we are all immersed *interacts* with quarks and electrons: it resists their accelerations much as a vat of molasses resists the motion of a Ping-Pong ball that's been submerged. And this resistance, this drag on particulate constituents, contributes to what you perceive as the mass of your arm and the bowling ball you are swinging, or as the mass of an object you're throwing, or as the mass of your entire body as you accelerate toward the finish line in a 100-meter race. And so we *do* feel the Higgs ocean. The forces we all exert thousands of times a day in order to change the velocity of one object or another—to impart an acceleration—are forces that fight against the drag of the Higgs ocean.[8]

The molasses metaphor captures well some aspects of the Higgs ocean. To accelerate a Ping-Pong ball submerged in molasses, you'd have to push it *much* harder than when playing with it on your basement table — it will resist your attempts to change its velocity more strongly than it does when not in molasses, and so it behaves as if being submerged in molasses has increased its mass. Similarly, as a result of their interactions with the ubiquitous Higgs ocean, elementary particles resist attempts to change their velocities — they acquire mass. However, the molasses metaphor has three misleading features that you should be aware of.

First, you can always reach into the molasses, pull out the Ping-Pong ball, and see how its resistance to acceleration diminishes. This isn't true for particles. We believe that, today, the Higgs ocean fills all of space, so there is no way to remove particles from its influence; all particles have the masses they do regardless of where they are. Second, molasses resists all motion, whereas the Higgs field resists only accelerated motion. Unlike a Ping-Pong ball moving through molasses, a particle moving through outer space at constant speed would not be slowed down by "friction" with the Higgs ocean. Instead, its motion would continue unchanged. Only when we try to speed the particle up or slow it down does the ocean of Higgs field make its presence known by the force we have to exert. Third, when it comes to familiar matter composed of conglomerates of fundamental particles, there is another important source of mass. The quarks constituting protons and neutrons are held together by the strong nuclear force: gluon particles (the messenger particles of the strong force) stream between quarks, "gluing" them together. Experiments have shown that these gluons are highly energetic, and since Einstein's $E=mc^2$ tells us that energy (E) can manifest itself as mass (m), we learn that the gluons inside protons and neutrons contribute a significant fraction of these particles' total mass. Thus, a more precise picture is to think of the molasseslike drag force of the Higgs ocean as giving mass to *fundamental* particles such as electrons and quarks, but when these particles combine into composite particles like protons, neutrons, and atoms, other (well understood) sources of mass also come into play.

Physicists assume that the degree to which the Higgs ocean resists a particle's acceleration varies with the particular species of particle. This is essential, because the known species of fundamental particles all have different masses. For example, while protons and neutrons are composed of two species of quarks (called *up-quarks* and *down-quarks*: a proton is made from two ups and a down; a neutron, from two downs and an up), over the years experimenters using atom smashers have discovered four other species of quark particles, whose masses span a wide range, from .0047 to 189 times the mass of a proton. Physicists believe the explanation for the variety of masses is that different kinds of particles interact more or less strongly with the Higgs ocean. If a particle moves smoothly through the Higgs ocean with little or no interaction, there will be little or no drag and the particle will have little or no mass. The photon is a good example. Photons pass completely unhindered through the Higgs ocean and so have no mass at all. If, to the contrary, a particle interacts significantly with the Higgs ocean, it will have a higher mass. The heaviest quark (it's called the *top-quark*), with a mass that's about 350,000 times an electron's, interacts 350,000 times more strongly with the Higgs ocean than does an electron; it has greater difficulty accelerating through the Higgs ocean, and that's why it has a greater mass. If we liken a particle's mass to a person's fame, then the Higgs ocean is like the paparazzi: those who are unknown pass through the swarming photographers with ease, but famous politicians and movie stars have to push much harder to reach their destination.[9]

This gives a nice framework for thinking about why one particle has a different mass from another, but, as of today, there is no fundamental explanation for the precise manner in which each of the known particle species interacts with the Higgs ocean. As a result, there is no fundamental explanation for why the known particles have the particular masses that have been revealed experimentally. However, most physicists do believe that were it not for the Higgs ocean, *all fundamental particles would be like the photon and have no mass whatsoever.* In fact, as we will now see, this may have been what things were like in the earliest moments of the universe.

Unification in a Cooling Universe

Whereas gaseous steam condenses into liquid water at 100 degrees Celsius, and liquid water freezes into solid ice at 0 degrees Celsius, theoretical studies have shown that the Higgs field condenses into a nonzero value at a million billion (10^{15}) degrees. That's almost 100 million times the temperature at the core of the sun, and it is the temperature to which the universe is believed to have dropped by about a hundredth of a billionth (10^{-11}) of a second after the big bang (ATB). Prior to 10^{-11} seconds

ATB, the Higgs field fluctuated up and down but had an average value of zero; as with water above 100 degrees Celsius, at such temperatures a Higgs ocean couldn't form because it was too hot. The ocean would have evaporated immediately. And without a Higgs ocean there was no resistance to particles undergoing accelerated motion (the paparazzi vanished), which implies that all the known particles (electrons, up-quarks, down-quarks, and the rest) had the same mass: zero.

This observation partly explains why the formation of the Higgs ocean is described as a cosmological phase transition. In the phase transitions from steam to water and from water to ice, two essential things happen. There is a significant qualitative change in appearance, and the phase transition is accompanied by a reduction in symmetry. We see the same two features in the formation of the Higgs ocean. First, there was a significant qualitative change: particle species that had been massless suddenly acquired nonzero masses—the masses that those particle species are now found to have. Second, this change was accompanied by a decrease in symmetry: before the formation of the Higgs ocean, all particles had the same mass—zero—a highly symmetric state of affairs. If you were to exchange one particle species' mass with another, no one would know, because the masses were all the same. But after the Higgs field condensed, the particle masses transmuted into nonzero—and nonequal— values, and so the symmetry between the masses was lost.

In fact, the reduction in symmetry arising from the formation of the Higgs ocean is more extensive still. Above 10^{15} degrees, when the Higgs field had yet to condense, not only were all species of fundamental matter particles massless, but also, without the resistive drag from a Higgs ocean, all species of force particles were massless as well. (Today, the W and Z messenger particles of the weak nuclear force have masses that are about 86 and 97 times the mass of the proton.) And, as originally discovered in the 1960s by Sheldon Glashow, Steven Weinberg, and Abdus Salam, the masslessness of all the force particles was accompanied by another, fantastically beautiful symmetry.

In the late 1800s Maxwell realized that electricity and magnetism, although once thought to be two completely separate forces, are actually different facets of the same force—the electromagnetic force (see Chapter 3). His work showed that electricity and magnetism complete each other; they are the yin and yang of a more symmetric, unified whole. Glashow, Salam, and Weinberg discovered the next chapter in this story of unification. They realized that before the Higgs ocean formed, not only did all the force particles have identical masses—zero—but the photons and W and Z particles were identical in essentially every other way as well.[10] Just as a snowflake is unaffected by the particular rotations that interchange the locations of its tips, physical processes in the absence of the Higgs ocean would have been unaffected by particular interchanges of electromagnetic and weak-nuclear-force particles—by particular interchanges of photons and W and Z particles. And just as the insensitivity of a snowflake to being rotated reflects a symmetry (rotational symmetry), the insensitivity to interchange of these force particles also reflects a symmetry, one that for technical reasons is called a *gauge symmetry*. It has a profound implication. Since these particles convey their respective forces—they are their force's messenger particles—the symmetry between them means there was symmetry between the forces. At high enough temperatures, therefore, temperatures that would vaporize today's Higgs-filled vacuum, there is no distinction between the weak nuclear force and the electromagnetic force. At high enough temperatures, that is, the Higgs ocean evaporates; as it does, the distinction between the weak and electromagnetic forces evaporates, too.

Glashow, Weinberg, and Salam had extended Maxwell's century-old discovery by showing that the electromagnetic and weak nuclear forces are actually part of one and the same force. They had *unified* the description of these two forces in what is now called the *electroweak* force.

The symmetry between the electromagnetic and weak forces is not apparent today because as the universe cooled, the Higgs ocean formed, and—this is vital—photons and W and Z particles interact with the condensed Higgs field differently. Photons zip through the Higgs ocean as easily as B-movie has-beens slip through the paparazzi, and therefore remain massless. W and Z particles, though, like Bill Clinton and Madonna, have to slog their way through, acquiring masses that are 86 and 97 times that of a proton, respectively. (Note: this metaphor is not to scale.) That's why the electromagnetic and weak nuclear forces appear so different in the world around us. The underlying symmetry between them is "broken," or obscured, by the Higgs ocean.

This is a truly breathtaking result. Two forces that look very different at today's temperatures—the electromagnetic force responsible for light, electricity, and magnetic attraction, and the weak nuclear force responsible for radioactive decay—are fundamentally part of the same force, and appear to be different only because the nonzero Higgs field obscures the symmetry between them. Thus, what we normally think of as empty space—the vacuum, nothingness—plays a central role in making things in the world appear as they do. Only by vaporizing the vacuum, by raising the temperature high enough so that the Higgs field evaporated—that is, acquired an average value of zero throughout space—would the full symmetry underlying nature's laws be made apparent.

When Glashow, Weinberg, and Salam were developing these ideas, the W and Z particles had yet to be discovered experimentally. It was the strong faith these physicists had in the power of theory and the beauty of symmetry that gave them the confidence to go forward. Their boldness proved well founded. In due course, the W and Z particles were discovered and the electroweak theory was confirmed experimentally. Glashow, Weinberg, and Salam had looked beyond superficial appearances—they had peered through the obscuring fog of nothingness—to reveal a deep and subtle symmetry entwining two of nature's four forces. They were awarded the 1979 Nobel Prize for the successful unification of the weak nuclear force and electromagnetism.

Grand Unification

When I was a freshman in college, I'd drop in every now and then on my adviser, the physicist Howard Georgi. I never had much to say, but it hardly mattered. There was always something that Georgi was excited to share with interested students. On one occasion in particular, Georgi was especially worked up and he spoke rapid fire for over an hour, filling the chalkboard a number of times over with symbols and equations. Throughout, I nodded enthusiastically. But frankly, I hardly understood a word. Years later I realized that Georgi had been telling me about plans to test a discovery he had made called *grand unification*.

Grand unification addresses a question that naturally follows the success of the electroweak unification: If two forces of nature were part of a unified whole in the early universe, might it be the case that, at even higher temperatures, at even earlier times in the history of the universe, the distinctions among three or possibly all four forces might similarly evaporate, yielding even greater symmetry? This raises the intriguing possibility that there might actually be a single fundamental force of nature that, through a series of cosmological phase transitions, has crystallized into the four seemingly different forces of which we are currently aware. In 1974, Georgi and Glashow put forward the first theory to go partway toward this goal of total unity. Their *grand unified theory*, together with later insights of Georgi, Helen Quinn, and Weinberg, suggested that three of the four forces—the strong, weak, and electromagnetic forces— were all part of one unified force when the temperature was above 10 billion billion billion (10^{28}) degrees—some thousand billion billion times the temperature at the center of the sun—extreme conditions that existed prior to 10^{-35} seconds after the bang. Above that temperature, these physicists suggested, photons, gluons of the strong force, as well as W and Z particles, could all be freely interchanged with one another—a more robust gauge symmetry than that of the electroweak theory—without any observable consequence. Georgi and Glashow thus suggested that at these high energies and temperatures there was complete symmetry among the three nongravitational-force particles, and hence complete symmetry among the three nongravitational forces.[11]

Glashow and Georgi's grand unified theory went on to say that we do not see this symmetry in the world around us—the strong nuclear force that keeps protons and neutrons tightly glued together in atoms seems completely separate from the weak and electromagnetic forces—because as the temperature dropped below 10^{28} degrees, another species of Higgs field entered the story. This Higgs field is called the *grand unified Higgs*. (Whenever they might be confused, the Higgs field involved in electroweak unification is called the *electroweak Higgs*.) Similar to its electroweak cousin, the grand unified Higgs fluctuated wildly above 10^{28} degrees, but calculations suggested that it condensed into a nonzero value when the universe dropped below this temperature. And, as with the electroweak Higgs, when this grand unified Higgs ocean formed, the universe went through a phase transition with an accompanying reduction in symmetry. In this case, because the grand unified Higgs ocean has a different effect on gluons than it does on the other force particles, the strong force splintered off from the electroweak force, yielding two distinct nongravitational forces where previously there was one. A fraction of a second and a drop of billions and billions of degrees later, the electroweak Higgs condensed, causing the weak and electromagnetic forces to split apart as well.

While a beautiful idea, grand unification (unlike electroweak unification) has not been confirmed experimentally. To the contrary, Georgi's and Glashow's original proposal predicted a trace, residual implication of the universe's early symmetry that should be apparent today, one that would allow protons to every so often transmute into other species of particles (such as anti-electrons and particles known as pions). But after years of painstaking search for such *proton decay* in elaborate underground experiments—the experiment Georgi had excitedly described to me in his office years ago—none were found; this ruled out Georgi and Glashow's proposal. Since then, however, physicists have developed variations on that original model that are not ruled out by such experiments; however, so far none of these alternative theories have been confirmed.

The consensus among physicists is that grand unification is one of the great, as yet unrealized, ideas in particle physics. Since unification and cosmological phase transitions have proven so potent for electromagnetism and the weak nuclear force, many feel that it is only a matter of time before other forces are also gathered within a unified framework. As we shall see in Chapter 12, great strides in this direction have recently been made using a different approach—*superstring theory*—that has, for the first time, brought *all* forces, including gravity, into a unified theory, albeit one which is still, as of this writing, under vigorous development. But what is already clear, even in just considering the electroweak theory, is that the universe we currently see exhibits but a remnant of the early universe's resplendent symmetry.

The Return of the Aether

The concept of symmetry's breaking, and its realization through the electroweak Higgs field, clearly plays a central role in particle physics and cosmology. But the discussion may have left you wondering about the following: If a Higgs ocean is an invisible something that fills what we ordinarily think of as empty space, isn't it just another incarnation of the long discredited notion of the aether? The answer: yes and no. The explanation: yes, indeed, in some ways a Higgs ocean does smack of the aether. Like the aether, a condensed Higgs field permeates space, surrounds us all, seeps right through everything material, and, as a nonremovable feature of empty space (unless we reheat the universe above 10^{15} degrees, which we can't actually do), it redefines our conception of nothingness. But unlike the original aether, which was introduced as an invisible medium to carry light waves in much the same way that air carries sound waves, a Higgs ocean has nothing to do with the motion of light; it does not affect light's speed in any way, and so experiments from the turn of the twentieth century that ruled out the aether by studying light's motion have no bearing on the Higgs ocean.

Moreover, since the Higgs ocean has no effect on anything moving with constant velocity, it does not pick out one observational vantage point as somehow being special, as the aether did. Instead, even with a Higgs ocean, all constant velocity observers remain on a completely equal footing, and hence a Higgs ocean does not conflict with special relativity. Of course, these observations do not prove that Higgs fields exist; instead, they show that despite certain similarities to the aether, Higgs fields are not in conflict with any theory or experiment.

If there is an ocean of Higgs field, though, it should yield other consequences that will be experimentally testable within the next few years. As a primary example, just as electromagnetic fields are composed of photons, Higgs fields are composed of particles that, not surprisingly, are called *Higgs particles*. Theoretical calculations have shown that if there is a Higgs ocean permeating space, Higgs particles should be among the debris from the high-energy collisions that will take place at the Large Hadron Collider, a giant atom smasher now under construction at Centre Européenne pour la Recherche Nucleaire (CERN) in Geneva, Switzerland, and slated to come online in 2007. Roughly speaking, enormously energetic head-on collisions between protons should be able to knock a Higgs particle out of the Higgs ocean somewhat as energetic underwater collisions can knock H_2O molecules out of the Atlantic. In due course, these experiments should allow us to determine whether this modern form of the aether exists or whether it will go the way of its earlier incarnation. This is a critical question to settle because, as we have seen, condensing Higgs fields play a deep and pivotal role in our current formulation of fundamental physics.

If the Higgs ocean is not found, it will require major rethinking of a theoretical framework that has been in place for more than thirty years. But if it is found, the event will be a triumph for theoretical physics: it will confirm the power of symmetry to correctly shape our mathematical reasoning as we venture forth into the unknown. Beyond this, confirmation of the Higgs ocean's existence would also do two more things. First, it would provide direct evidence of an ancient era when various aspects of today's universe that appear distinct were part of a symmetric whole. Second, it would establish that our intuitive notion of empty space—the end result of removing everything we can from a region so that its energy and temperature drop as low as possible—has, for a long time, been naïve. The emptiest empty space need not involve a state of absolute nothingness. Without invoking the spiritual, therefore, we may well closely brush

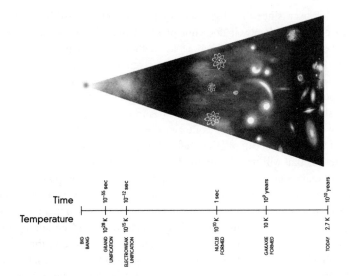

Time

10⁻³⁵ sec | 10⁻¹² sec | 1 sec | 10⁹ years | 10¹⁰ years

Temperature

10²⁸ K | 10¹⁵ K | 10¹⁰ K | 10 K | 2.7 K

BIG BANG | GRAND UNIFICATION | ELECTROWEAK UNIFICATION | NUCLEI FORMED | GALAXIES FORMED | TODAY

Figure 9.2 A time line schematically illustrating the standard big bang model of cosmology.

up against the thinking of Henry More (Chapter 2) in our scientific quest to understand space and time. To More, the usual concept of empty space was meaningless because space is always filled with divine spirit. To us, the usual concept of empty space may be similarly elusive, since the empty space we're privy to may always be filled with an ocean of Higgs field.

Entropy and Time

The time line in Figure 9.2 places the phase transitions we've discussed in historical context and hence gives us a firmer grasp of the sequence of events the universe has gone through from the big bang to the egg on your kitchen counter. But crucial information is still hidden within the fuzzy patch. Remember, knowing how things begin—the order of the stack of pages of *War and Peace*, the pressurized carbon dioxide molecules in your bottle of Coke, the state of the universe at the big bang—is essential to understanding how they evolve. Entropy can increase only if it is given room to increase. Entropy can increase only if it starts out low. If the pages of *War and Peace* begin thoroughly jumbled, further tosses will merely leave them jumbled; if the universe started out in a thoroughly disordered, high-entropy state, further cosmic evolution would merely maintain the disorder.

The history illustrated in Figure 9.2 is manifestly not a chronicle of eternal, unchanging disorder. Even though particular symmetries have been lost through cosmic phase transitions, the overall entropy of the universe has steadily increased. In the beginning, therefore, the universe must have been highly ordered. This fact allows us to associate "forward" in time with the direction of increasing entropy, but we still need to figure out an explanation for the incredibly low entropy—the incredibly high state of uniformity—of the newly born universe. This requires that we go even farther back than we have so far and try to understand more of what went on at the beginning—during the fuzzy patch in Figure 9.2—a task to which we now turn.

10

Deconstructing the Bang

WHAT BANGED?

A common misconception is that the big bang provides a theory of cosmic origins. It doesn't. The big bang is a theory, partly described in the last two chapters, that delineates cosmic evolution from a split second after whatever happened to bring the universe into existence, but *it says nothing at all about time zero itself*. And since, according to the big bang theory, the bang is what is supposed to have happened at the beginning, the big bang leaves out the bang. It tells us nothing about what banged, why it banged, how it banged, or, frankly, whether it ever really banged at all.[1] In fact, if you think about it for a moment, you'll realize that the big bang presents us with quite a puzzle. At the huge densities of matter and energy characteristic of the universe's earliest moments, gravity was by far the dominant force. But gravity is an *attractive* force. It impels things to come together. So what could possibly be responsible for the *outward* force that drove space to expand? It would seem that some kind of powerful repulsive force must have played a critical role at the time of the bang, but which of nature's forces could that possibly be?

For many decades this most basic of all cosmological questions went unanswered. Then, in the 1980s, an old observation of Einstein's was resurrected in a sparkling new form, giving rise to what has become known as *inflationary cosmology*. And with this discovery, credit for the bang could finally be bestowed on the deserving force: *gravity*. It's surprising, but physicists realized that in just the right environment gravity can be repulsive, and, according to the theory, the necessary conditions prevailed during the earliest moments of cosmic history. For a time interval that would make a nanosecond seem an eternity, the early universe provided an arena in which gravity exerted its repulsive side with a vengeance, driving every region of space away from every other with unrelenting ferocity. So powerful was the repulsive push of gravity that not only was the bang identified, it was revealed to be bigger — much bigger — than anyone had previously imagined. In the inflationary framework, the early universe expanded by an astonishingly huge factor compared with what is predicted by the standard big bang theory, enlarging our cosmological vista to a degree that dwarfed last century's realization that ours is but one galaxy among hundreds of billions.[2]

In this and the next chapter, we discuss inflationary cosmology. We will see that it provides a "front end" for the standard big bang model, offering critical modifications to the standard theory's claims about events during the universe's earliest moments. In doing so, inflationary cosmology resolves key issues that are beyond the reach of the standard big bang, makes a number of predictions that have been and in the near future will continue to be experimentally tested, and, perhaps most strikingly, shows how quantum processes can, through cosmological expansion, iron tiny wrinkles into the fabric of space that leave a visible imprint on the night sky. And beyond these achievements, inflationary cosmology gives significant insight into how the early universe may have acquired its exceedingly low entropy, taking us closer than ever to an explanation of the arrow of time.

Einstein and Repulsive Gravity

After putting the finishing touches on general relativity in 1915, Einstein applied his new equations for gravity to a variety of problems. One was the long-standing puzzle that Newton's equations couldn't account for the so-called precession of the perihelion of Mercury's orbit — the observed fact that Mercury does not trace the same path each time it orbits the sun: instead, each successive orbit shifts slightly relative to the previous. When Einstein redid the standard orbital calculations with his new equations, he derived the observed perihelion precession precisely, a result he found so thrilling that it gave him heart palpitations.[3] Einstein also applied general relativity to the question of how sharply the path of light emitted by a distant star would be bent by spacetime's curvature as it passed by the sun on its way to earth. In 1919, two teams of astronomers — one camped out on the island of Principe off the west coast of Africa, the other in Brazil—tested this prediction during a solar eclipse by comparing observations of starlight that just grazed the sun's surface (these are the light rays most affected by the sun's presence, and only during an eclipse are they visible) with photographs taken when the earth's orbit had placed it between these same stars and the sun, virtually eliminating the sun's gravitational impact on the starlight's trajectory. The comparison revealed a bending angle that, once again, confirmed Einstein's calculations. When the press caught wind of the result, Einstein became a world-renowned celebrity overnight. With general relativity, it's fair to say, Einstein was on a roll.

Yet, despite the mounting successes of general relativity, for years after he first applied his theory to the most immense of all challenges—understanding the entire universe—Einstein absolutely refused to accept the answer that emerged from the mathematics. Before the work of Friedmann and Lemaître discussed in Chapter 8, Einstein, too, had realized that the equations of general relativity showed that the universe could not be static; the fabric of space could stretch or it could shrink, but it could not maintain a fixed size. This suggested that the universe might have had a definite beginning, when the fabric was maximally compressed, and might even have a definite end. Einstein stubbornly balked at this consequence of general relativity, because he and everyone else "knew" that the universe was eternal and, on the largest of scales, fixed and unchanging. Thus, notwithstanding the beauty and the successes of general relativity, Einstein reopened his notebook and sought a modification of the equations that would allow for a universe that conformed to the prevailing prejudice. It didn't take him long. In 1917 he achieved the goal by introducing a new term into the equations of general relativity: the *cosmological constant*.[4]

Einstein's strategy in introducing this modification is not hard to grasp. The gravitational force between any two objects, whether they're baseballs, planets, stars, comets, or what have you, is attractive, and as a result, gravity constantly acts to draw objects toward one another. The gravitational attraction between the earth and a dancer leaping upward causes the dancer to slow down, reach a maximum height, and then head back down. If a choreographer wants a static configuration in which the dancer floats in midair, there would have to be a repulsive force between the dancer and the earth that would precisely balance their gravitational attraction: a static configuration can arise only when there is a perfect cancellation between attraction and repulsion. Einstein realized that exactly the same reasoning holds for the entire universe. In just the same way that the attractive pull of gravity acts to slow the dancer's ascent, it also acts to slow the expansion of space. And just as the dancer can't achieve stasis—it can't hover at a fixed height—without a repulsive force to balance the usual pull of gravity, space can't be static—space can't hover at a fixed overall size—without there also being some kind of balancing repulsive force. Einstein introduced the cosmological constant because he found that with this new term included in the equations, gravity could provide just such a repulsive force.

But what physics does this mathematical term represent? What is the cosmological constant, from what is it made, and how does it manage to go against the grain of usual attractive gravity and exert a repulsive outward push? Well, the modern reading of Einstein's work—one that goes back to Lemaître—interprets the cosmological constant as an exotic form of energy that uniformly and homogeneously fills all of space. I say "exotic" because Einstein's analysis didn't specify where this energy might come from and, as we'll shortly see, the mathematical description he invoked ensured that it could not be composed of anything familiar like protons, neutrons, electrons, or photons. Physicists today invoke phrases like "the energy of space itself" or "dark energy" when discussing the meaning of Einstein's cosmological constant, because if there were a cosmological constant, space would be filled with a transparent, amorphous presence that you wouldn't be able to see directly; space filled with a cosmological constant would still look dark. (This resembles the old notion of an aether and the newer notion of a Higgs field that has acquired a nonzero value throughout space. The latter similarity is more than mere coincidence since there is an important connection between a cosmological constant and Higgs fields, which we will come to shortly.) But even without specifying the origin or identity of the cosmological constant, Einstein was able to work out its gravitational implications, and the answer he found was remarkable.

To understand it, you need to be aware of one feature of general relativity that we have yet to discuss. In Newton's approach to gravity, the strength of attraction between two objects depends solely on two things: their masses and the distance between them. The more massive the objects and the closer they are, the greater the gravitational pull they exert on each other. The situation in general relativity is much the same, except that Einstein's equations show that Newton's focus on mass was too limited. According to general relativity, it is not just the mass (and the separation) of objects that contributes to the strength of the gravitational field. *Energy* and *pressure* also contribute. This is important, so let's spend a moment to see what it means.

Imagine that it's the twenty-fifth century and you're being held in the Hall of Wits, the newest Department of Corrections experiment employing a meritocratic approach to disciplining white-collar felons. The convicts are each given a puzzle, and they can regain their freedom only by solving it. The guy in the cell next to you has to figure out why *Gilligan's Island* reruns made a surprise comeback in the twenty-second century and have been the most popular show ever since, so he's likely to be calling the Hall home for quite some time. Your puzzle is simpler. You are given two identical solid gold cubes—they are the same size and each is made from precisely the same quantity of gold. Your challenge is to find a way to make the cubes register different weights when gently resting on a fixed, exquisitely accurate scale, subject to one stipulation: you're not allowed to change the amount of matter in either cube, so there's to be no chipping, scraping, soldering, shaving, etc. If you posed this puzzle to Newton, he'd immediately declare it to have no solution. According to Newton's laws, identical quantities of gold translate into identical masses. And since each cube will rest on the same, fixed scale, earth's gravitational pull on them will be identical. Newton would conclude that the two cubes must register an identical weight, no ifs, ands, or buts.

With your twenty-fifth-century high school knowledge of general relativity, though, you see a way out. General relativity shows that the strength of the gravitational attraction between two objects does not just depend on their masses[5] (and their separation), but also on any and all additional contributions to each object's total *energy*. And so far we have said nothing about the temperature of the golden cubes. Temperature is a measure of how quickly, on average, the atoms of gold that make up each cube are moving to and fro—it's a measure of how energetic the atoms are (it reflects their *kinetic* energy). Thus, you realize that if you heat up one cube, its atoms will be more energetic, so it will weigh a bit more than the cooler cube. This is a fact Newton was unaware of (an increase of 10 degrees Celsius would increase the weight of a one-pound cube of gold by about a millionth of a billionth of a pound, so the effect is minuscule), and with this solution you win release from the Hall.

Well, almost. Because your crime was particularly devious, at the last minute the parole board decides that you must solve a second puzzle. You are given two identical old-time Jack-in-the-box toys, and your new challenge is to find a way to make each have a different weight. But in this go-around, not only are you forbidden to change the amount of mass in either object, you are also required to keep both at exactly the same temperature. Again, were Newton given this puzzle, he would immediately resign himself to life in the Hall. Since the toys have identical masses, he would conclude that their weights are identical, and so the puzzle is insoluble. But once again, your knowledge of general relativity comes to the rescue: On one of the toys you compress the spring, tightly squeezing Jack under the closed lid, while on the other you leave Jack in his popped-up posture. Why? Well, a compressed spring has more energy than an uncompressed one; you had to exert energy to squeeze the spring down and you can see evidence of your labor because the compressed spring exerts pressure, causing the toy's lid to strain slightly outward. And, again, according to Einstein, *any* additional energy affects gravity, resulting in additional weight. Thus, the closed Jack-in-the-box, with its compressed spring exerting an outward pressure, weighs a touch more than the open Jack-in-the-box, with its uncompressed spring. This is a realization that would have escaped Newton, and with it you finally *do* earn back your freedom.

The solution to that second puzzle hints at the subtle but critical feature of general relativity that we're after. In his paper presenting general relativity, Einstein showed mathematically that the gravitational force depends not only on mass, and not only on energy (such as heat), but also on any *pressures* that may be exerted. And this is the essential physics we need if we are to understand the cosmological constant. Here's why. Outward-directed pressure, like that exerted by a compressed spring, is called *positive pressure*. Naturally enough, positive pressure makes a positive contribution to gravity. But, and this is the critical point, there are situations in which the pressure in a region, unlike mass and total energy, can be *negative*, meaning that the pressure sucks inward instead of pushing outward. And although that may not sound particularly exotic, negative pressure can result in something extraordinary from the point of view of general relativity: *whereas positive pressure contributes to ordinary attractive gravity, negative pressure contributes to "negative" gravity, that is, to repulsive gravity!*[6]

With this stunning realization, Einstein's general relativity exposed a loophole in the more than two-hundred-year-old belief that gravity is always an attractive force. Planets, stars, and galaxies, as Newton correctly showed, certainly do exert an attractive gravitational pull. But when pressure becomes important (for ordinary matter under everyday conditions, the gravitational contribution of pressure is negligible) and, in particular, when pressure is negative (for ordinary matter like protons and electrons, pressure is positive, which is why the cosmological constant can't be composed of anything familiar) there is a contribution to gravity that would have shocked Newton. *It's repulsive.*

This result is central to much of what follows and is easily misunderstood, so let me emphasize one essential point. Gravity and pressure are two related but separate characters in this story. Pressures, or more precisely, pressure differences, can exert their own, nongravitational forces. When you dive underwater, your eardrums can sense the pressure difference between the water pushing on them from the outside and the air pushing on them from the inside. That's all true. But the point we're now making about pressure and gravity is completely different. According to general relativity, pressure can indirectly exert another force—it can exert a gravitational force—because pressure contributes to the gravitational field. Pressure, like mass and energy, is a source of gravity. And remarkably, if the pressure in a region is negative, it contributes a gravitational *push* to the gravitational field permeating the region, not a gravitational pull.

This means that when pressure is negative, there is competition between ordinary attractive gravity, arising from ordinary mass and energy, and exotic repulsive gravity, arising from the negative pressure. If the negative pressure in a region is negative enough, repulsive gravity will dominate; gravity will push things apart rather than draw them together. Here is where the cosmological constant comes into the story. The cosmological term Einstein added to the equations of general relativity would mean that space is uniformly suffused with energy but, crucially, the equations show that this energy has a uniform, negative pressure. What's more, the gravitational repulsion of the cosmological constant's negative pressure overwhelms the gravitational attraction coming from its positive energy, and so repulsive gravity wins the competition: *a cosmological constant exerts an overall repulsive gravitational force.*[7]

For Einstein, this was just what the doctor ordered. Ordinary matter and radiation, spread throughout the universe, exert an attractive gravitational force, causing every region of space to *pull* on every other. The new cosmological term, which he envisioned as also being spread uniformly throughout the universe, exerts a repulsive gravitational force, causing every region of space to *push* on every other. By carefully choosing the size of the new term, Einstein found that he could precisely balance the usual attractive gravitational force with the newly discovered repulsive gravitational force, and produce a static universe.

Moreover, because the new repulsive gravitational force arises from the energy and pressure in space itself, Einstein found that its strength is cumulative; the force becomes stronger over larger spatial separations, since more intervening space means more outward pushing. On the distance scales of the earth or the entire solar system, Einstein showed that the new repulsive gravitational force is immeasurably tiny. It becomes important only over vastly larger cosmological expanses, thereby preserving all the successes of both Newton's theory and his own general relativity when they are applied closer to home. In short, Einstein found he could have his cake and eat it too: he could maintain all the appealing, experimentally confirmed features of general relativity while basking in the eternal serenity of an unchanging cosmos, one that was neither expanding nor contracting.

With this result, Einstein no doubt breathed a sigh of relief. How heart-wrenching it would have been if the decade of grueling research he had devoted to formulating general relativity resulted in a theory that was incompatible with the static universe apparent to anyone who gazed up at the night sky. But, as we have seen, a dozen years later the story took a sharp turn. In 1929, Hubble showed that cursory skyward gazes can be misleading. His systematic observations revealed that the universe is *not* static. It *is* expanding. Had Einstein trusted the original equations of general relativity, he would have predicted the expansion of the universe more than a decade before it was discovered observationally. That would certainly have ranked among the greatest discoveries—it might have been *the* greatest discovery—of all time. After learning of Hubble's results, Einstein rued the day he had thought of the cosmological constant, and he carefully erased it from the equations of general relativity. He wanted everyone to forget the whole sorry episode, and for many decades everyone did.

In the 1980s, however, the cosmological constant resurfaced in a surprising new form and ushered in one of the most dramatic upheavals in cosmological thinking since our species first engaged in cosmological thought.

Of Jumping Frogs and Supercooling

If you caught sight of a baseball flying upward, you could use Newton's law of gravity (or Einstein's more refined equations) to figure out its subsequent trajectory. And if you carried out the required calculations, you'd have a solid understanding of the ball's motion. But there would still be an unanswered question: Who or what threw the ball upward in the first place? How did the ball acquire the initial upward motion whose subsequent unfolding you've evaluated mathematically? In this example, a little further investigation is all it generally takes to find the answer (unless, of course, the aspiring big-leaguers realize that the ball just hit is on a collision course with the windshield of a parked Mercedes). But a more difficult version of a similar question dogs general relativity's explanation of the expansion of the universe.

The equations of general relativity, as originally shown by Einstein, the Dutch physicist Willem de Sitter, and, subsequently, Friedmann and Lemaître, allow for an expanding universe. But, just as Newton's equations tell us nothing about how a ball's upward journey got started, Einstein's equations tell us nothing about how the expansion of the universe got started. For many years, cosmologists took the initial outward expansion of space as an unexplained given, and simply worked the equations forward from there. This is what I meant earlier when I said that the big bang is silent on the bang.

Such was the case until one fateful night in December 1979, when Alan Guth, a physics postdoctoral fellow working at the Stanford Linear Accelerator Center (he is now a professor at MIT), showed that we can do better. Much better. Although there are details that today, more than two decades later, have yet to be resolved fully, Guth made a discovery that finally filled the cosmological silence by providing the big bang with a bang, and one that was bigger than anyone expected.

Guth was not trained as a cosmologist. His specialty was particle physics, and in the late 1970s, together with Henry Tye from Cornell University, he was studying various aspects of Higgs fields in grand unified theories. Remember from the last chapter's discussion of spontaneous symmetry breaking that a Higgs field contributes the least possible energy it can to a region of space when its value settles down to a particular nonzero number (a number that depends on the detailed shape of its potential energy bowl). In the early universe, when the temperature was extraordinarily high, we discussed how the value of a Higgs field would wildly fluctuate from one number to another, like the frog in the hot metal bowl whose legs were being singed, but as the universe cooled, the Higgs would roll down the bowl to a value that would minimize its energy.

Guth and Tye studied reasons why the Higgs field might be delayed in reaching the least energetic configuration (the bowl's valley in Figure 9.1c). If we apply the frog analogy to the question Guth and Tye asked, it was this: what if the frog, in one of its earlier jumps when the bowl was starting to cool, just happened to land on the central plateau? And what if, as the bowl continued to cool, the frog hung out on the central plateau (leisurely eating worms), rather than sliding down to the bowl's valley? Or, in physics terms, what if a fluctuating Higgs field's value should land on the energy bowl's central plateau and remain there as the universe continues to cool? If this happens, physicists say that the Higgs field has *supercooled*, indicating that even though the temperature of the universe has dropped to the point where you'd expect the Higgs value to approach the low-energy valley, it remains trapped in a higher-energy configuration. (This is analogous to highly purified water, which can be supercooled below 0 degrees Celsius, the temperature at which you'd expect it to turn into ice, and yet remain liquid because the formation of ice requires small impurities around which the crystals can grow.)

Guth and Tye were interested in this possibility because their calculations suggested it might be relevant to a problem (the *magnetic monopole* problem[8]) researchers had encountered with various attempts at grand unification. But Guth and Tye realized that there might be another implication and, in retrospect, that's why their work proved pivotal. They suspected that the energy associated with a supercooled Higgs field—remember, the height of the field represents its energy, so the field has zero energy only if its value lies in the bowl's valley—might have an effect on the expansion of the universe. In early December 1979, Guth followed up on this hunch, and here's what he found.

A Higgs field that has gotten caught on a plateau not only suffuses space with energy, but, of crucial importance, Guth realized that it also contributes a uniform *negative pressure*. In fact, he found that as far as energy and pressure are concerned, a Higgs field that's caught on a plateau has the same properties as a cosmological constant: it suffuses space with energy and negative pressure, and in exactly the same proportions as a cosmological constant. So Guth discovered that a supercooled Higgs field does have an important effect on the expansion of space: like a cosmological constant, it exerts a repulsive gravitational force that drives space to expand.[9]

At this point, since you are already familiar with negative pressure and repulsive gravity, you may be thinking, All right, it's nice that Guth found a specific physical mechanism for realizing Einstein's idea of a cosmological constant, but so what? What's the big deal? The concept of a cosmological constant had long been abandoned. Its introduction into physics was nothing but an embarrassment for Einstein. Why get excited over rediscovering something that had been discredited more than six decades earlier?

Inflation

Well, here's why. Although a supercooled Higgs field shares certain features with a cosmological constant, Guth realized that they are not completely identical. Instead, there are two key differences—differences that make all the difference.

First, whereas a cosmological constant is constant—it does not vary with time, so it provides a constant, unchanging outward push—a supercooled Higgs field need not be constant. Think of a frog perched on the bump in Figure 10.1a. It may hang out there for a while, but sooner or

(a)   (b)

Figure 10.1 (a) A supercooled Higgs field is one whose value gets trapped on the energy bowl's high-energy plateau, like the frog on a bump. (b) Typically, a supercooled Higgs field will quickly find its way off the plateau and drop to a value with lower energy, like the frog's jumping off the bump.

later a random jump this way or that—a jump taken not because the bowl is hot (it no longer is), but merely because the frog gets restless— will propel the frog beyond the bump, after which it will slide down to the bowl's lowest point, as in Figure 10.1b. A Higgs field can behave sim- ilarly. Its value throughout all of space may get stuck on its energy bowl's central bump while the temperature drops too low to drive significant thermal agitation. But quantum processes will inject random jumps into the Higgs field's value, and a large enough jump will propel it off the plateau, allowing its energy and pressure to relax to zero.[10] Guth's calcu- lations showed that, depending on the precise shape of the bowl's bump, this jump could have happened rapidly, perhaps in as short a time as .0000000000000000000000000000000001 (10^{-35}) seconds. Subsequently, Andrei Linde, then working at the Lebedev Physical Institute in Moscow, and Paul Steinhardt, then working with his student Andreas Albrecht at the University of Pennsylvania, discovered a way for the Higgs field's relaxation to zero energy and pressure throughout all of space to happen even more efficiently and significantly more uniformly (thereby curing certain technical problems inherent to Guth's original proposal[11]). They showed that if the potential energy bowl had been smoother and more gradually sloping, as in Figure 10.2, no quantum jumps would have been necessary: the Higgs field's value would quickly roll down to the valley, much like a ball rolling down a hill. The upshot is that if a Higgs field acted like a cosmological constant, it did so only for a brief moment.

The second difference is that whereas Einstein carefully and arbitrar- ily chose the value of the cosmological constant—the amount of energy

Figure 10.2 A smoother and more gradually sloping bump allows the Higgs field value to roll down to the zero-energy valley more easily and more uniformly throughout space.

and negative pressure it contributed to each volume of space—so that its outward repulsive force would precisely balance the inward attractive force arising from the ordinary matter and radiation in the cosmos, Guth was able to estimate the energy and negative pressure contributed by the Higgs fields he and Tye had been studying. And the answer he found was more than 10000000000000000000000000000000000000000000000000 00000000000000000000000000000000000000000000000000000 (10^{100}) times larger than the value Einstein had chosen. This number is huge, obviously, and so the outward push supplied by the Higgs field's repulsive gravity is *monumental* compared with what Einstein envisioned originally with the cosmological constant.

Now, if we combine these two observations—that the Higgs field will stay on the plateau, in the high-energy, negative-pressure state, only for the briefest of instants, and that while it is on the plateau, the repulsive outward push it generates is enormous—what do we have? Well, as Guth realized, we have a phenomenal, short-lived, outward burst. In other words, we have exactly what the big bang theory was missing: a *bang*, and a big one at that. That's why Guth's discovery is something to get excited about.[12]

The cosmological picture emerging from Guth's breakthrough is thus the following. A long time ago, when the universe was enormously dense, its energy was carried by a Higgs field perched at a value far from the lowest point on its potential energy bowl. To distinguish this particular Higgs field from others (such as the electroweak Higgs field responsible for giving mass to the familiar particle species, or the Higgs field that arises in grand unified theories[13]) it is usually called the *inflaton* field.* Because of its negative pressure, the inflaton field generated a gigantic gravitational repulsion that drove every region of space to rush away from every other; in Guth's language, the inflaton drove the universe to *inflate*. The repulsion lasted only about 10^{-35} seconds, but it was so powerful that even in that brief moment the universe swelled by a huge factor. Depending on details such as the precise shape of the inflaton field's potential energy, the universe could easily have expanded by a factor of 10^{30}, 10^{50}, or 10^{100} or more.

These numbers are staggering. An expansion factor of 10^{30}—a conservative estimate—would be like scaling up a molecule of DNA to

---

*You might think I've left out an "i" in the last syllable of "inflaton," but I haven't; physicists often give fields names, such as photon and gluon, which end with "on."

roughly the size of the Milky Way galaxy, and in a time interval that's much shorter than a billionth of a billionth of a billionth of the blink of an eye. By comparison, even this conservative expansion factor is billions and billions of times the expansion that would have occurred according to the standard big bang theory during the same time interval, and it exceeds the total expansion factor that has cumulatively occurred over the subsequent 14 billion years! In the many models of inflation in which the calculated expansion factor is much larger than 10^{30}, the resulting spatial expanse is so enormous that the region we are able to see, even with the most powerful telescope possible, is but a tiny fraction of the whole universe. According to these models, none of the light emitted from the vast majority of the universe could have reached us yet, and much of it won't arrive until long after the sun and earth have died out. If the entire cosmos were scaled down to the size of earth, the part accessible to us would be much smaller than a grain of sand.

Roughly 10^{-35} seconds after the burst began, the inflaton field found its way off the high-energy plateau and its value throughout space slid down to the bottom of the bowl, turning off the repulsive push. And as the inflaton value rolled down, it relinquished its pent-up energy to the production of ordinary particles of matter and radiation—like a foggy mist settling on the grass as morning dew—that uniformly filled the expanding space.[14] From this point on, the story is essentially that of the standard big bang theory: space continued to expand and cool in the aftermath of the burst, allowing particles of matter to clump into structures like galaxies, stars, and planets, which slowly arranged themselves into the universe we currently see, as illustrated in Figure 10.3.

Guth's discovery—dubbed *inflationary cosmology*—together with the important improvements contributed by Linde, and by Albrecht and Steinhardt, provided an explanation for what set space expanding in the first place. A Higgs field perched above its zero energy value can provide an outward blast driving space to swell. Guth provided the big bang with a bang.

The Inflationary Framework

Guth's discovery was quickly hailed as a major advance and has become a dominant fixture of cosmological research. But notice two things. First, in the standard big bang model, the bang supposedly happened at time zero,

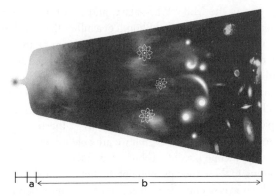

Figure 10.3 (a) Inflationary cosmology inserts a quick, enormous burst of spatial expansion early on in the history of the universe. **(b)** After the burst, the evolution of the universe merges into the standard evolution theorized in the big bang model.

at the very beginning of the universe, so it is viewed as the creation event. But just as a stick of dynamite explodes only when it's properly lit, in inflationary cosmology the bang happened only when conditions were right—when there was an inflaton field whose value provided the energy and negative pressure that fueled the outward burst of repulsive gravity—and that need not have coincided with the "creation" of the universe. For this reason, the inflationary bang is best thought of as *an* event that the preexisting universe experienced, but not necessarily as *the* event that created the universe. We denote this in Figure 10.3 by maintaining some of the fuzzy patch of Figure 9.2, indicating our continuing ignorance of fundamental origin: specifically, if inflationary cosmology is right, our ignorance of why there is an inflaton field, why its potential energy bowl has the right shape for inflation to have occurred, why there are space and time within which the whole discussion takes place, and, in Leibniz's more grandiose phrasing, why there is something rather than nothing.

A second and related observation is that inflationary cosmology is not a single, unique theory. Rather, it is a cosmological framework built around the realization that gravity can be repulsive and can thus drive a swelling of space. The precise details of the outward burst—when it happened, how long it lasted, the strength of the outward push, the factor by which the universe expanded during the burst, the amount of energy the inflaton deposited in ordinary matter as the burst drew to a close, and so on—depend on details, most notably the size and shape of the inflaton field's potential energy, that are presently beyond our ability to determine from theoretical considerations alone. So for many years physicists have studied all sorts of possibilities—various shapes for the potential energy, various numbers of inflaton fields that work in tandem, and so on—and determined which choices give rise to theories consistent with astronomical observations. The important thing is that there are aspects of inflationary cosmological theories that transcend the details and hence are common to essentially any realization. The outward burst itself, by definition, is one such feature, and hence any inflationary model comes with a bang. But there are a number of other features inherent to all inflationary models that are vital because they solve important problems that have stumped standard big bang cosmology.

Inflation and the Horizon Problem

One such problem is called the *horizon problem* and concerns the uniformity of the microwave background radiation that we came across previously. Recall that the temperature of the microwave radiation reaching us from one direction in space agrees with that coming from any other direction to fantastic accuracy (to better than a thousandth of a degree). This observational fact is pivotal, because it attests to homogeneity throughout space, allowing for enormous simplifications in theoretical models of the cosmos. In earlier chapters, we used this homogeneity to narrow down drastically the possible shapes for space and to argue for a uniform cosmic time. The problem arises when we try to explain *how* the universe became so uniform. How is it that vastly distant regions of the universe have arranged themselves to have nearly identical temperatures?

If you think back to Chapter 4, one possibility is that just as nonlocal quantum entanglement can correlate the spins of two widely separated particles, maybe it can also correlate the temperatures of two widely separated regions of space. While this is an interesting suggestion, the tremendous dilution of entanglement in all but the most controlled settings, as discussed at the end of that chapter, essentially rules it out. Okay, perhaps there is a simpler explanation. Maybe a long time ago when every region of space was nearer to every other, their temperatures equalized through their close contact much as a hot kitchen and a cool living room come to the same temperature when a door between them is opened for a while. In the standard big bang theory, though, this explanation also fails. Here's one way to think about it.

Imagine watching a film that depicts the full course of cosmic evolution from the beginning until today. Pause the film at some arbitrary moment and ask yourself: Could two particular regions of space, like the kitchen and the living room, have influenced each other's temperature? Could they have exchanged light and heat? The answer depends on two things: The distance between the regions and the amount of time that has elapsed since the bang. If their separation is less than the distance light could have traveled in the time since the bang, then the regions could have influenced each other; otherwise, they couldn't have. Now, you might think that *all* regions of the observable universe could have interacted with each other way back near the beginning because the farther back we wind the film, the closer the regions become and hence the easier it is for them to interact. But this reasoning is too quick; it doesn't take account of the fact that not only were regions of space closer, but there was also less time for them to have communicated.

To do a proper analysis, imagine running the cosmic film in reverse while focusing on two regions of space currently on opposite sides of the observable universe—regions that are so distant that they are currently beyond each other's spheres of influence. If in order to halve their separation we have to roll the cosmic film more than halfway back toward the beginning, then even though the regions of space were closer together, communication between them was still impossible: they were half as far apart, but the time since the bang was *less* than half of what it is today, and so light could travel only *less* than half as far. Similarly, if from that point in the film we have to run more than halfway back to the beginning in order to halve the separation between the regions once again, communication becomes more difficult still. With this kind of cosmic evolution, even though regions were closer together in the past, it becomes more puzzling—not less—that they somehow managed to equalize their temperatures. Relative to how far light can travel, the regions become increasingly cut off as we examine them ever farther back in time.

This is exactly what happens in the standard big bang theory. In the standard big bang, gravity acts only as an attractive force, and so, ever since the beginning, it has been acting to slow the expansion of space. Now, if something is slowing down, it will take more time to cover a given distance. For instance, imagine that Secretariat left the gate at a blistering pace and covered the first half of a racecourse in two minutes, but because it's not his best day, he slows down considerably during the second half and takes three more minutes to finish. When viewing a film of the race in reverse, we'd have to roll the film more than halfway back in order to see Secretariat at the course's halfway mark (we'd have to run the five-minute film of the race all the way back to the two-minute mark). Similarly, since in the standard big bang theory gravity slows the expansion of space, from any point in the cosmic film we have to wind more than halfway back in time in order to halve the separation between two regions. And, as above, this means that even though the regions of space were closer together at earlier times, it was more difficult—not less—for them to influence each other and hence more puzzling—not less—that they somehow reached the same temperature.

Physicists define a region's *cosmic horizon* (or *horizon* for short) as the most distant surrounding regions of space that are close enough to the given region for the two to have exchanged light signals in the time since the bang. The analogy is to the most distant things we can see on earth's surface from any particular vantage point.[15] The *horizon problem*, then, is the puzzle, inherent in the observations, that regions whose horizons have always been separate—regions that could never have interacted, communicated, or exerted any kind of influence on each other—somehow have nearly identical temperatures.

The horizon problem does not imply that the standard big bang model is wrong, but it does cry out for explanation. Inflationary cosmology provides one.

In inflationary cosmology, there was a brief instant during which gravity was repulsive and this drove space to expand faster and faster. During this part of the cosmic film, you would have to wind the film *less* than halfway back in order to halve the distance between two regions. Think of a race in which Secretariat covers the first half of the course in two minutes and, because he's having the run of his life, speeds up and blazes through the second half in one minute. You'd only have to wind the three-minute film of the race back to the two-minute mark—less than halfway back—to see him at the course's halfway point. Similarly, the increasingly rapid separation of any two regions of space during inflationary expansion implies that halving their separation requires winding the cosmic film less—*much less*—than halfway back toward the beginning. As we go farther back in time, therefore, it becomes *easier* for any two regions of space to influence each other, because, proportionally speaking, there is more time for them to communicate. Calculations show that if the inflationary-expansion phase drove space to expand by at least a factor of 10^{30}, an amount that is readily achieved in specific realizations of inflationary expansion, all the regions in space that we currently see—all the regions in space whose temperatures we have measured—were able to communicate as easily as the adjacent kitchen and living room and hence efficiently come to a common temperature in the earliest moments of the universe.[16] In a nutshell, space expands slowly enough in the very beginning for a uniform temperature to be broadly established and then, through an intense burst of ever more rapid expansion, the universe makes up for the sluggish start and widely disperses nearby regions.

That's how inflationary cosmology explains the otherwise mysterious uniformity of the microwave background radiation suffusing space.

Inflation and the Flatness Problem

A second problem addressed by inflationary cosmology has to do with the shape of space. In Chapter 8, we imposed the criterion of uniform spatial symmetry and found three ways in which the fabric of space can curve. Resorting to our two-dimensional visualizations, the possibilities are positive curvature (shaped like the surface of a ball), negative curvature (saddle-shaped), and zero curvature (shaped like an infinite flat tabletop or like a finite-sized video game screen). Since the early days of general relativity, physicists have realized that the total matter and energy in each volume of space—the *matter/energy density*—determine the curvature of space. If the matter/energy density is high, space will pull back on itself in the shape of a sphere; that is, there will be positive curvature. If the matter/energy density is low, space will flare outward like a saddle; that is, there will be negative curvature. Or, as mentioned in the last chapter, for a very special amount of matter/energy density—the critical density, equal to the mass of about five hydrogen atoms (about 10^{-23} grams) in each cubic meter—space will lie just between these two extremes, and will be perfectly flat: that is, there will be no curvature.

Now for the puzzle.

The equations of general relativity, which underlie the standard big bang model, show that if the matter/energy density early on was *exactly* equal to the critical density, then it would stay equal to the critical density as space expanded.[17] But if the matter/energy density was even slightly more or slightly less than the critical density, subsequent expansion would drive it enormously far from the critical density. Just to get a feel for the numbers, if at one second ATB, the universe was just shy of criticality, having 99.99 percent of the critical density, calculations show that by today its density would have been driven all the way down to .00000000001 of the critical density. It's kind of like the situation faced by a mountain climber who is walking across a razor-thin ledge with a steep drop off on either side. If her step is right on the mark, she'll make it across. But even a tiny misstep that's just a little too far left or right will be amplified into a significantly different outcome. (And, at the risk of having one too many analogies, this feature of the standard big bang model also reminds me of the shower years ago in my college dorm: if you managed to set the knob perfectly, you could get a comfortable water temperature. But if you were off by the slightest bit, one way or the other, the water would be either scalding or freezing. Some students just stopped showering altogether.)

For decades, physicists have been attempting to measure the matter/energy density in the universe. By the 1980s, although the measurements were far from complete, one thing was certain: the matter/energy density of the universe is not thousands and thousands of times smaller or larger than the critical density; equivalently, space is not substantially curved, either positively or negatively. This realization cast an awkward light on the standard big bang model. It implied that for the standard big bang to be consistent with observations, some mechanism—one that nobody could explain or identify—must have tuned the matter/energy density of the early universe *extraordinarily* close to the critical density. For example, calculations showed that at one second ATB, the matter/energy density of the universe needed to have been within a *millionth of a millionth of a percent* of the critical density; if the matter/energy density deviated from the critical value by any more than this minuscule amount, the standard big bang model predicts a matter/energy density today that is *vastly* different from what we observe. According to the standard big bang model, then, the early universe, much like the mountain climber, teetered along an extremely narrow ledge. A tiny deviation in conditions billions of years ago would have led to a present-day universe very different from the one revealed by astronomers' measurements. This is known as the *flatness problem*.

Although we've covered the essential idea, it's important to understand the sense in which the flatness problem is a problem. By no means does the flatness problem show that the standard big bang model is wrong. A staunch believer reacts to the flatness problem with a shrug of the shoulders and the curt reply "That's just how it was back then," taking the finely tuned matter/energy density of the early universe—which the standard big bang requires to yield predictions that are in the same ball-park as observations—as an unexplained given. But this answer makes most physicists recoil. Physicists feel that a theory is grossly unnatural if its success hinges on extremely precise tunings of features for which we lack a fundamental explanation. Without supplying a reason for why the matter/energy density of the early universe would have been so finely tuned to an acceptable value, many physicists have found the standard big bang model highly contrived. Thus, the flatness problem highlights the extreme sensitivity of the standard big bang model to conditions in the remote past of which we know very little; it shows how the theory must assume the universe was *just so*, in order to work.

By contrast, physicists long for theories whose predictions are insensitive to unknown quantities such as how things were a long time ago. Such theories feel robust and natural because their predictions don't depend delicately on details that are hard, or perhaps even impossible, to determine directly. This is the kind of theory provided by inflationary cosmology, and its solution to the flatness problem illustrates why.

The essential observation is that whereas attractive gravity amplifies any deviation from the critical matter/energy density, the repulsive gravity of the inflationary theory does the opposite: it *reduces* any deviation from the critical density. To get a feel for why this is the case, it's easiest to use the tight connection between the universe's matter/energy density and its curvature to reason geometrically. In particular, notice that even if the shape of the universe were significantly curved early on, after inflationary expansion a portion of space large enough to encompass today's observable universe looks very nearly flat. This is a feature of geometry we are all well aware of: The surface of a basketball is obviously curved, but it took both time and thinkers with chutzpah before everyone was convinced that the earth's surface was also curved. The reason is that, all else being equal, the larger something is, the more gradually it curves and the flatter a patch of a given size on its surface appears. If you draped the state of Nebraska over a sphere just a few hundred miles in diameter, as in Figure 10.4a, it would look curved, but on the earth's surface, as just about all Nebraskans concur, it looks flat. If you laid Nebraska out on a sphere a billion times larger than earth, it would look flatter still. In inflationary cos-

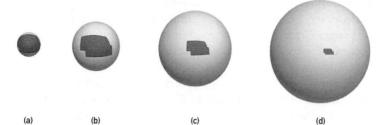

(a)   (b)   (c)   (d)

Figure 10.4 A shape of fixed size, such as that of the state of Nebraska, appears flatter and flatter when laid out on larger and larger spheres. In this analogy, the sphere represents the entire universe, while Nebraska represents the *observable* universe—the part within our cosmic horizon.

mology, space was stretched by such a colossal factor that the observable universe, the part we can see, is but a small patch in a gigantic cosmos. And so, like Nebraska laid out on a giant sphere as in Figure 10.4d, even if the entire universe were curved, the *observable* universe would be very nearly flat.[18]

It's as if there are powerful, oppositely oriented magnets embedded in the mountain climber's boots and the thin ledge she is crossing. Even if her step is aimed somewhat off the mark, the strong attraction between the magnets ensures that her foot lands squarely on the ledge. Similarly, even if the early universe deviated a fair bit from the critical matter/energy density and hence was far from flat, the inflationary expansion ensured that the part of space we have access to was *driven* toward a flat shape and that the matter/energy density we have access to was *driven* to the critical value.

Progress and Prediction

Inflationary cosmology's insights into the horizon and flatness problems represent tremendous progress. For cosmological evolution to yield a homogeneous universe whose matter/energy density is even remotely close to what we observe today, the standard big bang model requires precise, unexplained, almost eerie fine-tuning of conditions early on. This tuning can be assumed, as the staunch adherent to the standard big bang advocates, but the lack of an explanation makes the theory seem artificial.

To the contrary, regardless of the detailed properties of the early universe's matter/energy density, inflationary cosmological evolution *predicts* that the part we can see should be very nearly flat; that is, it *predicts* that the matter/energy density we observe should be very nearly 100 percent of the critical density.

Insensitivity to the detailed properties of the early universe is a wonderful feature of the inflationary theory, because it allows for definitive predictions irrespective of our ignorance of conditions long ago. But we must now ask: How do these predictions stand up to detailed and precise observations? Do the data support inflationary cosmology's prediction that we should observe a flat universe containing the critical density of matter/energy?

For many years the answer seemed to be "Not quite." Numerous astronomical surveys carefully measured the amount of matter/energy that could be seen in the cosmos, and the answer they came up with was about 5 percent of the critical density. This is far from the enormous or minuscule densities to which the standard big bang naturally leads— without artificial fine-tuning—and is what I alluded to earlier when I said that observations establish that the universe's matter/energy density is not thousands and thousands of times larger or smaller than the critical amount. Even so, 5 percent falls short of the 100 percent inflation predicts. But physicists have long realized that care must be exercised in evaluating the data. The astronomical surveys tallying 5 percent took account only of matter and energy that gave off light and hence could be seen with astronomers' telescopes. And for decades, even before the discovery of inflationary cosmology, there had been mounting evidence that the universe has a hefty dark side.

A Prediction of Darkness

During the early 1930s, Fritz Zwicky, a professor of astronomy at the California Institute of Technology (a famously caustic scientist whose appreciation for symmetry led him to call his colleagues spherical bastards because, he explained, they were bastards any way you looked at them[19]), realized that the outlying galaxies in the Coma cluster, a collection of thousands of galaxies some 370 million light-years from earth, were moving too quickly for their visible matter to muster an adequate gravitational force to keep them tethered to the group. Instead, his analysis showed that many of the fastest-moving galaxies should be flung clear of the cluster, like water droplets thrown off a spinning bicycle tire. And yet none were. Zwicky conjectured that there might be additional matter permeating the cluster that did not give off light but supplied the additional gravitational pull necessary to hold the cluster together. His calculations showed that if this explanation was right, the vast majority of the cluster's mass would comprise this nonluminous material. By 1936, corroborating evidence was found by Sinclair Smith of the Mount Wilson observatory, who was studying the Virgo cluster and came to a similar conclusion. But since both men's observations, as well as a number of subsequent others, had various uncertainties, many remained unconvinced that there was voluminous unseen matter whose gravitational pull was keeping the groups of galaxies together.

Over the next thirty years observational evidence for nonluminous matter continued to mount,[20] but it was the work of the astronomer Vera Rubin from the Carnegie Institution of Washington, together with Kent Ford and others, that really clinched the case. Rubin and her collaborators studied the movements of stars within numerous spinning galaxies and concluded that if what you see is what there is, then many of the galaxy's stars should be routinely flung outward. Their observations showed conclusively that the visible galactic matter could not exert a gravitational grip anywhere near strong enough to keep the fastest-moving stars from breaking free. However, their detailed analyses also showed that the stars *would* remain gravitationally tethered if the galaxies they inhabited were immersed in a giant ball of nonluminous matter (as in Figure 10.5), whose total mass far exceeded that of the galaxy's luminous material. And so, like an audience that infers the presence of a dark-robed mime even though it sees only his white-gloved hands flitting to and fro on the unlit stage, astronomers concluded that the universe must be suffused with *dark matter*—matter that does not clump together in stars and hence does not give off light, and that thus exerts a gravitational pull without revealing itself visibly. The universe's luminous constituents—stars— were revealed as but floating beacons in a giant ocean of dark matter.

But if dark matter must exist in order to produce the observed motions of stars and galaxies, what's it made of? So far, no one knows. The identity of the dark matter remains a major, looming mystery, although astronomers and physicists have suggested numerous possible constituents ranging from various kinds of exotic particles to a cosmic bath of miniature black holes. But even without determining its composition, by

Figure 10.5 A galaxy immersed in a ball of dark matter (with the dark matter artificially highlighted to make it visible in the figure).

closely analyzing its gravitational effects astronomers have been able to determine with significant precision how much dark matter is spread throughout the universe. And the answer they've found amounts to about 25 percent of the critical density.[21] Thus, together with the 5 percent found in visible matter, the dark matter brings our tally up to 30 percent of the amount predicted by inflationary cosmology.

Well, this is certainly progress, but for a long time scientists scratched their heads, wondering how to account for the remaining 70 percent of the universe, which, if inflationary cosmology was correct, had apparently gone AWOL. But then, in 1998, two groups of astronomers came to the same shocking conclusion, which brings our story full circle and once again reveals the prescience of Albert Einstein.

The Runaway Universe

Just as you may seek a second opinion to corroborate a medical diagnosis, physicists, too, seek second opinions when they come upon data or theories that point toward puzzling results. Of these second opinions, the most convincing are those that reach the same conclusion from a point of view that differs sharply from the original analysis. When the arrows of explanation converge on one spot from different angles, there's a good chance that they're pointing at the scientific bull's-eye. Naturally then, with inflationary cosmology strongly suggesting something totally bizarre—that 70 percent of the universe's mass/energy has yet to be measured or identified—physicists have yearned for independent confirmation. It has long been realized that measurement of the *deceleration parameter* would do the trick.

Since just after the initial inflationary burst, ordinary attractive gravity has been slowing the expansion of space. The rate at which this slowing occurs is called the deceleration parameter. A precise measurement of the parameter would provide independent insight into the total amount of matter in the universe: more matter, whether or not it gives off light, implies a greater gravitational pull and hence a more pronounced slowing of spatial expansion.

For many decades, astronomers have been trying to measure the deceleration of the universe, but although doing so is straightforward in principle, it's a challenge in practice. When we observe distant heavenly bodies such as galaxies or quasars, we are seeing them as they were a long time ago: the farther away they are, the farther back in time we are looking. So, if we could measure how fast they were receding from us, we'd have a measure of how fast the universe was expanding in the distant past. Moreover, if we could carry out such measurements for astronomical objects situated at a variety of distances, we would have measured the universe's expansion rate at a variety of moments in the past. By comparing these expansion rates, we could determine how the expansion of space is slowing over time and thereby determine the deceleration parameter.

Carrying out this strategy for measuring the deceleration parameter thus requires two things: a means of determining the distance of a given astronomical object (so that we know how far back in time we are looking) and a means of determining the speed with which the object is receding from us (so that we know the rate of spatial expansion at that moment in the past). The latter ingredient is easier to come by. Just as the pitch of a police car's siren drops to lower tones as it rushes away from us, the frequency of vibration of the light emitted by an astronomical source also drops as the object rushes away. And since the light emitted by atoms like hydrogen, helium, and oxygen—atoms that are among the constituents of stars, quasars, and galaxies—has been carefully studied under laboratory conditions, a precise determination of the object's speed can be made by examining how the light we receive differs from that seen in the lab.

But the former ingredient, a method for determining precisely how far away an object is, has proven to be the astronomer's headache. The farther away something is, the dimmer you expect it to appear, but turning this simple observation into a quantitative measure is difficult. To judge the distance to an object by its apparent brightness, you need to know its intrinsic brightness—how bright it would be were it right next to you. And it is difficult to determine the intrinsic brightness of an object billions of light-years away. The general strategy is to seek a species of heavenly bodies that, for fundamental reasons of astrophysics, always burn with a standard, dependable brightness. If space were dotted with glowing 100-watt lightbulbs, that would do the trick, since we could easily determine a given bulb's distance on the basis of how dim it appears (although it would be a challenge to see 100-watt bulbs from significantly far away). But, as space isn't so endowed, what can play the role of standard-brightness lightbulbs, or, in astronomy-speak, what can play the role of *standard candles*? Through the years astronomers have studied a variety of possibilities, but the most successful candidate to date is a particular class of supernova explosions.

When stars exhaust their nuclear fuel, the outward pressure from nuclear fusion in the star's core diminishes and the star begins to implode under its own weight. As the star's core crashes in on itself, its temperature rapidly rises, sometimes resulting in an enormous explosion that blows off the star's outer layers in a brilliant display of heavenly fireworks. Such an explosion is known as a supernova; for a period of weeks, a single exploding star can burn as bright as a billion suns. It's truly mind-boggling: a single star burning as bright as almost an entire galaxy! Different types of stars—of different sizes, with different atomic abundances, and so on—give rise to different kinds of supernova explosions, but for many years astronomers have realized that certain supernova explosions always seem to burn with the same intrinsic brightness. These are *type Ia* supernova explosions.

In a type Ia supernova, a white dwarf star—a star that has exhausted its supply of nuclear fuel but has insufficient mass to ignite a supernova explosion on its own—sucks the surface material from a nearby companion star. When the dwarf star's mass reaches a particular critical value, about 1.4 times that of the sun, it undergoes a runaway nuclear reaction that causes the star to go supernova. Since such supernova explosions occur when the dwarf star reaches the same critical mass, the characteristics of the explosion, including its overall intrinsic brightness, are largely the same from episode to episode. Moreover, since supernovae, unlike 100-watt lightbulbs, are so fantastically powerful, not only do they have a standard, dependable brightness but you can also see them clear across the universe. They are thus prime candidates for standard candles.[22]

In the 1990s, two groups of astronomers, one led by Saul Perlmutter at the Lawrence Berkeley National Laboratory, and the other led by Brian Schmidt at the Australian National University, set out to determine the deceleration—and hence the total mass/energy—of the universe by measuring the recession speeds of type Ia supernovae. Identifying a supernova as being of type Ia is fairly straightforward because the light their explosions generate follows a distinctive pattern of steeply rising then gradually falling intensity. But actually catching a type Ia supernova in the act is no small feat, since they happen only about once every few hundred years in a typical galaxy. Nevertheless, through the innovative technique of simultaneously observing thousands of galaxies with wide-field-of-view telescopes, the teams were able to find nearly four dozen type Ia supernovae at various distances from earth. After painstakingly determining the distance and recessional velocities of each, both groups came to a totally unexpected conclusion: ever since the universe was about 7 billion years old, its expansion rate has *not* been decelerating. Instead, the expansion rate has been *speeding up*.

The groups concluded that the expansion of the universe slowed down for the first 7 billion years after the initial outward burst, much like a car slowing down as it approaches a highway tollbooth. This was as expected. But the data revealed that, like a driver who hits the gas pedal after gliding through the EZ-Pass lane, the expansion of the universe has been accelerating ever since. The expansion rate of space 7 billion years ATB was *less* than the expansion rate 8 billion years ATB, which was *less* than the expansion rate 9 billion years ATB, and so on, all of which are *less* than the expansion rate today. The expected deceleration of spatial expansion has turned out to be an unexpected *acceleration*.

But how could this be? Well, the answer provides the corroborating second opinion regarding the missing 70 percent of mass/energy that physicists had been seeking.

The Missing 70 Percent

If you cast your mind back to 1917 and Einstein's introduction of a cosmological constant, you have enough information to suggest how it might be that the universe is accelerating. Ordinary matter and energy give rise to ordinary attractive gravity, which slows spatial expansion. But as the universe expands and things get increasingly spread out, this cosmic gravitational pull, while still acting to slow the expansion, gets weaker. And this sets us up for the new and unexpected twist. If the universe should have a cosmological constant—and if its magnitude should have just the right, small value—up until about 7 billion years ATB its gravitational repulsion would have been overwhelmed by the usual gravitational attraction of ordinary matter, yielding a net slowing of expansion, in keeping with the data. But then, as ordinary matter spread out and its gravitational pull diminished, the repulsive push of the cosmological constant (whose strength does not change as matter spreads out) would have gradually gained the upper hand, and *the era of decelerated spatial expansion would have given way to a new era of accelerated expansion.*

In the late 1990s, such reasoning and an in-depth analysis of the data led both the Perlmutter group and the Schmidt group to suggest that Einstein had not been wrong some eight decades earlier when he introduced a cosmological constant into the gravitational equations. The universe, they suggested, *does* have a cosmological constant.[23] Its magnitude is not what Einstein proposed, since he was chasing a static universe in which gravitational attraction and repulsion matched precisely, and these researchers found that for billions of years repulsion has dominated. But that detail notwithstanding, should the discovery of these groups continue to hold up under the close scrutiny and follow-up studies now under way, Einstein will have once again seen through to a fundamental feature of the universe, one that this time took more than eighty years to be confirmed experimentally.

The recession speed of a supernova depends on the difference between the gravitational pull of ordinary matter and the gravitational push of the "dark energy" supplied by the cosmological constant. Taking the amount of matter, both visible and dark, to be about 30 percent of the critical density, the supernova researchers concluded that the accelerated expansion they had observed required an outward push of a cosmological constant whose dark energy contributes about 70 percent of the critical density.

This is a remarkable number. If it's correct, then not only does ordinary matter—protons, neutrons, electrons—constitute a paltry 5 percent of the mass/energy of the universe, and not only does some currently unidentified form of dark matter constitute at least *five times* that amount, but also the *majority* of the mass/energy in the universe is contributed by a totally different and rather mysterious form of dark energy that is spread throughout space. If these ideas are right, they dramatically extend the Copernican revolution: not only are we not the center of the universe, but the stuff of which we're made is like flotsam on the cosmic ocean. If protons, neutrons, and electrons had been left out of the grand design, the total mass/energy of the universe would hardly have been diminished.

But there is a second, equally important reason why 70 percent is a remarkable number. A cosmological constant that contributes 70 percent of the critical density would, together with the 30 percent coming from ordinary matter and dark matter, bring the total mass/energy of the universe right up to the full 100 percent predicted by inflationary cosmology! Thus, the outward push demonstrated by the supernova data can be explained by just the right amount of dark energy to account for the unseen 70 percent of the universe that inflationary cosmologists had been scratching their heads over. The supernova measurements and inflationary cosmology are wonderfully complementary. They confirm each other. Each provides a corroborating second opinion for the other.[24]

Combining the observational results of supernovae with the theoretical insights of inflation, we thus arrive at the following sketch of cosmic evolution, summarized in Figure 10.6. Early on, the energy of the universe was carried by the inflaton field, which was perched away from its minimum energy state. Because of its negative pressure, the inflaton field drove an enormous burst of inflationary expansion. Then, some 10^{-35} seconds later, as the inflaton field slid down its potential energy bowl, the burst of expansion drew to a close and the inflaton released its pent-up energy to the production of ordinary matter and radiation. For many billions of years, these familiar constituents of the universe exerted an ordinary attractive gravitational pull that slowed the spatial expansion. But as the universe grew and thinned out, the gravitational pull diminished. About 7 billion years ago, ordinary gravitational attraction became weak enough for the gravitational repulsion of the universe's cosmological constant to become dominant, and since then the rate of spatial expansion has been continually increasing.

About 100 billion years from now, all but the closest of galaxies will be dragged away by the swelling space at faster-than-light speed and so would be impossible for us to see, regardless of the power of telescopes used. If these ideas are right, then in the far future the universe will be a vast, empty, and lonely place.

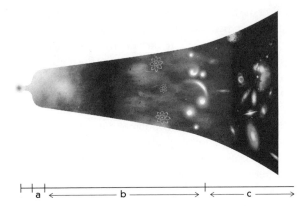

Figure 10.6 A time line of cosmic evolution. **(a)** Inflationary burst. **(b)** Standard Big Bang evolution. **(c)** Era of accelerated expansion.

Puzzles and Progress

With these discoveries, it thus seemed manifest that the pieces of the cosmological puzzle were falling into place. Questions left unanswered by the standard big bang theory—What ignited the outward swelling of space? Why is the temperature of the microwave background radiation so uniform? Why does space seem to have a flat shape?—were addressed by the inflationary theory. Even so, thorny issues regarding fundamental origins have continued to mount: Was there an era before the inflationary burst, and if so, what was it like? What introduced an inflaton field displaced from its lowest-energy configuration to initiate the inflationary expansion? And, the newest question of all, why is the universe apparently composed of such a mishmash of ingredients—5 percent familiar matter, 25 percent dark matter, 70 percent dark energy? Despite the immensely pleasing fact that this cosmic recipe agrees with inflation's prediction that the universe should have 100 percent of the critical density, and although it simultaneously explains the accelerated expansion found by supernova studies, many physicists view the hodgepodge composition as distinctly unattractive. Why, many have asked, has the universe's composition turned out to be so complicated? Why are there a handful of disparate ingredients in such seemingly random abundances? Is there some sensible underlying plan that theoretical studies have yet to reveal?

No one has advanced any convincing answers to these questions; they are among the pressing research problems driving current cosmological research and they serve to remind us of the many tangled knots we must still unravel before we can claim to have fully understood the birth of the universe. But despite the significant challenges that remain, inflation is far and away the front-running cosmological theory. To be sure, physicists' belief in inflation is grounded in the achievements we've so far discussed. But the confidence in inflationary cosmology has roots that run deeper still. As we'll see in the next chapter, a number of other considerations—coming from both observational and theoretical discoveries—have convinced many physicists who work in the field that the inflationary framework is our generation's most important and most lasting contribution to cosmological science.

11

Quanta in the Sky with Diamonds

INFLATION, QUANTUM JITTERS, AND THE ARROW OF TIME

The discovery of the inflationary framework launched a new era in cosmological research, and in the decades since, many thousands of papers have been written on the subject. Scientists have explored just about every nook and cranny of the theory you could possibly imagine. While many of these works have focused on details of technical importance, others have gone further and shown how inflation not only solves specific cosmological problems beyond the reach of the standard big bang, but also provides powerful new approaches to a number of age-old questions. Of these, there are three developments—having to do with the formation of clumpy structures such as galaxies; the amount of energy required to spawn the universe we see; and (of prime importance to our story) the origin of time's arrow—on which inflation has ushered in substantial and, some would say, spectacular progress.

Let's take a look.

Quantum Skywriting

Inflationary cosmology's solution to the horizon and flatness problems was its initial claim to fame, and rightly so. As we've seen, these were major accomplishments. But in the years since, many physicists have come to believe that another of inflation's achievements shares the top spot on the list of the theory's most important contributions.

The lauded insight concerns an issue that, to this point, I have encouraged you not to think about: How is it that there are galaxies, stars, planets, and other clumpy things in the universe? In the last three chapters, I asked you to focus on astronomically large scales—scales on which the universe appears homogeneous, scales so large that entire galaxies can be thought of as single H_2O molecules, while the universe itself is the whole, uniform glass of water. But sooner or later cosmology has to come to grips with the fact that when you examine the cosmos on "finer" scales you discover clumpy structures such as galaxies. And here, once again, we are faced with a puzzle.

If the universe is indeed smooth, uniform, and homogeneous on large scales—features that are supported by observation and that lie at the heart of all cosmological analyses—where could the smaller-scale lumpiness have come from? The staunch believer in standard big bang cosmology can, once again, shrug off this question by appealing to highly favorable and mysteriously tuned conditions in the early universe: "Near the very beginning," such a believer can say, "things were, by and large, smooth and uniform, but not *perfectly* uniform. How conditions got that way, I can't say. That's just how it was back then. Over time, this tiny lumpiness grew, since a lump has greater gravitational pull, being denser than its surroundings, and therefore grabs hold of more nearby material, growing larger still. Ultimately, the lumps got big enough to form stars and galaxies." This would be a convincing story were it not for two deficiencies: the utter lack of an explanation for either the initial overall homogeneity or these important tiny nonuniformities. That's where inflationary cosmology provides gratifying progress. We've already seen that inflation offers an explanation for the large-scale uniformity, and as we'll now learn, the explanatory power of the theory goes even further. According to inflationary cosmology, the initial nonuniformity that ultimately resulted in the formation of stars and galaxies came from *quantum mechanics*.

This magnificent idea arises from an interplay between two seemingly disparate areas of physics: the inflationary expansion of space and the quantum uncertainty principle. The uncertainty principle tells us that there are always trade-offs in how sharply various complementary physical features in the cosmos can be determined. The most familiar example (see Chapter 4) involves matter: the more precisely the position of a particle is determined, the less precisely its velocity can be determined. But the uncertainty principle also applies to fields. By essentially the same reasoning we used in its application to particles, the uncertainty principle implies that the more precisely the value of a field is determined at one location in space, the less precisely its rate of change at that location can be determined. (The position of a particle and the rate of change of its position — its velocity — play analogous roles in quantum mechanics to the value of a field and the rate of change of the field value, at a given location in space.)

I like to summarize the uncertainty principle by saying, roughly speaking, that quantum mechanics makes things jittery and turbulent. If the velocity of a particle can't be delineated with total precision, we also can't delineate where the particle will be located even a fraction of a second later, since velocity *now* determines position *then*. In a sense, the particle is free to take on this or that velocity, or more precisely, to assume a mixture of many different velocities, and hence it will jitter frantically, haphazardly going this way and that. For fields, the situation is similar. If a field's rate of change can't be delineated with total precision, then we also can't delineate what the value of the field will be, at any location, even a moment later. In a sense, the field will undulate up or down at this or that speed, or, more precisely, it will assume a strange mixture of many different rates of change, and hence its value will undergo a frenzied, fuzzy, random jitter.

In daily life we aren't directly aware of the jitters, either for particles or fields, because they take place on subatomic scales. But that's where inflation makes a big impact. The sudden burst of inflationary expansion stretched space by such an enormous factor that what initially inhabited the microscopic was drawn out to the macroscopic. As a key example, pioneers[1] of inflationary cosmology realized that random differences between the quantum jitters in one spatial location and another would have generated slight inhomogeneities in the microscopic realm; because of the indiscriminate quantum agitation, the amount of energy in one location would have been a bit different from what it was in another. Then, through the subsequent inflationary swelling of space, these tiny variations would have been stretched to scales far larger than the quantum domain, yielding a small amount of lumpiness, much as tiny wiggles drawn on a balloon with a Magic Marker are stretched clear across the balloon's surface when you blow it up. This, physicists believe, is the origin of the lumpiness that the staunch believer in the standard big bang model simply declares, without justification, to be "how it was back then." Through the enormous stretching of inevitable quantum fluctuations, inflationary cosmology provides an explanation: inflationary expansion stretches tiny, inhomogeneous quantum jitters and smears them clear across the sky.

Over the few billion years following the end of the brief inflationary phase, these tiny lumps continued to grow through gravitational clumping. Just as in the standard big bang picture, lumps have slightly higher gravitational pull than their surroundings, so they draw in nearby material, growing larger still. In time, the lumps grew large enough to yield the matter making up galaxies and the stars that inhabit them. Certainly, there are *numerous* steps of detail in going from a little lump to a galaxy, and many still need elucidation. But the overall framework is clear: in a quantum world, nothing is ever perfectly uniform because of the jitteriness inherent to the uncertainty principle. And, in a quantum world that experienced inflationary expansion, such nonuniformity can be stretched from the microworld to far larger scales, providing the seeds for the formation of large astrophysical bodies like galaxies.

That's the basic idea, so feel free to skip over the next paragraph. But for those who are interested, I'd like to make the discussion a bit more precise. Recall that inflationary expansion came to an end when the inflaton field's value slid down its potential energy bowl and the field relinquished all its pent-up energy and negative pressure. We described this as happening uniformly throughout space—the inflaton value here, there, and everywhere experienced the same evolution—as that's what naturally emerges from the governing equations. However, this is strictly true only if we ignore the effects of quantum mechanics. On *average*, the inflaton field value did indeed slide down the bowl, as we expect from thinking about a simple classical object like a marble rolling down an incline. But just as a frog sliding down the bowl is likely to jump and jiggle along the way, quantum mechanics tells us that the inflaton field experienced quivers and jitters. On its way down, the value may have suddenly jumped up a little bit over there or jiggled down a little bit over there. And because of this jittering, the inflaton reached the value of lowest energy at different places at slightly different moments. In turn, inflationary expansion shut off at slightly different times at different locations in space, so that the amount of spatial expansion at different locations varied slightly, giving rise to inhomogeneities—wrinkles—similar to the kind you see when the pizza maker stretches the dough a bit more in one place than another and creates a little bump. Now the normal intuition is that jitters arising from quantum mechanics would be too small to be relevant on astrophysical scales. But with inflation, space expanded at such a colossal rate, doubling in size every 10^{-37} seconds, that even a slightly different duration of inflation at nearby locations resulted in a significant wrinkle. In fact, calculations undertaken in specific realizations of inflation have shown that the inhomogeneities produced in this way have a tendency to be too large; researchers often have to adjust details in a given inflationary model (the precise shape of the inflaton field's potential energy bowl) to ensure that the quantum jitters don't predict a universe that's *too* lumpy. And so inflationary cosmology supplies a ready-made mechanism for understanding how the small-scale nonuniformity responsible for lumpy structures like stars and galaxies emerged in a universe that on the largest of scales appears thoroughly homogeneous.

According to inflation, the more than 100 billion galaxies, sparkling throughout space like heavenly diamonds, are nothing but quantum mechanics writ large across the sky. To me, this realization is one of the greatest wonders of the modern scientific age.

The Golden Age of Cosmology

Dramatic evidence supporting these ideas comes from meticulous satellite-based observations of the microwave background radiation's temperature. I have emphasized a number of times that the temperature of the radiation in one part of the sky agrees with that in another to high accuracy. But what I have yet to mention is that by the fourth digit after the decimal place, the temperatures in different locations *do* differ. Precision measurements, first accomplished in 1992 by COBE (the Cosmic Background Explorer satellite) and more recently by WMAP (the Wilkinson Microwave Anisotropy Probe), have determined that while the temperature might be 2.7249 Kelvin in one spot in space, it might be 2.7250 Kelvin in another, and 2.7251 Kelvin in still another.

The wonderful thing is that these extraordinarily small temperature variations follow a pattern on the sky that can be explained by attributing them to the same mechanism that has been suggested for seeding galaxy formation: quantum fluctuations stretched out by inflation. The rough idea is that when tiny quantum jitters are smeared across space, they make it slightly hotter in one region and slightly cooler in another (photons received from a slightly denser region expend more energy overcoming the slightly stronger gravitational field, and hence their energy and temperature are slightly lower than those of photons received from a less dense

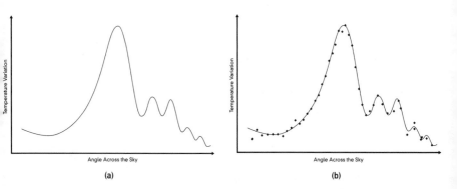

Figure 11.1 (**a**) Inflationary cosmology's prediction for temperature variations of the microwave background radiation from one point to another on the sky. (**b**) Comparison of those predictions with satellite-based observations.

region). Physicists have carried out precise calculations based on this proposal, and generated predictions for how the microwave radiation's temperature should vary from place to place across the sky, as illustrated in Figure 11.1a. (The details are not essential, but the horizontal axis is related to the angular separation of two points on the sky, and the vertical axis is related to their temperature difference.) In Figure 11.1b, these predictions are compared with satellite observations, represented by little diamonds, and as you can see there is *extraordinary agreement.*

I hope you're blown away by this concordance of theory and observation, because if not it means I've failed to convey the full wonder of the result. So, just in case, let me reemphasize what's going on here: satellite-borne telescopes have recently measured the temperature of microwave photons that have been traveling toward us, unimpeded, for nearly 14 billion years. They've found that photons arriving from different directions in space have nearly identical temperatures, differing by no more than a few ten-thousandths of a degree. Moreover, the observations have shown that these tiny temperature differences fill out a particular pattern on the sky, demonstrated by the orderly progression of diamonds in Figure 11.1b. And marvel of marvels, calculations done today, using the inflationary framework, are able to *explain* the pattern of these minuscule temperature variations—variations set down nearly 14 billion years ago—and, to top it off, the key to this explanation involves jitters arising from quantum uncertainty. Wow.

This success has convinced many physicists of the inflationary theory's validity. What is of equal importance, these and other precision astronomical measurements, which have only recently become possible, have allowed cosmology to mature from a field based on speculation and conjecture to one firmly grounded in observation—a coming of age that has inspired many in the field to call our era the golden age of cosmology.

Creating a Universe

With such progress, physicists have been motivated to see how much further inflationary cosmology can go. Can it, for example, resolve the ultimate mystery, encapsulated in Leibniz's question of why there is a universe at all? Well, at least with our current level of understanding, that's asking for too much. Even if a cosmological theory were to make headway on this question, we could ask why that particular theory—its assumptions, ingredients, and equations—was relevant, thus merely pushing the question of origin one step further back. If logic alone somehow required the universe to exist and to be governed by a unique set of laws with unique ingredients, then perhaps we'd have a convincing story. But, to date, that's nothing but a pipe dream.

A related but somewhat less ambitious question, one that has also been asked in various guises through the ages, is: Where did all the mass/energy making up the universe come from? Here, although inflationary cosmology does not provide a complete answer, it has cast the question in an intriguing new light.

To understand how, think of a huge but flexible box filled with many thousands of swarming children, incessantly running and jumping. Imagine that the box is completely impermeable, so no heat or energy can escape, but because it's flexible, its walls can move outward. As the children relentlessly slam into each of the box's walls—hundreds at a time, with hundreds more immediately to follow—the box steadily expands. Now, you might expect that because the walls are impermeable, the total energy embodied by the swarming children will stay fully within the expanding box. After all, where else could their energy go? Well, although a reasonable proposition, it's not quite right. There *is* some place for it to go. The children expend energy every time they slam into a wall, and much of this energy is transferred to the *wall's motion*. The very expansion of the box absorbs, and hence depletes, the children's energy.

Even though space doesn't have walls, a similar kind of energy transfer takes place as the universe expands. Just as the fast-moving children work against the inward force exerted by the box's walls as it expands, the fast-moving particles in our universe work against an inward force as space expands: They work against the inward force of gravity. And just as the total energy embodied by the children drops because it's continuously transferred to the energy of the walls as the box expands, the total energy carried by ordinary particles of matter and radiation drops becasue it is continually *transferred to gravity* as the universe expands. In short, by drawing an analogy between the inward force exerted by the box's walls and the inward force exerted by gravity (an analogy that can be established mathematically), we conclude that gravity depletes the energy in fast-moving particles of matter and radiation as space swells. The loss of energy from fast-moving particles from cosmic expansion has been confirmed by observations of the microwave background radiation.*

Let's now modify our analogy a bit to gain insight into how an inflaton field impacts our description of energy exchange as space expands. Imagine that a few pranksters among the children hook up a number of enormous rubber bands between each of the opposite, outward-moving walls of the box. The rubber bands exert an inward, negative pressure on the box walls, which has exactly the opposite effect of the children's outward, positive pressure; rather than transferring energy to the expansion of the box, the rubber bands' negative pressure "saps" energy from the expansion. As the box expands, the rubber bands get increasingly taut, which means *they* embody increasing amounts of energy.

This modified scenario is relevant to cosmology because, as we've learned, like the pranksters' rubber bands, a uniform inflaton field exerts a negative pressure within an exanding universe. And so, just as the total energy embodied by the rubber bands increases as the box expands because they extract energy from the box's walls, the total energy embod-

---

*As the universe expands, the energy loss of photons can be directly observed because their wavelengths stretch — they undergo *redshift* — and the longer a photon's wavelength, the less energy it has. The microwave background photons have undergone such redshift for nearly 14 billion years, explaining their long — microwave — wavelengths, and their low temperature. Matter undergoes a similar loss of its kinetic energy (energy from particle motion), but the total energy bound up in the mass of particles (their *rest energy* — the energy equivalent of their mass, when at rest) remains constant.

ied by the inflaton field increases as the universe expands because it *extracts energy from gravity.**

To summarize: *as the universe expands, matter and radiation lose energy to gravity while an inflaton field gains energy from gravity.*

The pivotal nature of these observations becomes clear when we try to explain the origin of the matter and radiation that make up galaxies, stars, and everything else inhabiting the cosmos. In the standard big bang theory, the mass/energy carried by matter and radiation has steadily decreased as the universe has expanded, and so the mass/energy in the early universe greatly exceeded what we see today. Thus, instead of offering an explanation for where all the mass/energy currently inhabiting the universe originated, the standard big bang fights an unending uphill battle: the farther back the theory looks, the *more* mass/energy it must somehow explain.

In inflationary cosmology, though, much the opposite is true. Recall that the inflationary theory argues that matter and radiation were produced at the end of the inflationary phase as the inflaton field released its pent-up energy by rolling from perch to valley in its potential-energy bowl. The relevant question, therefore, is whether, just as the inflationary phase was drawing to a close, the theory can account for the inflaton field embodying the *stupendous* quantity of mass/energy necessary to yield the matter and radiation in today's universe.

The answer to this question is that inflation can, without even breaking a sweat. As just explained, the inflaton field is a gravitational parasite—it feeds on gravity—and so the total energy the inflaton field carried increased as space expanded. More precisely, the mathematical analysis shows that the energy *density* of the inflaton field remained constant throughout the inflationary phase of rapid expansion, implying that the total energy it embodied grew in direct proportion to the volume of the space it filled. In the previous chapter, we saw that the size of the universe increased by at least a factor of 10^{30} during inflation, which means the volume of the universe increased by a factor of at least $(10^{30})^3 = 10^{90}$. Conse-

---

*While useful, the rubber-band analogy is not perfect. The inward, negative pressure exerted by the rubber bands impedes the expansion of the box, whereas the inflaton's negative pressure drives the expansion of space. This important difference illustrates the clarification emphasized on page 278: in cosmology, it is not that uniform negative pressure drives expansion (only pressure differences result in forces, so uniform pressure, whether positive or negative, exerts *no* force). Rather, pressure, like mass, gives rise to a gravitational force. And negative pressure gives rise to a repulsive gravitational force that drives expansion. This does not affect our conclusions.

quently, the energy embodied in the inflaton field *increased* by the same huge factor: as the inflationary phase drew to a close, a mere 10^{-35} or so seconds after it began, the energy in the inflaton field grew by a factor on the order of 10^{90}, if not more. *This means that at the onset of inflation, the inflaton field didn't need to have much energy, since the enormous expansion it was about to spawn would enormously amplify the energy it carried.* A simple calculation shows that a tiny nugget, on the order of 10^{-26} centimeters across, filled with a uniform inflaton field—and weighing a mere *twenty pounds*—would, through the ensuing inflationary expansion, acquire enough energy to account for all we see in the universe today.[2]

Thus, in stark contrast to the standard big bang theory in which the total mass/energy of the early universe was huge beyond words, inflationary cosmology, by "mining" gravity, can produce all the ordinary matter and radiation in the universe from a tiny, twenty-pound speck of inflaton-filled space. By no means does this answer Leibniz's question of why there is something rather than nothing, since we've yet to explain why there is an inflaton or even the space it occupies. But the something in need of explanation weighs a whole lot less than my dog Rocky, and that's certainly a very different starting point than envisaged in the standard big bang.*

Inflation, Smoothness, and the Arrow of Time

Perhaps my enthusiasm has already betrayed my bias, but of all the progress that science has achieved in our age, advances in cosmology fill me with the greatest awe and humility. I seem never to have lost the rush I initially felt years ago when I first read up on the basics of general relativity and realized that from our tiny little corner of spacetime we can apply Einstein's theory to learn about the evolution of the entire cosmos. Now, a few decades later, technological progress is subjecting these once abstract proposals for how the universe behaved in its earliest moments to observational tests, and the theories *really work*.

---

*Some researchers, including Alan Guth and Eddie Farhi, have investigated whether one might, hypothetically, create a new universe in the laboratory by synthesizing a nugget of inflaton field. Beyond the fact that we still don't have direct experimental verification that there is such a thing as an inflaton field, note that the twenty pounds of inflaton field would need to be crammed in a tiny space, roughly 10^{-26} or so centimeters on a side, and hence the density would be enormous—some 10^{67} times the density of an atomic nucleus—way beyond what we can produce, now or perhaps ever.

Recall, though, that besides cosmology's overall relevance to the story of space and time, Chapters 6 and 7 launched us into a study of the universe's early history with a specific goal: to find the origin of time's arrow. Remember from those chapters that the only convincing framework we found for explaining time's arrow was that the early universe had extremely high order, that is, extremely low entropy, which set the stage for a future in which entropy got ever larger. Just as the pages of *War and Peace* wouldn't have had the capacity to get increasingly jumbled if they had not been nice and ordered at some point, so too the universe wouldn't have had the capacity to get increasingly disordered—milk spilling, eggs breaking, people aging—unless it had been in a highly ordered configuration early on. The puzzle we encountered is to explain how this high-order, low-entropy starting point came to be.

Inflationary cosmology offers substantial progress, but let me first remind you more precisely of the puzzle, in case any of the relevant details have slipped your mind.

There is strong evidence and little doubt that, early in the history of the universe, matter was spread uniformly throughout space. Ordinarily, this would be characterized as a high-entropy configuration—like the carbon dioxide molecules from a bottle of Coke being spread uniformly throughout a room—and hence would be so commonplace that it would hardly require an explanation. But when gravity matters, as it does when considering the entire universe, a uniform distribution of matter is a rare, low-entropy, highly ordered configuration, because gravity drives matter to form clumps. Similarly, a smooth and uniform spatial curvature also has very low entropy; it is highly ordered compared with a wildly bumpy, nonuniform spatial curvature. (Just as there are many ways for the pages of *War and Peace* to be disordered but only one way for them to be ordered, so there are *many* ways for space to have a disordered, nonuniform shape, but very few ways in which it can be fully ordered, smooth, and uniform.) So we are left to puzzle: Why did the early universe have a low-entropy (highly ordered) uniform distribution of matter instead of a high-entropy (highly disordered) clumpy distribution of matter such as a diverse population of black holes? And why was the curvature of space smooth, ordered, and uniform to extremely high accuracy rather than being riddled with a variety of huge warps and severe curves, also like those generated by black holes?

As first discussed in detail by Paul Davies and Don Page,[3] inflationary cosmology gives important insight into these issues. To see how, bear in mind that an essential assumption of the puzzle is that once a clump forms here or there, its greater gravitational pull attracts yet more material, causing it to grow larger; correspondingly, once a wrinkle in space forms here or there, its greater gravitational pull tends to make the wrinkle yet more severe, leading to a bumpy, highly nonuniform spatial curvature. When gravity matters, ordinary, unremarkable, high-entropy configurations are lumpy and bumpy.

But note the following: this reasoning relies completely on the *attractive* nature of ordinary gravity. Lumps and bumps grow because they *pull* strongly on nearby material, coaxing such material to join the lump. During the brief inflationary phase, though, gravity was *repulsive* and that changed everything. Take the shape of space. The enormous outward push of repulsive gravity drove space to swell so swiftly that initial bumps and warps were stretched smooth, much as fully inflating a shriveled balloon stretches out its creased surface.* What's more, since the volume of space increased by a colossal factor during the brief inflationary period, the density of any clumps of matter was completely diluted, much as the density of fish in your aquarium would be diluted if the tank's volume suddenly increased to that of an Olympic swimming pool. Thus, although attractive gravity causes clumps of matter and creases of space to grow, repulsive gravity does the opposite: it causes them to diminish, leading to an ever smoother, ever more uniform outcome.

Thus, by the end of the inflationary burst, the size of the universe had grown fantastically, any nonuniformity in the curvature of space had been stretched away, and any initial clumps of anything at all had been diluted to the point of irrelevance. Moreover, as the inflaton field slid down to the bottom of its potential-energy bowl, bringing the burst of inflationary expansion to a close, it converted its pent-up energy into a nearly uniform bath of particles of ordinary matter throughout space (uniform up to the tiny but critical inhomogeneities coming from quantum jitters). In total, this all sounds like great progress. The outcome we've reached via inflation—*a smooth, uniform spatial expansion populated by a nearly uniform distribution of matter*—was exactly what we were trying to explain. It's exactly the low-entropy configuration that we need to explain time's arrow.

*Don't get confused here: The inflationary stretching of quantum jitters discussed in the last section still produced a minuscule, unavoidable nonuniformity of about 1 part in 100,000. But that tiny nonuniformity overlaid an otherwise smooth universe. We are now describing how the latter—the underlying smooth uniformity—came to be.

Entropy and Inflation

Indeed, this is significant progress. But two important issues remain.

First, we seem to be concluding that the inflationary burst smooths things out and hence lowers total entropy, embodying a physical mechanism—not just a statistical fluke—that appears to violate the second law of thermodynamics. Were that the case, either our understanding of the second law or our current reasoning would have to be in error. In actuality, though, we don't have to face either of these options, because total entropy does not go down as a result of inflation. What really happens during the inflationary burst is that the total entropy goes up, but it goes up *much less than it might have*. You see, by the end of the inflationary phase, space was stretched smooth and so the gravitational contribution to entropy—the entropy associated with the possible bumpy, nonordered, nonuniform shape of space—was minimal. However, when the inflaton field slid down its bowl and relinquished its pent-up energy, it is estimated to have produced about 10^{80} particles of matter and radiation. Such a huge number of particles, like a book with a huge number of pages, embodies a huge amount of entropy. Thus, even though the gravitational entropy went down, the increase in entropy from the production of all these particles more than compensated. The total entropy increased, just as we expect from the second law.

But, and this is the important point, the inflationary burst, by smoothing out space and ensuring a homogeneous, uniform, low-entropy gravitational field, created a huge *gap* between what the entropy contribution from gravity was and what it might have been. Overall entropy increased during inflation, but by a paltry amount compared with how much it *could* have increased. It's in this sense that inflation generated a low-entropy universe: by the end of inflation, entropy had increased, but by nowhere near the factor by which the spatial expanse had increased. If entropy is likened to property taxes, it would be as if New York City acquired the Sahara Desert. The total property taxes collected would go up, but by a tiny amount compared with the total increase in acreage.

Ever since the end of inflation, gravity has been trying to make up the entropy difference. Every clump—be it a galaxy, or a star in a galaxy, or a planet, or a black hole—that gravity has subsequently coaxed out of the uniformity (seeded by the tiny nonuniformity from quantum jitters) has increased entropy and has brought gravity one step closer to realizing its entropy potential. In this sense, then, inflation is a mechanism that yielded a large universe with relatively low gravitational entropy, and in that way set the stage for the subsequent billions of years of gravitational clumping whose effects we now witness. And so inflationary cosmology gives a direction to time's arrow by generating a past with exceedingly low gravitational entropy; the future is the direction in which this entropy grows.[4]

The second issue becomes apparent when we continue down the path to which time's arrow led us in Chapter 6. From an egg, to the chicken that laid it, to the chicken's feed, to the plant kingdom, to the sun's heat and light, to the big bang's uniformly distributed primordial gas, we followed the universe's evolution into a past that had ever greater order, at each stage pushing the puzzle of low entropy one step further back in time. We have just now realized that an even earlier stage of inflationary expansion can naturally explain the smooth and uniform aftermath of the bang. But what about inflation itself? Can we explain the initial link in this chain we've followed? Can we explain why conditions were right for an inflationary burst to happen at all?

This is an issue of paramount importance. No matter how many puzzles inflationary cosmology resolves in theory, if an era of inflationary expansion never took place, the approach will be rendered irrelevant. Moreover, since we can't go back to the early universe and determine directly whether inflation occurred, assessing whether we've made real progress in setting a direction to time's arrow requires that we determine the *likelihood* that the conditions necessary for an inflationary burst were achieved. That is, physicists bristle at the standard big bang's reliance on finely tuned homogeneous initial conditions that, while observationally motivated, are theoretically unexplained. It feels deeply unsatisfying for the low-entropy state of the early universe simply to be assumed; it feels hollow for time's arrow to be imposed on the universe, without explanation. At first blush, inflation offers progress by showing that what's assumed in the standard big bang emerges from inflationary evolution. But if the initiation of inflation requires yet other, highly special, exceedingly low-entropy conditions, we will pretty much find ourselves back at square one. We will merely have traded the big bang's special conditions for those necessary to ignite inflation, and the puzzle of time's arrow will remain just as puzzling.

What are the conditions necessary for inflation? We've seen that inflation is the inevitable result of the inflaton field's value getting stuck, for just a moment and within just a tiny region, on the high-energy plateau in its potential energy bowl. Our charge, therefore, is to determine how likely this starting configuration for inflation actually is. If initiating inflation proves easy, we'll be in great shape. But if the necessary conditions are extraordinarily unlikely to be attained, we will merely have shifted the question of time's arrow one step further back—to finding the explanation for the low-entropy inflaton field configuration that got the ball rolling.

I'll first describe current thinking on this issue in the most optimistic light, and then return to essential elements of the story that remain cloudy.

Boltzmann Redux

As mentioned in the previous chapter, the inflationary burst is best thought of as an event occurring in a preexisting universe, rather than being thought of as the creation of the universe itself. Although we don't have an unassailable understanding of what the universe was like during such a preinflationary era, let's see how far we can get if we assume that things were in a thoroughly ordinary, high-entropy state. Specifically, let's imagine that primordial, preinflationary space was riddled with warps and bumps, and that the inflaton field was also highly disordered, its value jumping to and fro like the frog in the hot metal bowl.

Now, just as you can expect that if you patiently play a fair slot machine, sooner or later the randomly spinning dials will land on triple diamonds, we expect that sooner or later a chance fluctuation within this highly energetic, turbulent arena of the primordial universe will cause the inflaton field's value to jump to the correct, uniform value in some small nugget of space, initiating an outward burst of inflationary expansion. As explained in the previous section, calculations show that the nugget of space need only have been tiny—on the order of 10^{-26} centimeters across—for the ensuing cosmological expansion (inflationary expansion followed by standard big bang expansion) to have stretched it larger than the universe we see today. Thus, rather than assuming or simply declaring that conditions in the early universe were right for inflationary expansion to take place, in this way of thinking about things an ultramicroscopic fluctuation weighing a mere twenty pounds, occurring within an ordinary, unremarkable environment of disorder, gave rise to the necessary conditions.

What's more, just as the slot machine will also generate a wide variety of nonwinning results, in other regions of primordial space other kinds of inflaton fluctuations would also have happened. In most, the fluctuation wouldn't have had the right value or have been sufficiently uniform for inflationary expansion to occur. (Even in a region that's a mere 10^{-26} centimeters across, a field's value can vary wildly.) But all that matters to us is that there was one nugget that yielded the space-smoothing inflationary burst that provided the first link in the low-entropy chain, ultimately leading to our familiar cosmos. As we see only our one big universe, we only need the cosmic slot machine to pay out once.[5]

Since we are tracing the universe back to a statistical fluctuation from primordial chaos, this explanation for time's arrow shares certain features with Boltzmann's original proposal. Remember from Chapter 6 that Boltzmann suggested that everything we now see arose as a rare but every so often expectable fluctuation from total disorder. The problem with Boltzmann's original formulation, though, was that it could not explain why the chance fluctuation had gone so far overboard and produced a universe hugely more ordered than it would need to be even to support life as we know it. Why is the universe so vast, having billions and billions of galaxies, each with billions and billions of stars, when it could have drastically cut corners by having, say, just a few galaxies, or even only one?

From the statistical point of view, a more modest fluctuation that produced some order but not as much as we currently see would be *far* more likely. Moreover, since on average entropy is on the rise, Boltzmann's reasoning suggests that it would be much more likely that everything we see today *just now* arose as a rare statistical jump to lower entropy. Recall the reason: the farther back the fluctuation happened, the lower the entropy it would have had to attain (entropy starts to rise after any dip to low entropy, as in Figure 6.4, so if the fluctuation happened yesterday, it must have dipped down to yesterday's lower entropy, and if it happened a billion years ago, it must have dipped down to that era's even lower entropy). Hence, the farther back in time, the more drastic and improbable the required fluctuation. Thus, it is much more likely that the jump just happened. But if we accept this conclusion, we can't trust memories, records, or even the laws of physics that underlie the discussion itself—a completely intolerable position.

The tremendous advantage of the inflationary incarnation of Boltzmann's idea is that a *small* fluctuation early on—a *modest* jump to the favorable conditions, within a *tiny* nugget of space—inevitably yields the huge and ordered universe we are aware of. Once inflationary expansion set in, the little nugget was *inexorably* stretched to scales at least as large as the universe we currently see, and hence there is no mystery as to why the universe didn't cut corners; there is no mystery why the universe is vast and is populated by a huge number of galaxies. From the get-go, inflation gave the universe an amazing deal. A jump to lower entropy within a tiny nugget of space was leveraged by inflationary expansion into the vast reaches of the cosmos. And, of utmost importance, the inflationary stretching didn't just yield any old large universe. It yielded *our* large universe—inflation explains the shape of space, it explains the large-scale uniformity, and it even explains the "smaller"-scale inhomogeneities such as galaxies and temperature variations in the background radiation. Inflation packages a wealth of explanatory and predictive power in a single fluctuation to low entropy.

And so Boltzmann may well have been right. Everything we see may have resulted from a chance fluctuation out of a highly disordered state of primeval chaos. In this realization of his ideas, though, we can trust our records and we can trust our memories: the fluctuation did not happen just now. The past really happened. Our records are records of things that took place. Inflationary expansion amplifies a tiny speck of order in the early universe—it "wound up" the universe to a huge expanse with minimal gravitational entropy—so the 14 billion years of subsequent unwinding, of subsequent clumping into galaxies, stars, and planets, presents no puzzle.

In fact, this approach even tells us a bit more. Just as it's possible to hit the jackpot on a number of slot machines on the floor of the Bellagio, in the primordial state of high entropy and overall chaos there was no reason why the conditions necessary for inflationary expansion would arise only in a single spatial nugget. Instead, as Andrei Linde has proposed, there could have been many nuggets scattered here and there that underwent space-smoothing inflationary expansion. If that were so, our universe would be but one among many that sprouted—and perhaps continue to sprout—when chance fluctuations made the conditions right for an inflationary burst, as illustrated in Figure 11.2. As these other universes would likely be forever separate from ours, it's hard to imagine how we would ever establish whether this "multiverse" picture is true. However, as a conceptual framework, it's both rich and tantalizing. Among other things, it suggests a possible shift in how we think about cosmology: In Chapter 10, I described inflation as a "front end" to the standard big bang theory, in which the bang is identified with a fleeting burst of rapid expansion. But

Figure 11.2 Inflation can occur repeatedly, sprouting new universes from older ones.

if we think of the inflationary sprouting of each new universe in Figure 11.2 as its own bang, then inflation itself is best viewed as the overarching cosmological framework within which big bang–like evolutions happen, bubble by bubble. Thus, rather than inflation's being incorporated into the standard big bang theory, in this approach the standard big bang would be incorporated into inflation.

Inflation and the Egg

So why do you see an egg splatter but not unsplatter? Where does the arrow of time that we all experience come from? Here is where this approach has taken us. Through a chance but every so often expectable fluctuation from an unremarkable primordial state with high entropy, a tiny, twenty-pound nugget of space achieved conditions that led to a brief burst of inflationary expansion. The tremendous outward swelling resulted in space's being stretched enormously large and extremely smooth, and, as the burst drew to a close, the inflaton field relinquished its hugely amplified energy by filling space nearly uniformly with matter and radiation. As the inflaton's repulsive gravity diminished, ordinary attractive gravity became dominant. And, as we've seen, attractive gravity exploits tiny inhomogeneities caused by quantum jitters to cause matter to clump, forming galaxies and stars and ultimately leading to the formation of the sun, the earth, the rest of the solar system, and the other features of our observed universe. (As discussed, some 7 billion or so years

ATB, repulsive gravity once again became dominant, but this is only relevant on the largest of cosmic scales and has no direct impact on smaller entities like individual galaxies or our solar system, where ordinary attractive gravity still reigns.) The sun's relatively low-entropy energy was used by low-entropy plant and animal life forms on earth to produce yet more low-entropy life forms, slowly raising the total entropy through heat and waste. Ultimately, this chain produced a chicken that produced an egg—and you know the rest of the story: the egg rolled off your kitchen counter and splattered on the floor as part of the universe's relentless drive to higher entropy. It's the low-entropy, highly ordered, uniformly smooth nature of the spatial fabric produced by inflationary stretching that is the analog of having the pages of *War and Peace* all in their proper numerical arrangement; it is this early state of order—the absence of severe bumps or warps or gargantuan black holes—that primed the universe for the subsequent evolution to higher entropy and hence provided the arrow of time we all experience. With our current level of understanding, this is the most complete explanation for time's arrow that has been given.

The Fly in the Ointment?

To me, this story of inflationary cosmology and time's arrow is lovely. From a wild and energetic realm of primordial chaos, there emerged an ultramicroscopic fluctuation of uniform inflaton field weighing far less than the limit for carry-on luggage. This initiated inflationary expansion, which set a direction to time's arrow, and the rest is history.

But in telling this story, we've made a pivotal assumption that's as yet unjustified. To assess the likelihood of inflation's being initiated, we've had to specify the characteristics of the preinflationary realm out of which inflationary expansion is supposed to have emerged. The particular realm we've envisioned—wild, chaotic, energetic—sounds reasonable, but delineating this intuitive description with mathematical precision proves challenging. Moreover, it is only a guess. The bottom line is that we don't know what conditions were like in the supposed preinflationary realm, in the fuzzy patch of Figure 10.3, and without that information we are unable to make a convincing assessment of the likelihood of inflation's initiating; any calculation of the likelihood depends sensitively on the assumptions we make.[6]

With this hole in our understanding, the most sensible summary is that inflation offers a powerful explanatory framework that bundles together seemingly disparate problems—the horizon problem, the flatness problem, the origin-of-structure problem, the low-entropy-of-the-early-universe problem—and offers a single solution that addresses them all. This feels right. But to go to the next step, we need a theory that can cope with the extreme conditions characteristic of the fuzzy patch—extremes of heat and colossal density—so that we will stand a chance of gaining sharp, unambiguous insight into the earliest moments of the cosmos.

As we will learn in the next chapter, this requires a theory that can overcome perhaps the greatest obstacle theoretical physics has faced during the last eighty years: a fundamental rift between general relativity and quantum mechanics. Many researchers believe that a relatively new approach called *superstring theory* may have accomplished this, but if superstring theory is right, the fabric of the cosmos is far stranger than almost anyone ever imagined.

IV

ORIGINS AND UNIFICATION

12

The World on a String

THE FABRIC ACCORDING TO STRING THEORY

Imagine a universe in which to understand anything you'd need to understand everything. A universe in which to say anything about why a planet orbits a star, about why a baseball flies along a particular trajectory, about how a magnet or a battery works, about how light and gravity operate—a universe in which to say anything about anything—you would need to uncover the most fundamental laws and determine how they act on the finest constituents of matter. Thankfully, this universe is not our universe.

If it were, it's hard to see how science would have made any progress at all. Over the centuries, the reason we've been able to make headway is that we've been able to work piecemeal; we've been able to unravel mysteries step by step, with each new discovery going a bit deeper than the previous. Newton didn't need to know about atoms to make great strides in understanding motion and gravity. Maxwell didn't need to know about electrons and other charged particles to develop a powerful theory of electromagnetism. Einstein didn't need to address the primordial incarnation of space and time to formulate a theory of how they curve in the service of the gravitational force. Instead, each of these discoveries, as well as the many others that underlie our current conception of the cosmos, proceeded within a limited context that unabashedly left many basic questions unanswered. Each discovery was able to contribute its own piece to the puzzle, even though no one knew—and we still don't know—what grand synthesizing picture comprises all the puzzle's pieces.

A closely related observation is that although today's science differs sharply from that of even fifty years ago, it would be simplistic to summarize scientific progress in terms of new theories overthrowing their predecessors. A more correct description is that each new theory refines its predecessor by providing a more accurate and more wide-reaching framework. Newton's theory of gravity has been superseded by Einstein's general relativity, but it would be naïve to say that Newton's theory was wrong. In the domain of objects that don't move anywhere near as fast as light and don't produce gravitational fields anywhere near as strong as those of black holes, Newton's theory is fantastically accurate. Yet this is not to say that Einstein's theory is a minor variant on Newton's; in the course of improving Newton's approach to gravity, Einstein invoked a whole new conceptual schema, one that radically altered our understanding of space and time. But the power of Newton's discovery within the domain he intended it for (planetary motion, commonplace terrestrial motion, and so on) is unassailable.

We envision each new theory taking us closer to the elusive goal of truth, but whether there is an ultimate theory—a theory that cannot be refined further, because it has finally revealed the workings of the universe at the deepest possible level—is a question no one can answer. Even so, the pattern traced out during the last three hundred years of discovery gives tantalizing evidence that such a theory can be developed. Broadly speaking, each new breakthrough has gathered a wider range of physical phenomena under fewer theoretical umbrellas. Newton's discoveries showed that the forces governing planetary motion are the same as those governing the motion of falling objects here on earth. Maxwell's discoveries showed that electricity and magnetism are two sides of the same coin. Einstein's discoveries showed that space and time are as inseparable as Midas' touch and gold. The discoveries of a generation of physicists in the early twentieth century established that myriad mysteries of microphysics could be explained with precision using quantum mechanics. More recently, the discoveries of Glashow, Salam, and Weinberg showed that electromagnetism and the weak nuclear force are two manifestations of a single force—the electroweak force—and there is even tentative, circumstantial evidence that the strong nuclear force may join the electroweak force in a yet grander synthesis.[1] Taking all this together, we see a pattern that goes from complexity to simplicity, a pattern that goes from diversity to unity. The explanatory arrows seem to be converging on a powerful, yet-to-be discovered framework that would unify all of nature's forces and all of matter within a single theory capable of describing all physical phenomena.

Albert Einstein, who for more than three decades sought to combine electromagnetism and general relativity in a single theory, is rightly credited with initiating the modern search for a unified theory. For long stretches during those decades, he was the sole searcher for such a unified theory, and his passionate yet solitary quest alienated him from the mainstream physics community. During the last twenty years, though, there has been a resurgence in the quest for a unified theory; Einstein's lonely dream has become the driving force for a whole generation of physicists. But with the discoveries since Einstein's time has come a shift in focus. Even though we don't yet have a successful theory combining the strong nuclear force and the electroweak force, all three of these forces (electromagnetic, weak, strong) have been described by a single uniform language based on quantum mechanics. But general relativity, our most refined theory of the fourth force, stands outside this framework. General relativity is a classical theory: it does not incorporate any of the probabilistic concepts of quantum theory. A primary goal of the modern unification program is therefore to combine general relativity and quantum mechanics, and to describe all four forces within the same quantum mechanical framework. This has proven to be one of the most difficult problems theoretical physics has ever encountered.

Let's see why.

Quantum Jitters and Empty Space

If I had to select the single most evocative feature of quantum mechanics, I'd choose the uncertainty principle. Probabilities and wavefunctions certainly provide a radically new framework, but it's the uncertainty principle that encapsulates the break from classical physics. Remember, in the seventeenth and eighteenth centuries, scientists believed that a complete description of physical reality amounted to specifying the positions and velocities of every constituent of matter making up the cosmos. And with the advent of the field concept in the nineteenth century, and its subsequent application to the electromagnetic and gravitational forces, this view was augmented to include the value of each field—the strength of each field, that is—and the rate of change of each field's value, at every location in space. But by the 1930s, the uncertainty principle dismantled this conception of reality by showing that you can't ever know both the position and the velocity of a particle; you can't ever know both the value of a field at some location in space and how quickly the field value is changing. Quantum uncertainty forbids it.

As we discussed in the last chapter, this quantum uncertainty ensures that the microworld is a turbulent and jittery realm. Earlier, we focused on uncertainty-induced quantum jitters for the inflaton field, but quantum uncertainty applies to all fields. The electromagnetic field, the strong and weak nuclear force fields, and the gravitational field are all subject to frenzied quantum jitters on microscopic scales. In fact, these field jitters exist even in space you'd normally think of as empty, space that would seem to contain no matter and no fields. This is an idea of critical importance, but if you haven't encountered it previously, it's natural to be puzzled. If a region of space contains nothing—if it's a vacuum—doesn't that mean there's nothing to jitter? Well, we've already learned that the concept of nothingness is subtle. Just think of the Higgs ocean that modern theory claims to permeate empty space. The quantum jitters I'm now referring to serve only to make the notion of "nothing" subtler still. Here's what I mean.

In prequantum (and pre-Higgs) physics, we'd declare a region of space completely empty if it contained no particles and the value of every field was uniformly zero.* Let's now think about this classical notion of emptiness in light of the quantum uncertainty principle. If a field were to have and maintain a vanishing value, we would know its value—zero—and also the rate of change of its value—zero, too. But according to the uncertainty principle, it's impossible for both these properties to be definite. Instead, if a field has a definite value at some moment, zero in the case at hand, the uncertainty principle tells us that its rate of change is completely random. And a random rate of change means that in subsequent moments the field's value will randomly jitter up and down, even in what we normally think of as completely empty space. So the intuitive notion of emptiness, one in which all fields have and maintain the value

---

*For ease of writing, we'll consider only fields that reach their lowest energy when their values are zero. The discussion for other fields—Higgs fields—is identical, except the jitters fluctuate about the field's *nonzero*, lowest-energy value. If you are tempted to say that a region of space is empty only if there is no matter present and all fields are *absent*, not just that they have the value zero, see notes section.[2]

zero, is incompatible with quantum mechanics. *A field's value can jitter around the value zero but it can't be uniformly zero throughout a region for more than a brief moment.*[3] In technical language, physicists say that fields undergo *vacuum fluctuations.*

The random nature of vacuum field fluctuations ensures that in all but the most microscopic of regions, there are as many "up" jitters as "down" and hence they average out to zero, much as a marble surface appears perfectly smooth to the naked eye even though an electron microscope reveals that it's jagged on minuscule scales. Nevertheless, even though we can't see them directly, more than half a century ago the reality of quantum field jitters, even in empty space, was conclusively established through a simple yet profound discovery.

In 1948, the Dutch physicist Hendrik Casimir figured out how vacuum fluctuations of the electromagnetic field could be experimentally detected. Quantum theory says that the jitters of the electromagnetic field in empty space will take on a variety of shapes, as illustrated in Figure 12.1a. Casimir's breakthrough was to realize that by placing two ordinary metal plates in an otherwise empty region, as in Figure 12.1b, he could induce a subtle modification to these vacuum field jitters. Namely, the quantum equations show that in the region between the plates there will be fewer fluctuations (only those electromagnetic field fluctuations whose values vanish at the location of each plate are allowed). Casimir analyzed the implications of this reduction in field jitters and found

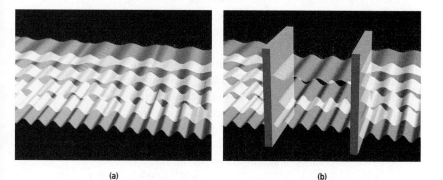

(a)   (b)

Figure 12.1 (a) Vacuum fluctuations of the electromagnetic field. (b) Vacuum fluctuations between two metal plates and those outside the plates.

something extraordinary. Much as a reduction in the amount of air in a region creates a pressure imbalance (for example, at high altitude you can feel the thinner air exerting less pressure on the outside of your eardrums), the reduction in quantum field jitters between the plates also yields a pressure imbalance: the quantum field jitters between the plates become a bit weaker than those outside the plates, and this imbalance *drives the plates toward each other*.

Think about how thoroughly odd this is. You place two plain, ordinary, uncharged metal plates into an *empty* region of space, facing one another. As their masses are tiny, the gravitational attraction between them is so small that it can be completely ignored. Since there is nothing else around, you naturally conclude that the plates will stay put. But this is *not* what Casimir's calculations predicted would happen. He concluded that the plates would be gently guided by the ghostly grip of quantum vacuum fluctuations to move toward one another.

When Casimir first announced these theoretical results, equipment sensitive enough to test his predictions didn't exist. Yet, within about a decade, another Dutch physicist, Marcus Spaarnay, was able to initiate the first rudimentary tests of this *Casimir force*, and increasingly precise experiments have been carried out ever since. In 1997, for example, Steve Lamoreaux, then at the University of Washington, confirmed Casimir's predictions to an accuracy of 5 percent.[4] (For plates roughly the size of playing cards and placed one ten-thousandth of a centimeter apart, the force between them is about equal to the weight of a single teardrop; this shows how challenging it is to measure the Casimir force.) There is now little doubt that the intuitive notion of empty space as a static, calm, eventless arena is thoroughly off base. Because of quantum uncertainty, empty space is teeming with quantum activity.

It took scientists the better part of the twentieth century to fully develop the mathematics for describing such quantum activity of the electromagnetic, and strong and weak nuclear forces. The effort was well spent: calculations using this mathematical framework agree with experimental findings to an unparalleled precision (e.g., calculations of the effect of vacuum fluctuations on the magnetic properties of electrons agree with experimental results to one part in a billion).[5]

Yet despite all this success, for many decades physicists have been aware that quantum jitters have been fomenting discontent within the laws of physics.

Jitters and Their Discontent[6]

So far, we've discussed only quantum jitters for fields that exist *within* space. What about the quantum jitters of space itself? While this might sound mysterious, it's actually just another example of quantum field jitters—an example, however, that proves particularly troublesome. In the general theory of relativity, Einstein established that the gravitational force can be described by warps and curves in the fabric of space; he showed that gravitational fields manifest themselves through the shape or geometry of space (and of spacetime, more generally). Now, just like any other field, the gravitational field is subject to quantum jitters: the uncertainty principle ensures that over tiny distance scales, the gravitational field fluctuates up and down. And since the gravitational field is synonymous with the shape of space, such quantum jitters mean that the shape of space fluctuates randomly. Again, as with all examples of quantum uncertainty, on everyday distance scales the jitters are too small to be sensed directly, and the surrounding environment appears smooth, placid, and predictable. But the smaller the scale of observation the larger the uncertainty, and the more tumultuous the quantum fluctuations become.

This is illustrated in Figure 12.2, in which we sequentially magnify the fabric of space to reveal its structure at ever smaller distances. The lowermost level of the figure shows the quantum undulations of space on familiar scales and, as you can see, there's nothing to see—the undulations are unobservably small, so space appears calm and flat. But as we home in by sequentially magnifying the region, we see that the undulations of space get increasingly frenetic. By the highest level in the figure, which shows the fabric of space on scales smaller than the *Planck length*—a millionth of a billionth of a billionth of a billionth (10^{-33}) of a centimeter—space becomes a seething, boiling cauldron of frenzied fluctuations. As the illustration makes clear, the usual notions of left/right, back/forth, and up/down become so jumbled by the ultramicroscopic tumult that they lose all meaning. Even the usual notion of before/after, which we've been illustrating by sequential slices in the spacetime loaf, is rendered meaningless by quantum fluctuations on time scales shorter than the *Planck time*, about a tenth of a millionth of a trillionth of a trillionth of a trillionth (10^{-43}) of a second (which is roughly the time it takes light to travel a Planck length). Like a blurry photograph, the wild undu-

Figure 12.2 Successive magnifications of space reveal that below the Planck length, space becomes unrecognizably tumultuous due to quantum jitters. (These are imaginary magnifying glasses, each of which magnifies between 10 million and 100 million times.)

lations in Figure 12.2 make it impossible to distinguish one time slice from another unambiguously when the time interval between them becomes shorter than the Planck time. The upshot is that on scales shorter than Planck distances and durations, quantum uncertainty renders the fabric of the cosmos so twisted and distorted that the usual conceptions of space and time are no longer applicable.

While exotic in detail, the broad-brush lesson illustrated by Figure 12.2 is one with which we are already familiar: concepts and conclusions relevant on one scale may not be applicable on all scales. This is a key principle in physics, and one that we encounter repeatedly even in far more prosaic contexts. Take a glass of water. Describing the water as a smooth, uniform liquid is both useful and relevant on everyday scales, but it's an approximation that breaks down if we analyze the water with submicroscopic precision. On tiny scales, the smooth image gives way to a completely different framework of widely separated molecules and atoms. Similarly, Figure 12.2 shows that Einstein's conception of a smooth, gently curving, geometrical space and time, although powerful and accurate for describing the universe on large scales, breaks down if we analyze the universe at extremely short distance and time scales. Physicists believe that, as with water, the smooth portrayal of space and time is an approximation that gives way to another, more fundamental framework when considered on ultramicroscopic scales. What that framework is—what constitutes the "molecules" and "atoms" of space and time—is a question currently being pursued with great vigor. It has yet to be resolved.

Even so, what is thoroughly clear from Figure 12.2 is that on tiny scales the smooth character of space and time envisioned by general relativity locks horns with the frantic, jittery character of quantum mechanics. The core principle of Einstein's general relativity, that space and time form a gently curving geometrical shape, runs smack into the core principle of quantum mechanics, the uncertainty principle, which implies a wild, tumultuous, turbulent environment on the tiniest of scales. The violent clash between the central ideas of general relativity and quantum mechanics has made meshing the two theories one of the most difficult challenges physicists have encountered during the last eighty years.

Does It Matter?

In practice, the incompatibility between general relativity and quantum mechanics rears its head in a very specific way. If you use the combined equations of general relativity and quantum mechanics, they almost always yield one answer: infinity. And that's a problem. It's nonsense. Experimenters never measure an infinite amount of anything. Dials never spin around to infinity. Meters never reach infinity. Calculators never register infinity. Almost always, an infinite answer is meaningless. All it tells us is that the equations of general relativity and quantum mechanics, when merged, go haywire.

Notice that this is quite unlike the tension between *special* relativity and quantum mechanics that came up in our discussion of quantum nonlocality in Chapter 4. There we found that reconciling the tenets of special relativity (in particular, the symmetry among all constant velocity observers) with the behavior of entangled particles requires a more complete understanding of the quantum measurement problem than has so far been attained (see pages 117–120). But this incompletely resolved issue does not result in mathematical inconsistencies or in equations that yield nonsensical answers. To the contrary, the combined equations of special relativity and quantum mechanics have been used to make the most precisely confirmed predictions in the history of science. The quiet tension between special relativity and quantum mechanics points to an area in need of further theoretical development, but it has hardly any impact on their combined predictive power. Not so with the explosive union between general relativity and quantum mechanics, in which all predictive power is lost.

Nevertheless, you can still ask whether the incompatibility between general relativity and quantum mechanics really matters. Sure, the combined equations may result in nonsense, but when do you ever really need to use them together? Years of astronomical observations have shown that general relativity describes the macro world of stars, galaxies, and even the entire expanse of the cosmos with impressive accuracy; decades of experiments have confirmed that quantum mechanics does the same for the micro world of molecules, atoms, and subatomic particles. Since each theory works wonders in its own domain, why worry about combining them? Why not keep them separate? Why not use general relativity for things that are large and massive, quantum mechanics for things that are tiny and light, and celebrate humankind's impressive achievement of successfully understanding such a wide range of physical phenomena?

As a matter of fact, this *is* what most physicists have done since the early decades of the twentieth century, and there's no denying that it's been a distinctly fruitful approach. The progress science has made by working in this disjointed framework is impressive. All the same, there are a number of reasons why the antagonism between general relativity and quantum mechanics must be reconciled. Here are two.

First, at a gut level, it is hard to believe that the deepest understanding of the universe consists of an uneasy union between two powerful theoretical frameworks that are mutually incompatible. It's not as though the universe comes equipped with a line in the sand separating things that are properly described by quantum mechanics from things properly described by general relativity. Dividing the universe into two separate realms seems both artificial and clumsy. To many, this is evidence that there must be a deeper, unified truth that overcomes the rift between general relativity and quantum mechanics and that can be applied to *everything*. We have one universe and therefore, many strongly believe, we should have one theory.

Second, although most things are either big and heavy or small and light, and therefore, as a practical matter, can be described using general relativity *or* quantum mechanics, this is not true of all things. Black holes provide a good example. According to general relativity, all the matter that makes up a black hole is crushed together at a single minuscule point at the black hole's center.[7] This makes the center of a black hole both enormously massive and incredibly tiny, and hence it falls on both sides of the purported divide: we need to use general relativity because the large mass creates a substantial gravitational field, and we also need to use quantum mechanics because all the mass is squeezed to a tiny size. But in combination, the equations break down, so no one has been able to determine what happens right at the center of a black hole.

That's a good example, but if you're a real skeptic, you might still wonder whether this is something that should keep anyone up at night. Since we can't see inside a black hole unless we jump in, and, moreover, were we to jump in we wouldn't be able to report our observations back to the outside world, our incomplete understanding of the black hole's interior may not strike you as particularly worrisome. For physicists, though, the existence of a realm in which the known laws of physics break down — no matter how esoteric the realm might seem — throws up red flags. If the known laws of physics break down under any circumstances, it is a clear signal that we have not reached the deepest possible understanding. After all, the universe works; as far as we can tell, the universe does not break down. The correct theory of the universe should, at the very least, meet the same standard.

Well, that surely seems reasonable. But for my money, the full urgency of the conflict between general relativity and quantum mechanics is revealed only through another example. Look back at Figure 10.6. As you can see, we have made great strides in piecing together a consistent and predictive story of cosmic evolution, but the picture remains incomplete because of the fuzzy patch near the inception of the universe. And within the foggy haze of those earliest moments lies insight into the most tantalizing of mysteries: the origin and fundamental nature of space and time. So what has prevented us from penetrating the haze? The blame rests squarely on the conflict between general relativity and quantum mechanics. The antagonism between the laws of the large and those of the small is the reason the fuzzy patch remains obscure and we still have no insight into what happened at the very beginning of the universe.

To understand why, imagine, as in Chapter 10, running a film of the expanding cosmos in reverse, heading back toward the big bang. In reverse, everything that is now rushing apart comes together, and so as we run the film farther back, the universe gets ever smaller, hotter, and denser. As we close in on time zero itself, the *entire* observable universe is compressed to the size of the sun, then further squeezed to the size of the earth, then crushed to the size of a bowling ball, a pea, a grain of sand— smaller and smaller the universe shrinks as the film rewinds toward its initial frames. There comes a moment in this reverse-run film when the entire known universe has a size close to the Planck length—the millionth of a billionth of a billionth of a billionth of a centimeter at which general relativity and quantum mechanics find themselves at loggerheads. At this moment, all the mass and energy responsible for spawning the observable universe is contained in a speck that's less than a hundredth of a billionth of a billionth of the size of a single atom.[8]

Thus, just as in the case of a black hole's center, the early universe falls on both sides of the divide: The enormous density of the early universe requires the use of general relativity. The tiny size of the early universe requires the use of quantum mechanics. But once again, in combination the laws break down. The projector jams, the cosmic film burns up, and we are unable to access the universe's earliest moments. Because of the conflict between general relativity and quantum mechanics, we remain ignorant about what happened at the beginning and are reduced to drawing a fuzzy patch in Figure 10.6.

If we ever hope to understand the origin of the universe—one of the deepest questions in all of science—the conflict between general relativity and quantum mechanics *must* be resolved. We must settle the differences between the laws of the large and the laws of the small and merge them into a single harmonious theory.

The Unlikely Road to a Solution*

As the work of Newton and Einstein exemplifies, scientific breakthroughs are sometimes born of a single scientist's staggering genius, pure and simple. But that's rare. Much more frequently, great breakthroughs represent the collective effort of many scientists, each building on the insights of

---

*The remainder of this chapter recounts the discovery of superstring theory and discusses the theory's essential ideas regarding unification and the structure of spacetime. Readers of *The Elegant Universe* (especially Chapters 6 through 8) will be familiar with much of this material, and should feel free to skim this chapter and move on to the next.

others to accomplish what no individual could have achieved in isolation. One scientist might contribute an idea that sets a colleague thinking, which leads to an observation that reveals an unexpected relationship that inspires an important advance, which starts anew the cycle of discovery. Broad knowledge, technical facility, flexibility of thought, openness to unanticipated connections, immersion in the free flow of ideas worldwide, hard work, and significant luck are all critical parts of scientific discovery. In recent times, there is perhaps no major breakthrough that better exemplifies this than the development of *superstring theory*.

Superstring theory is an approach that many scientists believe successfully merges general relativity and quantum mechanics. And as we will see, there is reason to hope for even more. Although it is still very much a work in progress, superstring theory may well be a fully unified theory of all forces and all matter, a theory that reaches Einstein's dream and beyond—a theory, I and many others believe, that is blazing the beginnings of a trail which will one day lead us to the deepest laws of the universe. Truthfully, though, superstring theory was not conceived as an ingenious means to reach these noble and long-standing goals. Instead, the history of superstring theory is full of accidental discoveries, false starts, missed opportunities, and nearly ruined careers. It is also, in a precise sense, the story of the discovery of the right solution for the wrong problem.

In 1968, Gabriele Veneziano, a young postdoctoral research fellow working at CERN, was one of many physicists trying to understand the strong nuclear force by studying the results of high-energy particle collisions produced in atom smashers around the world. After months of analyzing patterns and regularities in the data, Veneziano recognized a surprising and unexpected connection to an esoteric area of mathematics. He realized that a two-hundred-year-old formula discovered by the famous Swiss mathematician Leonhard Euler (the *Euler beta function*) seemed to match data on the strong nuclear force with precision. While this might not sound particularly unusual—theoretical physicists deal with arcane formulae all the time—it was a striking case of the cart's rolling miles ahead of the horse. More often than not, physicists first develop an intuition, a mental picture, a broad understanding of the physical principles underlying whatever they are studying and only then seek the equations necessary to ground their intuition in rigorous mathematics. Veneziano, to the contrary, jumped right to the equation; his brilliance was to recognize unusual patterns in the data and to make the unanticipated link to a formula devised centuries earlier for purely mathematical interest.

But although Veneziano had the formula in hand, he had no explanation for *why* it worked. He lacked a physical picture of why Euler's beta function should be relevant to particles influencing each other through the strong nuclear force. Within two years the situation completely changed. In 1970, papers by Leonard Susskind of Stanford, Holger Nielsen of the Niels Bohr Institute, and Yoichiro Nambu of the University of Chicago revealed the physical underpinnings of Veneziano's discovery. These physicists showed that if the strong force between two particles were due to a tiny, extremely thin, almost rubber-band-like strand that connected the particles, then the quantum processes that Veneziano and others had been poring over would be mathematically described using Euler's formula. The little elastic strands were christened *strings* and now, with the horse properly before the cart, string theory was officially born.

But hold the bubbly. To those involved in this research, it was gratifying to understand the physical origin of Veneziano's insight, since it suggested that physicists were on their way to unmasking the strong nuclear force. Yet the discovery was not greeted with universal enthusiasm; far from it. Very far. In fact, Susskind's paper was returned by the journal to which he submitted it with the comment that the work was of minimal interest, an evaluation Susskind recalls well: "I was stunned, I was knocked off my chair, I was depressed, so I went home and got drunk."[9] Eventually, his paper and the others that announced the string concept were all published, but it was not long before the theory suffered two more devastating setbacks. Close scrutiny of more refined data on the strong nuclear force, collected during the early 1970s, showed that the string approach failed to describe the newer results accurately. Moreover, a new proposal called *quantum chromodynamics*, which was firmly rooted in the traditional ingredients of particles and fields—no strings at all—*was* able to describe all the data convincingly. And so by 1974, string theory had been dealt a one-two knockout punch. Or so it seemed.

John Schwarz was one of the earliest string enthusiasts. He once told me that from the start, he had a gut feeling that the theory was deep and important. Schwarz spent a number of years analyzing its various mathematical aspects; among other things, this led to the discovery of *super-string theory—as we shall see, an important refinement of the original string proposal. But with the rise of quantum chromodynamics and the failure of the string framework to describe the strong force, the justification for continuing to work on string theory began to run thin. Nevertheless, there was one particular mismatch between string theory and the strong nuclear force that kept nagging at Schwarz, and he found that he just couldn't let it go. The quantum mechanical equations of string theory predicted that a particular, rather unusual, particle should be copiously produced in the high-energy collisions taking place in atom smashers. The particle would have zero mass, like a photon, but string theory predicted it would have *spin-two*, meaning, roughly speaking, that it would spin twice as fast as a photon. None of the experiments had ever found such a particle, so this appeared to be among the erroneous predictions made by string theory.

Schwarz and his collaborator Joël Scherk puzzled over this case of a missing particle, until in a magnificent leap they made a connection to a completely different problem. Although no one had been able to combine general relativity and quantum mechanics, physicists had determined certain features that would emerge from any successful union. And, as indicated in Chapter 9, one feature they found was that just as the electromagnetic force is transmitted microscopically by photons, the gravitational force should be microscopically transmitted by another class of particles, gravitons (the most elementary, quantum bundles of gravity). Although gravitons have yet to be detected experimentally, the theoretical analyses all agreed that gravitons must have two properties: they must be massless and have spin-two. For Schwarz and Scherk this rang a loud bell—these were just the properties of the rogue particle predicted by string theory—and inspired them to make a bold move, one that would transform a failing of string theory into a striking success.

They proposed that string theory shouldn't be thought of as a quantum mechanical theory of the strong nuclear force. They argued that even though the theory had been discovered in an attempt to understand the strong force, it was actually the solution to a different problem. It was actually the first ever quantum mechanical theory of the *gravitational* force. They claimed that the massless spin-two particle predicted by string theory was the graviton, and that the equations of string theory necessarily embodied a quantum mechanical description of gravity.

Schwarz and Scherk published their proposal in 1974 and expected a major reaction from the physics community. Instead, their work was ignored. In retrospect, it's not hard to understand why. It seemed to some that the string concept had become a theory in search of an application.

After the attempt to use string theory to explain the strong nuclear force had failed, it seemed as though its proponents wouldn't accept defeat and, instead, were flat out determined to find relevance for the theory elsewhere. Fuel was added to this view's fire when it became clear that Schwarz and Scherk needed to change the size of strings in their theory radically so that the force transmitted by the candidate gravitons would have the familiar, known strength of gravity. Since gravity is an extremely weak force* and since, it turns out, the longer the string the *stronger* the force transmitted, Schwarz and Scherk found that strings needed to be extremely tiny to transmit a force with gravity's feeble strength; they needed to be about the Planck length in size, a hundred billion billion times smaller than previously envisioned. So small, doubters wryly noted, that there was no equipment that would be able to see them, which meant that the theory could not be tested experimentally.[10]

By contrast, the 1970s witnessed one success after another for the more conventional, non-string-based theories, formulated with point particles and fields. Theorists and experimenters alike had their heads and hands full of concrete ideas to investigate and predictions to test. Why turn to speculative string theory when there was so much exciting work to be done within a tried-and-true framework? In much the same vein, although physicists knew in the backs of their minds that the problem of merging gravity and quantum mechanics remained unsolved using conventional methods, it was not a problem that commanded attention. Almost everyone acknowledged that it was an important issue and would need to be addressed one day, but with the wealth of work still to be done on the nongravitational forces, the problem of quantizing gravity was pushed to a barely burning back burner. And, finally, in the mid to late 1970s, string theory was far from having been completely worked out. Containing a candidate for the graviton was a success, but many conceptual and technical issues had yet to be addressed. It seemed thoroughly plausible that the theory would be unable to surmount one or more of these issues, so working on string theory meant taking a considerable risk. Within a few years, the theory might be dead.

Schwarz remained resolute. He believed that the discovery of string theory, the first plausible approach for describing gravity in the language

---

*Remember, as noted in Chapter 9, even a puny magnet can overpower the pull of the entire earth's gravity and pick up a paper clip. Numerically, the gravitational force has about 10^{-42} times the strength of the electromagnetic force.

of quantum mechanics, was a major breakthrough. If no one wanted to listen, fine. He would press on and develop the theory, so that when people were ready to pay attention, string theory would be that much further along. His determination proved prescient.

In the late 1970s and early 1980s, Schwarz teamed up with Michael Green, then of Queen Mary College in London, and set to work on some of the technical hurdles facing string theory. Primary among these was the problem of *anomalies*. The details are not of the essence, but, roughly speaking, an anomaly is a pernicious quantum effect that spells doom for a theory by implying that it violates certain sacred principles, such as energy conservation. To be viable, a theory must be free of all anomalies. Initial investigations had revealed that string theory was plagued by anomalies, which was one of the main technical reasons it had failed to generate much enthusiasm. The anomalies signified that although string theory appeared to provide a quantum theory of gravity, since it contained gravitons, on closer inspection the theory suffered from its own subtle mathematical inconsistencies.

Schwarz realized, however, that the situation was not clear-cut. There was a chance—it was a long shot—that a complete calculation would reveal that the various quantum contributions to the anomalies afflicting string theory, when combined correctly, cancelled each other out. Together with Green, Schwarz undertook the arduous task of calculating these anomalies, and by the summer of 1984 the two hit pay dirt. One stormy night, while working late at the Aspen Center for Physics in Colorado, they completed one of the field's most important calculations—a calculation proving that all of the potential anomalies, in a way that seemed almost miraculous, *did* cancel each other out. String theory, they revealed, was free of anomalies and hence suffered from no mathematical inconsistencies. String theory, they demonstrated convincingly, was quantum mechanically viable.

This time physicists listened. It was the mid-1980s, and the climate in physics had shifted considerably. Many of the essential features of the three nongravitational forces had been worked out theoretically and confirmed experimentally. Although important details remained unresolved—and some still do—the community was ready to tackle the next major problem: the merging of general relativity and quantum mechanics. Then, out of a little-known corner of physics, Green and Schwarz burst on the scene with a definite, mathematically consistent, and aesthetically pleasing proposal for how to proceed. Almost overnight, the number of researchers working on string theory leaped from two to over a thousand. The first string revolution was under way.

The First Revolution

I began graduate school at Oxford University in the fall of 1984, and within a few months the corridors were abuzz with talk of a revolution in physics. As the Internet was yet to be widely used, rumor was a dominant channel for the rapid spread of information, and every day brought word of new breakthroughs. Researchers far and wide commented that the atmosphere was charged in a way unseen since the early days of quantum mechanics, and there was serious talk that the end of theoretical physics was within reach.

String theory was new to almost everyone, so in those early days its details were not common knowledge. We were particularly fortunate at Oxford: Michael Green had recently visited to lecture on string theory, so many of us became familiar with the theory's basic ideas and essential claims. And impressive claims they were. In a nutshell, here is what the theory said:

Take any piece of matter—a block of ice, a chunk of rock, a slab of iron—and imagine cutting it in half, then cutting one of the pieces in half again, and on and on; imagine continually cutting the material into ever smaller pieces. Some 2,500 years ago, the ancient Greeks had posed the problem of determining the finest, uncuttable, indivisible ingredient that would be the end product of such a procedure. In our age we have learned that sooner or later you come to atoms, but atoms are not the answer to the Greeks' question, because they can be cut into finer constituents. Atoms can be split. We have learned that they consist of electrons that swarm around a central nucleus that is composed of yet finer particles—protons and neutrons. And in the late 1960s, experiments at the Stanford Linear Accelerator revealed that even neutrons and protons themselves are made up of more fundamental constituents: each proton and each neutron consists of three particles known as quarks, as mentioned in Chapter 9 and illustrated in Figure 12.3a.

Conventional theory, supported by state-of-the-art experiments, envisions electrons and quarks as dots with no spatial extent whatsoever; in this view, therefore, they mark the end of the line—the last of nature's matryoshka dolls to be found in the microscopic makeup of matter. Here

Figure 12.3 (a) Conventional theory is based on electrons and quarks as the basic constituents of matter. **(b)** String theory suggests that each particle is actually a vibrating string.

is where string theory makes its appearance. String theory challenges the conventional picture by proposing that electrons and quarks are *not* zero-sized particles. Instead, the conventional particle-as-dot model, according to string theory, is an approximation of a more refined portrayal in which each particle is actually a tiny, vibrating filament of energy, called a *string*, as you can see in Figure 12.3b. These strands of vibrating energy are envisioned to have no thickness, only length, and so strings are one-dimensional entities. Yet, because the strings are so small, some hundred billion billion times smaller than a single atomic nucleus (10^{-33} centimeters), they appear to be points even when examined with our most advanced atom smashers.

Because our understanding of string theory is far from complete, no one knows for sure whether the story ends here—whether, assuming the theory is correct, strings are truly the final Russian doll, or whether strings themselves might be composed of yet finer ingredients. We will come back to this issue, but for now we follow the historical development of the subject and imagine that strings are truly where the buck stops; we imagine that strings are *the* most elementary ingredient in the universe.

String Theory and Unification

That's string theory in brief, but to convey the power of this new approach, I need to describe conventional particle physics a little more fully. Over the past hundred years, physicists have prodded, pummeled, and pulverized matter in search of the universe's elementary constituents. And, indeed, they have found that in almost everything anyone has ever encountered, the fundamental ingredients are the electrons and quarks just mentioned—more precisely, as in Chapter 9, electrons and two kinds of quarks, up-quarks and down-quarks, that differ in mass and in electrical charge. But the experiments also revealed that the universe has other, more exotic particle species that don't arise in ordinary matter. In addition to up-quarks and down-quarks, experimenters have identified four other species of quarks (*charm-quarks, strange-quarks, bottom-quarks,* and *top-quarks*) and two other species of particles that are very much like electrons, only heavier (*muons* and *taus*). It is likely that these particles were plentiful just after the big bang, but today they are produced only as the ephemeral debris from high-energy collisions between the more familiar particle species. Finally, experimenters have also discovered three species of ghostly particles called *neutrinos* (*electron-neutrinos, muon-neutrinos,* and *tau-neutrinos*) that can pass through trillions of miles of lead as easily as we pass through air. These particles—the electron and its two heavier cousins, the six kinds of quarks, and the three kinds of neutrinos—constitute a modern-day particle physicist's answer to the ancient Greek question about the makeup of matter.[11]

The laundry list of particle species can be organized into three "families" or "generations" of particles, as in Table 12.1. Each family has two of the quarks, one of the neutrinos, and one of the electronlike particles; the only difference between corresponding particles in each family is that their masses increase in each successive family. The division into families certainly suggests an underlying pattern, but the barrage of particles can easily make your head spin (or, worse, make your eyes glaze over). Hang on, though, because one of the most beautiful features of string theory is that it provides a means for taming this apparent complexity.

According to string theory, there is only *one* fundamental ingredient—the string—and the wealth of particle species simply reflects the different vibrational patterns that a string can execute. It's just like what happens with more familiar strings like those on a violin or cello. A cello string can vibrate in many different ways, and we hear each pattern as a different musical note. In this way, one cello string can produce a range of different sounds. The strings in string theory behave similarly: they too can vibrate in different patterns. But instead of yielding different musical tones, *the different vibrational patterns in string theory correspond to different kinds of particles.* The key realization is that the detailed pattern of vibration executed by a string produces a specific mass, a specific electric charge, a specific spin, and so on—the specific list of properties, that is, which distinguish one kind of particle from another. A string vibrating in one particular pattern might have the properties of an electron, while a string

Family 1		Family 2		Family 3	
Particle	Mass	Particle	Mass	Particle	Mass
Electron	.00054	Muon	.11	Tau	1.9
Electron-neutrino	$<10^{-9}$	Muon-neutrino	$<10^{-4}$	Tau-neutrino	$<10^{-3}$
Up-quark	.0047	Charm-quark	1.6	Top-quark	189
Down-quark	.0074	Strange-quark	.16	Bottom-quark	5.2

Table 12.1 The three families of fundamental particles and their masses (in multiples of the proton mass). The values of the neutrino masses are known to be nonzero but their exact values have so far eluded experimental determination.

vibrating in a different pattern might have the properties of an up-quark, a down-quark, or any of the other particle species in Table 12.1. It is not that an "electron string" makes up an electron, or an "up-quark string" makes up an up-quark, or a "down-quark string" makes up a down-quark. Instead, the *single* species of string can account for a great variety of particles because the string can execute a great variety of vibrational patterns.

As you can see, this represents a potentially giant step toward unification. If string theory is correct, the head-spinning, eye-glazing list of particles in Table 12.1 manifests the vibrational repertoire of a single basic ingredient. Metaphorically, the different notes that can be played by a single species of string would account for all of the different particles that have been detected. At the ultramicroscopic level, the universe would be akin to a string symphony vibrating matter into existence.

This is a delightfully elegant framework for explaining the particles in Table 12.1, yet string theory's proposed unification goes even further. In Chapter 9 and in our discussion above, we discussed how the forces of nature are transmitted at the quantum level by other particles, the messenger particles, which are summarized in Table 12.2. String theory accounts for the messenger particles exactly as it accounts for the matter particles. Namely, each messenger particle is a string that's executing a particular vibrational pattern. A photon is a string vibrating in one particular pattern, a W particle is a string vibrating in a different pattern, a gluon is a string vibrating in yet another pattern. And, of prime importance, what Schwarz and Scherk showed in 1974 is that there is a parti-

Force	Force particle	Mass
Strong	Gluon	0
Electromagnetic	Photon	0
Weak	W, Z	86, 97
Gravity	Graviton	0

Table 12.2 The four forces of nature, together with their associated force particles and their masses in multiples of the proton mass. (There are actually two W particles—one with charge +1 and one with charge −1—that have the same mass; for simplicity we ignore this detail and refer to each as a W particle.

cular vibrational pattern that has all the properties of a graviton, so that the gravitational force is included in string theory's quantum mechanical framework. Thus, not only do matter particles arise from vibrating strings, but so do the messenger particles—even the messenger particle for gravity.

And so, beyond providing the first successful approach for merging gravity and quantum mechanics, string theory revealed its capacity to provide a unified description of all matter and all forces. That's the claim that knocked thousands of physicists off their chairs in the mid-1980s; by the time they got up and dusted themselves off, many were converts.

Why Does String Theory Work?

Before the development of string theory, the path of scientific progress was strewn with unsuccessful attempts to merge gravity and quantum mechanics. So what is it about string theory that has allowed it to succeed thus far? We've described how Schwarz and Scherk realized, much to their surprise, that one particular string vibrational pattern had just the right properties to be the graviton particle, and how they then concluded that string theory provided a ready-made framework for merging the two theories. Historically, that is indeed how the power and promise of string theory was fortuitously realized, but as an explanation for why the string approach succeeded where all other attempts failed, it leaves us wanting. Figure 12.2 encapsulates the conflict between general relativity and quantum mechanics—on ultrashort distance (and time) scales, the frenzy of quantum uncertainty becomes so violent that the smooth geometrical model of spacetime underlying general relativity is destroyed—so the question is, How does string theory solve the problem? How does string theory calm the tumultuous fluctuations of spacetime at ultramicroscopic distances?

The main new feature of string theory is that its basic ingredient is not a point particle—a dot of no size—but instead is an object that has spatial extent. This difference is the key to string theory's success in merging gravity and quantum mechanics.

The wild frenzy depicted in Figure 12.2 arises from applying the uncertainty principle to the gravitational field; on smaller and smaller scales, the uncertainty principle implies that fluctuations in the gravitational field get larger and larger. On such extremely tiny distance scales, though, we should describe the gravitational field in terms of its fundamental constituents, gravitons, much as on molecular scales we should describe water in terms of H_2O molecules. In this language, the frenzied gravitational field undulations should be thought of as large numbers of gravitons wildly flitting this way and that, like bits of dirt and dust caught up in a ferocious tornado. Now, if gravitons were point particles (as envisioned in all earlier, failed attempts to merge general relativity and quantum mechanics), Figure 12.2 would accurately reflect their collective behavior: ever shorter distance scales, ever greater agitation. But string theory changes this conclusion.

In string theory, each graviton is a vibrating string—something that is not a point, but instead is roughly a Planck length (10^{-33} centimeters) in size.[12] And since the gravitons are the finest, most elementary constituents of a gravitational field, it makes no sense to talk about the behavior of gravitational fields on sub–Planck length scales. Just as resolution on your TV screen is limited by the size of individual pixels, resolution of the gravitational field in string theory is limited by the size of gravitons. Thus, the nonzero size of gravitons (and everything else) in string theory sets a limit, at roughly the Planck scale, to how finely a gravitational field can be resolved.

That is the vital realization. The uncontrollable quantum fluctuations illustrated in Figure 12.2 arise only when we consider quantum uncertainty on arbitrarily short distance scales—scales shorter than the Planck length. In a theory based on zero-sized point particles, such an application of the uncertainty principle is warranted and, as we see in the figure, this leads us to a wild terrain beyond the reach of Einstein's general relativity. A theory based on strings, however, includes a built-in fail-safe. In string theory, strings are the smallest ingredient, so our journey into the ultramicroscopic comes to an end when we reach the Planck length—the size of strings themselves. In Figure 12.2, the Planck scale is represented by the second highest level; as you can see, on such scales there are still undulations in the spatial fabric because the gravitational field is still subject to quantum jitters. But the jitters are mild enough to avoid irreparable conflict with general relativity. The precise mathematics underlying general relativity must be modified to incorporate these quantum undulations, but this can be done and the math remains sensible.

Thus, by limiting how small you can get, string theory limits how violent the jitters of the gravitational field become—and the limit is just big enough to avoid the catastrophic clash between quantum mechanics and general relativity. In this way, string theory quells the antagonism between the two frameworks and is able, for the first time, to join them.

Cosmic Fabric in the Realm of the Small

What does this mean for the ultramicroscopic nature of space and space-time more generally? For one thing, it forcefully challenges the conventional notion that the fabric of space and time is continuous—that you can always divide the distance between here and there or the duration between now and then in half and in half again, endlessly partitioning space and time into ever smaller units. Instead, when you get down to the Planck length (the length of a string) and Planck time (the time it would take light to travel the length of a string) and try to partition space and time more finely, you find you can't. The concept of "going smaller" ceases to have meaning once you reach the size of the *smallest* constituent of the cosmos. For zero-sized point particles this introduces no constraint, but since strings have size, it does. If string theory is correct, the usual concepts of space and time, the framework within which all of our daily experiences take place, simply don't apply on scales finer than the Planck scale—the scale of strings themselves.

As for what concepts take over, there is as yet no consensus. One possibility that jibes with the explanation above for how string theory meshes quantum mechanics and general relativity is that the fabric of space on the Planck scale resembles a lattice or a grid, with the "space" between the grid lines being outside the bounds of physical reality. Just as a microscopic ant walking on an ordinary piece of fabric would have to leap from thread to thread, perhaps motion through space on ultramicroscopic scales similarly requires discrete leaps from one "strand" of space to another. Time, too, could have a grainy structure, with individual moments being packed closely together but not melding into a seamless continuum. In this way of thinking, the concepts of ever smaller space and time intervals would sharply come to an end at the Planck scale. Just as there is no such thing as an American coin value smaller than a penny, if ultramicroscopic spacetime has a grid structure, there would be no such thing as a distance shorter than the Planck length or a duration shorter than the Planck time.

Another possibility is that space and time do not abruptly cease to have meaning on extremely small scales, but instead gradually morph into other, more fundamental concepts. Shrinking smaller than the Planck scale would be off limits not because you run into a fundamental grid, but because the concepts of space and time segue into notions for which "shrinking smaller" is as meaningless as asking whether the number nine is happy. That is, we can envision that as familiar, macroscopic space and time gradually transform into their unfamiliar ultramicroscopic counterparts, many of their usual properties—such as length and duration—become irrelevant or meaningless. Just as you can sensibly study the temperature and viscosity of liquid water—concepts that apply to the macroscopic properties of a fluid—but when you get down to the scale of individual H_2O molecules, these concepts cease to be meaningful, so, perhaps, although you can divide regions of space and durations of time in half and in half again on everyday scales, as you pass the Planck scale they undergo a transformation that renders such division meaningless.

Many string theorists, including me, strongly suspect that something along these lines actually happens, but to go further we need to figure out the more fundamental concepts into which space and time transform.* To date, this is an unanswered question, but cutting-edge research (described in the final chapter) has suggested some possibilities with far-reaching implications.

---

*I might note that the proponents of another approach for merging general relativity and quantum mechanics, *loop quantum gravity*, to be briefly discussed in Chapter 16, take a viewpoint that is closer to the former conjecture—that spacetime has a discrete structure on the smallest of scales.

The Finer Points

With the description I've given so far, it might seem baffling that any physicist would resist the allure of string theory. Here, finally, is a theory that promises to realize Einstein's dream and more; a theory that could quell the hostility between quantum mechanics and general relativity; a theory with the capacity to unify all matter and all forces by describing everything in terms of vibrating strings; a theory that suggests an ultramicroscopic realm in which familiar space and time might be as quaint as a rotary telephone; a theory, in short, that promises to take our understanding of the universe to a whole new level. But bear in mind that no one has ever seen a string and, except for some maverick ideas discussed in the next chapter, it is likely that even if string theory is right, no one ever will. Strings are so small that a direct observation would be tantamount to reading the text on this page from a distance of 100 light-years: it would require resolving power nearly a billion billion times finer than our current technology allows. Some scientists argue vociferously that a theory so removed from direct empirical testing lies in the realm of philosophy or theology, but not physics.

I find this view shortsighted, or, at the very least, premature. While we may never have technology capable of seeing strings directly, the history of science is replete with theories that were tested experimentally through indirect means.[13] String theory isn't modest. Its goals and promises are big. And that's exciting and useful, because if a theory is to be *the* theory of our universe, it must match the real world not just in the broad-brush outline discussed so far, but also in minute detail. As we'll now discuss, therein lie potential tests.

During the 1960s and 1970s, particle physicists made great strides in understanding the quantum structure of matter and the nongravitational forces that govern its behavior. The framework to which they were finally led by experimental results and theoretical insights is called the *standard model* of particle physics and is based on quantum mechanics, the matter particles in Table 12.1, and the force particles in Table 12.2 (ignoring the graviton, since the standard model does not incorporate gravity, and including the Higgs particle, which is not listed in the tables), all viewed as point particles. The standard model is able to explain essentially all data produced by the world's atom smashers, and over the years its inventors have been deservedly lauded with the highest of honors. Even so, the standard model has significant limitations. We've already discussed how it, and every other approach prior to string theory, failed to merge gravity and quantum mechanics. But there are other shortcomings as well.

The standard model failed to explain *why* the forces are transmitted by the precise list of particles in Table 12.2 and *why* matter is composed of the precise list of particles in Table 12.1. Why are there three families of matter particles and why does each family have the particles it does? Why not two families or just one? Why does the electron have three times the electric charge of the down-quark? Why does the muon weigh 23.4 times as much as the up-quark, and why does the top-quark weigh about 350,000 times as much as an electron? Why is the universe constructed with this range of seemingly random numbers? The standard model takes the particles in Tables 12.1 and 12.2 (again, ignoring the graviton) as *input,* then makes impressively accurate predictions for how the particles will interact and influence each other. But the standard model can't explain the input—the particles and their properties—any more than your calculator can explain the numbers you input the last time you used it.

Puzzling over the properties of these particles is not an academic question of why this or that esoteric detail happens to be one way or another. Over the last century, scientists have realized that the universe has the familiar features of common experience only because the particles in Tables 12.1 and 12.2 have precisely the properties they do. Even fairly minor changes to the masses or electric charges of some of the particles would, for example, make them unable to engage in the nuclear processes that power stars. And without stars, the universe would be a completely different place. Thus, the detailed features of the elementary particles are entwined with what many view as the deepest question in all of science: *Why do the elementary particles have just the right properties to allow nuclear processes to happen, stars to light up, planets to form around stars, and on at least one such planet, life to exist?*

The standard model can't offer any insight into this question since the particle properties are part of its required input. The theory won't start to chug along and produce results until the particle properties are specified. But string theory is different. In string theory, particle properties are *determined* by string vibrational patterns and so the theory holds the promise of providing an explanation.

Particle Properties in String Theory

To understand string theory's new explanatory framework, we need to have a better feel for how string vibrations produce particle properties, so let's consider the simplest property of a particle, its mass.

From $E = mc^2$, we know that mass and energy are interchangeable; like dollars and euros, they are convertible currencies (but unlike monetary currencies, they have a fixed exchange rate, given by the speed of light times itself, c^2). Our survival depends on Einstein's equation, since the sun's life-sustaining heat and light are generated by the conversion of 4.3 million tons of matter into energy every second; one day, nuclear reactors on earth may emulate the sun by safely harnessing Einstein's equation to provide humanity with an essentially limitless supply of energy.

In these examples, energy is produced from mass. But Einstein's equation works perfectly well in reverse—the direction in which mass is produced from energy—and that's the direction in which string theory uses Einstein's equation. The *mass* of a particle in string theory is nothing but the *energy* of its vibrating string. For instance, the explanation string theory offers for why one particle is heavier than another is that the string constituting the heavier particle is vibrating faster and more furiously than the string constituting the lighter particle. Faster and more furious vibration means higher energy, and higher energy translates, via Einstein's equation, into greater mass. Conversely, the lighter a particle is, the slower and less frenetic is the corresponding string vibration; a massless particle like a photon or a graviton corresponds to a string executing the most placid and gentle vibrational pattern that it possibly can.*[14]

Other properties of a particle, such as its electric charge and its spin, are encoded through more subtle features of the string's vibrations. Compared with mass, these features are harder to describe nonmathematically, but they follow the same basic idea: the vibrational pattern is the particle's fingerprint; all the properties that we use to distinguish one particle from another are determined by the vibrational pattern of the particle's string.

In the early 1970s, when physicists analyzed the vibrational patterns arising in the first incarnation of string theory—the *bosonic string theory*—

---

*The relationship to mass arising from a Higgs ocean will be discussed later in the chapter.

to determine the kinds of particle properties the theory predicted, they hit a snag. Every vibrational pattern in the bosonic string theory had a whole-number amount of spin: spin-0, spin-1, spin-2, and so on. This was a problem, because although the messenger particles have spin values of this sort, particles of matter (like electrons and quarks) don't. They have a fractional amount of spin, spin-½. In 1971, Pierre Ramond of the University of Florida set out to remedy this deficiency; in short order, he found a way to modify the equations of the bosonic string theory to allow for half-integer vibrational patterns as well.

In fact, on closer inspection, Ramond's research, together with results found by Schwarz and his collaborator André Neveu and later insights of Ferdinando Gliozzi, Joël Scherk, and David Olive, revealed a perfect balance—a novel symmetry—between the vibrational patterns with different spins in the modified string theory. These researchers found that the new vibrational patterns arose in pairs whose spin values differed by half a unit. For every vibrational pattern with spin-½ there was an associated vibrational pattern with spin-0. For every vibrational pattern of spin-1, there was an associated vibrational pattern of spin-½, and so on. The relationship between integer and half-integer spin values was named *supersymmetry*, and with these results the *supersymmetric string theory*, or *superstring theory*, was born. Nearly a decade later, when Schwarz and Green showed that all the potential anomalies that threatened string theory canceled each other out, they were actually working in the framework of superstring theory, and so the revolution their paper ignited in 1984 is more appropriately called the first *superstring* revolution. (In what follows, we will often refer to strings and to string theory, but that's just a shorthand; we always mean superstrings and superstring theory.)

With this background, we can now state what it would mean for string theory to reach beyond broad-brush features and explain the universe in detail. It comes down to this: among the vibrational patterns that strings can execute, there must be patterns whose properties agree with those of the known particle species. The theory has vibrational patterns with spin-½, but it must have spin-½ vibrational patterns that match *precisely* the known matter particles, as summarized in Table 12.1. The theory has spin-1 vibrational patterns, but it must have spin-1 vibrational patterns that match *precisely* the known messenger particles, as summarized in Table 12.2. Finally, if experiments do indeed discover spin-0 particles, such as are predicted for Higgs fields, string theory must yield vibrational patterns that match *precisely* the properties of these particles as well. In short, for string theory to be viable, its vibrational patterns must yield and explain the particles of the standard model.

Here, then, is string theory's grand opportunity. If string theory is right, there *is* an explanation for the particle properties that experimenters have measured, and it's to be found in the resonant vibrational patterns that strings can execute. If the properties of these vibrational patterns match the particle properties in Tables 12.1 and 12.2, I think that would convince even the diehard skeptics of string theory's veracity, whether or not anyone had directly seen the extended structure of a string itself. And beyond establishing itself as the long-sought unified theory, with such a match between theory and experimental data, string theory would provide the first fundamental explanation for why the universe is the way it is.

So how does string theory fare on this critical test?

Too Many Vibrations

Well, at first blush, string theory fails. For starters, there are an infinite number of different string vibrational patterns, with the first few of an endless series schematically illustrated in Figure 12.4. Yet Tables 12.1 and 12.2 contain only a finite list of particles, and so from the get-go we appear to have a vast mismatch between string theory and the real world. What's more, when we analyze mathematically the possible energies—and hence masses—of these vibrational patterns, we come upon another significant mismatch between theory and observation. The masses of the permissible string vibrational patterns bear no resemblance to the experimentally measured particle masses recorded in Tables 12.1 and 12.2. It's not hard to see why.

Since the early days of string theory, researchers have realized that the stiffness of a string is inversely proportional to its length (its length squared, to be more precise): while long strings are easy to bend, the shorter the string the more rigid it becomes. In 1974, when Schwarz and Scherk proposed decreasing the size of strings so that they'd embody a gravitational force of the right strength, they therefore also proposed increasing the tension of the strings—all the way, it turns out, to about a thousand trillion trillion trillion (10^{39}) tons, about 1000000000000000000000000000000000000000 (10^{41}) times the tension on an average piano string. Now, if you imagine bending a tiny,

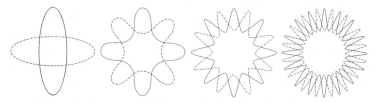

Figure 12.4 The first few examples of string vibrational patterns.

extremely stiff string into one of the increasingly elaborate patterns in Figure 12.4, you'll realize that the more peaks and troughs there are, the more energy you'll have to exert. Conversely, once a string is vibrating in such an elaborate pattern, it embodies a huge amount of energy. Thus, all but the simplest string vibrational patterns are highly energetic and hence, via $E = mc^2$, correspond to particles with huge masses.

And by huge, I really mean huge. Calculations show that the masses of the string vibrations follow a series analogous to musical harmonics: they are all multiples of a fundamental mass, the *Planck mass*, much as overtones are all multiples of a fundamental frequency or tone. By the standards of particle physics, the Planck mass is colossal—it is some 10 billion billion (10^{19}) times the mass of a proton, roughly the mass of a dust mote or a bacterium. Thus, the possible masses of string vibrations are 0 times the Planck mass, 1 times the Planck mass, 2 times the Planck mass, 3 times the Planck mass, and so on, showing that the masses of all but the 0-mass string vibrations are gargantuan.[15]

As you can see, some of the particles in Tables 12.1 and 12.2 are indeed massless, but most aren't. And the nonzero masses in the tables are farther from the Planck mass than the Sultan of Brunei is from needing a loan. Thus, we see clearly that the known particle masses do not fit the pattern advanced by string theory. Does this mean that string theory is ruled out? You might think so, but it doesn't. Having an endless list of vibrational patterns whose masses become ever more remote from those of known particles is a challenge the theory must overcome. Years of research have revealed promising strategies for doing so.

As a start, note that experiments with the known particle species have taught us that heavy particles tend to be unstable; typically, heavy particles disintegrate quickly into a shower of lower-mass particles, ultimately generating the lightest and most familiar species in Tables 12.1 and 12.2.

(For instance, the top-quark disintegrates in about 10^{-24} seconds.) We expect this lesson to hold true for the "superheavy" string vibrational patterns, and that would explain why, even if they were copiously produced in the hot, early universe, few if any would have survived until today. Even if string theory is right, our only chance to see the superheavy vibrational patterns would be to produce them through high-energy collisions in particle accelerators. However, as current accelerators can reach only energies equivalent to roughly 1,000 times the mass of a proton, they are far too feeble to excite any but string theory's most placid vibrational patterns. Thus, string theory's prediction of a tower of particles with masses starting some million billion times greater than that achievable with today's technology is not in conflict with observations.

This explanation also makes clear that contact between string theory and particle physics will involve only the lowest-energy—the massless— string vibrations, since the others are way beyond what we can reach with today's technology. But what of the fact that most of the particles in Tables 12.1 and 12.2 are not massless? It's an important issue, but less troubling than it might at first appear. Since the Planck mass is huge, even the most massive particle known, the top-quark, weighs in at only .0000000000000000116 (about 10^{-17}) times the Planck mass. As for the electron, it weighs in at .00000000000000000000000034 (about 10^{-23}) times the Planck mass. So, to a first approximation—*valid to better than 1 part in 10^{17}*—all the particles in Tables 12.1 and 12.2 *do* have masses equal to zero times the Planck mass (much as most earthlings' wealth, to a first approximation, is 0 times that of the Sultan of Brunei), just as "predicted" by string theory. Our goal is to better this approximation and show that string theory explains the tiny deviations from 0 times the Planck mass characteristic of the particles in Tables 12.1 and 12.2. But massless vibrational patterns are not as grossly at odds with the data as you might have initially thought.

This is encouraging, but detailed scrutiny reveals yet further challenges. Using the equations of superstring theory, physicists have listed every massless string vibrational pattern. One entry is the spin-2 graviton, and that's the great success which launched the whole subject; it ensures that gravity is a part of quantum string theory. But the calculations also show that there are *many* more massless spin-1 vibrational patterns than there are particles in Table 12.2, and there are *many* more massless spin-½ vibrational patterns than there are particles in Table 12.1. Moreover, the list of spin-½ vibrational patterns shows no trace of any repetitive groupings like the family structure of Table 12.1. With a less cursory inspection, then, it seems increasingly difficult to see how string vibrations will align with the known particle species.

Thus, by the mid-1980s, while there were reasons to be excited about superstring theory, there were also reasons to be skeptical. Undeniably, superstring theory presented a bold step toward unification. By providing the first consistent approach for merging gravity and quantum mechanics, it did for physics what Roger Bannister did for the four-minute mile: it showed the seemingly impossible to be possible. Superstring theory established definitively that we could break through the seemingly impenetrable barrier separating the two pillars of twentieth-century physics.

Yet, in trying to go further and show that superstring theory could explain the detailed features of matter and nature's forces, physicists encountered difficulties. This led the skeptics to proclaim that superstring theory, despite all its potential for unification, was merely a mathematical structure with no direct relevance for the physical universe.

Even with the problems just discussed, at the top of the skeptics' list of superstring theory's shortcomings was a feature I've yet to introduce. Superstring theory does indeed provide a successful merger of gravity and quantum mechanics, one that is free of the mathematical inconsistencies that plagued all previous attempts. However, strange as it may sound, in the early years after its discovery, physicists found that the equations of superstring theory do *not* have these enviable properties if the universe has three spatial dimensions. Instead, the equations of superstring theory are mathematically consistent only if the universe has *nine* spatial dimensions, or, including the time dimension, they work only in a universe with *ten* spacetime dimensions!

In comparison to this bizarre-sounding claim, the difficulty in making a detailed alignment between string vibrational patterns and known particle species seems like a secondary issue. Superstring theory requires the existence of *six* dimensions of space that no one has ever seen. That's not a fine point—*that's* a problem.

Or is it?

Theoretical discoveries made during the early decades of the twentieth century, long before string theory came on the scene, suggested that extra dimensions need not be a problem at all. And, with a late-twentieth-century updating, physicists showed that these extra dimensions have the capacity to bridge the gap between string theory's vibrational patterns and the elementary particles experimenters have discovered.

This is one of the theory's most gratifying developments; let's see how it works.

Unification in Higher Dimensions

In 1919, Einstein received a paper that could easily have been dismissed as the ravings of a crank. It was written by a little-known German mathematician named Theodor Kaluza, and in a few brief pages it laid out an approach for unifying the two forces known at the time, gravity and electromagnetism. To achieve this goal, Kaluza proposed a radical departure from something so basic, so completely taken for granted, that it seemed beyond questioning. He proposed that the universe does not have three space dimensions. Instead, Kaluza asked Einstein and the rest of the physics community to entertain the possibility that the universe has *four* space dimensions so that, together with time, it has a total of five spacetime dimensions.

First off, what in the world does that mean? Well, when we say that there are three space dimensions we mean that there are three independent directions or axes along which you can move. From your current position you can delineate these as left/right, back/forth, and up/down; in a universe with three space dimensions, any motion you undertake is some combination of motion along these three directions. Equivalently, in a universe with three space dimensions you need precisely three pieces of information to specify a location. In a city, for example, you need a building's street, its cross street, and a floor number to specify the whereabouts of a dinner party. And if you want people to show up while the food is still hot, you also need to specify a fourth piece of data: a time. That's what we mean by spacetime's being four-dimensional.

Kaluza proposed that in addition to left/right, back/forth, and up/down, *the universe actually has one more spatial dimension that, for some reason, no one has ever seen.* If correct, this would mean that there is another independent direction in which things can move, and therefore that we need to give four pieces of information to specify a precise location in space, and a total of five pieces of information if we also specify a time.

Okay; that's what the paper Einstein received in April 1919 proposed.

The question is, Why didn't Einstein throw it away? We don't see another space dimension—we never find ourselves wandering aimlessly because a street, a cross street, and a floor number are somehow insufficient to specify an address—so why contemplate such a bizarre idea? Well, here's why. Kaluza realized that the equations of Einstein's general theory of relativity could fairly easily be extended mathematically to a universe that had one more space dimension. Kaluza undertook this extension and found, naturally enough, that the higher-dimensional version of general relativity not only included Einstein's original gravity equations but, because of the extra space dimension, also had extra equations. When Kaluza studied these extra equations, he discovered something extraordinary: the extra equations were none other than the equations Maxwell had discovered in the nineteenth century for describing the electromagnetic field! By imagining a universe with one new space dimension, Kaluza had proposed a solution to what Einstein viewed as one of the most important problems in all of physics. *Kaluza had found a framework that combined Einstein's original equations of general relativity with those of Maxwell's equations of electromagnetism.* That's why Einstein didn't throw Kaluza's paper away.

Intuitively, you can think of Kaluza's proposal like this. In general relativity, Einstein awakened space and time. As they flexed and stretched, Einstein realized that he'd found the geometrical embodiment of the gravitational force. Kaluza's paper suggested that the geometrical reach of space and time was greater still. Whereas Einstein realized that gravitational fields can be described as warps and ripples in the usual three space and one time dimensions, Kaluza realized that in a universe with an additional space dimension there would be additional warps and ripples. And those warps and ripples, his analysis showed, would be just right to describe electromagnetic fields. In Kaluza's hands, Einstein's own geometrical approach to the universe proved powerful enough to unite gravity and electromagnetism.

Of course, there was still a problem. Although the mathematics worked, there was—and still is—no evidence of a spatial dimension beyond the three we all know about. So was Kaluza's discovery a mere curiosity, or was it somehow relevant to our universe? Kaluza had a powerful trust in theory—he had, for example, learned to swim by studying a treatise on swimming and then diving into the sea—but the idea of an invisible space dimension, no matter how compelling the theory, still sounded outrageous. Then, in 1926, the Swedish physicist Oskar Klein injected a new twist into Kaluza's idea, one that suggested where the extra dimension might be hiding.

The Hidden Dimensions

To understand Klein's idea, picture Philippe Petit walking on a long, rubber-coated tightrope stretched between Mount Everest and Lhotse. Viewed from a distance of many miles, as in Figure 12.5, the tightrope appears to be a one-dimensional object like a line—an object that has extension only along its length. If we were told that a tiny worm was slithering along the tightrope in front of Philippe, we'd wildly cheer it on because it needs to stay ahead of Philippe's step to avoid disaster. Of course, with a moment's reflection we all realize that there is more to the surface of the tightrope than the left/right dimension we can directly perceive. Although difficult to see with the naked eye from a great distance, the surface of the tightrope has a second dimension: the clockwise/counterclockwise dimension that is "wrapped" around it. With the aid of a modest telescope, this circular dimension becomes visible and we see that the worm can move not only in the long, unfurled left/right direction but also in the short, "curled-up" clockwise/counterclockwise direction. That is, at every point on the tightrope, the worm has two independent directions in which it can move (that's what we mean when we say the tightrope's surface is two-dimensional*), so it can safely stay out of Philippe's way either by slithering ahead of him, as we initially envisioned, or by crawling around the tiny circular dimension and letting Philippe pass above.

The tightrope illustrates that dimensions—the independent directions in which anything can move—come in two qualitatively distinct varieties. They can be big and easy to see, like the left/right dimension of the tightrope's surface, or they can be tiny and more difficult to see, like the clockwise/counterclockwise dimension that circles around the tightrope's surface. In this example, it was not a major challenge to see the small circular girth of the tightrope's surface. All we needed was a reasonable magnifying instrument. But as you can imagine, the smaller a

---

*Were you to count left, right, clockwise, and counterclockwise all separately, you'd conclude that the worm can move in four directions. But when we speak of "independent" directions, we always group those that lie along the same geometrical axis—like left and right, and also clockwise and counterclockwise.

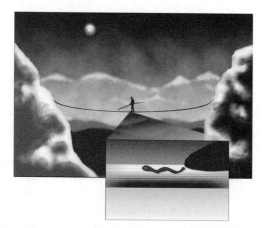

Figure 12.5 From a distance, a tightrope wire looks one-dimensional, although with a strong enough telescope, its second, curled-up dimension becomes visible.

curled-up dimension, the more difficult it is to detect. At a distance of a few miles, it's one thing to reveal the circular dimension of a tightrope's surface; it would be quite another to reveal the circular dimension of something as thin as dental floss or a narrow nerve fiber.

Klein's contribution was to suggest that what's true for an object *within* the universe might be true for the fabric of the universe itself. Namely, just as the tightrope's surface has both large and small dimensions, so does the fabric of space. Maybe the three dimensions we all know about—left/right, back/forth, and up/down—are like the horizontal extent of the tightrope, dimensions of the big, easy-to-see variety. But just as the surface of the tightrope has an additional, small, curled-up, circular dimension, maybe the fabric of space also has a small, curled-up, circular dimension, one so small that no one has powerful enough magnifying equipment to reveal its existence. Because of its tiny size, Klein argued, the dimension would be hidden.

How small is small? Well, by incorporating certain features of quantum mechanics into Kaluza's original proposal, Klein's mathematical analysis revealed that the radius of an extra circular spatial dimension would likely be roughly the Planck length,[16] certainly way too small for experimental accessibility (current state-of-the-art equipment cannot resolve anything smaller than about a thousandth the size of an atomic nucleus, falling short of the Planck length by more than a factor of a million billion). Yet, to an imaginary, Planck-sized worm, this tiny, curled-up circular dimension would provide a new direction in which it could roam just as freely as an ordinary worm negotiates the circular dimension of the tightrope in Figure 12.5. Of course, just as an ordinary worm finds that there isn't much room to explore in the clockwise direction before it finds itself back at its starting point, a Planck-sized worm slithering along a curled-up dimension of space would also repeatedly circle back to its starting point. But aside from the length of the travel it permitted, a curled-up dimension would provide a direction in which the tiny worm could move just as easily as it does in the three familiar unfurled dimensions.

To get an intuitive sense of what this looks like, notice that what we've been referring to as the tightrope's curled-up dimension—the clockwise/counterclockwise direction—*exists at each point along its extended dimension.* The earthworm can slither around the circular girth of the tightrope at any point along its outstretched length, and so the tightrope's surface can be described as having one long dimension, with a tiny, circular direction tacked on at each point, as in Figure 12.6. This is a useful image to have in mind because it also applies to Klein's proposal for hiding Kaluza's extra dimension of space.

To see this, let's again examine the fabric of space by sequentially showing its structure on ever smaller distance scales, as in Figure 12.7. At the first few levels of magnification, nothing new is revealed: the fabric of space still appears three-dimensional (which, as usual, we schematically represent on the printed page by a two-dimensional grid). However, when we get down to the Planck scale, the highest level of magnification in the figure, Klein suggested that a new, curled-up dimension becomes visible.

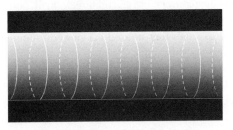

Figure 12.6 The surface of a tightrope has one long dimension with a circular dimension tacked on at each point.

Figure 12.7 The Kaluza-Klein proposal is that on very small scales, space has an extra circular dimension tacked on to each familiar point.

Just as the circular dimension of the tightrope exists at each point along its big, extended dimension, the circular dimension in this proposal exists at each point in the familiar three extended dimensions of daily life. In Figure 12.7, we illustrate this by drawing the additional circular dimension at various points along the extended dimensions (since drawing the circle at every point would obscure the image) and you can immediately see the similarity with the tightrope in Figure 12.6. In Klein's proposal, therefore, space should be envisioned as having three unfurled dimensions (of which we show only two in the figure) with an additional circular dimension tacked on to each point. Notice that the extra dimension is not a bump or a loop within the usual three spatial dimensions, as the graphic limitations of the figure might lead you to think. Instead, the extra dimension is a new direction, completely distinct from the three we know about, which exists at every point in our ordinary three-dimensional space, but is so small that it escapes detection even with our most sophisticated instruments.

With this modification to Kaluza's original idea, Klein provided an answer to how the universe might have more than the three space dimensions of common experience that could remain hidden, a framework that has since become known as *Kaluza-Klein theory*. And since an extra dimension of space was all Kaluza needed to merge general relativity and electromagnetism, Kaluza-Klein theory would seem to be just what Einstein was looking for. Indeed, Einstein and many others became quite excited about unification through a new, hidden space dimension, and a vigorous effort was launched to see whether this approach would work in complete detail. But it was not long before Kaluza-Klein theory encountered its own problems. Perhaps most glaring of all, attempts to incorporate the electron into the extra-dimensional picture proved unworkable.[17] Einstein continued to dabble in the Kaluza-Klein framework until at least the early 1940s, but the initial promise of the approach failed to materialize, and interest gradually died out.

Within a few decades, though, Kaluza-Klein theory would make a spectacular comeback.

String Theory and Hidden Dimensions

In addition to the difficulties Kaluza-Klein theory encountered in trying to describe the microworld, there was another reason scientists were hesitant about the approach. Many found it both arbitrary and extravagant to postulate a hidden spatial dimension. It is not as though Kaluza was led to the idea of a new spatial dimension by a rigid chain of deductive reasoning. Instead, he pulled the idea out of a hat, and upon analyzing its implications discovered an unexpected link between general relativity and electromagnetism. Thus, although it was a great discovery in its own right, it lacked a sense of inevitability. If you asked Kaluza and Klein *why* the universe had five spacetime dimensions rather than four, or six, or seven, or 7,000 for that matter, they wouldn't have had an answer much more convincing than "Why not?"

More than three decades later, the situation changed radically. String theory is the first approach to merge general relativity and quantum mechanics; moreover, it has the potential to unify our understanding of all forces and all matter. But the quantum mechanical equations of string theory don't work in four spacetime dimensions, nor in five, six, seven, or 7,000. Instead, for reasons discussed in the next section, the equations of string theory work only in ten spacetime dimensions—nine of space, plus time. String theory *demands* more dimensions.

This is a fundamentally different kind of result, one never before encountered in the history of physics. Prior to strings, no theory said anything at all about the number of spatial dimensions in the universe. Every theory from Newton to Maxwell to Einstein assumed that the universe had three space dimensions, much as we all assume the sun will rise tomorrow. Kaluza and Klein proffered a challenge by suggesting that there were four space dimensions, but this amounted to yet another assumption—a different assumption, but an assumption nonetheless. Now, for the first time, string theory provided equations that *predicted* the number of space dimensions. A calculation—not an assumption, not a hypothesis, not an inspired guess—determines the number of space dimensions according to string theory, and the surprising thing is that the calculated number is not three, but nine. String theory leads us, *inevitably*, to a universe with six extra space dimensions and hence provides a compelling, ready-made context for invoking the ideas of Kaluza and Klein.

The original proposal of Kaluza and Klein assumed only one hidden dimension, but it's easily generalized to two, three, or even the six extra dimensions required by string theory. For example, in Figure 12.8a we replace the additional circular dimension of Figure 12.7, a one-dimensional shape, with the surface of a sphere, a two-dimensional shape (recall from the discussion in Chapter 8 that the surface of a sphere is two-dimensional because you need two pieces of information—like latitude

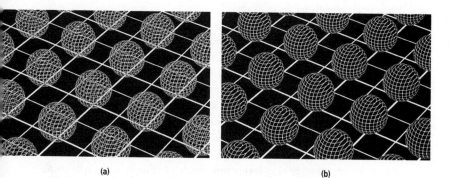

(a)   (b)

Figure 12.8 A close-up of a universe with the three usual dimensions, represented by the grid, and **(a)** two curled-up dimensions, in the form of hollow spheres, and **(b)** three curled-up dimensions in the form of solid balls.

and longitude on the earth's surface—to specify a location). As with the circle, you should envision the sphere tacked on to every point of the usual dimensions, even though in Figure 12.8a, to keep the image clear, we draw only those that lie on the intersections of grid lines. In a universe of this sort, you would need a total of five pieces of information to locate a position in space: three pieces to locate your position in the big dimensions (street, cross street, floor number) and two pieces to locate your position on the sphere (latitude, longitude) tacked on at that point. Certainly, if the sphere's radius were tiny—billions of times smaller than an atom—the last two pieces of information wouldn't matter much for comparatively large beings like ourselves. Nevertheless, the extra dimension would be an integral part of the ultramicroscopic makeup of the spatial fabric. An ultramicroscopic worm would need all five pieces of information and, if we include time, it would need six pieces of information in order to show up at the right dinner party at the right time.

Let's go one dimension further. In Figure 12.8a, we considered only the surface of the spheres. Imagine now that, as in Figure 12.8b, the fabric of space also includes the interior of the spheres—our little Planck-sized worm can burrow into the sphere, as ordinary worms do with apples, and freely move throughout its interior. To specify the worm's location would now require *six* pieces of information: three to locate its position in the usual extended spatial dimensions, and three more to locate its position in the ball tacked on to that point (latitude, longitude, depth of penetration). Together with time, this is therefore an example of a universe with *seven* spacetime dimensions.

Now comes a leap. Although it is impossible to draw, imagine that at every point in the three extended dimensions of everyday life, the universe has not one extra dimension as in Figure 12.7, not two extra dimensions as in Figure 12.8a, not three extra dimensions as in Figure 12.8b, but six extra space dimensions. I certainly can't visualize this and I've never met anyone who can. But its meaning is clear. To specify the spatial location of a Planck-sized worm in such a universe requires *nine* pieces of information: three to locate its position in the usual extended dimensions and six more to locate its position in the curled-up dimensions tacked on to that point. When time is also taken into account, this is a ten-spacetime-dimensional universe, as required by the equations of string theory. If the extra six dimensions are curled up small enough, they would easily have escaped detection.

The Shape of Hidden Dimensions

The equations of string theory actually determine more than just the number of spatial dimensions. They also determine the kinds of shapes the extra dimensions can assume.[18] In the figures above, we focused on the simplest of shapes—circles, hollow spheres, solid balls—but the equations of string theory pick out a significantly more complicated class of six-dimensional shapes known as Calabi-Yau shapes or Calabi-Yau spaces. These shapes are named after two mathematicians, Eugenio Calabi and Shing-Tung Yau, who discovered them mathematically long before their relevance to string theory was realized; a rough illustration of one example is given in Figure 12.9a. Bear in mind that in this figure a two-dimensional graphic illustrates a six-dimensional object, and this results in a variety of significant distortions. Even so, the picture gives a rough sense of what these shapes look like. If the particular Calabi-Yau shape in Figure 12.9a constituted the extra six dimensions in string theory, on ultramicroscopic scales space would have the form illustrated in Figure 12.9b. As the Calabi-Yau shape would be tacked on to every point in the usual three dimensions, you and I and everyone else would right now be surrounded by and filled with these little shapes. Literally, as you walk

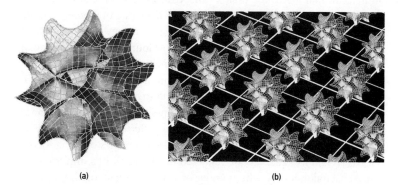

(a)   (b)

Figure 12.9: (a) One example of a Calabi-Yau shape. (b) A highly magnified portion of space with additional dimensions in the form of a tiny Calabi-Yau shape.

from one place to another, your body would move through all nine dimensions, rapidly and repeatedly circumnavigating the entire shape, on average making it seem as if you weren't moving through the extra six dimensions at all.

If these ideas are right, the ultramicroscopic fabric of the cosmos is embroidered with the richest of textures.

String Physics and Extra Dimensions

The beauty of general relativity is that the physics of gravity is controlled by the geometry of space. With the extra spatial dimensions proposed by string theory, you'd naturally guess that the power of geometry to determine physics would substantially increase. And it does. Let's first see this by taking up a question that I've so far skirted. Why does string theory require ten spacetime dimensions? This is a tough question to answer nonmathematically, but let me explain enough to illustrate how it comes down to an interplay of geometry and physics.

Imagine a string that's constrained to vibrate only on the two-dimensional surface of a flat tabletop. The string will be able to execute a variety of vibrational patterns, but only those involving motion in the left/right and back/forth directions of the table's surface. If the string is then released to vibrate in the third dimension, motion in the up/down dimension that leaves the table's surface, additional vibrational patterns become accessible. Now, although it is hard to picture in more than three dimensions, this conclusion—more dimensions means more vibrational patterns—is general. If a string can vibrate in a fourth spatial dimension, it can execute more vibrational patterns than it could in only three; if a string can vibrate in a fifth spatial dimension, it can execute more vibrational patterns than it could in only four; and so on. This is an important realization, because there is an equation in string theory that demands that the number of independent vibrational patterns meet a very precise constraint. If the constraint is violated, the mathematics of string theory falls apart and its equations are rendered meaningless. In a universe with three space dimensions, the number of vibrational patterns is too small and the constraint is not met; with four space dimensions, the number of vibrational patterns is still too small; with five, six, seven, or eight dimensions it is still too small; but with nine space dimensions, the constraint on the number of vibrational patterns is satisfied perfectly. And that's how string theory determines the number of space dimensions.*[19]

While this illustrates well the interplay of geometry and physics, their association within string theory goes further and, in fact, provides a way to address a critical problem encountered earlier. Recall that, in trying to make detailed contact between string vibrational patterns and the known particle species, physicists ran into trouble. They found that there were far too many massless string vibrational patterns and, moreover, the detailed properties of the vibrational patterns did not match those of the known matter and force particles. But what I didn't mention earlier, because we hadn't yet discussed the idea of extra dimensions, is that although those calculations took account of the *number* of extra dimensions (explaining, in part, why so many string vibrational patterns were found), they did not take account of the small size and complex *shape* of the extra dimensions—they assumed that all space dimensions were flat and fully unfurled—and that makes a substantial difference.

Strings are so small that even when the extra six dimensions are crumpled up into a Calabi-Yau shape, the strings still vibrate into those directions. For two reasons, that's extremely important. First, it ensures that the strings always vibrate in all nine space dimensions, and hence the constraint on the number of vibrational patterns continues to be satisfied, even when the extra dimensions are tightly curled up. Second, just as the vibrational patterns of air streams blown through a tuba are affected by the twists and turns of the instrument, the vibrational patterns of strings are influenced by the twists and turns in the geometry of the extra six dimensions. If you were to change the shape of a tuba by making a passageway narrower or by making a chamber longer, the air's vibrational patterns and hence the sound of the instrument would change. Similarly, if the shape and size of the extra dimensions were modified, the precise properties of

---

*Let me prepare you for one relevant development we will encounter in the next chapter. String theorists have known for decades that the equations they generally use to mathematically analyze string theory are approximate (the exact equations have proven difficult to identify and understand). However, most thought that the approximate equations were sufficiently accurate to determine the required number of extra dimensions. More recently (and to the shock of most physicists in the field), some string theorists showed that the approximate equations *missed* one dimension; it is now accepted that the theory needs *seven* extra dimensions. As we will see, this does not compromise the material discussed in this chapter, but shows that it fits within a larger, in fact more unified, framework.[20]

each possible vibrational pattern of a string would also be significantly affected. And since a string's vibrational pattern determines its mass and charge, this means that the extra dimensions play a pivotal role in determining particle properties.

This is a key realization. *The precise size and shape of the extra dimensions has a profound impact on string vibrational patterns and hence on particle properties.* As the basic structure of the universe—from the formation of galaxies and stars to the existence of life as we know it—depends sensitively on the particle properties, the code of the cosmos may well be written in the geometry of a Calabi-Yau shape.

We saw one example of a Calabi-Yau shape in Figure 12.9, but there are at least hundreds of thousands of other possibilities. The question, then, is which Calabi-Yau shape, if any, constitutes the extra-dimensional part of the spacetime fabric. This is one of the most important questions string theory faces since only with a definite choice of Calabi-Yau shape are the detailed features of string vibrational patterns determined. To date, the question remains unanswered. The reason is that the current understanding of string theory's equations provides no insight into how to pick one shape from the many; from the point of view of the known equations, each Calabi-Yau shape is as valid as any other. The equations don't even determine the size of the extra dimensions. Since we don't see the extra dimensions, they must be small, but precisely how small remains an open question.

Is this a fatal flaw of string theory? Possibly. But I don't think so. As we will discuss more fully in the next chapter, the exact equations of string theory have eluded theorists for many years and so much work has used *approximate* equations. These have afforded insight into a great many features of string theory, but for certain questions—including the exact size and shape of the extra dimensions—the approximate equations fall short. As we continue to sharpen our mathematical analysis and improve these approximate equations, determining the form of the extra dimensions is a prime—and in my opinion attainable—objective. So far, this goal remains beyond reach.

Nevertheless, we can still ask whether *any* choice of Calabi-Yau shape yields string vibrational patterns that closely approximate the known particles. And here the answer is quite gratifying.

Although we are far from having investigated every possibility, examples of Calabi-Yau shapes have been found that give rise to string vibrational patterns in rough agreement with Tables 12.1 and 12.2. For instance, in the mid-1980s Philip Candelas, Gary Horowitz, Andrew Strominger, and Edward Witten (the team of physicists who realized the relevance of Calabi-Yau shapes for string theory) discovered that each hole—the term is used in a precisely defined mathematical sense—contained within a Calabi-Yau shape gives rise to a *family* of lowest-energy string vibrational patterns. A Calabi-Yau shape with three holes would therefore provide an explanation for the repetitive structure of three families of elementary particles in Table 12.1. Indeed, a number of such three-holed Calabi-Yau shapes have been found. Moreover, among these preferred Calabi-Yau shapes are ones that also yield just the right number of messenger particles as well as just the right electric charges and nuclear force properties to match the particles in Tables 12.1 and 12.2.

This is an extremely encouraging result; by no means was it ensured. In merging general relativity and quantum mechanics, string theory might have achieved one goal only to find it impossible to come anywhere near the equally important goal of explaining the properties of the known matter and force particles. Researchers take heart in the theory's having blazed past that disappointing possibility. Going further and calculating the precise masses of the particles is significantly more challenging. As we discussed, the particles in Tables 12.1 and 12.2 have masses that deviate from the lowest-energy string vibrations—zero times the Planck mass—by less than one part in a million billion. Calculating such infinitesimal deviations requires a level of precision way beyond what we can muster with our current understanding of string theory's equations.

As a matter of fact, I suspect, as do many other string theorists, that the tiny masses in Tables 12.1 and 12.2 arise in string theory much as they do in the standard model. Recall from Chapter 9 that in the standard model, a Higgs field takes on a nonzero value throughout all space, and the mass of a particle depends on how much drag force it experiences as it wades through the Higgs ocean. A similar scenario likely plays out in string theory. If a huge collection of strings all vibrate in just the right coordinated way throughout all of space, they can provide a uniform background that for all intents and purposes would be indistinguishable from a Higgs ocean. String vibrations that initially yielded zero mass would then acquire tiny nonzero masses through the drag force they experience as they move and vibrate through the string theory version of the Higgs ocean.

Notice, though, that in the standard model, the drag force experienced by a given particle—and hence the mass it acquires—is determined by experimental measurement and specified as an input to the theory. In the string theory version, the drag force—and hence the masses of the vibrational patterns—would be traced back to interactions between strings (since the Higgs ocean would be made of strings) and should be *calculable*. String theory, at least in principle, allows all particle properties to be determined by the theory itself.

No one has accomplished this, but as emphasized, string theory is still very much a work in progress. In time, researchers hope to realize fully the vast potential of this approach to unification. The motivation is strong because the potential payoff is big. With hard work and substantial luck, string theory may one day explain the fundamental particle properties and, in turn, explain why the universe is the way it is.

The Fabric of the Cosmos According to String Theory

Even though much about string theory still lies beyond the bounds of our comprehension, it has already exposed dramatic new vistas. Most strikingly, in mending the rift between general relativity and quantum mechanics, string theory has revealed that the fabric of the cosmos may have many more dimensions than we perceive directly—dimensions that may be the key to resolving some of the universe's deepest mysteries. Moreover, the theory intimates that the familiar notions of space and time do not extend into the sub-Planckian realm, which suggests that space and time as we currently understand them may be mere approximations to more fundamental concepts that still await our discovery.

In the universe's initial moments, these features of the spacetime fabric that, today, can be accessed only mathematically, would have been manifest. Early on, when the three familiar spatial dimensions were also small, there would likely have been little or no distinction between what we now call the big and the curled-up dimensions of string theory. Their current size disparity would be due to cosmological evolution which, in a way that we don't yet understand, would have had to pick three of the spatial dimensions as special, and subject only them to the 14 billion years of expansion discussed in earlier chapters. Looking back in time even further, the entire observable universe would have shrunk into the sub-Planckian domain, so that what we've been referring to as the fuzzy patch (in Figure 10.6), we can now identify as the realm where familiar space and time have yet to emerge from the more fundamental entities—whatever they may be—that current research is struggling to comprehend.

Further progress in understanding the primordial universe, and hence in assessing the origin of space, time, and time's arrow, requires a significant honing of the theoretical tools we use to understand string theory—a goal that, not too long ago, seemed noble yet distant. As we'll now see, with the development of M-theory, progress has exceeded many of even the optimists' most optimistic predictions.

13

The Universe on a Brane

SPECULATIONS ON SPACE AND TIME IN M-THEORY

String theory has one of the most twisted histories of any scientific breakthrough. Even today, more than three decades after its initial articulation, most string practitioners believe we still don't have a comprehensive answer to the rudimentary question, What is string theory? We know a lot *about* string theory. We know its basic features, we know its key successes, we know the promise it holds, and we know the challenges it faces; we can also use string theory's equations to make detailed calculations of how strings should behave and interact in a wide range of circumstances. But most researchers feel that our current formulation of string theory still lacks the kind of core principle we find at the heart of other major advances. Special relativity has the constancy of the speed of light. General relativity has the equivalence principle. Quantum mechanics has the uncertainty principle. String theorists continue to grope for an analogous principle that would capture the theory's essence as completely.

To a large extent, this deficiency exists because string theory developed piecemeal instead of emerging from a grand, overarching vision. The *goal* of string theory—the unification of all forces and all matter in a quantum mechanical framework—is about as grand as it gets, but the theory's evolution has been distinctly fragmented. After its serendipitous discovery more than three decades ago, string theory has been cobbled together as one group of theorists has uncovered key properties by studying *these* equations, while another group has revealed critical implications by examining *those*.

String theorists can be likened to a primitive tribe excavating a buried spacecraft onto which they've stumbled. By tinkering and fiddling, the tribe would slowly establish aspects of the spacecraft's operation, and this would nurture a sense that all the buttons and toggles work together in a coordinated and unified manner. A similar feeling prevails among string theorists. Results found over many years of research are dovetailing and converging. This has instilled a growing confidence among researchers that string theory is closing in on one powerful, coherent framework— which has yet to be unearthed fully, but ultimately will expose nature's inner workings with unsurpassed clarity and comprehensiveness.

In recent times, nothing illustrates this better than the realization that sparked the *second superstring revolution*—a revolution that has, among other things, exposed another hidden dimension entwined in the spatial fabric, opened new possibilities for experimental tests of string theory, suggested that our universe may be brushing up against others, revealed that black holes may be created in the next generation of high-energy accelerators, and led to a novel cosmological theory in which time and its arrow, like the graceful arc of Saturn's rings, may cycle around and around.

The Second Superstring Revolution

There's an awkward detail regarding string theory that I've yet to divulge, but that readers of my previous book, *The Elegant Universe*, may recall. Over the last three decades, not one but *five* distinct versions of string theory have been developed. While their names are not of the essence, they are called *Type I, Type IIA, Type IIB, Heterotic-O,* and *Heterotic-E*. All share the essential features introduced in the last chapter—the basic ingredients are strands of vibrating energy—and, as calculations in the 1970s and 1980s revealed, each theory requires six extra space dimensions; but when they are analyzed in detail, significant differences appear. For example, the Type I theory includes the vibrating string loops discussed in the last chapter, so-called *closed strings*, but unlike the other string theories, it also contains *open strings*, vibrating string snippets that have two loose ends. Furthermore, calculations show that the list of string vibrational patterns and the way each pattern interacts and influences others differ from one formulation to another.

The most optimistic of string theorists envisioned that these differences would serve to eliminate four of the five versions when detailed comparisons to experimental data could one day be carried out. But, frankly, the mere existence of five different formulations of string theory was a source of quiet discomfort. The dream of unification is one in which scientists are led to a unique theory of the universe. If research established that only one theoretical framework could embrace both quantum mechanics and general relativity, theorists would reach unification nirvana. They would have a strong case for the framework's validity even in the absence of direct experimental verification. After all, a wealth of experimental support for both quantum mechanics and general relativity already exists, and it seems plain as day that the laws governing the universe should be mutually compatible. If a particular theory were the unique, mathematically consistent arch spanning the two experimentally confirmed pillars of twentieth-century physics, that would provide powerful, albeit indirect, evidence for the theory's inevitability.

But the fact that there are five versions of string theory, superficially similar yet distinct in detail, would seem to mean that string theory fails the uniqueness test. Even if the optimists are some day vindicated and only one of the five string theories is confirmed experimentally, we would still be vexed by the nagging question of why there are four other consistent formulations. Would the other four simply be mathematical curiosities? Would they have any significance for the physical world? Might their existence be the tip of a theoretical iceberg in which clever scientists would subsequently show that there are actually five other versions, or six, or seven, or perhaps even an endless number of distinct mathematical variations on a theme of strings?

During the late 1980s and early 1990s, with many physicists hotly pursuing an understanding of one or another of the string theories, the enigma of the five versions was not a problem researchers typically dealt with on a day-to-day basis. Instead, it was one of those quiet questions that everyone assumed would be addressed in the distant future, when the understanding of each individual string theory had become significantly more refined.

But in the spring of 1995, with little warning, these modest hopes were wildly exceeded. Drawing on the work of a number of string theorists (including Chris Hull, Paul Townsend, Ashoke Sen, Michael Duff, John Schwarz, and many others), Edward Witten—who for two decades has been the world's most renowned string theorist—uncovered a hidden unity that tied all five string theories together. Witten showed that rather than being distinct, the five theories are actually just five different ways of mathematically analyzing a *single* theory. Much as the translations of a book into five different languages might seem, to a monolingual reader, to be five distinct texts, the five string formulations appeared distinct only because Witten had yet to write the dictionary for translating among them. But once revealed, the dictionary provided a convincing demonstration that—like a single master text from which five translations have been made—a single master theory links all five string formulations. The unifying master theory has tentatively been called *M-theory*, M being a tantalizing placeholder whose meaning—Master? Majestic? Mother? Magic? Mystery? Matrix?—awaits the outcome of a vigorous worldwide research effort now seeking to complete the new vision illuminated by Witten's powerful insight.

This revolutionary discovery was a gratifying leap forward. String theory, Witten demonstrated in one of the field's most prized papers (and in important follow-up work with Petr Hořava), *is* a single theory. No longer did string theorists have to qualify their candidate for the unified theory Einstein sought by adding, with a tinge of embarrassment, that the proposed unified framework lacked unity because it came in five different versions. How fitting, by contrast, for the farthest-reaching proposal for a unified theory to be, itself, the subject of a meta-unification. Through Witten's work, the unity embodied by each individual string theory was extended to the whole string framework.

Figure 13.1 sketches the status of the five string theories before and after Witten's discovery, and is a good summary image to keep in mind. It illustrates that M-theory is not a new approach, per se, but that, by clearing the clouds, it promises a more refined and complete formulation of physical law than is provided by any one of the individual string theories. M-theory links together and embraces equally all five string theories by showing that each is part of a grander theoretical synthesis.

The Power of Translation

Although Figure 13.1 schematically conveys the essential content of Witten's discovery, expressed in this way it might strike you like a bit of inside baseball. Before Witten's breakthrough, researchers thought there were

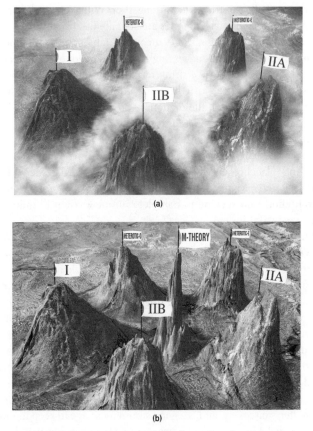

(a)

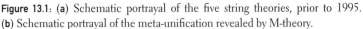

(b)

Figure 13.1: (a) Schematic portrayal of the five string theories, prior to 1995. (b) Schematic portrayal of the meta-unification revealed by M-theory.

five separate versions of string theory; after his breakthrough, they didn't. But if you'd never known that there were five purportedly distinct string theories, why should you care that the cleverest of all string theorists showed they aren't distinct after all? Why, in other words, was Witten's discovery revolutionary as opposed to a modest insight correcting a previous misconception?

Here's why. Over the past few decades, string theorists have been stymied repeatedly by a mathematical problem. Because the exact equations describing any one of the five string theories have proven so difficult to extract and analyze, theorists have based much of their research on approximate equations that are far easier to work with. While there are good reasons to believe that the approximate equations should, in many circumstances, give answers close to those given by the true equations, approximations—like translations—always miss something. For this reason, certain key problems have proved beyond the approximate equations' mathematical reach, significantly impeding progress.

For the imprecision inherent in textual translations, readers have a couple of immediate remedies. The best option, if the reader's linguistic skills are up to the task, is to consult the original manuscript. At the moment, the analog of this option is not available to string theorists. By virtue of the consistency of the dictionary developed by Witten and others, we have strong evidence that all five string theories are different descriptions of a single master theory, M-theory, but researchers have yet to develop a complete understanding of this theoretical nexus. We have learned much about M-theory in the last few years, but we still have far to go before anyone could sensibly claim that it is properly or completely understood. In string theory, it's as if we have five translations of a yet-to-be-discovered master text.

Another helpful remedy, well known to readers of translations who either don't have the original (as in string theory) or, more commonly, don't understand the language in which it's written, is to consult several translations of the master text into languages with which they are familiar. Passages on which the translations agree give confidence; passages on which they differ flag possible inaccuracies or highlight different interpretations. It is this approach that Witten made available with his discovery that the five string theories are different translations of the same underlying theory. In fact, his discovery provided an extremely powerful version of this line of attack that is best understood through a slight extension of the translation analogy.

Imagine a master manuscript infused with such an enormous range of puns, rhymes, and offbeat, culture-sensitive jokes, that the complete text cannot be expressed gracefully in any single one of five given languages into which it is being translated. Some passages might translate into Swahili with ease, while other portions might prove thoroughly impenetrable in this tongue. Much insight into some of the latter passages might emerge from the Inuit translation; in yet other sections that translation might be completely opaque. Sanskrit might capture the essence of some of these tricky passages, but for other, particularly troublesome sections, all five translations might leave you dumbfounded and only the master text will be intelligible. This is much closer to the situation with the five string theories. Theorists have found that for certain questions, one of the five may give a transparent description of the physical implications, while the descriptions given by the other four are too mathematically complex to be useful. And therein lies the power of Witten's discovery. Prior to his breakthrough, string theory researchers who encountered intractably difficult equations would be stuck. But Witten's work showed that each such question admits four mathematical translations—four mathematical reformulations—and sometimes one of the reformulated questions proves far simpler to answer. Thus, *the dictionary for translating between the five theories can sometimes provide a means for translating impossibly difficult questions into comparatively simple ones.*

It's not foolproof. Just as all five translations of certain passages in that master text might be equally incomprehensible, sometimes the mathematical descriptions given by all five string theories are equally difficult to understand. In such cases, just as we would need to consult the original text itself, we would need full comprehension of the elusive M-theory to make progress. Even so, in a wealth of circumstances, Witten's dictionary has provided a powerful new tool for analyzing string theory.

Hence, just as each translation of a complex text serves an important purpose, each string formulation does too. By combining insights gained from the perspective of each, we are able to answer questions and reveal features that are completely beyond the reach of any single string formulation. Witten's discovery thus gave theorists five times the firepower for advancing string theory's front line. And that, in large part, is why it sparked a revolution.

Eleven Dimensions

So, with our newfound power to analyze string theory, what insights have emerged? There have been many. I will focus on those that have had the greatest impact on the story of space and time.

Of primary importance, Witten's work revealed that the approximate string theory equations used in the 1970s and 1980s to conclude that the universe must have nine space dimensions *missed the true number by one.* The exact answer, his analysis showed, is that the universe according to M-theory has ten space dimensions, that is, eleven spacetime dimensions. Much as Kaluza found that a universe with five spacetime dimensions provided a framework for unifying electromagnetism and gravity, and much as string theorists found that a universe with ten spacetime dimensions provided a framework for unifying quantum mechanics and general relativity, Witten found that a universe with eleven spacetime dimensions provided a framework for unifying all string theories. Like five villages that appear, viewed from ground level, to be completely separate but, when viewed from a mountaintop—making use of an additional, vertical dimension—are seen to be connected by a web of paths and roadways, the additional space dimension emerging from Witten's analysis was crucial to his finding connections between all five string theories.

While Witten's discovery surely fit the historical pattern of achieving unity through more dimensions, when he announced the result at the annual international string theory conference in 1995, it shook the foundations of the field. Researchers, including me, had thought long and hard about the approximate equations being used, and everyone was confident that the analyses had given the final word on the number of dimensions. But Witten revealed something startling.

He showed that all of the previous analyses had made a mathematical simplification tantamount to *assuming* that a hitherto unrecognized tenth spatial dimension would be extremely small, much smaller than all others. So small, in fact, that the approximate string theory equations that all researchers were using lacked the resolving power to reveal even a mathematical hint of the dimension's existence. And that led everyone to conclude that string theory had only nine space dimensions. But with the new insights of the unified M-theoretic framework, Witten was able to go beyond the approximate equations, probe more finely, and demonstrate that one space dimension had been overlooked all along. Thus, Witten showed that the five ten-dimensional frameworks that string theorists had developed for more than a decade were actually five approximate descriptions of a single, underlying eleven-dimensional theory.

You might wonder whether this unexpected realization invalidated previous work in string theory. By and large, it didn't. The newfound tenth spatial dimension added an unanticipated feature to the theory, but if string/M-theory is correct, and should the tenth spatial dimension turn out to be much smaller than all others—as, for a long time, had been unwittingly assumed—previous work would remain valid. However, because the known equations are still unable to nail down the sizes or shapes of extra dimensions, string theorists have expended much effort over the last few years investigating the new possibility of a not-so-small tenth spatial dimension. Among other things, the wide-ranging results of these studies have put the schematic illustration of the unifying power of M-theory, Figure 13.1, on a firm mathematical foundation.

I suspect that the updating from ten to eleven dimensions—regardless of its great importance to the mathematical structure of string/M-theory—doesn't substantially alter your mind's-eye picture of the theory. To all but the cognoscenti, trying to imagine seven curled-up dimensions is an exercise that's pretty much the same as trying to imagine six.

But a second and closely related insight from the second superstring revolution does alter the basic intuitive picture of string theory. The collective insights of a number of researchers—Witten, Duff, Hull, Townsend, and many others—established that *string theory is not just a theory of strings.*

Branes

A natural question, which may have occurred to you in the last chapter, is *Why strings?* Why are one-dimensional ingredients so special? In reconciling quantum mechanics and general relativity, we found it crucial that strings are not dots, that they have nonzero size. But that requirement can be met with two-dimensional ingredients shaped like miniature disks or Frisbees, or by three-dimensional bloblike ingredients, shaped like base-balls or lumps of clay. Or, since the theory has such an abundance of space dimensions, we can even imagine blobs with more dimensions still. Why don't these ingredients play any role in our fundamental theories?

In the 1980s and early 1990s, most string theorists had what seemed like a convincing answer. They argued that there *had* been attempts to formulate a fundamental theory of matter based on bloblike constituents by, among others, such icons of twentieth-century physics as Werner Heisenberg and Paul Dirac. But their work, as well as many subsequent studies, showed that it was extremely difficult to develop a theory based on tiny blobs that met the most basic of physical requirements—for example, ensuring that all quantum mechanical probabilities lie between 0 and 1 (no sense can be made of negative probabilities or of probabilities greater than 1), and debarring faster-than-light communication. For point parti-cles, a half-century of research initiated in the 1920s showed that these conditions could be met (as long as gravity was ignored). And, by the 1980s, more than a decade of investigation by Schwarz, Scherk, Green, and others established, to the surprise of most researchers, that the conditions could also be met for one-dimensional ingredients, strings (which necessarily *included* gravity). But it seemed impossible to proceed to fundamental ingredients with two or more spatial dimensions. The reason, briefly put, is that the number of symmetries respected by the equations peaks enormously for one-dimensional objects (strings) and drops off precipitously thereafter. The symmetries in question are more abstract than the ones discussed in Chapter 8 (they have to do with how equations change if, while studying the motion of a string or a higher dimensional ingredient, we were to zoom in or out, suddenly and arbitrarily changing the resolution of our observations). These transformations prove critical to formulating a physically sensible set of equations, and beyond strings it seemed that the required fecundity of symmetries was absent.[1]

It was thus another shock to most string theorists when Witten's paper and an avalanche of subsequent results[2] led to the realization that string theory, and the M-theoretic framework to which it now belongs, *does* contain ingredients besides strings. The analyses showed that there are two-dimensional objects called, naturally enough, *membranes* (another possible meaning for the "M" in M-theory) or—in deference to systematically naming their higher-dimensional cousins—*two-branes*. There are objects with three spatial dimensions called *three-branes*. And, although increasingly difficult to visualize, the analyses showed that there are also objects with p spatial dimensions, where p can be any whole number less than 10, known—with no derogation intended—as *p-branes*. Thus, strings are but one ingredient in string theory, not *the* ingredient.

These other ingredients escaped earlier theoretical investigation for much the same reason the tenth space dimension did: the approximate string equations proved too coarse to reveal them. In the theoretical contexts that string researchers had investigated mathematically, it turns out that all p-branes are significantly heavier than strings. And the more massive something is, the more energy is required to produce it. But a limitation of the approximate string equations—a limitation embedded in the equations and well known to all string theorists—is that they become less and less accurate when describing entities and processes involving more and more energy. At the extreme energies relevant for p-branes, the approximate equations lacked the precision to expose the branes lurking in the shadows, and that's why decades passed without their presence being noticed in the mathematics. But with the various rephrasings and new approaches provided by the unified M-theoretic framework, researchers were able to skirt some of the previous technical obstacles, and there, in full mathematical view, they found a whole panoply of higher-dimensional ingredients.[3]

The revelation that there are other ingredients besides strings in string theory does not invalidate or make obsolete earlier work any more than the discovery of the tenth spatial dimension did. Research shows that if the higher-dimensional branes are much more massive than strings—as had been unknowingly assumed in previous studies—they have minimal impact on a wide range of theoretical calculations. But just as the tenth space dimension does not have to be much smaller than all others, so the higher-dimensional branes do not have to be much heavier. There are a variety of circumstances, still hypothetical, in which the mass of a higher-dimensional brane can be on a par with the lowest-mass string vibrational patterns, and in such cases the brane *does* have a significant impact on the resulting physics. For example, my own work with Andrew Strominger and David Morrison showed that a brane can wrap itself around a spherical portion of a Calabi-Yau shape, much like plastic wrap vacuum-sealed around a grapefruit; should that portion of space shrink, the wrapped brane would also shrink, causing its mass to decrease. This decrease in mass, we were able to show, allows the portion of space to collapse fully and tear open—space itself can rip apart—while the wrapped brane ensures that there are no catastrophic physical consequences. I discussed this development in detail in *The Elegant Universe* and will briefly return to it when we discuss time travel in Chapter 15, so I won't elaborate further here. But this snippet makes clear how higher-dimensional branes can have a significant effect on the physics of string theory.

For our current focus, though, there is another profound way that branes impact the view of the universe according to string/M-theory. The grand expanse of the cosmos—the entirety of the spacetime of which we are aware—may itself be nothing but an enormous brane. Ours may be a braneworld.

Braneworlds

Testing string theory is a challenge because strings are ultrasmall. But remember the physics that determined the string's size. The messenger particle of gravity—the graviton—is among the lowest-energy string vibrational patterns, and the strength of the gravitational force it communicates is proportional to the length of the string. Since gravity is such a weak force, the string's length must be tiny; calculations show that it must be within a factor of a hundred or so of the Planck length for the string's graviton vibrational pattern to communicate a gravitational force with the observed strength.

Given this explanation, we see that a highly energetic string is not constrained to be tiny, since it no longer has any direct connection to the graviton particle (the graviton is a *low*-energy, zero-mass vibrational pattern). In fact, as more and more energy is pumped into a string, at first it will vibrate more and more frantically. But after a certain point, additional energy will have a different effect: it will cause the string's length to increase, and there's no limit to how long it can grow. By pumping enough energy into a string, you could even make it grow to macroscopic size. With today's technology we couldn't come anywhere near achieving this, but it's possible that in the searingly hot, extremely energetic aftermath of the big bang, long strings were produced. If some have managed to survive until today, they could very well stretch clear across the sky. Although a long shot, it's even possible that such long strings could leave tiny but detectable imprints on the data we receive from space, perhaps allowing string theory to be confirmed one day through astronomical observations.

Higher-dimensional *p*-branes need not be tiny, either, and because they have more dimensions than strings do, a qualitatively new possibility opens up. When we picture a long—perhaps infinitely long—string, we envision a long one-dimensional object that exists within the three large space dimensions of everyday life. A power line stretched as far as the eye can see provides a reasonable image. Similarly, if we picture a large—perhaps infinitely large—two-brane, we envision a large two-dimensional surface that exists within the three large space dimensions of common experience. I don't know of a realistic analogy, but a ridiculously huge drive-in movie screen, extremely thin but as high and as wide as the eye can see, offers a visual image to latch on to. When it comes to a large three-brane, though, we find ourselves in a qualitatively new situation. A three-brane has three dimensions, so if it were large—perhaps infinitely large—it would *fill* all three big spatial dimensions. Whereas a one-brane and a two-brane, like the power line and movie screen, are objects that exist *within* our three large space dimensions, a large three-brane would occupy all the space of which we're aware.

This raises an intriguing possibility. Might we, right now, be living within a three-brane? Like Snow White, whose world exists within a two-dimensional movie screen—a two-brane—that itself resides within a higher-dimensional universe (the three space dimensions of the movie theater), might everything we know exist within a three-dimensional screen—a three-brane—that itself resides within the higher-dimensional universe of string/M-theory? Could it be that what Newton, Leibniz, Mach, and Einstein called three-dimensional space is actually a particular three-dimensional entity in string/M-theory? Or, in more relativistic language, could it be that the four-dimensional spacetime developed by Minkowski and Einstein is actually the wake of a three-brane as it evolves through time? In short, might the universe as we know it be a brane?[4]

The possibility that we are living within a three-brane—the so-called *braneworld scenario*—is the latest twist in string/M-theory's story. As we will see, it provides a qualitatively new way of thinking about string/M-theory, with numerous and far-reaching ramifications. The essential physics is that branes are rather like cosmic Velcro; in a particular way we'll now discuss, they are very sticky.

Sticky Branes and Vibrating Strings

One of the motivations for introducing the term "M-theory" is that we now realize that "string theory" highlights but one of the theory's many ingredients. Theoretical studies revealed one-dimensional strings decades before more refined analyses discovered the higher-dimensional branes, so "string theory" is something of an historical artifact. But, even though M-theory exhibits a democracy in which extended objects of a variety of dimensions are represented, strings still play a central role in our current formulation. In one way this is immediately clear. When all the higher-dimensional p-branes are much heavier than strings, they can be ignored, as researchers had done unknowingly since the 1970s. But there is another, more general way in which strings are first among equals.

In 1995, shortly after Witten announced his breakthrough, Joe Polchinski of the University of California at Santa Barbara got to thinking. Years earlier, in a paper he had written with Robert Leigh and Jin Dai, Polchinski had discovered an interesting though fairly obscure feature of string theory. Polchinski's motivation and reasoning were somewhat technical and the details are not essential to our discussion, but his results are. He found that in certain situations the endpoints of open strings—

remember, these are string segments with two loose ends—would not be able to move with complete freedom. Instead, just as a bead on a wire is free to move, but must follow the wire's contour, and just as a pinball is free to move, but must follow the contours of the pinball table's surface, the endpoints of an open string would be free to move but would be restricted to particular shapes or contours in space. While the string would still be free to vibrate, Polchinski and his collaborators showed that its endpoints would be "stuck" or "trapped" within certain regions.

In some situations, the region might be one-dimensional, in which case the string's endpoints would be like two beads sliding on a wire, with the string itself being like a cord connecting them. In other situations, the region might be two-dimensional, in which case the endpoints of the string would be very much like two pinballs connected by a cord, rolling around a pinball table. In yet other situations, the region might have three, four, or any other number of spatial dimensions less than ten. These results, as shown by Polchinski and also by Petr Hořava and Michael Green, helped resolve a long-standing puzzle in the comparison of open and closed strings, but over the years, the work attracted limited attention.[5] In October 1995, when Polchinski finished rethinking these earlier insights in light of Witten's new discoveries, that changed.

A question that Polchinski's earlier paper left without a complete answer is one that may have occurred to you while reading the last paragraph: If the endpoints of open strings are stuck within a particular region of space, *what is it that they are stuck to?* Wires and pinball machines have a tangible existence independent of the beads or balls that are constrained to move along them. What about the regions of space to which the endpoints of open strings are constrained? Are they filled with some independent and fundamental ingredient of string theory, one that jealously clutches open string endpoints? Prior to 1995, when string theory was thought to be a theory of strings only, there didn't seem to be any candidate for the job. But after Witten's breakthrough and the torrent of results it inspired, the answer became obvious to Polchinski: if the endpoints of open strings are restricted to move within some p-dimensional region of space, then that region of space must be occupied by a p-brane.* His calculations showed that the newly discovered p-branes had exactly the right properties to be the objects that exert an unbreakable grip on open string

---

*The more precise name for these sticky entities is *Dirichlet-p-branes*, or *D-p-branes* for short. We will stick with the shorter *p-brane*.

endpoints, constraining them to move within the *p*-dimensional region of space they fill.

To get a better sense for what this means, look at Figure 13.2. In (a), we see a couple of two-branes with a slew of open strings moving around and vibrating, all with their endpoints restricted to motion along their respective branes. Although it is increasingly difficult to draw, the situation with higher-dimensional branes is identical. Open string endpoints can move freely on and within the *p*-brane, but they can't leave the brane itself. When it comes to the possibility of motion off a brane, branes are the stickiest things imaginable. It's also possible for one end of an open string to be stuck to one *p*-brane and its other end to be stuck to a different *p*-brane, one that may have the same dimension as the first (Figure 13.2b) or may not (Figure 13.2c).

To Witten's discovery of the connection between the various string theories, Polchinski's new paper provided a companion manifesto for the second superstring revolution. While some of the great minds of twentieth-century theoretical physics had struggled and failed to formulate a theory containing fundamental ingredients with more dimensions than dots (zero dimensions) or strings (one dimension), the results of Witten and Polchinski, together with important insights of many of today's leading researchers, revealed the path to progress. Not only did these physicists establish that string/M-theory contains higher-dimensional ingredients,

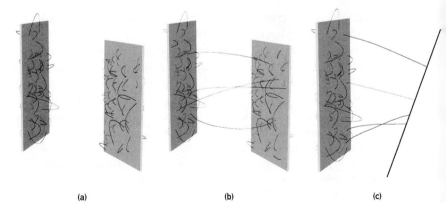

(a)   (b)   (c)

Figure 13.2 (a) Open strings with endpoints attached to two-dimensional branes, or two-branes. **(b)** Strings stretching from one two-brane to another. **(c)** Strings stretching from a two-brane to a one-brane.

but Polchinski's insights in particular provided a means for analyzing their detailed physical properties theoretically (should they prove to exist). The properties of a brane, Polchinski argued, are to a large extent captured by the properties of the vibrating open strings whose endpoints it contains. Just as you can learn a lot about a carpet by running your hand through its pile—the snippets of wool whose endpoints are attached to the carpet backing—many qualities of a brane can be determined by studying the strings whose endpoints it clutches.

That was a paramount result. It showed that decades of research that produced sharp mathematical methods to study one-dimensional objects—strings—could be used to study higher-dimensional objects, p-branes. Wonderfully, then, Polchinski revealed that the analysis of higher-dimensional objects was reduced, to a large degree, to the thoroughly familiar, if still hypothetical, analysis of strings. It's in this sense that strings are special among equals. If you understand the behavior of strings, you're a long way toward understanding the behavior of p-branes.

With these insights, let's now return to the braneworld scenario—the possibility that we're all living out our lives within a three-brane.

Our Universe as a Brane

If we are living within a three-brane—if our four-dimensional spacetime is nothing but the history swept out by a three-brane through time—then the venerable question of whether spacetime is a something would be cast in a brilliant new light. Familiar four-dimensional spacetime would arise from a real physical entity in string/M-theory, a three-brane, not from some vague or abstract idea. In this approach, the reality of our four-dimensional spacetime would be on a par with the reality of an electron or a quark. (Of course, you could still ask whether the larger spacetime within which strings and branes exist—the eleven dimensions of string/M-theory—is itself an entity; the reality of the spacetime arena we directly experience, though, would be rendered obvious.) But if the universe we're aware of really is a three-brane, wouldn't even a casual glance reveal that we are immersed within something—within the three-brane interior?

Well, we've already learned of things within which modern physics suggests we may be immersed—a Higgs ocean; space filled with dark energy; myriad quantum field fluctuations—none of which make themselves directly apparent to unaided human perceptions. So it shouldn't be a shock to learn that string/M-theory adds another candidate to the list of invisible things that may fill "empty" space. But let's not get cavalier. For each of the previous possibilities, we understand its impact on physics and how we might establish that it truly exists. Indeed, for two of the three— dark energy and quantum fluctuations—we've seen that strong evidence supporting their existence has already been gathered; evidence for the Higgs field is being sought at current and future accelerators. So what is the corresponding situation for life within a three-brane? If the brane-world scenario is correct, why don't we see the three-brane, and how would we establish that it exists?

The answer highlights how the physical implications of string/M-theory in the braneworld context differ radically from the earlier "brane-free" (or, as they're sometimes affectionately called, no-braner) scenarios. Consider, as an important example, the motion of light—the motion of photons. In string theory, a photon, as you now know, is a particular string vibrational pattern. More specifically, mathematical studies have shown that in the braneworld scenario, only open string vibrations, not closed ones, produce photons, and this makes a big difference. Open string end-points are constrained to move within the three-brane, but are otherwise completely free. This implies that photons (open strings executing the photon mode of vibration) would travel without any constraint or obstruc-tion throughout our three-brane. And that would make the brane appear *completely transparent—completely invisible*—thus preventing us from seeing that we are immersed within it.

Of equal importance, because open string endpoints cannot leave a brane, they are unable to move into the extra dimensions. Just as the wire constrains its beads and the pinball machine constrains its balls, our sticky three-brane would permit photons to move *only* within our three spatial dimensions. Since photons are the messenger particles for electromagnet-ism, this implies that the electromagnetic force—light—would be trapped within our three dimensions, as illustrated (in two dimensions so we can draw it) in Figure 13.3.

That's an intense realization with important consequences. Earlier, we required the extra dimensions of string/M-theory to be tightly curled up. The reason, clearly, is that we don't see the extra dimensions and so they must be hidden away. And one way to hide them is to make them smaller than we or our equipment can detect. But let's now reexamine

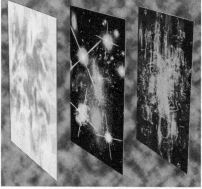

Figure 13.3 (a) In the braneworld scenario, photons are open strings with endpoints trapped within the brane, so they—light—cannot leave the brane itself. **(b)** Our braneworld could be floating in a grand expanse of additional dimensions that remain invisible to us, because the light we see cannot leave our brane. There might also be other braneworlds floating nearby.

this issue in the braneworld scenario. How do we detect things? Well, when we use our eyes, we use the electromagnetic force; when we use powerful instruments like electron microscopes, we also use the electromagnetic force; when we use atom smashers, one of the forces we use to probe the ultrasmall is, once again, the electromagnetic force. But if the electromagnetic force is confined to our three-brane, our three space dimensions, it is *unable* to probe the extra dimensions, regardless of their size. Photons cannot escape our dimensions, enter the extra dimensions, and then travel back to our eyes or equipment allowing us to detect the extra dimensions, *even if they were as large as the familiar space dimensions.*

So, if we live in a three-brane, there is an alternative explanation for why we're not aware of the extra dimensions. It is not necessarily that the extra dimensions are extremely small. They could be big. We don't see them because of the *way* we see. We see by using the electromagnetic force, which is unable to access any dimensions beyond the three we know about. Like an ant walking along a lily pad, completely unaware of the deep waters lying just beneath the visible surface, we could be floating within a grand, expansive, higher-dimensional space, as in Figure 13.3b, but the electromagnetic force—eternally trapped within our dimensions—would be unable to reveal this.

Okay, you might say, but the electromagnetic force is only one of nature's four forces. What about the other three? Can they probe into the extra dimensions, thus enabling us to reveal their existence? For the strong and weak nuclear forces, the answer is, again, no. In the braneworld scenario, calculations show that the messenger particles for these forces—gluons and W and Z particles—also arise from open-string vibrational patterns, so they are just as trapped as photons, and processes involving the strong and weak nuclear forces are just as blind to the extra dimensions. The same goes for particles of matter. Electrons, quarks, and all other particle species also arise from the vibrations of open strings with trapped endpoints. *Thus, in the braneworld scenario, you and I and everything we've ever seen are permanently imprisoned within our three-brane.* Taking account of time, everything is trapped within our four-dimensional slice of spacetime.

Well, almost everything. For the force of gravity, the situation is different. Mathematical analyses of the braneworld scenario have shown that gravitons arise from the vibrational pattern of closed strings, much as they do in the previously discussed no-braner scenarios. And closed strings—strings with no endpoints—are not trapped by branes. They are as free to leave a brane as they are to roam on or through it. So, if we were living in a brane, we would not be completely cut off from the extra dimensions. Through the gravitational force, we could both influence and be influenced by the extra dimensions. Gravity, in such a scenario, would provide our sole means for interacting beyond our three space dimensions.

How big could the extra dimensions be before we'd become aware of them through the gravitational force? This is an interesting and critical question, so let's take a look.

Gravity and Large Extra Dimensions

Back in 1687, when Newton proposed his universal law of gravity, he was actually making a strong statement about the number of space dimensions. Newton didn't just say that the force of attraction between two objects gets weaker as the distance between them gets larger. He proposed a formula, the *inverse square law*, which describes precisely how the gravitational attraction will diminish as two objects are separated. According to this formula, if you double the distance between the objects, their gravitational attraction will fall by a factor of 4 (2^2); if you triple the distance, it will fall by a factor of 9 (3^2); if you quadruple the distance, it will fall by a factor of 16 (4^2); and more generally, the gravitational force drops in proportion to the square of the separation. As has become abundantly evident over the last few hundred years, this formula works.

But *why* does the force depend on the square of the distance? Why doesn't the force drop like the cube of the separation (so that if you double the distance, the force diminishes by a factor of 8) or the fourth power (so that if you double the distance, the force diminishes by a factor of 16), or perhaps, even more simply, why doesn't the gravitational force between two objects drop in direct proportion to the separation (so that if you double the distance, the force diminishes by a factor of 2)? The answer is tied directly to the number of dimensions of space.

One way to see this is to think about how the number of gravitons emitted and absorbed by the two objects depends on their separation, or by thinking about how the curvature of spacetime that each object experiences diminishes as the distance between them increases. But let's take a simpler, more old-fashioned approach, which gets us quickly and intuitively to the correct answer. Let's draw a figure (Figure 13.4a) that schematically illustrates the gravitational field produced by a massive object—let's say the sun—much as Figure 3.1 schematically illustrates the magnetic field produced by a bar magnet. Whereas magnetic field lines sweep around from the magnet's north pole to its south pole, notice that gravitational field lines emanate radially outward in all directions and just keep on going. The strength of the gravitational pull another object—imagine it's an orbiting satellite—would feel at a given distance is proportional to the density of field lines at that location. The more field lines penetrate the satellite, as in Figure 13.4b, the greater the gravitational pull to which it is subject.

We can now explain the origin of Newton's inverse square law. An imaginary sphere centered on the sun and passing through the satellite's location, as in Figure 13.4c, has a surface area that—like the surface of any sphere in three-dimensional space—is proportional to the *square* of its radius, which in this case is the *square* of the distance between the sun and the satellite. This means that the density of field lines passing through the sphere—the total number of field lines divided by the sphere's area—decreases as the square of sun-satellite separation. If you double the dis-

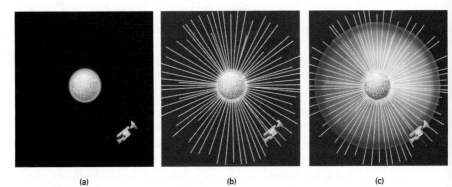

(a)   (b)   (c)

Figure 13.4 (a) The gravitational force exerted by the sun on an object, such as a satellite, is inversely proportional to the square of the distance between them. The reason is that the sun's gravitational field lines spread out uniformly as in (b) and hence have a density at a distance d that is inversely proportional to the area of an imaginary sphere of radius d—schematically drawn in (c)—an area which basic geometry shows to be proportional to d^2.

tance, the same number of field lines are now uniformly spread out on a sphere with four times the surface area, and hence the gravitational pull at that distance will drop by a factor of four. Newton's inverse square law for gravity is thus a reflection of a geometrical property of spheres in three space dimensions.

By contrast, if the universe had two or even just one space dimension, how would Newton's formula change? Well, Figure 13.5a shows a two-dimensional version of the sun and its orbiting satellite. As you can see, at any given distance the sun's gravitational field lines uniformly spread out on a circle, the analog of a sphere in one lower dimension. Since the circle's circumference is proportional to its radius (not to the square of its radius), if you double the sun–satellite separation, the density of field lines will decrease by a factor of 2 (not 4) and so the strength of the sun's gravitational pull will drop only by a factor of 2 (not 4). If the universe had only two space dimensions, then, gravitational pull would be inversely proportional to separation, not the square of separation.

If the universe had only one space dimension, as in Figure 13.5b, the law of gravity would be simpler still. Gravitational field lines would have no room to spread out, and so the force of gravity would not decrease with separation. If you were to double the distance between the sun and the

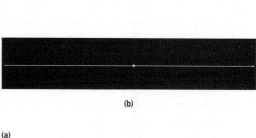

(b)

(a)

Figure 13.5 (a) In a universe with only two spatial dimensions, the gravitational force drops in proportion to separation, because gravitational field lines uniformly spread on a circle whose circumference is proportional to its radius. **(b)** In a universe with one space dimension, gravitational field lines do not have any room to spread, so the gravitational force is constant, regardless of separation.

satellite (assuming that versions of such objects could exist in such a universe), the same number of field lines would penetrate the satellite and hence the force of gravity acting between them would not change at all.

Although it is impossible to draw, the pattern illustrated by Figures 13.4 and 13.5 extends directly to a universe with four or five or six or any number of space dimensions. The more space dimensions there are, the more room gravitational lines of force have to spread out. And the more they spread out, the more precipitously the force of gravity drops with increasing separation. In four space dimensions, Newton's law would be an inverse cube law (double the separation, force drops by a factor of 8); in five space dimensions, it would be an inverse fourth-power law (double the separation, force drops by a factor of 16); in six space dimensions, it would be an inverse fifth-power law (double the separation, force drops by a factor of 32); and so on for ever higher-dimensional universes.

You might think that the success of the inverse square version of Newton's law in explaining a wealth of data—from the motion of planets to the paths of comets—confirms that we live in a universe with precisely three space dimensions. But that conclusion would be hasty. We know that the inverse square law works on astronomical scales,[6] and we know that it works on terrestrial scales, and that jibes well with the fact that on such scales we see three space dimensions. But do we know that it works on smaller scales? How far down into the microcosmos has gravity's inverse square law been tested? As it turns out, experimenters have confirmed it down to only about a tenth of a millimeter; if two objects are brought to within a separation of a tenth of a millimeter, the data verify that the strength of their gravitational attraction follows the predictions of the inverse square law. But so far, it has proven a significant technical challenge to test the inverse square law on shorter scales (quantum effects and the weakness of gravity complicate the experiments). This is a critical issue, because deviations from the inverse square law would be a convincing signal of extra dimensions.

To see this explicitly, let's work with a lower-dimensional toy example that we can easily draw and analyze. Imagine we lived in a universe with one space dimension—or so we thought, because only one space dimension was visible and, moreover, centuries of experiments had shown that the force of gravity does not vary with the separation between objects. But also imagine that in all those years experimenters had been able only to test the law of gravity down to distances of about a tenth of a millimeter. For distances shorter than that, no one had any data. Now, imagine further that, unbeknownst to everyone but a handful of fringe theoretical physicists, the universe actually had a second, curled-up space dimension making its shape like the surface of Philippe Petit's tightrope, as in Figure 12.5. How would this affect future, more refined gravitational tests? We can deduce the answer by looking at Figure 13.6. As two tiny objects are brought close enough together—much closer than the circumference of the curled-up dimension—the two-dimensional character of space would become apparent immediately, because on those scales gravitational field lines *would* have room to spread out (Figure 13.6a). Rather than being

(a)   (b)

Figure 13.6 (a) When objects are close, the gravitational pull varies as it does in two space dimensions. **(b)** When objects are farther apart, the gravitational pull behaves as it does in one space dimension—it is constant.

independent of distance, the force of gravity would vary *inversely* with separation when objects were close enough together.

Thus, if you were an experimenter in this universe, and you developed exquisitely accurate methods for measuring gravitational attraction, here's what you would find. When two objects were extremely close, much closer than the size of the curled-up dimension, their gravitational attraction would diminish in proportion to their separation, just as you expect for a universe with two space dimensions. But then, when the objects were about as far apart as the circumference of the curled-up dimension, things would change. Beyond this distance, the gravitational field lines would be unable to spread any further. They would have spilled out as far as they could into the second curled-up dimension—they would have saturated that dimension—and so from this distance onward the gravitational force would no longer diminish, as illustrated in Figure 13.6b. You can compare this saturation with the plumbing in an old house. If someone opens the faucet in the kitchen sink when you're just about to rinse the shampoo out of your hair, the water pressure can drop because the water spreads between the two outlets. The pressure will diminish yet again should someone open the faucet in the laundry room, since the water will spread even more. But once all the faucets in the house are open, the pressure will remain constant. Although it might not provide the relaxing, high-water-pressure experience you'd anticipated, the pressure in the shower will not drop any further once the water has completely spread throughout all "extra" outlets. Similarly, once the gravitational field has completely spread throughout the extra curled-up dimension, it will not diminish with further separation.

From your data you would deduce two things. First, from the fact that the gravitational force diminished in proportion to distance when objects are very close, you'd realize that the universe has *two* space dimensions, not one. Second, from the crossover to a gravitational force that is constant—the result known from hundreds of years of previous experiments—you'd conclude that one of these dimensions is curled up, with a size about equal to the distance at which the crossover takes place. And with this result, you'd overturn centuries, if not millennia, of belief regarding something so basic, the number of space dimensions, that it seemed almost beyond questioning.

Although I set this story in a lower-dimensional universe, for visual ease, our situation could be much the same. Hundreds of years of experiments have confirmed that gravity varies inversely with the square of distance, giving strong evidence that there are three space dimensions. But as of 1998, no experiment had ever probed gravity's strength on separations smaller than a millimeter (today, as mentioned, this has been pushed to a tenth of a millimeter). This led Savas Dimopoulos, of Stanford, Nima Arkani-Hamed, now of Harvard, and Gia Dvali, of New York University, to propose that *in the braneworld scenario extra dimensions could be as large as a millimeter and would still have been undetected.* This radical suggestion inspired a number of experimental groups to initiate a study of gravity at submillimeter distances in hopes of finding violations of the inverse square law; so far, none have been found, down to a tenth of a millimeter. Thus, even with today's state-of-the-art gravity experiments, *if we are living within a three-brane, the extra dimensions could be as large as a tenth of a millimeter, and yet we wouldn't know it.*

This is one of the most striking realizations of the last decade. Using the three nongravitational forces, we can probe down to about a billionth of a billionth (10^{-18}) of a meter, and no one has found any evidence of extra dimensions. But in the braneworld scenario, the nongravitational forces are impotent in searching for extra dimensions since they are trapped on the brane itself. Only gravity can give insight into the nature of the extra dimensions, and, as of today, the extra dimensions could be as thick as a human hair and yet they'd be completely invisible to our most sophisticated instruments. Right now, right next to you, right next to me, and right next to everyone else, there could be another spatial dimension—a dimension beyond left/right, back/forth, and up/down, a dimension that's curled up but still large enough to swallow something as thick as this page—that remains beyond our grasp.*

Large Extra Dimensions and Large Strings

By trapping three of the four forces, the braneworld scenario significantly relaxes experimental constraints on how big the extra dimensions can be, but the extra dimensions aren't the only thing this approach allows to get bigger. Drawing on insights of Witten, Joe Lykken, Constantin Bachas,

---

*There is even a proposal, from Lisa Randall, of Harvard, and Raman Sundrum, of Johns Hopkins, in which gravity too can be trapped, not by a sticky brane, but by extra dimensions that curve in just the right way, relaxing even further the constraints on their size.

and others, Ignatios Antoniadis, together with Arkani-Hamed, Dimopoulos, and Dvali, realized that in the braneworld scenario even unexcited, low-energy strings can be *much* larger than previously thought. In fact, the two scales—the size of extra dimensions and the size of strings—are closely related.

Remember from the previous chapter that the basic size of string is determined by requiring that its graviton vibrational pattern communicate a gravitational force of the observed strength. The weakness of gravity translates into the string's being very short, about the Planck length (10^{-33} centimeters). But this conclusion is highly dependent on the size of the extra dimensions. The reason is that in string/M-theory, the strength of the gravitational force we observe in our three extended dimensions represents an interplay between two factors. One factor is the intrinsic, fundamental strength of the gravitational force. The second factor is the size of the extra dimensions. The larger the extra dimensions, the more gravity can spill into them and the weaker its force will *appear* in the familiar dimensions. Just as larger pipes yield weaker water pressure because they allow water more room to spread out, so larger extra dimensions yield weaker gravity, because they give gravity more room to spread out.

The original calculations that determined the string's length assumed that the extra dimensions were so tiny, on the order of the Planck length, that gravity couldn't spill into them at all. Under this assumption, gravity appears weak because it *is* weak. But now, if we work in the braneworld scenario and allow the extra dimensions to be much larger than had previously been considered, the observed feebleness of gravity no longer means that it's intrinsically weak. Instead, gravity could be a relatively powerful force that appears weak only because the relatively large extra dimensions, like large pipes, dilute its intrinsic strength. Following this line of reasoning, if gravity is much stronger than once thought, the strings can be much longer than once thought, too.

As of today, the question of exactly how long doesn't have a unique, definite answer. With the newfound freedom to vary both the size of strings and the size of the extra dimensions over a much wider range than previously envisioned, there are a number of possibilities. Dimopoulos and his collaborators have argued that existing experimental results, both from particle physics and from astrophysics, show that unexcited strings can't be larger than about a billionth of a billionth of a meter (10^{-18} meters). While small by everyday standards, this is about a hundred million billion (10^{17}) times larger than the Planck length—*nearly a hundred million billion times larger than previously thought.* As we'll now see, that would be large enough that signs of strings could be detected by the next generation of particle accelerators.

String Theory Confronts Experiment?

The possibility that we are living within a large three-brane is, of course, just that: a possibility. And, within the braneworld scenario, the possibility that the extra dimensions could be much larger than once thought—and the related possibility that strings could also be much larger than once thought—are also just that: possibilities. *But they are tremendously exciting possibilities.* True, even if the braneworld scenario is right, the extra dimensions and the string size could still be Planckian. But the possibility within string/M-theory for strings and the extra dimensions to be much larger—to be just beyond the reach of today's technology—is fantastic. It means that there is at least a chance that in the next few years, string/M-theory will make contact with observable physics and become an experimental science.

How big a chance? I don't know, and nor does anyone else. My intuition tells me it's unlikely, but my intuition is informed by a decade and a half of working within the conventional framework of Planck-sized strings and Planck-sized extra dimensions. Perhaps my instincts are old-fashioned. Thankfully, the question will be settled without the slightest concern for anyone's intuition. If the strings are big, or if some of the extra dimensions are big, the implications for upcoming experiments are spectacular.

In the next chapter, we'll consider a variety of experiments that will test, among other things, the possibilities of comparatively large strings and large extra dimensions, so here I will just whet your appetite. If strings are as large as a billionth of a billionth (10^{-18}) of a meter, the particles corresponding to the higher harmonic vibrations in Figure 12.4 will not have enormous masses, in excess of the Planck mass, as in the standard scenario. Instead, their masses will be only a thousand to a few thousand times that of a proton, and that's low enough to be within reach of the Large Hadron Collider now being built at CERN. If these string vibrations were to be excited through high-energy collisions, the accelerator's detectors would light up like the Times Square crystal ball on New Year's Eve. A whole host of never-before-seen particles would be produced, and their masses would be related to one another's much as the various harmonics are related on a cello. String theory's signature would be etched across the data with a flourish that would have impressed John Hancock. Researchers wouldn't be able to miss it, even without their glasses.

Moreover, in the braneworld scenario, high-energy collisions might even produce—get this—miniature black holes. Although we normally think of black holes as gargantuan structures out in deep space, it's been known since the early days of general relativity that if you crammed enough matter together in the palm of your hand, you'd create a tiny black hole. This doesn't happen because no one's grip—and no mechanical device—is even remotely strong enough to exert a sufficient compression force. Instead, the only accepted mechanism for black hole production involves the gravitational pull of an enormously massive star's overcoming the outward pressure normally exerted by the star's nuclear fusion processes, causing the star to collapse in on itself. But if gravity's intrinsic strength on small scales is far greater than previously thought, tiny black holes could be produced with significantly less compression force than previously believed. Calculations show that the Large Hadron Collider may have just enough squeezing power to create a cornucopia of microscopic black holes through high-energy collisions between protons.[7] Think about how amazing that would be. The Large Hadron Collider might turn out to be a factory for producing microscopic black holes! These black holes would be so small and would last for such a short time that they wouldn't pose us the slightest threat (years ago, Stephen Hawking showed that all black holes disintegrate via quantum processes—big ones very slowly, tiny ones very quickly), but their production would provide confirmation of some of the most exotic ideas ever contemplated.

Braneworld Cosmology

A primary goal of current research, one that is being hotly pursued by scientists worldwide (including me), is to formulate an understanding of cosmology that incorporates the new insights of string/M-theory. The reason is clear: not only does cosmology grapple with the big, gulp-in-the-throat questions, and not only have we come to realize that aspects of familiar experience—such as the arrow of time—are bound up with conditions at the universe's birth, but cosmology also provides a theorist with what New York provided Sinatra: a proving ground par excellence. If a theory can make it in the extreme conditions characteristic of the universe's earliest moments, it can make it anywhere.

As of today, cosmology according to string/M-theory is a work in progress, with researchers heading down two main pathways. The first and more conventional approach imagines that just as inflation provided a brief but profound front end to the standard big bang theory, string/M-theory provides a yet earlier and perhaps yet more profound front end to inflation. The vision is that string/M-theory will unfuzz the fuzzy patch we've used to denote our ignorance of the universe's earliest moments, and after that, the cosmological drama will unfold according to inflationary theory's remarkably successful script, recounted in earlier chapters.

While there has been progress on specific details required by this vision (such as trying to understand why only three of the universe's spatial dimensions underwent expansion, as well as developing mathematical methods that may prove relevant to analyzing the spaceless/timeless realm that may precede inflation), the eureka moments have yet to occur. The intuition is that whereas inflationary cosmology imagines the observable universe getting ever smaller at ever earlier times—and hence being ever hotter, denser, and energetic—string/M-theory tames this unruly (in physics-speak, "singular") behavior by introducing a minimal size (as in our discussion on pages 350–351) below which new and less singular physical quantities become relevant. This reasoning lies at the heart of string/M-theory's successful merger of general relativity and quantum mechanics, and many researchers expect that we will shortly determine how to apply the same reasoning in the context of cosmology. But, as of now, the fuzzy patch still looks fuzzy, and it's anybody's guess when clarity will be achieved.

The second approach employs the braneworld scenario, and in its most radical incarnation posits a completely new cosmological framework. It is far from clear whether this approach will survive detailed mathematical scrutiny, but it does provide a good example of how breakthroughs in fundamental theory can suggest novel trails through well-trodden territory. The proposal is called the *cyclic model.*

Cyclic Cosmology

From the standpoint of time, ordinary experience confronts us with two types of phenomena: those that have a clearly delineated beginning, middle, and end (this book, a baseball game, a human life) and those that are cyclic, happening over and over again (the changing seasons, the rising and setting of the sun, Larry King's weddings). Of course, on closer scrutiny we learn that cyclic phenomena also have a beginning and end, since cycles do not generally go on forever. The sun has been rising and setting—that is, the earth has been spinning on its axis while revolving around the sun—every day for some 5 billion years. But before that, the sun and the solar system had yet to form. And one day, some 5 billion years from now, the sun will turn into a red giant star, engulfing the inner planets, including earth, and there will no longer even be a notion of a rising and setting sun, at least not here.

But these are modern scientific recognitions. To the ancients, cyclic phenomena seemed eternally cyclic. And to many, the cyclic phenomena, running their course and continuously returning to begin anew, were the primary phenomena. The cycles of days and seasons set the rhythm of work and life, so it is no wonder that some of the oldest recorded cosmologies envisioned the unfolding of the world as a cyclical process. Rather than positing a beginning, a middle, and an end, a cyclic cosmology imagines that the world changes through time much as the moon changes through phases: after it has passed through a complete sequence, conditions are ripe for everything to start afresh and initiate yet another cycle.

Since the discovery of general relativity, a number of cyclic cosmological models have been proposed; the best-known was developed in the 1930s by Richard Tolman of the California Institute of Technology. Tolman suggested that the observed expansion of the universe might slow down, someday stop, and then be followed by a period of contraction in which the universe got ever smaller. But instead of reaching a fiery finale in which it implodes on itself and comes to an end, the universe might, Tolman proposed, undergo a *bounce*: space might shrink down to some small size and then rebound, initiating a new cycle of expansion followed once again by contraction. A universe eternally repeating this cycle—expansion, contraction, bounce, expansion again—would elegantly avoid the thorny issues of origin: in such a scenario, the very concept of origin would be inapplicable since the universe always was and would always be.

But Tolman realized that looking back in time from today, the cycles could have repeated for a while, but not indefinitely. The reason is that during each cycle, the second law of thermodynamics dictates that entropy would, on average, rise.[8] And according to general relativity, the amount of entropy at the beginning of each new cycle determines how long that cycle will last. More entropy means a longer period of expansion before the outward motion grinds to a halt and the inward motion takes over. Each successive cycle would therefore last much longer than its predecessor; equivalently, earlier cycles would be shorter and shorter. When analyzed mathematically, the constant shortening of the cycles implies that they cannot stretch infinitely far into the past. Even in Tolman's cyclic framework, the universe would have a beginning.

Tolman's proposal invoked a spherical universe, which, as we've seen, has been ruled out by observations. But a radically new incarnation of cyclic cosmology, involving a flat universe, has recently been developed within string/M-theory. The idea comes from Paul Steinhardt and his collaborator Neil Turok of Cambridge University (with heavy use of results discovered in their collaborations with Burt Ovrut, Nathan Seiberg, and Justin Khoury) and proposes a new mechanism for driving cosmic evolution.[9] Briefly put, they suggest that we are living within a three-brane that violently collides every few trillion years with another nearby, parallel three-brane. And the "bang" from the collision initiates each new cosmological cycle.

The basic setup of the proposal is illustrated in Figure 13.7 and was suggested some years ago by Hořava and Witten in a noncosmological

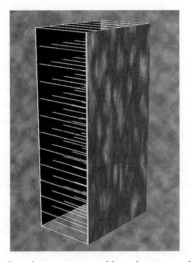

Figure 13.7 Two three-branes, separated by a short interval.

context. Hořava and Witten were trying to complete Witten's proposed unity among all five string theories and found that if one of the seven extra dimensions in M-theory had a very simple shape—not a circle, as in Figure 12.7, but a little segment of a straight line, as in Figure 13.7—and was bounded by so-called end-of-the-world branes attached like bookends, then a direct connection could be made between the Heterotic-E string theory and all others. The details of how they drew this connection are neither obvious nor of the essence (if you are interested, see, for example, *The Elegant Universe*, Chapter 12); what matters here is that it's a starting point that naturally emerges from the theory itself. Steinhardt and Turok enlisted it for their cosmological proposal.

Specifically, Steinhardt and Turok imagine that each brane in Figure 13.7 has three space dimensions, with the line segment between them providing a fourth space dimension. The remaining six space dimensions are curled up into a Calabi-Yau space (not shown in the figure) that has the right shape for string vibrational patterns to account for the known particle species.[10] The universe of which we are directly aware corresponds to one of these three-branes; if you like, you can think of the second three-brane as another universe, whose inhabitants, if any, would also be aware of only three space dimensions, assuming that their experimental technology and expertise did not greatly exceed ours. In this setup, then, another three-brane—another universe—is right next door. It's hovering no more than a fraction of a millimeter away (the separation being in the fourth spatial dimension, as in Figure 13.7), but because our three-brane is so sticky and the gravity we experience so weak, we have no direct evidence of its existence, nor its hypothetical inhabitants any evidence of ours.

But, according to the cyclic cosmological model of Steinhardt and Turok, Figure 13.7 isn't how it's always been or how it will always be. Instead, in their approach, the two three-branes are attracted to each other—almost as though connected by tiny rubber bands—and this implies that each drives the cosmological evolution of the other: the branes engage in an endless cycle of collision, rebound, and collision once again, eternally regenerating their expanding three-dimensional worlds. To see how this goes, look at Figure 13.8, which illustrates one complete cycle, step by step.

At Stage 1, the two three-branes have just rushed toward each other and slammed together, and are now rebounding. The tremendous energy of the collision deposits a significant amount of high-temperature radiation and matter on each of the rebounding three-branes, and—this is

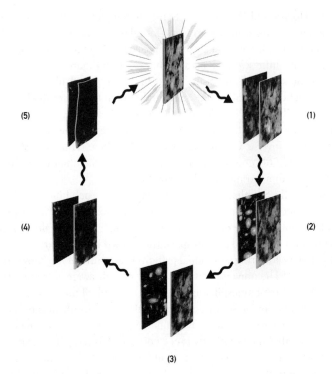

Figure 13.8 Various stages in the cyclic braneworld cosmological model.

key—Steinhardt and Turok argue that *the detailed properties of this matter and radiation have a nearly identical profile to what's produced in the inflationary model.* Although there is still some controversy on this point, Steinhardt and Turok therefore claim that the collision between the two three-branes results in physical conditions extremely close to what they'd be a moment after the burst of inflationary expansion in the more conventional approach discussed in Chapter 10. Not surprisingly, then, to a hypothetical observer within our three-brane, the next few stages in the cyclic cosmological model are essentially the same as those in the standard approach as illustrated in Figure 9.2 (where that figure is now interpreted as depicting evolution on one of the three-branes). Namely, as our three-brane rebounds from the collision, it expands and cools, and cosmic structures such as stars and galaxies gradually coalesce from the primordial plasma, as you can see in Stage 2. Then, inspired by the recent supernova observations discussed in Chapter 10, Steinhardt and Turok configure their model so that about 7 billion years into the cycle—Stage 3—the energy in ordinary matter and radiation becomes sufficiently diluted by the expansion of the brane so that a dark energy component gains the upper hand and, through its negative pressure, drives an era of accelerated expansion. (This requires an arbitrary tuning of details, but it allows the model to match observation, and so, the cyclic model's proponents argue, is well motivated.) About 7 billion years later, we humans find ourselves here on earth, at least in the current cycle, experiencing the early stages of the accelerated phase. Then, for roughly the next *trillion* years, not much new happens beyond our three-brane's continued accelerated expansion. That's long enough for our three-dimensional space to have stretched by such a colossal factor that matter and radiation are diluted almost completely away, leaving the braneworld looking almost completely empty and completely uniform: Stage 4.

By this point, our three-brane has completed its rebound from the initial collision and has started to approach the second three-brane once again. As we get closer and closer to another collision, quantum jitters of the strings attached to our brane overlie its uniform emptiness with tiny ripples, Stage 5. As we continue to pick up speed, the ripples continue to grow; then, in a cataclysmic collision, we smack into the second three-brane, we bounce off, and the cycle starts anew. The quantum ripples imprint tiny inhomogeneities in the radiation and matter produced during the collision and, much as in the inflationary scenario, these deviations from perfect uniformity grow into clumps that ultimately generate stars and galaxies.

These are the major stages in the cyclic model (also known tenderly as the *big splat*). Its premise—colliding braneworlds—is very different from that of the successful inflationary theory, but there are, nevertheless, significant points of contact between the two approaches. That both rely on quantum agitation to generate initial nonuniformities is one essential similarity. In fact, Steinhardt and Turok argue that the equations governing the quantum ripples in the cyclic model are nearly identical to those in the inflationary picture, so the resulting nonuniformities predicted by the two theories are nearly identical as well.[11] Moreover, while there isn't an inflationary burst in the cyclic model, there is a trillion-year period (beginning at Stage 3) of milder accelerated expansion. But it's really just a matter of haste versus patience; what the inflationary model accomplishes in a flash, the cyclic model accomplishes in a comparative eternity. Since the collision in the cyclic model is not the beginning of the universe, there is the luxury of slowly resolving cosmological issues (like the flatness and horizon problems) during the final trillion years of each *previous* cycle. Eons of gentle but steady accelerated expansion at the end of each cycle stretch our three-brane nice and flat, and, except for tiny but important quantum fluctuations, make it thoroughly uniform. And so the long, final stage of each cycle, followed by the splat at the beginning of the next cycle, yields an environment very close to that produced by the short surge of expansion in the inflationary approach.

A Brief Assessment

At their present levels of development, both the inflationary and the cyclic models provide insightful cosmological frameworks, but neither offers a complete theory. Ignorance of the prevailing conditions during the universe's earliest moments forces proponents of inflationary cosmology to simply assume, without theoretical justification, that the conditions required for initiating inflation arose. If they did, the theory resolves numerous cosmological conundrums and launches time's arrow. But such successes hinge on inflation's happening in the first place. What's more, inflationary cosmology has not been seamlessly embedded within string theory and so is not yet part of a consistent merger of quantum mechanics and general relativity.

The cyclic model has its own share of shortcomings. As with Tolman's model, consideration of entropy buildup (and also of quantum mechanics[12]) ensures that the cyclic model's cycles could not have gone on forever. Instead, the cycles began at some definite time in the past, and so, as with inflation, we need an explanation of how the first cycle got started. If it did, then the theory, also like inflation, resolves the key cosmological problems and sets time's arrow pointing from each low-entropy splat forward through the ensuing stages of Figure 13.8. But, as it's currently conceived, the cyclic model offers no explanation of how or why the universe finds itself in the necessary configuration of Figure 13.8. Why, for instance, do six dimensions curl themselves up into a particular Calabi-Yau shape while one of the extra dimensions dutifully takes the shape of a spatial segment separating two three-branes? How is it that the two end-of-the-world three-branes line up so perfectly and attract each other with just the right force so that the stages in Figure 13.8 proceed as we've described? And, of critical importance, what actually happens when the two three-branes collide in the cyclic model's version of a bang?

On this last question, there is hope that the cyclic model's splat is less problematic than the singularity encountered at time zero in inflationary cosmology. Instead of all of space being infinitely compressed, in the cyclic approach only the single dimension between the branes gets squeezed down; the branes themselves experience overall expansion, not contraction, during each cycle. And this, Steinhardt, Turok, and their collaborators have argued, implies *finite* temperature and *finite* densities on the branes themselves. But this is a highly tentative conclusion because, so far, no one has been able to get the better of the equations and figure out what would happen should branes slam together. In fact, the analyses so far completed point toward the splat being subject to the same problem that afflicts the inflationary theory at time zero: the mathematics breaks down. Thus, cosmology is still in need of a rigorous resolution of its singular start—be it the true start of the universe, or the start of our current cycle.

The most compelling feature of the cyclic model is the way it incorporates dark energy and the observed accelerated expansion. In 1998, when it was discovered that the universe is undergoing accelerated expansion, it was quite a surprise to most physicists and astronomers. While it can be incorporated into the inflationary cosmological picture by assuming that the universe contains precisely the right amount of dark energy, accelerated expansion seems like a clumsy add-on. In the cyclic model, by contrast, dark energy's role is natural and pivotal. The trillion-year period of slow but steadily accelerated expansion is crucial for wiping the slate clean, for diluting the observable universe to near nothingness, and for resetting conditions in preparation for the next cycle. From this point of view, both the inflationary model and the cyclic model rely on accelerated expansion—the inflationary model near its beginning and the cyclic model at the end of each of its cycles—but only the latter has direct observational support. (Remember, the cyclic approach is designed so that we are just entering the trillion-year phase of accelerated expansion, and such expansion has been recently observed.) That's a tick in the cyclic model's column, but it also means that should accelerated expansion fail to be confirmed by future observations, the inflationary model could survive (although the puzzle of the missing 70 percent of the universe's energy budget would emerge anew) but the cyclic model could not.

New Visions of Spacetime

The braneworld scenario and the cyclic cosmological model it spawned are both highly speculative. I have discussed them here not so much because I feel certain that they are correct, as because I want to illustrate the striking new ways of thinking about the space we inhabit and the evolution it has experienced that have been inspired by string/M-theory. If we are living within a three-brane, the centuries-old question regarding the corporeality of three-dimensional space would have its most definite answer: space would be a brane, and hence would most definitely be a something. It might also not be anything particularly special as there could be many other branes, of various dimensions, floating within string/M-theory's higher dimensional expanse. And if cosmological evolution on our three-brane is driven by repeated collisions with a nearby brane, time as we know it would span only one of the universe's many cycles, with one big bang followed by another, and then another.

To me, it's a vision that's both exciting and humbling. There may be much more to space and time than we anticipated; if there is, what we consider to be "everything" may be but a small constituent of a far richer reality.

V

REALITY AND
IMAGINATION

Up in the Heavens and Down in the Earth

EXPERIMENTING WITH SPACE AND TIME

We've come a long way since Empedocles of Agrigento explained the universe using earth, air, fire, and water. And much of the progress we've made, from Newton through the revolutionary discoveries of the twentieth century, has been borne out spectacularly by experimental confirmation of detailed and precise theoretical predictions. But since the mid-1980s, we've been the victims of our own success. With the incessant urge to push the limits of understanding ever further, our theories have entered realms beyond the reach of our current technology.

Nevertheless, with diligence and luck, many forefront ideas will be tested during the next few decades. As we'll discuss in this chapter, experiments either planned or under way have the potential to give much insight into the existence of extra dimensions, the composition of dark matter and dark energy, the origin of mass and the Higgs ocean, aspects of early-universe cosmology, the relevance of supersymmetry, and, possibly, the veracity of string theory itself. And so, with a fair bit more luck, some imaginative and innovative ideas regarding unification, the nature of space and time, and our cosmic origins may finally be tested.

Einstein in Drag

In his decade-long struggle to formulate the general theory of relativity, Einstein sought inspiration from a variety of sources. Most influential of all were insights into the mathematics of curved shapes developed in the nineteenth century by mathematical luminaries including Carl Friedrich Gauss, János Bolyai, Nikolai Lobachevsky, and Georg Bernhard Riemann. As we discussed in Chapter 3, Einstein was also inspired by the ideas of Ernst Mach. Remember that Mach advocated a relational conception of space: for him, space provided the language for specifying the location of one object relative to another but was not itself an independent entity. Initially, Einstein was an enthusiastic champion of Mach's perspective, because it was the most relative that a theory espousing relativity could be. But as Einstein's understanding of general relativity deepened, he realized that it did not incorporate Mach's ideas fully. According to general relativity, the water in Newton's bucket, spinning in an otherwise empty universe, would take on a concave shape, and this conflicts with Mach's purely relational perspective, since it implies an absolute notion of acceleration. Even so, general relativity does incorporate some aspects of Mach's viewpoint, and within the next few years a more than $500 million experiment that has been in development for close to forty years will test one of the most prominent Machian features.

The physics to be studied has been known since 1918, when the Austrian researchers Joseph Lense and Hans Thirring used general relativity to show that just as a massive object warps space and time—like a bowling ball resting on a trampoline—so a rotating object drags space (and time) around it, like a spinning stone immersed in a bucket of syrup. This is known as *frame dragging* and implies, for example, that an asteroid freely falling toward a rapidly rotating neutron star or black hole will get caught up in a whirlpool of spinning space and be whipped around as it journeys downward. The effect is called frame dragging because from the point of view of the asteroid—from its frame of reference—it isn't being whipped around at all. Instead, it's falling straight down along the spatial grid, but because space is swirling (as in Figure 14.1) the grid gets twisted, so the meaning of "straight down" differs from what you'd expect based on a distant, nonswirling perspective.

To see the connection to Mach, think about a version of frame drag-

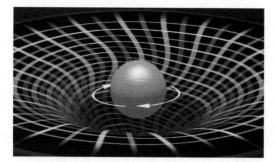

Figure 14.1 A massive spinning object drags space—the freely falling frames—around with it.

ging in which the massive rotating object is a huge, hollow sphere. Calculations initiated in 1912 by Einstein (even before he completed general relativity), which were significantly extended in 1965 by Dieter Brill and Jeffrey Cohen, and finally completed in 1985 by the German physicists Herbert Pfister and K. Braun, showed that space inside the hollow sphere would be dragged by the rotational motion and set into a whirlpool-like spin.[1] If a stationary bucket filled with water—stationary as viewed from a distant vantage point—were placed inside such a rotating sphere, the calculations show that the spinning space would exert a force on the stationary water, causing it to rise up the bucket walls and take on a concave shape.

This result would have pleased Mach no end. Although he might not have liked the description in terms of "spinning space"—since this phrase portrays spacetime as a something—he would have found it extremely gratifying that *relative* spinning motion between the sphere and the bucket causes the water's shape to change. In fact, for a shell that contains enough mass, an amount on a par with that contained in the entire universe, the calculations show that it doesn't matter one bit whether you think the hollow sphere is spinning around the bucket, or the bucket is spinning within the hollow sphere. Just as Mach advocated, the only thing that matters is the relative spinning motion between the two. And since the calculations I've referred to make use of nothing but general relativity, this is an explicit example of a distinctly Machian feature of Einstein's theory. (Nevertheless, whereas standard Machian reasoning would claim that the water would stay flat if the bucket spun in an infinite, empty universe, general relativity disagrees. What the Pfister and Braun results show is that a sufficiently massive rotating sphere is able to completely block the usual influence of the space that lies beyond the sphere itself.)

In 1960, Leonard Schiff of Stanford University and George Pugh of the U.S. Department of Defense independently suggested that general relativity's prediction of frame dragging might be experimentally tested using the rotational motion of the earth. Schiff and Pugh realized that according to Newtonian physics, a spinning gyroscope—a spinning wheel that's attached to an axis—floating in orbit high above the earth's surface would point in a fixed and unchanging direction. But, according to general relativity, its axis would rotate ever so slightly because of the earth's dragging of space. Since the earth's mass is puny in comparison with the hypothetical hollow sphere used in the Pfister and Braun calculation above, the degree of frame dragging caused by the earth's rotation is tiny. The detailed calculations showed that if the gyroscope's spin axis were initially directed toward a chosen reference star, a year later, slowly swirling space would shift the direction of its axis by about a hundred-thousandth of a degree. That's the angle the second hand on a clock sweeps through in roughly two millionths of a second, so its detection presents a major scientific, technological, and engineering challenge.

Four decades of development and nearly a hundred doctoral dissertations later, a Stanford team led by Francis Everitt and funded by NASA is ready to give the experiment a go. During the next few years, their *Gravity Probe B* satellite, floating 400 miles out in space and outfitted with four of the most stable gyroscopes ever built, will attempt to measure frame dragging caused by the earth's rotation. If the experiment is successful, it will be one of the most precise confirmations of general relativity ever achieved, and will provide the first direct evidence of a Machian effect.[2] Equally exciting is the possibility that the experiments will detect a deviation from what general relativity predicts. Such a tiny crack in general relativity's foundation might be just what we need to gain an experimental glimpse into hitherto hidden features of spacetime.

Catching the Wave

An essential lesson of general relativity is that mass and energy cause the fabric of spacetime to warp; we illustrated this in Figure 3.10 by showing the curved environment surrounding the sun. One limitation of a still figure, though, is that it fails to illustrate how the warps and curves in space evolve when mass and energy move or in some way change their configuration.[3] General relativity predicts that, just as a trampoline assumes a fixed, warped shape if you stand perfectly still, but heaves when you jump up and down, space can assume a fixed, warped shape if matter is perfectly still, as assumed in Figure 3.10, but ripples undulate through its fabric when matter moves to and fro. Einstein came to this realization between 1916 and 1918, when he used the newly fashioned equations of general relativity to show that—much as electric charges racing up and down a broadcast antenna produce electromagnetic waves (this is how radio and television waves are produced)—matter racing this way and that (as in a supernova explosion) produces gravitational waves. And since gravity is curvature, a gravitational wave is a wave of curvature. Just as tossing a pebble into a pond generates outward-spreading water ripples, gyrating matter generates outward-spreading spatial ripples; according to general relativity, a distant supernova explosion is like a cosmic pebble that's been tossed into a spacetime pond, as illustrated in Figure 14.2. The figure highlights an important distinguishing feature of gravitational waves: unlike electromagnetic, sound, and water waves—waves that travel *through* space—gravitational waves travel *within* space. They are traveling distortions in the geometry of space itself.

While gravitational waves are now an accepted prediction of general relativity, for many years the subject was mired in confusion and contro-

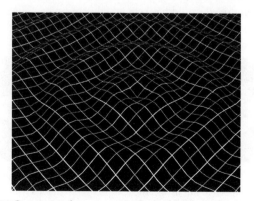

Figure 14.2 Gravitational waves are ripples in the fabric of spacetime.

versy, at least in part because of overadherence to Machian philosophy. If general relativity fully incorporated Mach's ideas, then the "geometry of space" would merely be a convenient language for expressing the location and motion of one massive object with respect to another. Empty space, in this way of thinking, would be an empty concept, so how could it be sensible to speak of empty space wiggling? Many physicists tried to prove that the supposed waves in space amounted to a misinterpretation of the mathematics of general relativity. But in due course, the theoretical analyses converged on the correct conclusion: gravitational waves are real, and space *can* ripple.

With every passing peak and trough, a gravitational wave's distorted geometry would stretch space—and everything in it—in one direction, and then compress space—and everything in it—in a perpendicular direction, as in the highly exaggerated depiction in Figure 14.3. In principle, you could detect a gravitational wave's passing by repeatedly measuring distances between a variety of locations and finding that the ratios between these distances had momentarily changed.

In practice, no one has been able to do this, so no one has directly detected a gravitational wave. (However, there is compelling, indirect evidence for gravitational waves.[4]) The difficulty is that the distorting influence of a passing gravitational wave is typically minute. The atomic bomb tested at Trinity on July 16, 1945, packed a punch equivalent to 20,000 tons of TNT and was so bright that witnesses miles away had to wear eye protection to avoid serious damage from the electromagnetic waves it generated. Yet, even if you were standing right under the hundred-foot steel tower on which the bomb was hoisted, the gravitational waves its explosion produced would have stretched your body one way or another

Figure 14.3 A passing gravitational wave stretches an object one way and then the other. (In this image, the scale of distortion for a typical gravitational wave is hugely exaggerated.)

only by a minuscule fraction of an atomic diameter. That's how comparatively feeble gravitational disturbances are, and it gives an inkling of the technological challenges involved in detecting them. (Since a gravitational wave can also be thought of as a huge number of gravitons traveling in a coordinated manner—just as an electromagnetic wave is composed of a huge number of coordinated photons—this also gives an inkling of how difficult it is to detect a *single* graviton.)

Of course, we're not particularly interested in detecting gravitational waves produced by nuclear weapons, but the situation with astrophysical sources is not much easier. The closer and more massive the astrophysical source and the more energetic and violent the motion involved, the stronger the gravitational waves we would receive. But even if a star at a distance of 10,000 light-years were to go supernova, as the resulting gravitational wave passed by earth it would stretch a one-meter-long rod by only a millionth of a billionth of a centimeter, barely a hundredth the size of an atomic nucleus. So, unless some highly unexpected astrophysical event of truly cataclysmic proportions were to happen relatively nearby, detecting a gravitational wave will require an apparatus capable of responding to fantastically small length changes.

The scientists who designed and built the *Laser Interferometer Gravitational Wave Observatory* (LIGO) (being run jointly by the California Institute of Technology and the Massachusetts Institute of Technology and funded by the National Science Foundation) have risen to the challenge. LIGO is impressive and the expected sensitivity is astounding. It consists of two hollow tubes, each *four kilometers* long and a bit over a meter wide, which are arranged in a giant L. Laser light simultaneously shot down vacuum tunnels inside each tube, and reflected back by highly polished mirrors, is used to measure the relative length of each to fantastic accuracy. The idea is that should a gravitational wave roll by, it will stretch one tube relative to the other, and if the stretching is big enough, scientists will be able to detect it.

The tubes are long because the stretching and compressing accomplished by a gravitational wave is cumulative. If a gravitational wave were to stretch something four meters long by, say, 10^{-20} meters, it would stretch something four kilometers long by a thousand times as much, 10^{-17} meters. So, the longer the span being monitored, the easier it is to detect a change in its length. To capitalize on this, the LIGO experimenters actually direct the laser beams to bounce back and forth between mirrors at opposite ends of each tube more than a hundred times on each run, increasing the roundtrip distance being monitored to about 800 kilometers per beam. With such clever tricks and engineering feats, LIGO should be able to detect any change in the tube lengths that exceeds a trillionth of the thickness of a human hair—a hundred millionth the size of an atom.

Oh, and there are actually two of these L-shaped devices. One is in Livingston, Louisiana, and the other is about 2,000 miles away in Hanford, Washington. If a gravity wave from some distant astrophysical hullabaloo rolls by earth, it should affect each detector identically, so any wave caught by one experiment had better also show up in the other. This is an important consistency check, since for all the precautions that have been taken to shield the detectors, the disturbances of everyday life (the rumble of a passing truck, the grinding of a chainsaw, the impact of a falling tree, and so on) could masquerade as gravitational waves. Requiring coincidence between distant detectors serves to rule out these false positives.

Researchers have also carefully calculated the gravitational wave frequencies—the number of peaks and troughs that should pass by their detector each second—that they expect to be produced by a range of astrophysical phenomena including supernova explosions, the rotational motion of nonspherical neutron stars, and collisions between black holes. Without this information the experimenters would be looking for a needle in a haystack; with it, they can focus the detectors on a sharply defined frequency band of physical interest. Curiously, the calculations reveal that some gravitational wave frequencies should be in the range of a few thousand cycles per second; if these were sound waves, they'd be right in the range of human audibility. Coalescing neutron stars would sound like a chirp with a rapidly rising pitch, while a pair of colliding black holes would mimic the trill of a sparrow that's received a sharp blow to the chest. There's a junglelike cacophony of gravitational waves oscillating through the spacetime fabric, and if all goes according to plan, LIGO will be the first instrument to tune in.[5]

What makes this all so exciting is that gravitational waves maximize the utility of gravity's two main features: its weakness and its ubiquity. Of all four forces, gravity interacts with matter most feebly. This implies that gravitational waves can pass through material that's opaque to light, giving access to astrophysical realms previously hidden. What's more, because *everything* is subject to gravity (whereas, for example, the electromagnetic force only affects objects carrying an electric charge), everything has the capacity to generate gravitational waves and hence produce an observable signature. LIGO thereby marks a significant turning point in the way we examine the cosmos.

There was a time when all we could do was raise our eyes and gaze skyward. In the seventeenth century, Hans Lippershey and Galileo Galilei changed that; with the aid of the telescope, the grand vista of the cosmos came within humanity's purview. But in time, we realized that visible light represented a narrow band of electromagnetic waves. In the twentieth century, with the aid of infrared, radio, X-ray, and gamma ray telescopes, the cosmos opened up to us anew, revealing wonders invisible in the wavelengths of light that our eyes have evolved to see. Now, in the twenty-first century, we are opening up the heavens once again. With LIGO and its subsequent improvements,* we will view the cosmos in a completely new way. Rather than using electromagnetic waves, we will use gravitational waves; rather than using the electromagnetic force, we will use the gravitational force.

To appreciate how revolutionary this new technology may be, imagine a world on which alien scientists were just now discovering how to detect electromagnetic waves—light—and think about how their view of the universe would, in short order, profoundly change. We are on the cusp of our first detection of gravitational waves and so may well be in a similar position. For millennia we have looked into the cosmos; now it's as if, for the first time in human history, we will listen to it.

The Hunt for Extra Dimensions

Before 1996, most theoretical models that incorporated extra dimensions imagined that their spatial extent was roughly Planckian (10^{-33} centimeters). As this is seventeen orders of magnitude smaller than anything resolvable using currently available equipment, without the discovery of miraculous new technology Planckian physics will remain out of reach.

---

*One of these is the planned Laser Interferometer Space Antenna (LISA), a space-based version of LIGO comprising multiple spacecraft, separated by millions of kilometers, playing the role of LIGO's four-kilometer tubes. There are also other detectors that are playing a critical role in the search for gravitational waves, including the German-British detector GEO600, the French-Italian detector VIRGO, and the Japanese detector TAMA300.

But if the extra dimensions are "large," meaning larger than a hundredth of a billionth of a billionth (10^{-20}) of a meter, about a millionth the size of an atomic nucleus, there is hope.

As we discussed in Chapter 13, if any of the extra dimensions are "very large"—within a few orders of magnitude of a millimeter—precision measurements of gravity's strength should reveal their existence. Such experiments have been under way for a few years and the techniques are being rapidly refined. So far, no deviations from the inverse square law characteristic of three space dimensions have been found, so researchers are pressing on to smaller distances. A positive signal would, to say the least, rock the foundations of physics. It would provide compelling evidence of extra dimensions accessible only to gravity, and that would give strong circumstantial support for the braneworld scenario of string/M-theory.

If the extra dimensions are large but not very large, precision gravity experiments will be unlikely to detect them, but other indirect approaches remain available. For example, we mentioned earlier that large extra dimensions would imply that gravity's intrinsic strength is greater than previously thought. The observed weakness of gravity would be attributed to its leaking out into the extra dimensions, not to its being fundamentally feeble; on short distance scales, before such leakage occurs, gravity would be strong. Among other implications, this means that the creation of tiny black holes would require far less mass and energy than it would in a universe in which gravity is intrinsically far weaker. In Chapter 13, we discussed the possibility that such microscopic black holes might be produced by high-energy proton-proton collisions at the Large Hadron Collider, the particle accelerator now under construction in Geneva, Switzerland, and slated for completion by 2007. That is an exciting prospect. But there is another tantalizing possibility that was raised by Alfred Shapere, of the University of Kentucky, and Jonathan Feng, of the University of California at Irvine. These researchers noted that cosmic rays—elementary particles that stream through space and continually bombard our atmosphere—might also initiate production of microscopic black holes.

Cosmic ray particles were discovered in 1912 by the Austrian scientist Victor Hess; more than nine decades later, they still present many mysteries. Every second, cosmic rays slam into the atmosphere and initiate a cascade of billions of downward-raining particles that pass through your body and mine; some of them are detected by a variety of dedicated instruments worldwide. But no one is completely sure what kinds of particles constitute the impinging cosmic rays (although there is a growing consensus that they are protons), and despite the fact that some of these high-energy particles are believed to come from supernova explosions, no one has any idea of where the highest-energy cosmic ray particles originate. For example, on October 15, 1991, the Fly's Eye cosmic ray detector, in the Utah desert, measured a particle streaking across the sky with an energy equivalent to 30 billion proton masses. That's almost as much energy in a single subatomic particle as in a Mariano Rivera fastball, and is about 100 million times the size of the particle energies that will be produced by the Large Hadron Collider.[6] The puzzling thing is that no known astrophysical process could produce particles with such high energy; experimenters are gathering more data with more sensitive detectors in hopes of solving the mystery.

For Shapere and Feng, the origin of super-energetic cosmic ray particles was of secondary concern. They realized that regardless of where such particles come from, if gravity on microscopic scales is far stronger than formerly thought, the highest-energy cosmic ray particles might have just enough oomph to create tiny black holes when they violently slam into the upper atmosphere.

As with their production in atom smashers, such tiny black holes would pose absolutely no danger to the experimenters or the world at large. After their creation, they would quickly disintegrate, sending off a characteristic cascade of other, more ordinary particles. In fact, the microscopic black holes would be so short-lived that experimenters would not search for them directly; instead, they would look for evidence of black holes through detailed studies of the resulting particle showers raining down on their detectors. The most sensitive of the world's cosmic ray detectors, the Pierre Auger Observatory—with an observing area the size of Rhode Island—is now being built on a vast stretch of land in western Argentina. Shapere and Feng estimate that if all of the extra dimensions are as large as 10^{-14} meters, then after a year's worth of data collection, the Auger detector will see the characteristic particle debris from about a dozen tiny black holes produced in the upper atmosphere. If such black hole signatures are not found, the experiment will conclude that extra dimensions are smaller. Finding the remains of black holes produced in cosmic ray collisions is certainly a long shot, but success would open the first experimental window on extra dimensions, black holes, string theory, and quantum gravity.

Beyond black hole production, there is another, accelerator-based way that researchers will be looking for extra dimensions during the next decade. The idea is a sophisticated variant on the "space-between-the-cushions" explanation for the loose coins missing from your pocket.

A central principle of physics is conservation of energy. Energy can manifest itself in many forms—the kinetic energy of a ball's motion as it flies off a baseball bat, gravitational potential energy as the ball flies upward, sound and heat energy when the ball hits the ground and excites all sorts of vibrational motion, the mass energy that's locked inside the ball itself, and so on—but when all carriers of energy have been accounted for, the amount with which you end always equals the amount with which you began.[7] To date, no experiment contradicts this law of perfect energy balance.

But depending on the precise size of the hypothesized extra dimensions, high-energy experiments to be carried out at the newly upgraded facility at Fermilab and at the Large Hadron Collider may reveal processes that appear to violate energy conservation: the energy at the end of a collision may be less than the energy at the beginning. The reason is that, much like your missing coins, energy (carried by gravitons) can seep into the cracks—the tiny additional space—provided by the extra dimensions and hence be inadvertently overlooked in the energy accounting calculation. The possibility of such a "missing energy signal" provides yet another means for establishing that the fabric of the cosmos has complexity well beyond what we can see directly.

No doubt, when it comes to extra dimensions, I'm biased. I've worked on aspects of extra dimensions for more than fifteen years, so they hold a special place in my heart. But, with that confession as a qualifier, it's hard for me to imagine a discovery that would be more exciting than finding evidence for dimensions beyond the three with which we're all familiar. To my mind, there is currently no other serious proposal whose confirmation would so thoroughly shake the foundation of physics and so thoroughly establish that we must be willing to question basic, seemingly self-evident, elements of reality.

The Higgs, Supersymmetry, and String Theory

Beyond the scientific challenges of searching into the unknown, and the chance of finding evidence of extra dimensions, there are a couple of specific motivations for recent upgrades on the accelerator at Fermilab and for building the mammoth Large Hadron Collider. One is to find Higgs particles. As we discussed in Chapter 9, the elusive Higgs particles would be the smallest constituents of a Higgs field—a field, physicists hypothesize, that forms the Higgs ocean and thereby gives mass to the other fundamental particle species. Current theoretical and experimental studies suggest that the Higgs should have a mass in the range of a hundred to a thousand times the mass of the proton. If the lower end of this range turns out to be right, Fermilab stands a reasonably good chance of discovering a Higgs particle in the near future. And certainly, if Fermilab fails and if the estimated mass range is nonetheless correct, the Large Hadron Collider should produce Higgs particles galore by the end of the decade. The detection of Higgs particles would be a major milestone, as it would confirm the existence of a species of field that theoretical particle physicists and cosmologists have invoked for decades, without any supporting experimental evidence.

Another major goal of both Fermilab and the Large Hadron Collider is to detect evidence of supersymmetry. Recall from Chapter 12 that supersymmetry pairs particles whose spins differ by half a unit and is an idea that originally arose from studies of string theory in the early 1970s. If supersymmetry is relevant to the real world, then for every known particle species with spin-$\frac{1}{2}$ there should be a partner species with spin-0; for every known particle species of spin-1, there should be a partner species with spin-$\frac{1}{2}$. For example, for the spin-$\frac{1}{2}$ electron there should be a spin-0 species called the *supersymmetric electron,* or *selectron* for short; for the spin-$\frac{1}{2}$ quarks there should be *supersymmetric quarks,* or *squarks*; for spin-$\frac{1}{2}$ neutrinos there should be spin-0 *sneutrinos*; for spin-1 gluons, photons, and W and Z particles there should be spin-$\frac{1}{2}$ *gluinos, photinos,* and *winos* and *zinos*. (Yes, physicists get carried away.)

No one has ever detected any of these purported doppelgängers, and the explanation, physicists hope with fingers crossed, is that the supersymmetric particles are substantially heavier than their known counterparts. Theoretical considerations suggest that the supersymmetric particles could be a thousand times as massive as a proton, and in that case their failure to appear in experimental data wouldn't be mysterious: existing atom smashers don't have adequate power to produce them. In the coming decade this will change. Already, the newly upgraded accelerator at Fermilab has a shot at discovering some supersymmetric particles. And, as with the Higgs, should Fermilab fail to find evidence of supersymmetry and if the expected mass range of the supersymmetric particles is fairly accurate, the Large Hadron Collider should produce them with ease.

The confirmation of supersymmetry would be the most important development in elementary particle physics in more than two decades. It would establish the next step in our understanding beyond the successful standard model of particle physics and would provide circumstantial evidence that string theory is on the right track. But note that it wouldn't prove string theory itself. Even though supersymmetry was discovered in the course of developing string theory, physicists have long since realized that supersymmetry is a more general principle that can easily be incorporated in traditional point-particle approaches. Confirmation of supersymmetry would establish a vital element of the string framework and would guide much subsequent research, but it wouldn't be string theory's smoking gun.

On the other hand, if the braneworld scenario is correct, upcoming accelerator experiments *do* have the potential of confirming string theory. As mentioned briefly in Chapter 13, should the extra dimensions in the braneworld scenario be as large as 10^{-16} centimeters, not only would gravity be intrinsically stronger than previously thought, but strings would be significantly longer as well. Since longer strings are less stiff, they require less energy to vibrate. Whereas in the conventional string framework, string vibrational patterns would have energies that are more than a million billion times beyond our experimental reach, in the braneworld scenario the energies of string vibrational patterns could be as low as a *thousand* times the proton's mass. Should this be the case, high-energy collisions at the Large Hadron Collider will be akin to a well-hit golf ball ricocheting around the inside of a piano; the collisions will have enough energy to excite many "octaves" of string vibrational patterns. Experimenters would detect a panoply of new, never before seen particles— new, never before seen string vibrational patterns, that is—whose energies would correspond to the harmonic resonances of string theory.

The properties of these particles and the relationships between them would show unmistakably that they're all part of the same cosmic score, that they're all different but related notes, that they're all distinct vibrational patterns of a single kind of object—a string. For the foreseeable future, this is the most likely scenario for a direct confirmation of string theory.

Cosmic Origins

As we saw in earlier chapters, the cosmic microwave background radiation has played a dominant role in cosmological research since its discovery in the mid-1960s. The reason is clear: in the early stages of the universe, space was filled with a bath of electrically charged particles—electrons and protons—which, through the electromagnetic force, incessantly buffeted photons this way and that. But by a mere 300,000 years after the bang (ATB), the universe cooled off just enough for electrons and protons to combine into electrically neutral atoms—and from this moment onward, the radiation has traveled throughout space, mostly undisturbed, providing a sharp snapshot of the early universe. There are roughly 400 million of these primordial cosmic microwave photons streaming through every cubic meter of space, pristine relics of the early universe.

Initial measurements of the microwave background radiation revealed its temperature to be remarkably uniform, but as we discussed in Chapter 11, closer inspection, first achieved in 1992 by the Cosmic Background Explorer (COBE) and since improved by a number of observational undertakings, found evidence of small temperature variations, as illustrated in Figure 14.4a. The data are gray-scale coded, with light and dark patches indicating temperature variations of about a few ten-thousandths of a degree. The figure's splotchiness shows the minute but undeniably real unevenness of the radiation's temperature across the sky.

While an impressive discovery in its own right, the COBE experiment also marked a fundamental change in the character of cosmological research. Before COBE, cosmological data were coarse. In turn, a cosmological theory was deemed viable if it could match the broad-brush features of astronomical observations. Theorists could propose scheme after scheme with only minimal consideration for satisfying observational constraints. There simply weren't many observational constraints, and the ones that existed weren't particularly precise. But COBE initiated a new era in which the standards have tightened considerably. There is now a growing body of precision data with which any theory must reckon successfully even to be considered. In 2001, the Wilkinson Microwave Anisotropy Probe (WMAP) satellite, a joint venture of NASA and Princeton University, was launched to measure the microwave background radiation with about forty times COBE's resolution and sensitivity. By

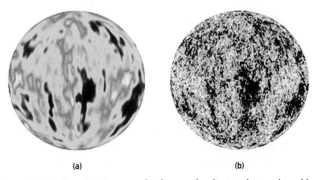

(a)   (b)

Figure 14.4 (a) Cosmic microwave background radiation data gathered by the COBE satellite. The radiation has been traveling through space unimpeded since about 300,000 years after the big bang, so this picture renders the tiny temperature variations present in the universe nearly 14 billion years ago. **(b)** Improved data collected by the WMAP satellite.

comparing WMAP's initial results, Figure 14.4b, with COBE's, Figure 14.4a, you can immediately see how much finer and more detailed a picture WMAP is able to provide. Another satellite, *Planck*, which is being developed by the European Space Agency, is scheduled for launch in 2007, and if all goes according to plan, will better WMAP's resolution by a factor of ten.

The influx of precision data has winnowed the field of cosmological proposals, with the inflationary model being, far and away, the leading contender. But as we mentioned in Chapter 10, inflationary cosmology is not a unique theory. Theorists have proposed *many* different versions (old inflation, new inflation, warm inflation, hybrid inflation, hyperinflation, assisted inflation, eternal inflation, extended inflation, chaotic inflation, double inflation, weak-scale inflation, hypernatural inflation, to name just a few), each involving the hallmark brief burst of rapid expansion, but all differing in detail (in the number of fields and their potential energy shapes, in which fields get perched on plateaus, and so on). These differences give rise to slightly different predictions for the properties of the microwave background radiation (different fields with different energies have slightly different quantum fluctuations). Comparison with the WMAP and Planck data should be able to rule out many proposals, substantially refining our understanding.

In fact, the data may be able to thin the field even further. Although quantum fluctuations stretched by inflationary expansion provide a compelling explanation for the observed temperature variations, this model has a competitor. The cyclic cosmological model of Steinhardt and Turok, described in Chapter 13, offers an alternative proposal. As the two three-branes of the cyclic model slowly head toward each other, quantum fluctuations cause different parts to approach at slightly different rates. When they finally slam together roughly a trillion years later, different locations on the branes will make contact at slightly different moments, rather as if two pieces of coarse sandpaper were being slapped together. The tiny deviations from a perfectly uniform impact yield tiny deviations from a perfectly uniform evolution across each brane. Since one of these branes is supposed to be our three-dimensional space, the deviations from uniformity are deviations we should be able to detect. Steinhardt, Turok, and their collaborators have argued that the inhomogeneities generate temperature deviations of the same form as those emerging from the inflationary framework, and hence, with today's data, the cyclic model offers an equally viable explanation of the observations.

However, the more refined data being gathered over the next decade may be able to distinguish between the two approaches. In the inflationary framework, not only are quantum fluctuations of the inflaton field stretched by the burst of exponential expansion, but tiny quantum ripples in the spatial fabric are also generated by the intense outward stretching. Since ripples in space are nothing but gravitational waves (as in our earlier discussion of LIGO), the inflationary framework predicts that gravitational waves were produced in the earliest moments of the universe.[8] These are often called *primordial gravitational waves*, to distinguish them from those generated more recently by violent astrophysical events. In the cyclic model, by contrast, the deviation from perfect uniformity is built up gently, over the course of an almost unfathomable length of time, as the branes spend a trillion years slowly heading toward their next splat. The absence of a brisk and vigorous change in the geometry of the branes, and in the geometry of space, means that spatial ripples are *not* generated, so the cyclic model predicts an absence of primordial gravitational waves. Thus, if primordial cosmological gravitational waves should be detected, it will be yet another triumph for the inflationary framework and will definitively rule out the cyclic approach.

It is unlikely that LIGO will be sensitive enough to detect inflation's predicted gravitational waves, but it is possible that they will be observed indirectly either by Planck or by another satellite experiment called the

Cosmic Microwave Background Polarization experiment (CMBPol) that is now being planned. Planck, and CMBPol in particular, will not focus solely on temperature variations of the microwave background radiation, but will also measure *polarization*, the average spin directions of the microwave photons detected. Through a chain of reasoning too involved to cover here, it turns out that gravitational waves from the bang would leave a distinct imprint on the polarization of the microwave background radiation, perhaps an imprint large enough to be measured.

So, within a decade, we may get sharp insight into whether the bang was really a splat, and whether the universe we're aware of is really a three-brane. In the golden age of cosmology, some of the wildest ideas may actually be testable.

Dark Matter, Dark Energy, and the Future of the Universe

In Chapter 10 we went through the strong theoretical and observational evidence indicating that a mere 5 percent of the universe's heft comes from the constituents found in familiar matter—protons and neutrons (electrons account for less than .05 percent of ordinary matter's mass)— while 25 percent comes from dark matter and 70 percent from dark energy. But there is still significant uncertainty regarding the detailed identity of all this dark stuff. A natural guess is that the dark matter is also composed of protons and neutrons, ones that somehow avoided clumping together to form light-emitting stars. But another theoretical considera-tion makes this possibility very unlikely.

Through detailed observations, astronomers have a clear knowledge of the average relative abundances of light elements—hydrogen, helium, deuterium, and lithium—that are scattered throughout the cosmos. To a high degree of accuracy, the abundances agree with theoretical calcula-tions of the processes believed to have synthesized these nuclei during the first few minutes of the universe. This agreement is one of the great suc-cesses of modern theoretical cosmology. However, these calculations assume that the bulk of the dark matter is *not* composed of protons and neutrons; if, on cosmological scales, protons and neutrons were a domi-nant constituent, the cosmic recipe is thrown off and the calculations yield results that are ruled out by observations.

So, if not protons and neutrons, what constitutes the dark matter? As of today, no one knows, but there is no shortage of proposals. The candidates' names run the gamut from axions to zinos, and whoever finds the answer will surely pay a visit to Stockholm. That no one has yet detected a dark matter particle places significant constraints on any proposal. The reason is that dark matter is not only situated out in space; it is distributed throughout the universe and so is also wafting by us here on earth. According to many of the proposals, right now billions of dark matter particles are shooting through your body every second, so viable candidates are only those particles that can pass through bulky matter without leaving a significant trace.

Neutrinos are one possibility. Calculations estimate their relic abundance since they were produced in the big bang, at about 55 million per cubic meter of space, so if any one of the three neutrino species weighed about a hundredth of a millionth (10^{-8}) as much as a proton, they would supply the dark matter. Although recent experiments have given strong evidence that neutrinos do have mass, according to current data they are too light to supply the dark matter; they fall short of the mark by a factor of more than a hundred.

Another promising proposal involves supersymmetric particles, especially the *photino*, the *zino*, and the *higgsino* (the partners of the photon, the Z, and the Higgs). These are the most standoffish of the supersymmetric particles—they could nonchalantly pass through the entire earth without the slightest effect on their motion—and hence could easily have escaped detection.[9] From calculations of how many of these particles would have been produced in the big bang and survived until today, physicists estimate that they would need to have mass on the order of 100 to 1,000 times that of the proton to supply the dark matter. This is an intriguing number, because various studies of supersymmetric-particle models as well as of superstring theory have arrived at the same mass range for these particles, without any concern for dark matter or cosmology. This would be a puzzling and completely unexplained confluence, unless, of course, the dark matter is indeed composed of supersymmetric particles. Thus, the search for supersymmetric particles at the world's current and pending accelerators may also be viewed as searches for the heavily favored dark matter candidates.

More direct searches for the dark matter particles streaming through the earth have also been under way for some time, although these are extremely challenging experiments. Of the million or so dark matter particles that should be passing through an area the size of a quarter each second, at most one per day would leave any evidence in the specially designed equipment that various experimenters have built to detect them. To date, no confirmed detection of a dark matter particle has been achieved.[10] With the prize still very much up in the air, researchers are pressing ahead with much intensity. It is quite possible that within the next few years, the identity of the dark matter will be settled.

Definitive confirmation that dark matter exists, and direct determination of its composition, would be a major advance. For the first time in history, we would learn something that is at once thoroughly basic and surprisingly elusive: the makeup of the vast majority of the universe's material content.

All the same, as we saw in Chapter 10, recent data suggest strongly that even with the identification of the dark matter, there would still be a significant plot twist in need of experimental vetting: the supernova observations that give evidence of an outward-pushing cosmological constant accounting for 70 percent of the total energy in the universe. As the most exciting and unexpected discovery of the last decade, the evidence for a cosmological constant—an energy that suffuses space—needs vigorous, airtight confirmation. A number of approaches are planned or already under way.

The microwave background experiments play an important role here as well. The size of the splotches in Figure 14.4—where, again, each splotch is a region of uniform temperature—reflects the overall shape of the spatial fabric. If space were shaped like a sphere, as in Figure 8.6a, the outward bloating would cause the splotches to be a bit bigger than they are in Figure 14.4b; if space were shaped like a saddle, as in Figure 8.6c, the inward shrinking would cause the splotches to be a bit smaller; and if space were flat, as in Figure 8.6b, the splotch size would be in between. The precision measurements initiated by COBE and since bettered by WMAP strongly support the proposition that space is *flat*. Not only does this match the theoretical expectations coming from inflationary models, but it also jibes perfectly with the supernova results. As we've seen, a spatially flat universe requires the total mass/energy density to equal the critical density. With ordinary and dark matter contributing about 30 percent and dark energy contributing about 70 percent, everything hangs together impressively.

A more direct confirmation of the supernova results is the goal of the SuperNova/Acceleration Probe (SNAP). Proposed by scientists at the

Lawrence Berkeley Laboratory, SNAP would be a satellite-borne orbiting telescope with the capacity to observe and measure more than twenty times the number of supernovae studied so far. Not only would SNAP be able to confirm the earlier result that 70 percent of the universe is dark energy, but it should also be able to determine the nature of the dark energy more precisely.

You see, although I have described the dark energy as being a version of Einstein's cosmological constant—a constant, unchanging energy that pushes space to expand—there is a closely related but alternative possibility. Remember from our discussion of inflationary cosmology (and the jumping frog) that a field whose value is perched above its lowest energy configuration can act like a cosmological constant, driving an accelerated expansion of space, but will typically do so only for a short time. Sooner or later, the field will find its way to the bottom of its potential energy bowl, and the outward push will disappear. In inflationary cosmology, this happens in a tiny fraction of a second. But by introducing a new field and by carefully choosing its potential energy shape, physicists have found ways for the accelerated expansion to be far milder in its outward push but to last far longer—for the field to drive a comparatively slow and steady accelerated phase of spatial expansion that lasts not for a fraction of a second, but for billions of years, as the field slowly rolls to the lowest energy value. This raises the possibility that, right now, we may be experiencing an extremely gentle version of the inflationary burst believed to have happened during the universe's earliest moments.

The difference between a true cosmological constant and the latter possibility, known as *quintessence,* is of minimal importance today, but has a profound effect on the long-term future of the universe. A cosmological constant is *constant*—it provides a never-ending accelerated expansion, so the universe will expand ever more quickly and become ever more spread out, diluted, and barren. But quintessence provides accelerated expansion that at some point draws to a close, suggesting a far future less bleak and desolate than that following from accelerated expansion that's eternal. By measuring changes in the acceleration of space over long time spans (through observations of supernovae at various distances and hence at various times in the past), SNAP may be able to distinguish between the two possibilities. By determining whether the dark energy truly is a cosmological constant, SNAP will give insight into the long-term fate of the universe.

Space, Time, and Speculation

The journey to discover the nature of space and time has been long and filled with many surprises; no doubt it is still in its early stages. During the last few centuries, we've seen one breakthrough after another radically reshape our conceptions of space and time and reshape them again. The theoretical and experimental proposals we've covered in this book represent our generation's sculpting of these ideas, and will likely be a major part of our scientific legacy. In Chapter 16, we will discuss some of the most recent and speculative advances in an effort to cast light on what might be the next few steps of the journey. But first, in Chapter 15, we will speculate in a different direction.

While there is no set pattern to scientific discovery, history shows that deep understanding is often the first step toward technological control. Understanding of the electromagnetic force in the 1800s ultimately led to the telegraph, radio, and television. With that knowledge, in conjunction with subsequent understanding of quantum mechanics, we were able to develop computers, lasers, and electronic gadgets too numerous to mention. Understanding of the nuclear forces led to dangerous mastery over the most powerful weapons the world has ever known, and to the development of technologies that might one day meet all the world's energy needs with nothing but vats of salt water. Could our ever deepening understanding of space and time be the first step in a similar pattern of discovery and technological development? Will we one day be masters of space and time and do things that for now are only part of science fiction?

No one knows. But let's see how far we've gotten and what it might take to succeed.

15

Teleporters
and Time Machines

Perhaps I just lacked imagination back in the 1960s, but what really struck me as unbelievable was the computer on board the *Enterprise*. My grade-school sensibilities granted poetic license to warp drive and to a universe populated by aliens fluent in English. But a machine that could—on demand—immediately display a picture of any historical figure who ever lived, give technical specifications for any piece of equipment ever built, or provide access to any book ever written? *That* strained my ability to suspend disbelief. In the late 1960s, this preteen was certain that there'd never be a way to gather, store, and give ready access to such a wealth of information. And yet, less than half a century later, I can sit here in my kitchen with laptop, wireless Internet connection, and voice recognition software and play Kirk, thumbing through a vast storehouse of knowledge—from the pivotal to the puerile—without lifting a finger. True, the speed and efficiency of computers depicted in the twenty-third-century world of *Star Trek* is still enviable, but it's easy to envisage that when that era arrives, our technology will have exceeded the imagined expectations.

This example is but one of many that have made a cliché of science fiction's ability to presage the future. But what of the most tantalizing of all plot devices—the one in which someone enters a chamber, flips a switch, and is transported to a faraway place or a different time? Is it possible we will one day break free from the meager spatial expanse and temporal epoch to which we have been so far confined and explore the farthest reaches of space and time? Or will this distinction between science fiction and reality remain forever sharply drawn? Having already been exposed to my childhood failure to anticipate the information revolution, you might question my ability to divine future technological breakthroughs. So, rather than speculating on the likelihood of what may be, in this chapter I'll describe how far we've actually gone, in both theory and practice, toward realizing teleporters and time machines, and what it would take to go further and attain control over space and time.

Teleportation in a Quantum World

In conventional science fiction depictions, a *teleporter* (or, in *Star Trek* lingo, a *transporter*) scans an object to determine its detailed composition and sends the information to a distant location, where the object is reconstituted. Whether the object itself is "dematerialized," its atoms and molecules being sent along with the blueprint for putting them back together, or whether atoms and molecules located at the receiving end are used to build an exact replica of the object, varies from one fictional incarnation to another. As we'll see, the scientific approach to teleportation developed over the last decade is closer in spirit to the latter category, and this raises two essential questions. The first is a standard but thorny philosophical conundrum: When, if ever, should an exact replica be identified, called, considered, or treated as if it were the original? The second is the question of whether it's possible, even in principle, to examine an object and determine its composition with complete accuracy so that we can draw up a perfect blueprint with which to reconstitute it.

In a universe governed by the laws of classical physics, the answer to the second question would be yes. In principle, the attributes of every particle making up an object—each particle's identity, position, velocity, and so on—could be measured with total precision, transmitted to a distant location, and used as an instruction manual for recreating the object. Doing this for an object composed of more than just a handful of elementary particles would be laughably beyond reach, but in a classical universe, the obstacle would be complexity, not physics.

In a universe governed by the laws of quantum physics—our universe—the situation is far more subtle. We've learned that the act of measurement coaxes one of the myriad potential attributes of an object to snap out of the quantum haze and take on a definite value. When we observe a particle, for example, the definite features we see do not generally reflect the fuzzy quantum mixture of attributes it had a moment before we looked.[1] Thus, if we want to replicate an object, we face a quantum Catch-22. To replicate we must observe, so we know what to replicate. But the act of observation causes change, so if we replicate what we see, we will not replicate what was there before we looked. This suggests that teleportation in a quantum universe is unattainable, not merely because of practical limitations arising from complexity, but because of fundamental limitations inherent in quantum physics. Nevertheless, as we'll see in the next section, in the early 1990s an international team of physicists found an ingenious way to circumvent this conclusion.

As for the first question, regarding the relationship between replica and original, quantum physics gives an answer that's both precise and encouraging. According to quantum mechanics, every electron in the universe is identical to every other, in that they all have exactly the same mass, exactly the same electric charge, exactly the same weak and strong nuclear force properties, and exactly the same total spin. Moreover, our well-tested quantum mechanical description says that these *exhaust* the attributes that an electron can possess; electrons are all identical with regard to these properties, and there are no other properties to consider. In the same sense, every up-quark is the same as every other, every down-quark is the same as every other, every photon is the same as every other, and so on for all other particle species. As recognized by quantum practitioners many decades ago, particles may be thought of as the smallest possible packets of a field (e.g., photons are the smallest packets of the electromagnetic field), and quantum physics shows that such smallest constituents of the same field are always identical. (Or, in the framework of string theory, particles of the same species have identical properties because they are identical vibrations of a single species of string.)

What can differ between two particles of the same species are the probabilities that they are located at various positions, the probabilities that their spins are pointing in particular directions, and the probabilities that they have particular velocities and energies. Or, as physicists say more succinctly, the two particles can be in different *quantum states*. But if two particles of the same species are in the same quantum state—except, possibly, for one particle having a high likelihood of being *here* while the other particle has a high likelihood of being over *there*—the laws of quantum mechanics ensure that they are indistinguishable, not just in practice but in principle. They are perfect twins. If someone were to exchange the particles' positions (more precisely, exchange the two particles' probabilities of being located at any given position), there'd be absolutely no way to tell.

Thus, if we imagine starting with a particle located here,* and somehow put another particle of the same species into exactly the same quantum state (same probabilities for spin orientation, energy, and so on) at some distant location, the resulting particle would be indistinguishable from the original and the process would rightly be called quantum teleportation. Of course, were the original particle to survive the process intact, you might be tempted to call the process quantum cloning or, perhaps, quantum faxing. But as we'll see, the scientific realization of these ideas does not preserve the original particle—it is unavoidably modified during the teleportation process—so we won't be faced with this taxonomic dilemma.

A more pressing concern, and one that philosophers have considered closely in various forms, is whether what's true for an individual particle is true for an agglomeration. If you were able to teleport from one location to another every single particle that makes up your DeLorean, ensuring that the quantum state of each, including its relationship to all others, was reproduced with 100% fidelity, would you have succeeded in teleporting the vehicle? Although we have no empirical evidence to guide us, the theoretical case in support of having teleported the car is strong. Atomic and molecular arrangements determine how an object looks and feels, sounds and smells, and even tastes, so the resulting vehicle should be identical to the original DeLorean—bumps, nicks, squeaky left wing-door, musty smell from the family dog, all of it—and the car should take a sharp turn and respond to flooring the gas pedal exactly as the original did. The question of whether the vehicle actually is the original or, instead, is an exact duplicate, is of no concern. If you'd asked United Quantum Van Lines to ship your car by boat from New York to London but, unbeknownst to you,

---

*Since teleportation starts with something here and seeks to make it appear at a distant location, in this section I will often speak as if particles have definite positions. To be more precise, I should always say, "starting with a particle that has a high likelihood of being located here" or "starting with a particle with a 99 percent chance of being located here," with similar language used where the particle is teleported, but for brevity's sake I will use the looser language.

they teleported it in the manner described, you could never know the difference—even in principle.

But what if the moving company did the same to your cat, or, having sated your appetite for airplane gastronomy, what if you decided on teleportation for your own transatlantic travel? Would the cat or person who steps out of the receiving chamber be the same as the one who stepped into the teleporter? Personally, I think so. Again, since we have no relevant data, the best that I or anyone can do is speculate. But to my way of thinking, a living being whose constituent atoms and molecules are in exactly the same quantum state as mine *is* me. Even if the "original" me still existed after the "copy" had been made, I (we) would say without hesitation that each was me. We'd be of the same mind—literally—in asserting that neither would have priority over the other. Thoughts, memories, emotions, and judgments have a physical basis in the human body's atomic and molecular properties; an identical quantum state of these elementary constituents should entail an identical conscious being. As time went by, our experiences would cause us to differentiate, but I truly believe that henceforth there'd be two of me, not an original that was somehow "really" me and a copy that somehow wasn't.

In fact, I'm willing to be a bit looser. Our physical composition goes through numerous transformations all the time—some minor, some drastic—but we remain the same person. From the Häagen-Dazs that inundates the bloodstream with fat and sugar, to the MRI that flips the spin axes of various atomic nuclei in the brain, to heart transplants and liposuction, to the trillion atoms in the average human body that are replaced every millionth of a second, we undergo constant change, yet our personal identity remains unaffected. So, even if a teleported being did not match my physical state with perfect accuracy, it could very well be fully indistinguishable from me. In my book, it could very well *be* me.

Certainly, if you believe that there is more to life, and conscious life in particular, than its physical makeup, your standards for successful teleportation will be more stringent than mine. This tricky issue—to what extent is our personal identity tied to our physical being?—has been debated for years in a variety of guises without being answered to everyone's satisfaction. While I believe identity all resides in the physical, others disagree, and no one can claim to have the definitive answer.

But irrespective of your point of view on the hypothetical question of teleporting a living being, scientists have now established that, through the wonders of quantum mechanics, *individual particles can be—and have been—teleported.*

Let's see how.

Quantum Entanglement and Quantum Teleportation

In 1997, a group of physicists led by Anton Zeilinger, then at the University of Innsbruck, and another group led by A. Francesco De Martini at the University of Rome,[2] each carried out the first successful teleportation of a single photon. In both experiments, an initial photon in a particular quantum state was teleported a short distance across a laboratory, but there is every reason to expect that the procedures would have worked equally well over any distance. Each group used a technique based on theoretical insights reported in 1993 by a team of physicists—Charles Bennett of IBM's Watson Research Center; Gilles Brassard, Claude Crepeau, and Richard Josza of the University of Montreal; the Israeli physicist Asher Peres; and William Wootters of Williams College—that rely on quantum entanglement (Chapter 4).

Remember, two entangled particles, say two photons, have a strange and intimate relationship. While each has only a certain probability of spinning one way or another, and while each, when measured, seems to "choose" randomly between the various possibilities, whatever "choice" one makes the other immediately makes too, regardless of their spatial separation. In Chapter 4, we explained that there is no way to use entangled particles to send a message from one location to another faster than the speed of light. If a succession of entangled photons were each measured at widely separated locations, the data collected at either detector would be a random sequence of results (with the overall frequency of spinning one way or another being consistent with the particles' probability waves). The entanglement would become evident only on comparing the two lists of results, and seeing, remarkably, that they were identical. But that comparison requires some kind of ordinary, slower-than-light-speed communication. And since before the comparison no trace of the entanglement could be detected, no faster than light-speed signal could be sent.

Nevertheless, even though entanglement can't be used for superluminal communication, one can't help feeling that long-distance correlations between particles are so bizarre that they've got to be useful for something extraordinary. In 1993, Bennett and his collaborators discovered one such possibility. They showed that quantum entanglement could be used for quantum teleportation. You might not be able to send a message at a speed greater than that of light, but if you'll settle for slower-than-light teleportation of a particle from here to there, entanglement's the ticket.

The reasoning behind this conclusion, while mathematically straightforward, is cunning and ingenious. Here's the flavor of how it goes.

Imagine I want to teleport a particular photon, one I'll call Photon A, from my home in New York to my friend Nicholas in London. For simplicity, let's see how I'd teleport the exact quantum state of the photon's spin—that is, how I'd ensure that Nicholas would acquire a photon whose probabilities of spinning one way or another were identical to Photon A's.

I can't just measure the spin of Photon A, call Nicholas, and have him manipulate a photon on his end so its spin matches my observation; the result I find would be affected by the observation I make, and so would not reflect the true state of Photon A before I looked. So what can I do? Well, according to Bennett and colleagues, the first step is to ensure that Nicholas and I each have one of two additional photons, let's call them Photons B and C, which are entangled. How we get these photons is not particularly important. Let's just assume that Nicholas and I are certain that even though we are on opposite sides of the Atlantic, if I were to measure Photon B's spin about any given axis, and he were to do the same for Photon C, we would find exactly the same result.

The next step, according to Bennett and coworkers, is *not* to directly measure Photon A—the photon I hope to teleport—since that turns out to be too drastic an intervention. Instead, I should measure a *joint* feature of Photon A and the entangled Photon B. For instance, quantum theory allows me to measure whether Photons A and B have the same spin about a vertical axis, without measuring their spins individually. Similarly, quantum theory allows me to measure whether Photons A and B have the same spin about a horizontal axis, without measuring their spins individually. With such a joint measurement, I do not learn Photon A's spin, but I do learn how Photon A's spin is related to Photon B's. And that's important information.

The distant Photon C is entangled with Photon B, so if I know how Photon A is related to Photon B, I can deduce how Photon A is related to Photon C. If I now phone this information to Nicholas, communicating how Photon A is spinning relative to his Photon C, he can determine how Photon C must be manipulated so that its quantum state will match Photon A's. Once he carries out the necessary manipulation, the quantum state of the photon in his possession will be identical to that of Photon A, and that's all we need to declare that Photon A has been successfully teleported. In the simplest case, for example, should my measurement reveal that Photon B's spin is identical to Photon A's, we would conclude that Photon C's spin is also identical to Photon A's, and without further ado, the teleportation would be complete. Photon C would be in the same quantum state as Photon A, as desired.

Well, almost. That's the rough idea, but to explain quantum teleportation in manageable steps, I've so far left out an absolutely crucial element of the story, one I'll now fill in. When I carry out the joint measurement on Photons A and B, I do indeed learn how the spin of Photon A is related to that of Photon B. But, as with all observations, the measurement itself affects the photons. Therefore, I do *not* learn how Photon A's spin was related to Photon B's before the measurement. Instead, I learn how they are related after they've both been disrupted by the act of measurement. So, at first sight, we seem to face the same quantum obstacle to replicating Photon A that I described at the outset: the unavoidable disruption caused by the measurement process. That's where Photon C comes to the rescue. Because Photons B and C are entangled, the disruption I cause to Photon B in New York *will also be reflected in the state of Photon C in London*. That is the wondrous nature of quantum entanglement, as elaborated in Chapter 4. In fact, Bennett and his collaborators showed mathematically that through its entanglement with Photon B, the disruption caused by my measurement *is imprinted on the distant Photon C*.

And that's fantastically interesting. Through my measurement, we are able to learn how Photon A's spin is related to Photon B's, but with the prickly problem that both photons were disrupted by my meddling. Through entanglement, however, Photon C is tied in to my measurement—even though it's thousands of miles away—and this allows us to isolate the effect of the disruption and thereby have access to information ordinarily lost in the measurement process. If I now call Nicholas with the result of my measurement, he will learn how the spins of Photons A and B are related after the disruption, and, via Photon C, he will *have access to the impact of the disruption itself*. This allows Nicholas to use Photon C to, roughly speaking, subtract out the disruption caused by my measurement and thus skirt the obstacle to duplicating Photon A. In fact, as Bennett and collaborators show in detail, by at most a simple manipulation of Photon C's spin (based on my phone call informing him how Photon A is spinning relative to Photon B) Nicholas will ensure that Photon C, as far as its spin goes, exactly replicates the quantum state of Photon A *prior to my measurement*. Moreover, although spin is only one characteristic of a photon, other features of Photon A's quantum state (such as the probability that it has one energy or another) can be replicated similarly. Thus, by using this procedure, we could teleport Photon A from New York to London.[3]

As you can see, quantum teleportation involves two stages, each of which conveys critical and complementary information. First, we undertake a joint measurement on the photon we want to teleport with one member of an entangled pair of photons. The disruption associated with the measurement is imprinted on the distant partner of the entangled pair through the weirdness of quantum nonlocality. That's Stage 1, the distinctly quantum part of the teleportation process. In Stage 2, the result of the measurement itself is communicated to the distant reception location by more standard means (telephone, fax, e-mail . . .) in what might be called the classical part of the teleportation process. In combination, Stage 1 and Stage 2 allow the exact quantum state of the photon we want to teleport to be reproduced by a straightforward operation (such as a rotation by a certain amount about particular axes) on the distant member of the entangled pair.

Notice, as well, a couple of key features of quantum teleportation. Since Photon A's original quantum state was disrupted by my measurement, *Photon C in London is now the only one in that original state*. There aren't two copies of the original Photon A and so, rather than calling this quantum faxing, it is indeed more accurate to call this quantum teleportation.[4] Furthermore, even though we teleported Photon A from New York to London—even though the photon in London becomes indistinguishable from the original photon we had in New York—we do not learn Photon A's quantum state. The photon in London has exactly the same probability of spinning in one direction or another as Photon A did before my meddling, but we do not know what that probability is. In fact, that's the trick underlying quantum teleportation. The disruption caused by measurement prevents us from determining Photon A's quantum state, but in the approach described, *we don't need to know the photon's quantum state in order to teleport it*. We need to know only an aspect of its quantum state—what we learn from the joint measurement with Photon

B. Quantum entanglement with distant Photon C fills in the rest.

Implementing this strategy for quantum teleportation was no small feat. By the early 1990s, creating an entangled pair of photons was a standard procedure, but carrying out a joint measurement of two photons (the joint measurement on Photons A and B described above, technically called a *Bell-state measurement*) had never been attained. The achievement of both Zeilinger's and De Martini's groups was to invent ingenious experimental techniques for the joint measurement and to realize them in the laboratory.[5] By 1997 they had achieved this goal, becoming the first groups to achieve the teleportation of a single particle.

Realistic Teleportation

Since you and I and a DeLorean and everything else are composed of many particles, the natural next step is to imagine applying quantum teleportation to such large collections of particles, allowing us to "beam" macroscopic objects from one place to another. But the leap from teleporting a single particle to teleporting a macroscopic collection of particles is staggering, and enormously far beyond what researchers can now accomplish and what many leaders in the field imagine achieving even in the distant future. But for kicks, here's how Zeilinger fancifully dreams we might one day go about it.

Imagine I want to teleport my DeLorean from New York to London. Instead of providing Nicholas and me with one member each of an entangled pair of photons (what we needed to teleport a single photon), we must each have a chamber of particles containing enough protons, neutrons, electrons, and so on to build a DeLorean, with all the particles in my chamber being quantum entangled with all those in Nicholas's chamber (see Figure 15.1). I also need a device that measures joint properties of all the particles making up my DeLorean with those particles flitting to and fro within my chamber (the analog of measuring joint features of Photons A and B). Through the entanglement of the particles in the two chambers, the impact of the joint measurements I carry out in New York will be imprinted on Nicholas's chamber of particles in London (the analog of Photon C's state reflecting the joint measurement of A and B). If I call Nicholas and communicate the results of my measurements (it'll be an expensive call, as I'll be giving Nicholas some 10^{30} results), the data will instruct him on how to manipulate the particles in his chamber

Figure 15.1 A fanciful approach to teleportation envisions having two chambers of quantum entangled particles at distant locations, and a means of carrying out appropriate joint measurements of the particles making up the object to be teleported with the particles in one of the chambers. The results of these measurements would then provide the necessary information to manipulate the particles in the second chamber to replicate the object, and complete the teleportation.

(much as my earlier phone call instructed him on how to manipulate Photon C). When he finishes, each particle in his chamber will be in precisely the same quantum state as each particle in the DeLorean (before it was subjected to any measurements) and so, as in our earlier discussion, Nicholas will now *have* the DeLorean.* Its teleportation from New York to London will be complete.

Note, though, that as of today, every step in this macroscopic version of quantum teleportation is fantasy. An object like a DeLorean has in excess of a billion billion billion particles. While experimenters are gaining facility with entangling more than a single pair of particles, they are extremely far from reaching numbers relevant for macroscopic entities.[6] Setting up the two chambers of entangled particles is thus absurdly beyond current reach. Moreover, the joint measurement of *two* photons was, in itself, a difficult and impressive feat. Extending this to a joint measurement of billions and billions of particles is, as of today, unimaginable. From our current vantage point, a dispassionate assessment would con-

*For collections of particles—as opposed to individual particles—the quantum state also encodes the relationship of each particle in the collection to every other. So, by exactly reproducing the quantum state of the particles making up the DeLorean, we ensure that they all stand in the same relation to each other; the only change they experience is that their overall location would have been shifted from New York to London.

clude that teleporting a macroscopic object, at least in the manner so far employed for a single particle, is eons—if not an eternity—away.

But, as the one constant in science and technology is the transcendence of naysaying prophesies, I'll simply note the obvious: teleportation of macroscopic bodies looks unlikely. Yet, who knows? Forty years ago, the *Enterprise*'s computer looked pretty unlikely too.[7]

The Puzzles of Time Travel

There's no denying that life would be different if teleporting macroscopic objects were as easy as calling FedEx or hopping on a subway. Impractical or impossible journeys would become available, and the concept of travel through space would be revolutionized to that rare degree at which a leap in convenience and practicality marks a fundamental shift in worldview.

Even so, teleportation's impact on our sense of the universe would pale in comparison to the upheaval wrought by achieving volitional travel through time. Everyone knows that with enough effort and dedication we can, at least in principle, get from here to there. Although there are technological limitations on our travels through space, within those constraints our travels are guided by choice and whim. But to get from now to then? Our experiences overwhelmingly attest to there being at most one route: we must wait it out—second must follow second as tick by tock now methodically gives way to then. And this assumes that "then" is later than "now." If then precedes now, experience dictates that there is no route at all; traveling to the past seems not to be an option. Unlike travels through space, travels through time appear to be anything but a matter of choice and whim. When it comes to time, we get dragged along in one direction, whether we like it or not.

Were we able to navigate time as easily as we navigate space, our worldview would not just change, it would undergo the single most dramatic shift in the history of our species. In light of such undeniable impact, I am often struck by how few people realize that the theoretical underpinnings for one kind of time travel—time travel to the future—have been in place since early last century.

When Einstein discovered the nature of special relativistic spacetime, he laid out a blueprint for fast-forwarding to the future. If you want to see what's happening on planet earth 1,000, or 10,000, or 10 million years in the future, the laws of Einsteinian physics tell you how to go about it. You build a vehicle whose speed can reach, say 99.9999999996 percent of light speed. At full throttle, you head off into deep space for a day, or ten days, or a little over twenty-seven years according to your ship's clock, then abruptly turn around and head back to earth, again at full throttle. On your return, 1,000, or 10,000, or 10 million years of earth time *will* have elapsed. This is an undisputed and experimentally verified prediction of special relativity; it is an example of the slowing of time with the increasing of speed described in Chapter 3.[8] Of course, since vehicles of such speed are beyond what we can build, no one has tested these predictions literally. But as we discussed earlier, researchers have confirmed the predicted slowing of time for a commercial airliner, traveling at a small fraction of light speed, as well as that of elementary particles like muons racing through accelerators at very nearly the speed of light (stationary muons decay into other particles in about two millionths of a second, but the faster they travel the slower their internal clock's tick, and so the longer the muons appear to live). There is every reason to believe, and no reason not to believe, that special relativity is correct, and its strategy for reaching the future would work as predicted. Technology, not physics, keeps each of us tethered to this epoch.*

Thornier issues arise, though, when we think about the other kind of time travel, travel to the past. No doubt you are familiar with some of these. For example, there's the standard scenario in which you travel to the past and prevent your own birth. In many fictional descriptions this is achieved with violence; however, any less drastic but equally effective intervention—such as preventing your parents from meeting—would do just as well. The paradox is clear: if you were never born, how did you come to be, and, in particular, how did you travel to the past and keep your parents from meeting? To travel to the past and keep your parents

*The fragility of the human body is another practical limitation: the acceleration required to reach such high speeds in a reasonable length of time is well beyond what the body can withstand. Note, too, that the slowing of time gives a strategy, in principle, for reaching distant locations in space. If a rocket were to leave earth and head for the Andromeda galaxy, traveling at 99.999999999999999999 percent of light speed, we'd have to wait nearly 6 million years for it to return. But at that speed, time on the rocket slows down relative to time on earth so dramatically that upon returning the astronaut would have aged only eight hours (setting aside the fact that he or she couldn't have survived the accelerations to get up to speed, turn back, and finally stop).

apart, you had to have been born; but if you were born, traveled to the past, and kept your parents apart, you *wouldn't* have been born. We run headlong into a logical impasse.

A similar paradox, suggested by the Oxford philosopher Michael Dummett and highlighted by his colleague David Deutsch, teases the brain in a slightly different, perhaps even more baffling way. Here's one version. Imagine I build a time machine and travel ten years into the future. After a quick lunch at Tofu-4-U (the chain that overtook McDonald's after the great mad-cow pandemic put a dent in the public enthusiasm for cheeseburgers), I find the nearest Internet café and get online to see what advances have been made in string theory. And do I get a splendid surprise. I read that all open issues in string theory have been resolved. The theory has been completely worked out and successfully used to explain all known particle properties. Incontrovertible evidence for the extra dimensions has been found, and the theory's predictions of supersymmetric partner particles—their masses, electric charges, and so on—have just been confirmed, spot on, by the Large Hadron Collider. There is no longer any doubt: string theory is the unified theory of the universe.

When I dig a little deeper to see who is responsible for these great advances, I get an even bigger surprise. The breakthrough paper was written a year earlier by none other than Rita Greene. My mother. I'm shocked. No disrespect intended: my mother is a wonderful person, but she's not a scientist, can't understand why anybody would be a scientist, and, for example, read only a few pages of *The Elegant Universe* before putting it down, saying it gave her a headache. So how in the world could she have written *the* key paper in string theory? Well, I read her paper online, am blown away by the simple yet deeply insightful reasoning, and see at the end that she's thanked *me* for years of intense instruction in mathematics and physics after a Tony Robbins seminar persuaded her to overcome her fears and pursue her inner physicist. Yikes, I think. She'd just enrolled in that seminar when I embarked on my trip to the future. I'd better head back to my own time to begin the instruction.

Well, I go back in time and begin to tutor my mother in string theory. But it's not going well. A year goes by. Then two. And although she's trying hard, she's just not getting it. I'm starting to worry. We stay at it for another couple of years, but progress is minimal. Now I'm really worried. There is not much time left before her paper is supposed to appear. How is she going to write it? Finally, I make the big decision. When I read her paper in the future, it left such an impression on me that I remember it clear as day. And so, instead of having her discover it on her own—something that's looking less and less likely—I tell her what to write, making sure she includes everything exactly as I remember reading it. She releases the paper, and in short order it sets the physics world on fire. All that I read about during my time in the future comes to pass.

Now here's the puzzling issue. Who should get the credit for my mother's groundbreaking paper? I certainly shouldn't. I learned of the results by reading them in her paper. Yet how can my mother take credit, when she wrote only what I told her to? Of course, the issue here is not really one of credit—it's the issue of where the new knowledge, new insights, and new understanding presented in my mother's paper came from. To what can I point and say, "This person or this computer came up with the new results"? I didn't have the insights, nor did my mother, there wasn't anyone else involved, and we didn't use a computer. Nevertheless, somehow these brilliant results are all in her paper. Apparently, in a world that allows time travel both to the future and to the past, knowledge can materialize out of thin air. Although not quite as paradoxical as preventing your own birth, this is positively weird.

What should we make of such paradox and weirdness? Should we conclude that while time travel to the future is allowed by the laws of physics, any attempt to return to the past must fail? Some have certainly thought so. But, as we'll now see, there are ways around the tricky issues we've come upon. This doesn't mean that travel to the past is possible—that's a separate issue we'll consider shortly—but it does show that travel back in time can't be ruled out merely by invoking the puzzles we've just discussed.

Rethinking the Puzzles

Recall that in Chapter 5 we discussed the flow of time, from the perspective of classical physics, and came upon an image that differs substantially from our intuitive picture. Careful thought led us to envision spacetime as a block of ice with every moment forever frozen in place, as opposed to the familiar image of time as a river sweeping us forward from one moment to the next. These frozen moments are grouped into *nows*—into events that happen at the same time—in different ways by observers in different states of motion. And to accommodate this flexibility of slicing the spacetime block into different notions of now, we also invoked an equivalent metaphor in which spacetime is viewed as a loaf of bread that can be sliced at different angles.

But regardless of the metaphor, Chapter 5's lesson is that moments—the events making up the spacetime loaf—just are. They are timeless. Each moment—each event or happening—exists, just as each point in space exists. Moments don't momentarily come to life when illuminated by the "spotlight" of an observer's present; that image aligns well with our intuition but fails to stand up to logical analysis. Instead, once illuminated, always illuminated. Moments don't change. Moments are. Being illuminated is simply one of the many unchanging features that constitute a moment. This is particularly evident from the insightful though imaginary perspective of Figure 5.1, in which all events making up the history of the universe are on view; they are all there, static and unchanging. Different observers don't agree on which of the events happen at the same time—they time-slice the spacetime loaf at different angles—but the total loaf and its constituent events are universal, literally.

Quantum mechanics offers certain modifications to this classical perspective on time. For example, we saw in Chapter 12 that on extremely short scales, space and spacetime become unavoidably wavy and bumpy. But (Chapter 7), a full assessment of quantum mechanics and time requires a resolution of the quantum measurement problem. One of the proposals for doing so, the Many Worlds interpretation, is particularly relevant for coping with paradoxes arising from time travel, and we will take that up in the next section. But in this section, let's stay classical and bring the block-of-ice/loaf-of-bread depiction of spacetime to bear on these puzzles.

Take the paradoxical example of your having gone back in time and having prevented your parents from meeting. Intuitively, we all know what that's supposed to mean. Before you time-traveled to the past, your parents had met—say, at the stroke of midnight, December 31, 1965,* at a New Year's party—and, in due course, your mother gave birth to you. Then, many years later, you decided to travel to the past—back to December 31, 1965—and once there, you changed things; in particular, you kept your parents apart, preventing your own conception and birth. But let's now counter this intuitive description with the more fully reasoned spacetime-loaf depiction of time.

At its core, the intuitive description fails to make sense because it

---

*Of course, I really should say January 1, 1966, but let's not worry about that.

assumes moments can change. The intuitive picture envisions the stroke of midnight, December 31, 1965 (using standard earthling time-slicing), as "initially" being the moment of your parents meeting, but envisions further that your interference "subsequently" changes things so that at the stroke of midnight, December 31, 1965, your parents are miles, if not continents, apart. The problem with this recounting of events, though, is that moments don't change; as we've seen, they just are. The spacetime loaf exists, fixed and unchanging. There is no meaning to a moment's "initially" being one way and "subsequently" being another way.

If you time-traveled back to December 31, 1965, then you were there, you were always there, you will always be there, you were never not there. December 31, 1965, did not happen twice, with your missing the debut but attending the encore. From the timeless perspective of Figure 5.1, you exist—static and unchanging—at various locations in the spacetime loaf. If today you set the dials on your time machine to send you to 11:50 p.m., December 31, 1965, then this latter moment will be among the locations in the spacetime loaf at which you can be found. But your presence on New Year's Eve, 1965, will be an *eternal* and *immutable* feature of spacetime.

This realization still leads us to some quirky conclusions, but it avoids paradox. For example, you would appear in the spacetime loaf at 11:50 p.m., December 31, 1965, but before that moment there would be no record of your existence. This is strange, but not paradoxical. If a guy saw you pop in at 11:50 p.m. and asked you, with fear in his eyes, where you came from, you could calmly answer, "The future." In this scenario, at least so far, we are not caught in a logical impasse. Where things get more interesting, of course, is if you then try to carry out your mission and keep your parents from meeting. What happens? Well, carefully maintaining the "spacetime block" perspective, we inescapably conclude that you can't succeed. No matter what you do on that fateful New Year's Eve, you'll fail. Keeping your parents apart—while seeming to be within the realm of things you can do—actually amounts to logical gobbledygook. Your parents met at the stroke of midnight. You were there. And you will "always" be there. Each moment just is; it doesn't change. Applying the concept of change to a moment makes as much sense as subjecting a rock to psychoanalysis. Your parents met at the stroke of midnight, December 31, 1965, and *nothing* can change that because their meeting is an immutable, unchangeable event, eternally occupying its spot in spacetime.

In fact, now that you think about it, you remember that sometime in your teens, when you asked your dad what it was like to propose to your mother, he told you that he hadn't planned to propose at all. He had barely met your mother before asking the big question. But about ten minutes before midnight at a New Year's party, he got so freaked by seeing a man pop in from nowhere — a man who claimed to be from the future — that when he met your mother he decided to propose, right on the spot.

The point is that the complete and unchanging set of events in space-time necessarily fits together into a coherent, self-consistent whole. The universe makes sense. If you time-travel back to December 31, 1965, you are actually fulfilling your own destiny. In the spacetime loaf, there is someone present at 11:50 p.m. on December 31, 1965, who is not there at any earlier time. From the imaginary, outside perspective of Figure 5.1, we would be able to see this directly; we would also see, undeniably, that the person is *you* at your current age. For these events, situated decades ago, to make sense, you *must* time-travel back to 1965. What's more, from our outside perspective we can see your father asking you a question just after 11:50 p.m. on December 31, 1965, looking frightened, rushing away, and meeting your mother at midnight; a little further along the loaf, we can see your parents' wedding, your birth, your ensuing childhood, and, later on, your stepping into the time machine. If time travel to the past were possible, we could no longer explain events at one time solely in terms of events at earlier times (from any given perspective); but the total-ity of events would necessarily constitute a sensible, coherent, noncontra-dictory story.

As emphasized in the last section, this doesn't, by any stretch of the imagination, signify that time travel to the past is possible. But it does sug-gest strongly that the purported paradoxes, such as preventing your own birth, are themselves born of logical flaws. If you time-travel to the past, you can't change it any more than you can change the value of pi. If you travel to the past, you are, will be, and always were part of the past, the very same past that leads to your traveling to it.

From the outside perspective of Figure 5.1, this explanation is both tight and coherent. Surveying the totality of events in the spacetime loaf, we see that they interlock with the rigid economy of a cosmic crossword puzzle. Yet, from your perspective on December 31, 1965, things are still puzzling. I declared above that even though you may be determined to keep your parents from meeting, you can't succeed in the classical approach to this problem. You can watch them meet. You can even facilitate their meeting, perhaps inadvertently as in the story I've told. You can travel back in time repeatedly, so there are many of you present, each intent on preventing your parents' union. But to succeed in preventing your parents from meeting would be to change something with respect to which the concept of change is meaningless.

But, even with the insight of these abstract observations, we can't help asking: What stops you from succeeding? If you are standing at the party at 11:50 p.m. and see your young mother, what stops you from whisking her away? Or, if you see your young father, what stops you from—oh, what the heck, let's just say it—shooting him? Don't you have free will? Here is where, some suspect, quantum mechanics may enter the story.

Free Will, Many Worlds, and Time Travel

Free will is a tricky issue, even absent the complicating factor of time travel. The laws of classical physics are deterministic. As we saw earlier, if you were to know precisely how things are now (the position and velocity of every particle in the universe), the laws of classical physics would tell you exactly how things were or would be at any other moment you specified. The equations are indifferent to the supposed freedom of human will. Some have taken this to mean that in a classical universe, free will would be an illusion. You are made of a collection of particles, so if the laws of classical physics could determine everything about your particles at any moment—where they'd be, how they'd be moving and so on—your willful ability to determine your own actions would appear fully compromised. This reasoning convinces me, but those who believe we are more than the sum of our particles may disagree.

Anyway, the relevance of these observations is limited, since ours is a quantum, not a classical, universe. In quantum physics, real-world physics, there are resemblances to this classical perspective; there are also potentially pivotal differences. As you read in Chapter 7, if you know the quantum wavefunction right now for every particle in the universe, Schrödinger's equation tells you how the wavefunction was or will be at any other moment you specify. This component of quantum physics is fully deterministic, just as in classical physics. However, the act of observation complicates the quantum mechanical story and, as we've seen, heated debate over the quantum measurement problem still rages. If physicists one day conclude that Schrödinger's equation is all there is to quantum mechanics, then quantum physics, in its entirety, would be every bit as deterministic as classical physics. As with classical determinism, some would say this means free will is an illusion; others would not. But if we're currently missing part of the quantum story—if the passage from probabilities to definite outcomes requires something beyond the standard quantum framework—it's at least possible that free will might find a concrete realization within physical law. We might one day find, as some physicists have speculated, that the act of conscious observation is an integral element of quantum mechanics, being the catalyst that coaxes one outcome from the quantum haze to be realized.[9] Personally, I find this extremely unlikely, but I know of no way to rule it out.

The upshot is that the status of free will and its role within fundamental physical law remain unresolved. So let's consider both possibilities, free will that's illusory and free will that's real.

If free will is an illusion, and if time travel to the past is possible, then your inability to prevent your parents from meeting poses no puzzle. Although you feel as if you have control over your actions, the laws of physics are really pulling the strings. When you go to whisk away your mother or shoot your father, the laws of physics get in the way. The time machine lands you on the wrong side of town, and you arrive after your parents have met; or you try to pull the trigger and the gun jams; or you do pull the trigger, but you miss the target and instead knock off your father's only competitor for your mother's hand, clearing the way for their union; or, perhaps, when you step out of the time machine you no longer have the desire to prevent your parents from meeting. Regardless of your intention when you enter the time machine, your actions when you exit are part of spacetime's consistent story. The laws of physics trump all attempts to thwart logic. Everything you do fits in perfectly. It always has and always will. You can't change the unchangeable.

If free will is not an illusion, and if time travel to the past is possible, quantum physics gives alternative suggestions for what might happen, and is distinctly different from the formulation based on classical physics. One particularly compelling proposal, championed by Deutsch, makes use of the Many Worlds interpretation of quantum mechanics. Remember from Chapter 7 that in the Many Worlds framework, every potential outcome embodied in a quantum wavefunction—a particle's spinning this way or that, another particle's being here or there—is realized in its own separate, parallel universe. The universe we're aware of at any given moment is but one of an infinite number in which every possible evolution allowed by quantum physics is separately realized. In this framework, it's tempting to suggest that the freedom we feel to make this or that choice reflects the possibility we have to enter this or that parallel universe in a subsequent moment. Of course, since infinitely many copies of you and me are sprinkled across the parallel universes, the concepts of personal identity and free will need to be interpreted in this broadened context.

As far as time travel and the potential paradoxes go, the Many Worlds interpretation suggests a novel resolution. When you travel to 11:50 p.m. on December 31, 1965, pull out your weapon, aim at your father, and pull the trigger, the gun works and you hit the intended target. But since this is not what happened in the universe from which you embarked on your time travel odyssey, your journey must not only have been through time, *it must have been also from one parallel universe to another.* The parallel universe in which you now find yourself is one in which your parents never did meet—a universe which the Many Worlds interpretation assures us is out there (since every possible universe consistent with the laws of quantum physics is out there). And so, in this approach, we face no logical paradox, because there are various versions of a given moment, each situated in a different parallel universe; in the Many Worlds interpretations, it's as if there are infinitely many spacetime loaves, not just one. In the universe of origination, your parents met on December 31, 1965, you were born, you grew up, you held a grudge against your father, you became fascinated with time travel, and you embarked on a journey to December 31, 1965. In the universe in which you arrive, your father is killed on December 31, 1965, before meeting your mother, by a gunman claiming to be his son from the future. A version of you is never born in this universe, but that's okay, since the you who pulled the trigger *does* have parents. It's just that they happen to live in a different parallel universe. Whether anyone in this universe believes your story or, instead, views you as delusional, I can't say. But what's clear is that in each universe—the one you left and the one you entered—we avoid self-contradictory circumstances.

What's more, even in this broadened context, your time travel expedition doesn't change the past. In the universe you left, that's manifest, since you never visit its past. In the universe you enter, your presence at 11:50 p.m. on December 31, 1965, does not change that moment: in that universe you were, and always will be, present at that moment. Again, in the Many Worlds interpretation, every physically consistent sequence of events happens in one of the parallel universes. The universe you enter is one in which the murderous actions you choose to undertake are realized. Your presence on December 31, 1965, and all the mayhem you create, are part of the unchangeable fabric of that universe's reality.

The Many Worlds interpretation offers a similar resolution to the issue of knowledge seemingly materializing from nowhere, as in the scenario of my mother's writing a decisive paper in string theory. According to the Many Worlds interpretation, in one of the myriad parallel universes my mother *does* develop quickly into a string theory expert, and on her own discovers all that I read in her paper. When I undertake my excursion to the future, my time machine takes me to *that* universe. The results I read in my mother's paper while I'm there were indeed discovered by the version of my mother in that world. Then, when I travel back in time, I enter a different one of the parallel universes, one in which my mother has difficulty understanding physics. After years of trying to teach her, I give up and finally tell her what to write in the paper. But in this scenario there is no puzzle regarding who is responsible for the breakthroughs. The discoverer is the version of my mother in the universe in which she's a physics whiz. All that's happened as a result of my various time travels is that her discoveries are communicated to a version of herself in another parallel universe. Assuming you find parallel universes easier to swallow than authorless discoveries—a debatable proposition—this provides a less baffling explanation of the interplay of knowledge and time travel.

None of the proposals we've discussed in this or the previous section are necessarily *the* resolution to the puzzles and paradoxes of time travel. Instead, these proposals are meant to show that puzzles and paradoxes do not rule out time travel to the past since, with our current state of understanding, physics provides possible avenues for end runs around the problems. But failing to rule something out is a far cry from declaring it possible. So we are now led to ask the main question:

Is Time Travel to the Past Possible?

Most sober physicists would answer no. I would say no. But unlike the definitive no you'd get if you asked whether special relativity allows a massive object to accelerate up to and then exceed the speed of light, or whether Maxwell's theory allows a particle with one unit of electric charge to disintegrate into particles with two units of electric charge, this is a qualified no.

The fact is, no one has shown that the laws of physics absolutely rule out past-directed time travel. To the contrary, some physicists have even laid out hypothetical instructions for how a civilization with unlimited technological prowess, operating fully within the known laws of physics, might go about building a time machine (when we speak of time machines, we will always mean something that is able to travel both to the future and to the past). The proposals bear no resemblance to the spinning gizmo described by H. G. Wells or Doc Brown's souped-up DeLorean. And the design elements all brush right up against the limits of known physics, leading many researchers to suspect that with subsequent refinements in our grasp of nature's laws, existing and future proposals for time machines will be deemed beyond the bounds of what's physically possible. But as of today, this suspicion is based on gut feeling and circumstantial evidence, not solid proof.

Einstein himself, during the decade of intense research leading to the publication of his general theory of relativity, pondered the question of travel to the past.[10] Frankly, it would have been strange if he hadn't. As his radical reworkings of space and time discarded long-accepted dogma, an ever-present question was how far the upheaval would go. Which features, if any, of familiar, everyday, intuitive time would survive? Einstein never wrote much on the issue of time travel because, by his own account, he never made much progress. But in the decades following the release of his paper on general relativity, slowly but surely, other physicists did.

Among the earliest general relativity papers with relevance for time machines were those written in 1937 by the Scottish physicist W. J. van Stockum[11] and in 1949 by a colleague of Einstein's at the Institute for Advanced Study, Kurt Gödel. Van Stockum studied a hypothetical problem in general relativity in which a very dense and infinitely long cylinder is set into spinning motion about its (infinitely) long axis. Although an infinite cylinder is physically unrealistic, van Stockum's analysis led to an interesting revelation. As we saw in Chapter 14, massive spinning objects drag space into a whirlpool-like swirl. In this case, the swirl is so significant that, mathematical analysis shows, not only space but also time would get caught up in the whirlpool. Roughly speaking, the spinning twists the time direction on its side, so that circular motion around the cylinder takes you to the past. If your rocket ship encircles the cylinder, you can return to your starting point in space *before* you embark on your journey. Certainly, no one can build an infinitely long spinning cylinder, but this work was an early hint that general relativity might not prohibit time travel to the past.

Gödel's paper also investigated a situation involving rotational motion. But rather than focusing on an object rotating within space, Gödel studied what happens if all of space undergoes rotational motion. Mach would have thought this meaningless. If the whole universe is rotating, then there's nothing with respect to which the purported rotation is happening. Mach would conclude, a rotating universe and a stationary universe are one and the same. But this is another example in which general relativity fails to fully conform to Mach's relational conception of space. According to general relativity, it does make sense to speak of the entire universe's rotating, and with this possibility come simple observational consequences. For example, if you fire a laser beam in a rotating universe, general relativity shows that it will appear to travel along a spiral path rather than a straight line (somewhat like the path you'd see a slow-moving bullet follow if you fired a toy gun upward while riding a merry-go-round). The surprising feature of Gödel's analysis was his realization that if your rocket ship were to follow appropriate trajectories in a spinning universe, you could also return to your place of origin in space *before* the time of your departure. A rotating universe would thus itself be a time machine.

Einstein congratulated Gödel on his discovery, but suggested that further investigation might show that solutions to the equations of general relativity permitting travel to the past run afoul of other essential physical requirements, making them no more than mathematical curiosities. As far as Gödel's solution goes, increasingly precise observations have minimized the direct relevance of his work by establishing that our universe is not rotating. But van Stockum and Gödel had let the genie out of the bottle; within a couple of decades, yet more solutions to Einstein's equations permitting time travel to the past were found.

In recent decades, interest in hypothetical time machine designs has revived. In the 1970s, Frank Tipler reanalyzed and refined van Stockum's solution, and in 1991, Richard Gott of Princeton University discovered another method for building a time machine making use of so-called cosmic strings (hypothetical, infinitely long, filamentary remnants of phase transitions in the early universe). These are all important contributions, but the proposal that's simplest to describe, using concepts we've developed in previous chapters, was found by Kip Thorne and his students at the California Institute of Technology. It makes use of wormholes.

Blueprint for a Wormhole Time Machine

I'll first lay out the basic strategy for constructing Thorne's wormhole time machine, and in the next section I'll discuss the challenges faced by any contractor Thorne might hire to execute the plans.

A *wormhole* is a hypothetical tunnel through space. A more familiar kind of tunnel, such as one that's been bored through the side of a mountain, provides a shortcut from one location to another. Wormholes serve a similar function, but they differ from conventional tunnels in one important respect. Whereas conventional tunnels provide a new route through existing space—the mountain and the space it occupies exist before a tunnel is constructed—a wormhole provides a tunnel from one point in space to another along a new, previously nonexistent tube of space. Were you to remove the tunnel through the mountain, the space it occupied would still exist. Were you to remove a wormhole, the space it occupied would vanish.

Figure 15.2a illustrates a wormhole connecting the Kwik-E-Mart and the Springfield Nuclear Power Plant, but the drawing is misleading because the wormhole appears to stretch across Springfield airspace. More accurately, the wormhole should be thought of as a new region of space that interfaces with ordinary, familiar space only at its ends—its mouths. If while walking along the streets of Springfield, you scoured the

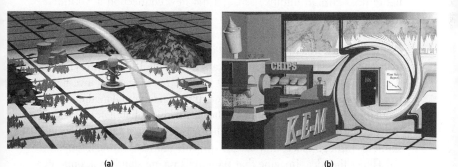

(a)   (b)

Figure 15.2 **(a)** A wormhole extending from the Kwik-E-Mart to the nuclear power plant. **(b)** The view through the wormhole, looking from the mouth at the Kwik-E-Mart and into the mouth in the power plant.

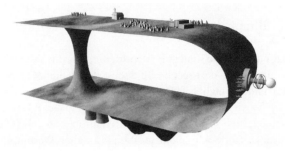

Figure 15.3 Geometry which more clearly shows that the wormhole is a shortcut. (Wormhole mouths are really inside Kwik-E-Mart and the nuclear power plant, although that is difficult to show in this representation.)

skyline in search of the wormhole, you'd see nothing. The only way to see it would be to hop on over to the Kwik-E-Mart, where you would find an opening in ordinary space—one wormhole mouth. Looking through the opening, you'd see the inside of the power plant, the location of the second mouth, as in Figure 15.2b. Another misleading feature of Figure 15.2a is that the wormhole doesn't appear to be a shortcut. We can fix this by modifying the illustration as in Figure 15.3. As you can see, the usual route from the power plant to the Kwik-E-Mart is indeed longer than the wormhole's new spatial passage. The contortions in Figure 15.3 reflect the difficulties in drawing general relativistic geometry on a page, but the figure does give an intuitive sense of the new connection a wormhole would provide.

No one knows whether wormholes exist, but many decades ago physicists established that they are allowed by the mathematics of general relativity and so are fair game for theoretical study. In the 1950s, John Wheeler and his coworkers were among the earliest researchers to investigate wormholes, and they discovered many of their fundamental mathematical properties. More recently, though, Thorne and his collaborators revealed the full richness of wormholes by realizing that not only can they provide shortcuts through space, they can also provide shortcuts through time.

Here's the idea. Imagine that Bart and Lisa are standing at opposite ends of Springfield's wormhole—Bart at the power plant, Lisa at the Kwik-E-Mart—idly chatting with each other about what to get Homer for his birthday, when Bart decides to take a short transgalactic jaunt (to get

Homer some of his favorite Andromedean fish fingers). Lisa doesn't feel up for the ride but, as she's always wanted to see Andromeda, she persuades Bart to load his wormhole mouth on his ship and take it along, so she can have a look. You might expect this to mean that Bart will have to keep stretching the wormhole longer as his journey progresses, but that assumes the wormhole connects the Kwik-E-Mart and Bart's ship through ordinary space. It doesn't. And, as illustrated in Figure 15.4, through the wonders of general relativistic geometry, the wormhole's length can remain fixed throughout the entire voyage. This is a key point. Even though Bart rockets off to Andromeda, his distance to Lisa through the wormhole does not change. This makes manifest the wormhole's role as a shortcut through space.

For definiteness, let's say that Bart heads off at 99.999999999999999999 percent of light speed and travels four hours outbound to Andromeda, all the while continuing to chat with Lisa through the wormhole, just as they'd been doing before the flight. When

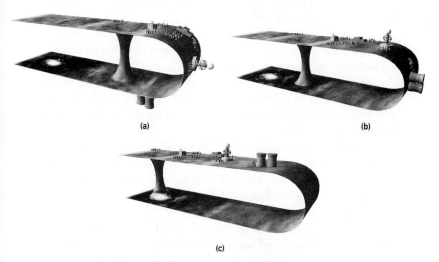

(a)   (b)

(c)

Figure 15.4 (a) A wormhole connecting the Kwik-E-Mart and the nuclear power plant. **(b)** The lower wormhole opening transported (from the nuclear power plant) to outer space (on spaceship, not shown). The wormhole length remains fixed. **(c)** The wormhole opening arrives at the Andromeda galaxy; the other opening is still at the Kwik-E-Mart. The length of the wormhole is unchanged throughout the entire voyage.

the ship reaches Andromeda, Lisa tells Bart to pipe down so she can take in the view without disturbance. She's exasperated by his insistence on quickly grabbing the takeout at the Fish Finger Flythrough and heading back to Springfield, but agrees to keep on chatting until he returns. Four hours and a few dozen rounds of tic-tac-toe later, Bart safely sets his ship down on the lawn of Springfield High.

When he looks out the ship window, though, Bart gets a bit of a shock. The buildings look completely different, and the scoreboard floating high above the rollerball stadium gives a date some 6 million years after his departure. "Dude!?!" he says to himself, but a moment later it all becomes clear. Special relativity, he remembers from a heart-to-heart he'd recently had with Sideshow Bob, ensures that the faster you travel the slower your clock ticks. If you travel out into space at high speed and then return, only a few hours might have elapsed aboard your ship while thousands or millions of years, if not more, will have elapsed according to someone stationary. With a quick calculation, Bart confirms that at the speed he was traveling, eight hours elapsed on the ship would mean 6 million years elapsed on earth. The date on the scoreboard is right; Bart realizes he has traveled far into earth's future.

". . . Bart! Hello, Bart!" Lisa yells through the wormhole. "Have you been listening to me? Step on it. I want to get home in time for dinner." Bart looks into his wormhole mouth and tells Lisa he's already landed on the lawn of Springfield High. Looking more closely through the wormhole, Lisa sees that Bart is telling the truth, but looking out of the Kwik-E-Mart toward Springfield High, she doesn't see his ship on the lawn. "I don't get it," she says.

"Actually, it makes perfect sense," Bart proudly answers. "I've landed at Springfield High, but 6 million years into the future. You can't see me by looking out the Kwik-E-Mart window, because you're looking at the right place, but you're not looking at the right time. You're looking 6 million years too early."

"Oh, right, that time-dilation thing of special relativity," Lisa agrees. "Cool. Anyway, I want to get home in time for dinner, so climb through the wormhole, because we've got to hurry." "Okay," Bart says, crawling through the wormhole. He buys a Butterfinger from Apu, and he and Lisa head home.

Notice that although Bart's passage through the wormhole took him but a moment, *it transported him 6 million years back in time.* He and his ship and the wormhole mouth had landed far into earth's future. Had he gotten out, spoken with people, and checked the newspaper, everything would have confirmed this. Yet, when he passed through the wormhole and rejoined Lisa, he found himself back in the present. The same holds true for anyone else who might follow Bart through the wormhole mouth: he would also travel 6 million years back in time. Similarly, anyone who climbs into the wormhole mouth at the Kwik-E-Mart, and out of the mouth Bart left in his ship, would travel 6 million years into the future. The important point is that Bart did not just take one of the wormhole mouths on a journey through space. His journey also transported the wormhole mouth through time. *Bart's voyage took him and the wormhole's mouth into earth's future. In short, Bart transformed a tunnel through space into a tunnel through time; he turned a wormhole into a time machine.*

A rough way to visualize what's going on is depicted in Figure 15.5. In Figure 15.5a we see a wormhole connecting one spatial location with another, with the wormhole configuration drawn so as to emphasize that it lies outside of ordinary space. In Figure 15.5b, we show the time evolution of this wormhole, assuming both its mouths are kept stationary. (The time slices are those of a stationary observer.) In Figure 15.5c, we show what happens when one wormhole mouth is loaded onto a spaceship and taken on a round-trip journey. Time for the moving mouth, just like time on a moving clock, slows down, so that the moving mouth is transported to the future. (If an hour elapses on a moving clock but a thousand years elapse on stationary clocks, the moving clock will have jumped into the stationary clocks' future.) Thus, instead of the stationary wormhole mouth's connecting, via the wormhole tunnel, to a mouth on the same time slice, it connects to a mouth on a *future* time slice, as in Figure 15.5c. Unless the wormhole mouths are moved further, the time difference between them will remain locked in. At any moment, should you enter one mouth and exit the other, you will have become a time traveler.

Building a Wormhole Time Machine

One blueprint for building a time machine is now clear. Step 1: find or create a wormhole wide enough for you, or anything you want to send through time, to pass. Step 2: establish a time difference between the wormhole mouths—say, by moving one relative to the other. That's it. In principle.

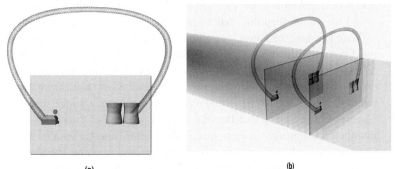

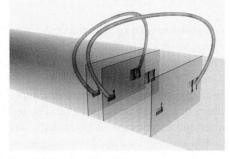

Figure 15.5 (a) A wormhole, created at some moment in time, connects one location in space with another. **(b)** If the wormhole mouths do not move relative to one another, they "pass" through time at the same rate, so the tunnel connects the two regions at the same time. **(c)** If one wormhole mouth is taken on a round-trip journey (not shown), less time will elapse for that mouth, and hence the tunnel will connect the two regions of space at different moments of time. The wormhole has become a time machine.

How about in practice? Well, as I mentioned at the outset, no one knows whether wormholes even exist. Some physicists have suggested that tiny wormholes might be plentiful in the microscopic makeup of the spatial fabric, being continually produced by quantum fluctuations of the gravitational field. If so, the challenge would be to enlarge one to macroscopic size. Proposals have been made for how this might be done, but they're barely beyond theoretical flights of fancy. Other physicists have envisioned the creation of large wormholes as an engineering project in applied general relativity. We know that space responds to the distribution of matter and energy, so with sufficient control over matter and energy, we might cause a region of space to spawn a wormhole. This approach presents an additional complication, because just as we must tear open the side of a mountain to attach the mouth of a tunnel, we must tear open the fabric of space to attach the mouth of a wormhole.[12] No one knows whether such tears in space are allowed by the laws of physics. Work with which I've been involved in string theory (see page 386) has shown that certain kinds of spatial tears are possible, but so far we have no idea whether these rips might be relevant to the creation of wormholes. The bottom line is that intentional acquisition of a macroscopic wormhole is a fantasy that, at best, is a *very* long way from being realized.

Morever, even if we somehow managed to get our hands on a macroscopic wormhole, we wouldn't be done; we'd still face a couple of significant obstacles. First, in the 1960s, Wheeler and Robert Fuller showed, using the equations of general relativity, that wormholes are unstable. Their walls tend to collapse inward in a fraction of a second, which eliminates their utility for any kind of travel. More recently, though, physicists (including Thorne and Morris, and also Matt Visser) have found a potential way around the collapse problem. If the wormhole is not empty, but instead contains material—so-called *exotic matter*—that can exert an outward push on its walls, then it might be possible to keep the wormhole open and stable. Although similar in its effect to a cosmological constant, exotic matter would generate outward-pushing repulsive gravity by virtue of having negative energy (not just the negative pressure characteristic of a cosmological constant[13]). Under highly specialized conditions, quantum mechanics allows for negative energy,[14] but it would be a monumental challenge to generate enough exotic matter to hold a macroscopic wormhole open. (For example, Visser has calculated that the amount of negative energy needed to keep open a one-meter-wide wormhole is roughly equal in magnitude to the total energy produced by the sun over about 10 billion years.[15])

Second, even if we somehow found or created a macroscopic wormhole, and even if we somehow were able to buttress its walls against immediate collapse, and even if we were able to induce a time difference between the wormhole mouths (say, by flying one mouth around at high speed), there would remain another hurdle to acquiring a time machine. A number of physicists, including Stephen Hawking, have raised the possibility that vacuum fluctuations—the jitters arising from the quantum uncertainty experienced by all fields, even in empty space, discussed in Chapter 12—might destroy a wormhole just as it was getting into position to be a time machine. The reason is that, just at the moment when time travel through the wormhole becomes possible, a devastating feedback mechanism, somewhat like the screeching noise generated when microphone and speaker levels in a sound system are not adjusted appropriately, may come into play. Vacuum fluctuations from the future can travel through the wormhole to the past, where they can then travel through ordinary space and time to the future, enter the wormhole, and travel back to the past again, creating an endless cycle through the wormhole and filling it with ever-increasing energy. Presumably, such an intense energy buildup would destroy the wormhole. Theoretical research suggests this as a real possibility, but the necessary calculations strain our current understanding of general relativity and quantum mechanics in curved spacetime, so there is no conclusive proof.

The challenges to building a wormhole time machine are clearly immense. But the final word won't be given until our facility with quantum mechanics and gravity is refined further, perhaps through advances in superstring theory. Although at an intuitive level physicists generally agree that time travel to the past is impossible, as of today the question has yet to be fully closed.

Cosmic Rubbernecking

In thinking about time travel, Hawking has raised an interesting point. Why, he asks, if time travel is possible, haven't we been inundated with visitors from the future? Well, you might answer, maybe we have. And you might go further and say we've put so many time travelers in locked wards that most of the others don't dare identify themselves. Of course, Hawking is half joking, and so am I, but he does raise a serious question. If you believe, as I do, that we have not been visited from the future, is that tantamount to believing time travel impossible? Surely, if people succeed in building time machines in the future, some historian is bound to get a grant to study, up close and personal, the building of the first atomic bomb, or the first voyage to the moon, or the first foray into reality television. So, if we believe no one has visited us from the future, perhaps we are implicitly saying that we believe no such time machine will ever be built.

Actually, though, this is not a necessary conclusion. *The time machines that have thus far been proposed do not allow travel to a time prior to the construction of the first time machine itself.* For the wormhole time machine, this is easy to see by examining Figure 15.5. Although there is a time difference between the wormhole mouths, and although that difference allows travel forward and backward in time, you can't reach a time before the time difference was established. The wormhole itself does not exist on the far left of the spacetime loaf, so there is no way you can use it to get there. Thus, if the first time machine is built, say, 10,000 years from now, *that* moment will no doubt attract many time-traveling tourists, but all previous times, such as ours, will remain inaccessible.

I find it curious and compelling that our current understanding of nature's laws not only suggests how to avoid the seeming paradoxes of time travel but also offers proposals for how time travel might actually be accomplished. Don't get me wrong: I count myself among the sober physicists who feel intuitively that we will one day rule out time travel to the past. But until there's definitive proof, I think it justified and appropriate to keep an open mind. At the very least, researchers focusing on these issues are substantially deepening our understanding of space and time in extreme circumstances. At the very best, they may be taking the first critical steps toward integrating us into the spacetime superhighway. After all, every moment that goes by without our having succeeded in building a time machine is a moment that will be forever beyond our reach and the reach of all who follow.

16

The Future
of an Allusion

PROSPECTS FOR SPACE AND TIME

P hysicists spend a large part of their lives in a state of confusion. It's an occupational hazard. To excel in physics is to embrace doubt while walking the winding road to clarity. The tantalizing discomfort of perplexity is what inspires otherwise ordinary men and women to extraordinary feats of ingenuity and creativity; nothing quite focuses the mind like dissonant details awaiting harmonious resolution. But en route to explanation—during their search for new frameworks to address outstanding questions—theorists must tread with considered step through the jungle of bewilderment, guided mostly by hunches, inklings, clues, and calculations. And as the majority of researchers have a tendency to cover their tracks, discoveries often bear little evidence of the arduous terrain that's been covered. But don't lose sight of the fact that nothing comes easily. Nature does not give up her secrets lightly.

In this book we've looked at numerous chapters in the story of our species' attempt to understand space and time. And although we have encountered some deep and astonishing insights, we've yet to reach that ultimate eureka moment when all confusion abates and total clarity prevails. We are, most definitely, still wandering in the jungle. So, where from here? What is the next chapter in spacetime's story? Of course, no one knows for sure. But in recent years a number of clues have come to light, and although they've yet to be integrated into a coherent picture, many physicists believe they are hinting at the next big upheaval in our understanding of the cosmos. In due course, space and time as currently conceived may be recognized as mere allusions to more subtle, more profound, and more fundamental principles underlying physical reality. In the final chapter of this account, let's consider some of these clues and catch a glimpse of where we may be headed in our continuing quest to grasp the fabric of the cosmos.

Are Space and Time Fundamental Concepts?

The German philosopher Immanuel Kant suggested that it would be not merely difficult to do away with space and time when thinking about and describing the universe, it would be downright impossible. Frankly, I can see where Kant was coming from. Whenever I sit, close my eyes, and try to think about things while somehow not depicting them as occupying space or experiencing the passage of time, I fall short. Way short. Space, through context, or time, through change, always manages to seep in. Ironically, the closest I come to ridding my thoughts of a direct spacetime association is when I'm immersed in a mathematical calculation (often having to do with spacetime!), because the nature of the exercise seems able to engulf my thoughts, if only momentarily, in an abstract setting that seems devoid of space and time. But the thoughts themselves and the body in which they take place are, all the same, very much part of familiar space and time. Truly eluding space and time makes escaping your shadow a cakewalk.

Nevertheless, many of today's leading physicists suspect that space and time, although pervasive, may not be truly fundamental. Just as the hardness of a cannonball emerges from the collective properties of its atoms, and just as the smell of a rose emerges from the collective properties of its molecules, and just as the swiftness of a cheetah emerges from the collective properties of its muscles, nerves, and bones, so too, the properties of space and time — our preoccupation for much of this book — may also emerge from the collective behavior of some other, more fundamental constituents, which we've yet to identify.

Physicists sometimes sum up this possibility by saying that spacetime may be an illusion — a provocative depiction, but one whose meaning requires proper interpretation. After all, if you were to be hit by a speeding cannonball, or inhale the alluring fragrance of a rose, or catch sight of a blisteringly fast cheetah, you wouldn't deny their existence simply because each is composed of finer, more basic entities. To the contrary, I think most of us would agree that these agglomerations of matter exist, and moreover, that there is much to be learned from studying how their familiar characteristics emerge from their atomic constituents. But because they are composites, what we wouldn't try to do is build a theory of the universe based on cannonballs, roses, or cheetahs. Similarly, if space and time turn out to be composite entities, it wouldn't mean that their familiar manifestations, from Newton's bucket to Einstein's gravity, are illusory; there is little doubt that space and time will retain their all-embracing positions in experiential reality, regardless of future developments in our understanding. Instead, composite spacetime would mean that an even more elemental description of the universe—one that is spaceless and timeless—has yet to be discovered. The illusion, then, would be one of our own making: the erroneous belief that the deepest understanding of the cosmos would bring space and time into the sharpest possible focus. Just as the hardness of a cannonball, the smell of the rose, and the speed of the cheetah disappear when you examine matter at the atomic and subatomic level, space and time may similarly dissolve when scrutinized with the most fundamental formulation of nature's laws.

That spacetime may not be among the fundamental cosmic ingredients may strike you as somewhat far-fetched. And you may well be right. But rumors of spacetime's impending departure from deep physical law are not born of zany theorizing. Instead, this idea is strongly suggested by a number of well-reasoned considerations. Let's take a look at some of the most prominent.

Quantum Averaging

In Chapter 12 we discussed how the fabric of space, much like everything else in our quantum universe, is subject to the jitters of quantum uncertainty. It is these fluctuations, you'll recall, that run roughshod over point-particle theories, preventing them from providing a sensible quantum theory of gravity. By replacing point particles with loops and snippets, string theory spreads out the fluctuations—substantially reducing their magnitude—and this is how it yields a successful unification of quantum mechanics and general relativity. Nevertheless, the diminished spacetime fluctuations certainly still exist (as illustrated in the next-to-last level of magnification in Figure 12.2), and within them we can find important clues regarding the fate of spacetime.

First, we learn that the familiar space and time that suffuse our thoughts and support our equations emerge from a kind of averaging process. Think of the pixelated image you see when your face is a few inches from a television screen. This image is very different from what you see at a more comfortable distance, because once you can no longer resolve individual pixels, your eyes combine them into an average that looks smooth. But notice that it's only through the averaging process that the pixels produce a familiar, continuous image. In a similar vein, the microscopic structure of spacetime is riddled with random undulations, but we aren't directly aware of them because we lack the ability to resolve spacetime on such minute scales. Instead, our eyes, and even our most powerful equipment, combine the undulations into an average, much like what happens with television pixels. Because the undulations are random, there are typically as many "up" undulations in a small region as there are "down," so when averaged they tend to cancel out, yielding a placid spacetime. But, as in the television analogy, *it's only because of the averaging process that a smooth and tranquil form for spacetime emerges.*

Quantum averaging provides a down-to-earth interpretation of the assertion that familiar spacetime may be illusory. Averages are useful for many purposes but, by design, they do not provide a sharp picture of underlying details. Although the average family in the U.S. has 2.2 children, you'd be in a bind were I to ask to visit such a family. And although the national average price for a gallon of milk is $2.783, you're unlikely to find a store selling it for exactly this price. So, too, familiar spacetime, itself the result of an averaging process, may not describe the details of something we'd want to call fundamental. Space and time may only be approximate, collective conceptions, extremely useful in analyzing the universe on all but ultramicroscopic scales, yet as illusory as a family with 2.2 children.

A second and related insight is that the increasingly intense quantum jitters that arise on decreasing scales suggest that the notion of being able to divide distances or durations into ever smaller units likely comes to an end at around the Planck length (10^{-33} centimeters) and Planck time (10^{-43} seconds). We encountered this idea in Chapter 12, where we emphasized that, although the notion is thoroughly at odds with our usual experiences of space and time, it is not particularly surprising that a property relevant to the everyday fails to survive when pushed into the micro-realm. And since the arbitrary divisibility of space and time is one of their most familiar everyday properties, the inapplicability of this concept on ultrasmall scales gives another hint that there is something else lurking in the microdepths—something that might be called the bare-bones sub-strate of spacetime—the entity to which the familiar notion of spacetime alludes. We expect that this *ur*-ingredient, this most elemental spacetime stuff, does not allow dissection into ever smaller pieces because of the violent fluctuations that would ultimately be encountered, and hence is quite unlike the large-scale spacetime we directly experience. It seems likely, therefore, that the appearance of the fundamental spacetime con-stituents—whatever they may be—is altered significantly through the aver-aging process by which they yield the spacetime of common experience.

Thus, looking for familiar spacetime in the deepest laws of nature may be like trying to take in Beethoven's Ninth Symphony solely note by single note or one of Monet's haystack paintings solely brushstroke by sin-gle brushstroke. Like these masterworks of human expression, nature's spacetime whole may be so different from its parts that nothing resem-bling it exists at the most fundamental level.

Geometry in Translation

Another consideration, one physicists call *geometrical duality*, also sug-gests that spacetime may not be fundamental, but suggests it from a very different viewpoint. Its description is a little more technical than quantum averaging, so feel free to go into skim mode if at any point this section gets too heavy. But because many researchers consider this material to be among string theory's most emblematic features, it's worth trying to get the gist of the ideas.

In Chapter 13 we saw how the five supposedly distinct string theories are actually different translations of one and the same theory. Among other things, we emphasized that this is a powerful realization because, when translated, supremely difficult questions sometimes become far simpler to answer. But there is a feature of the translation dictionary uni-fying the five theories that I've so far neglected to mention. Just as a ques-tion's degree of difficulty can be changed radically by the translation from one string formulation to another, so, too, can the description of the geo-metrical form of spacetime. Here's what I mean.

Because string theory requires more than the three space dimensions and one time dimension of common experience, we were motivated in Chapters 12 and 13 to take up the question of where the extra dimensions might be hiding. The answer we found is that they may be curled up into a size that, so far, has eluded detection because it's smaller than we are able to probe experimentally. We also found that physics in our familiar big dimensions is dependent on the precise size and shape of the extra dimensions because their geometrical properties affect the vibrational patterns strings can execute. Good. Now for the part I left out.

The dictionary that translates questions posed in one string theory into different questions posed in another string theory *also translates the geometry of the extra dimensions in the first theory into a different extra-dimensional geometry in the second theory*. If, for example, you are studying the physical implications of, say, the Type IIA string theory with extra dimensions curled up into a particular size and shape, then every conclusion you reach can, at least in principle, be deduced by considering appropriately translated questions in, say, the Type IIB string theory. But the dictionary for carrying out the translation *demands* that the extra dimensions in the Type IIB string theory be curled up into a precise geometrical form that depends on—*but generally differs from*—the form given by the Type IIA theory. In short, a given string theory with curled-up dimensions in one geometrical form is equivalent to—is a translation of—another string theory with curled-up dimensions in a *different* geometrical form.

And the differences in spacetime geometry need not be minor. For example, if one of the extra dimensions of, say, the Type IIA string theory should be curled up into a circle, as in Figure 12.7, the translation dictionary shows that this is absolutely equivalent to the Type IIB string theory with one of its extra dimensions also curled up into a circle, but one whose radius is *inversely* proportional to the original. If one circle is tiny, the other is big, and vice versa—and yet there is absolutely no way to distinguish between the two geometries. (Expressing lengths as multiples of the Planck length, if one circle has radius R, the mathematical dictionary shows that the other circle has radius $1/R$). You might think that you could easily and immediately distinguish between a big and a small dimension, but in string theory this is not always the case. All observations derive from the interactions of strings, and these two theories, the Type IIA with a big circular dimension and the Type IIB with a small circular dimension, are merely different translations of—different ways of expressing—the same physics. Every observation you describe within one string theory has an alternative and equally viable description within the other string theory, even though the language of each theory and the interpretation it gives may differ. (This is possible because there are two qualitatively different configurations for strings moving on a circular dimension: those in which the string is wrapped around the circle like a rubber band around a tin can, and those in which the string resides on a portion of the circle but does not wrap around it. The former have energies that are *proportional* to the radius of the circle [the larger the radius, the longer the wrapped strings are stretched, so the more energy they embody], while the latter have energies that are *inversely proportional* to the radius [the smaller the radius, the more hemmed in the strings are, so the more energetically they move because of quantum uncertainty]. Notice that if we were to replace the original circle by one of *inverted* radius, while also exchanging "wrapped" and "not wrapped" strings, physical energies—and, it turns out, physics more generally—would remain unaffected. This is exactly what the dictionary translating from the Type IIA theory to the Type IIB theory requires, and why two seemingly different geometries—a big and a small circular dimension—can be equivalent.)

A similar idea also holds when circular dimensions are replaced with the more complicated Calabi-Yau shapes introduced in Chapter 12. A given string theory with extra dimensions curled up into a particular Calabi-Yau shape gets translated by the dictionary into a different string theory with extra dimensions curled up into a different Calabi-Yau shape (one that is called the *mirror* or *dual* of the original). In these cases, not only can the sizes of the Calabi-Yaus differ, but so can their shapes, including the number and variety of their holes. But the translation dictionary ensures that they differ in just the right way, so that even though the extra dimensions have different sizes and shapes, the physics following from each theory is absolutely identical. (There are two types of holes in a given Calabi-Yau shape, but it turns out that string vibrational patterns—and hence physical implications—are sensitive only to the *difference* between the number of holes of each type. So if one Calabi-Yau has, say, two holes of the first kind and five of the second, while another Calabi-Yau has five holes of the first kind and two of the second, then even though they differ as geometrical shapes, they can give rise to identical physics.*)

*For details on geometrical duality involving both circles and Calabi-Yau shapes, see *The Elegant Universe*, Chapter 10.

From another perspective, then, this bolsters the suspicion that space is not a foundational concept. Someone describing the universe using one of the five string theories would claim that space, including the extra dimensions, has a particular size and shape, while someone else using one of the other string theories would claim that space, including the extra dimensions, has a different size and shape. Because the two observers would simply be using alternative *mathematical* descriptions of the same *physical* universe, it is not that one would be right and the other wrong. They would both be right, even though their conclusions about space— its size and shape—would differ. Note too, that it's not that they would be slicing up spacetime in different, equally valid ways, as in special relativity. These two observers would fail to agree on the overall structure of space-time itself. And that's the point. If spacetime were really fundamental, most physicists expect that everyone, regardless of perspective—regardless of the language or theory used—would agree on its geometrical properties. But the fact that, at least within string theory, this need not be the case, suggests that spacetime may be a secondary phenomenon.

We are thus led to ask: if the clues described in the last two sections are pointing us in the right direction, and familiar spacetime is but a large-scale manifestation of some more fundamental entity, what is that entity and what are its essential properties? As of today, no one knows. But in the search for answers, researchers have found yet further clues, and the most important have come from thinking about black holes.

Wherefore the Entropy of Black Holes?

Black holes have the universe's most inscrutable poker faces. From the outside, they appear just about as simple as you can get. The three distinguishing features of a black hole are its mass (which determines how big it is—the distance from its center to its event horizon, the enshrouding surface of no return), its electric charge, and how fast it's spinning. That's it. There are no more details to be gleaned from scrutinizing the visage that a black hole presents to the cosmos. Physicists sum this up with the saying "Black holes have no hair," meaning that they lack the kinds of detailed features that allow for individuality. When you've seen one black hole with a given mass, charge, and spin (though you've learned these indirectly, through their effect on surrounding gas and stars, since black holes are black), you've definitely seen them all.

Nevertheless, behind their stony countenances, black holes harbor the greatest reservoirs of mayhem the universe has ever known. Among all physical systems of a given size with *any* possible composition, black holes contain the highest possible entropy. Recall from Chapter 6 that one rough way to think about this comes directly from entropy's definition as a measure of the number of rearrangements of an object's internal constituents that have no effect on its appearance. When it comes to black holes, even though we can't say what their constituents actually are— since we don't know what happens when matter is crushed at the black hole's center—we can say confidently that rearranging these constituents will no more affect a black hole's mass, charge, or spin than rearranging the pages in *War and Peace* will affect the weight of the book. And since mass, charge, and spin fully determine the face that a black hole shows the external world, *all* such manipulations go unnoticed and we can say a black hole has maximal entropy.

Even so, you might suggest one-upping the entropy of a black hole in the following simple way. Build a hollow sphere of the same size as a given black hole and fill it with gas (hydrogen, helium, carbon dioxide, whatever) that you allow to spread through its interior. The more gas you pump in, the greater the entropy, since more constituents means more possible rearrangements. You might guess, then, that if you keep on pumping and pumping, the entropy of the gas will steadily rise and so will eventually exceed that of the given black hole. It's a clever strategy, but general relativity shows that it fails. The more gas you pump in, the more massive the sphere's contents become. And before you reach the entropy of an equal-sized black hole, the increasingly large mass within the sphere will reach a critical value that causes the sphere and its contents to *become a black hole*. There's just no way around it. Black holes have a monopoly on maximal disorder.

What if you try to further increase the entropy in the space inside the black hole itself by continuing to pump in yet more gas? Entropy will indeed continue to rise, but you'll have changed the rules of the game. As matter takes the plunge across a black hole's ravenous event horizon, not only does the black hole's entropy increase, but *its size increases as well.* The size of a black hole is proportional to its mass, so as you dump more matter into the hole, it gets heavier and bigger. Thus, once you max out the entropy in a region of space by creating a black hole, any attempt to further increase the entropy in that region will fail. The region just can't support more disorder. It's entropy-sated. Whatever you do, whether you pump in gas or toss in a Hummer, you will necessarily cause the black hole to grow and hence surround a larger spatial region. Thus, the amount of entropy contained within a black hole not only tells us a fundamental feature of the black hole, it also tells us something fundamental about space itself: *the maximum entropy that can be crammed into a region of space—any region of space, anywhere, anytime—is equal to the entropy contained within a black hole whose size equals that of the region in question.*

So, how much entropy does a black hole of a given size contain? Here is where things get interesting. Reasoning intuitively, start with something more easily visualized, like air in a Tupperware container. If you were to join together two such containers, doubling the total volume and number of air molecules, you might guess that you'd double the entropy. Detailed calculations confirm[1] this conclusion and show that, all else being equal (unchanging temperature, density, and so on), the entropies of familiar physical systems are proportional to their volumes. A natural next guess is that the same conclusion would also apply to less familiar things, like black holes, leading us to expect that a black hole's entropy is also proportional to its volume.

But in the 1970s, Jacob Bekenstein and Stephen Hawking discovered that this isn't right. Their mathematical analyses showed that the entropy of a black hole is not proportional to its volume, but instead is proportional to the *area* of its event horizon—roughly speaking, to its surface area. This is a very different answer. Were you to double the radius of a black hole, its volume would increase by a factor of 8 (2^3) while its surface area would increase by only a factor of 4 (2^2); were you to increase its radius by a factor of a hundred, its volume would increase by a factor of a million (100^3), while its surface area would increase only by a factor of $10{,}000$ (100^2). Big black holes have much more volume than they do surface area.[2] Thus, even though black holes contain the greatest entropy among all things of a given size, Bekenstein and Hawking showed that the amount of entropy they contain is less than what we'd naïvely guess.

That entropy is proportional to surface area is not merely a curious distinction between black holes and Tupperware, about which we can take note and swiftly move on. We've seen that black holes set a limit to the amount of entropy that, even in principle, can be crammed into a region of space: take a black hole whose size precisely equals that of the region in question, figure out how much entropy the black hole has, and that *is* the absolute limit on the amount of entropy the region of space can contain. Since this entropy, as the works of Bekenstein and Hawking showed, is proportional to the black hole's surface area—which equals the surface area of the region, since we chose them to have the same size—we conclude that the maximal entropy any given region of space can contain is proportional to the *region's* surface area.[3]

The discrepancy between this conclusion and that found from thinking about air trapped in Tupperware (where we found the amount of entropy to be proportional to the Tupperware's *volume*, not its surface area) is easy to pinpoint: Since we assumed the air was uniformly spread, the Tupperware reasoning ignored gravity; remember, when gravity matters, things clump. To ignore gravity is fine when densities are low, but when you are considering large entropy, densities are high, gravity matters, and the Tupperware reasoning is no longer valid. Instead, such extreme conditions require the gravity-based calculations of Bekenstein and Hawking, with the conclusion that the maximum entropy potential for a region of space is proportional to its surface area, not its volume.

All right, but why should we care? There are two reasons.

First, the entropy bound gives yet another clue that ultramicroscopic space has an atomized structure. In detail, Bekenstein and Hawking found that if you imagine drawing a checkerboard pattern on the event horizon of a black hole, with each square being one Planck length by one Planck length (so each such "Planck square" has an area of about 10^{-66} square centimeters), then the black hole's entropy equals the number of such squares that can fit on its surface.[4] It's hard to miss the conclusion to which this result strongly hints: each Planck square is a minimal, fundamental unit of space, and each carries a minimal, single unit of entropy. This suggests that there is nothing, even in principle, that can take place *within* a Planck square, because any such activity could support disorder and hence the Planck square could contain more than the single unit of entropy found by Bekenstein and Hawking. Once again, then, from a completely different perspective we are led to the notion of an elemental spatial entity.[5]

Second, for a physicist, the upper limit to the entropy that can exist in a region of space is a critical, almost sacred quantity. To understand why, imagine you're working for a behavioral psychiatrist, and your job is to keep a detailed, moment-to-moment record of the interactions between groups of intensely hyperactive young children. Every morning you pray that the day's group will be well behaved, because the more bedlam the children create, the more difficult your job. The reason is intuitively obvious, but it's worth saying explicitly: the more disorderly the children are, the more things you have to keep track of. The universe presents a physicist with much the same challenge. A fundamental physical theory is meant to describe everything that goes on—or could go on, even in principle—in a given region of space. And, as with the children, the more disorder the region can contain—even in principle—the more things the theory must be capable of keeping track of. Thus, the maximum entropy a region can contain provides a simple but incisive litmus test: physicists expect that a truly fundamental theory is one that is perfectly matched to the maximum entropy in any given spatial region. The theory should be so tightly in tune with nature that its maximum capacity to keep track of disorder *exactly* equals the maximum disorder a region can possibly contain, not more and not less.

The thing is, if the Tupperware conclusion had had unlimited validity, a fundamental theory would have needed the capacity to account for a volume's worth of disorder in any region. But since that reasoning fails when gravity is included—and since a fundamental theory must include gravity—we learn that a fundamental theory need only be able to account for a surface area's worth of disorder in any region. And as we showed with a couple of numerical examples a few paragraphs ago, for large regions the latter is much smaller than the former.

Thus, the Bekenstein and Hawking result tells us that a theory that includes gravity is, in some sense, simpler than a theory that doesn't. There are fewer "degrees of freedom"—fewer things that can change and hence contribute to disorder—that the theory must describe. This is an interesting realization in its own right, but if we follow this line of reasoning one step further, it seems to tell us something exceedingly bizarre. If the maximum entropy in any given region of space is proportional to the region's surface area and not its volume, then perhaps the true, fundamental degrees of freedom—the attributes that have the potential to give rise to that disorder—*actually reside on the region's surface and not within its volume.* Maybe, that is, the universe's real physical processes take place on a thin, distant surface that surrounds us, and all we see and experience is merely a projection of those processes. Maybe, that is, the universe is rather like a hologram.

This is an odd idea, but as we'll now discuss, it has recently received substantial support.

Is the Universe a Hologram?

A hologram is a two-dimensional piece of etched plastic, which, when illuminated with appropriate laser light, projects a three-dimensional image.[6] In the early 1990s, the Dutch Nobel laureate Gerard 't Hooft and Leonard Susskind, the same physicist who coinvented string theory, suggested that the universe itself might operate in a manner analogous to a hologram. They put forward the startling idea that the comings and goings we observe in the three dimensions of day-to-day life might themselves be holographic projections of physical processes taking place on a distant, two-dimensional surface. In their new and peculiar-sounding vision, we and everything we do or see would be akin to holographic images. Whereas Plato envisioned common perceptions as revealing a mere shadow of reality, the holographic principle concurs, but turns the metaphor on its head. The shadows—the things that are flattened out and hence live on a lower-dimensional surface—are real, while what seem to be the more richly structured, higher-dimensional entities (us; the world around us) are evanescent projections of the shadows.*

Again, while it is a fantastically strange idea, and one whose role in the final understanding of spacetime is far from clear, 't Hooft and Susskind's so-called *holographic principle* is well motivated. For, as we discussed in the last section, the maximum entropy that a region of space can contain scales with the area of its surface, not with the volume of its interior. It's natural to guess, then, that the universe's most fundamental ingredients, its most basic degrees of freedom—the entities that can carry the universe's entropy much as the pages of *War and Peace* carry its entropy—would reside on a bounding surface and not in the universe's interior. What we experience in the "volume" of the universe—in the *bulk*, as physicists often call it—would be determined by what takes place on the bounding surface, much as what we see in a holographic projection is determined by information encoded on a bounding piece of plastic. The laws of physics would act as the universe's laser, illuminating the real

---

*If you're reluctant to rewrite Plato, the braneworld scenario gives a version of holography in which shadows are put back in their proper place. Imagine that we live on a three-brane that surrounds a region with four space dimensions (much as the two-dimensional skin of an apple surrounds the apple's three-dimensional interior). The holographic principle in this setting would say that our three-dimensional perceptions would be the shadows of four-dimensional physics taking place in the region surrounded by our brane.

processes of the cosmos—processes taking place on a thin, distant surface—and generating the holographic illusions of daily life.

We have not yet figured out how this holographic principle might be realized in the real world. One challenge is that in conventional descriptions the universe is imagined either to go on forever, or if not, to wrap back on itself like a sphere or a video game screen (as in Chapter 8), and hence it wouldn't have any edges or boundaries. So, where would the supposed "bounding holographic surface" be located? Moreover, physical processes certainly seem to be under our control, right here, deep in the universe's interior. It doesn't seem that something on a hard-to-locate boundary is somehow calling the shots regarding what happens here in the bulk. Does the holographic principle imply that *that* sense of control and autonomy is illusory? Or is it better to think of holography as articulating a kind of duality in which, on the basis of taste—not of physics—one can choose a familiar description in which the fundamental laws operate here in the bulk (which aligns with intuition and perception) or an unfamiliar description in which fundamental physics takes place on some kind of boundary of the universe, with each viewpoint being equally valid? These are essential questions that remain controversial.

But in 1997, building on earlier insights of a number of string theorists, the Argentinian physicist Juan Maldacena had a breakthrough that dramatically advanced thinking on these matters. His discovery is not directly relevant to the question of holography's role in our real universe, but in the time-honored fashion of physics, he found a hypothetical context—a hypothetical universe—in which abstract musings on holography could be made both concrete and precise using mathematics. For technical reasons, Maldacena studied a hypothetical universe with four large space dimensions and one time dimension that have uniform negative curvature—a higher dimensional version of the Pringle's potato chip, Figure 8.6c. Standard mathematical analysis reveals that this five-dimensional spacetime has a boundary[7] that, like all boundaries, has one dimension less than the shape it bounds: three space dimensions and one time dimension. (As always, higher-dimensional spaces are hard to envision, so if you want a mental picture, think of a can of tomato soup—the three-dimensional liquid soup is analogous to the five-dimensional spacetime, while the two-dimensional surface of the can is analogous to the four-dimensional spacetime boundary.) After including additional curled-up dimensions as required by string theory, Maldacena convincingly argued that the physics witnessed by an observer living within this universe (an observer in the "soup") could be completely described in terms of physics taking place on the universe's boundary (physics on the surface of the can).

Although it is not realistic, this work provided the first concrete and mathematically tractable example in which the holographic principle was explicitly realized.[8] In doing so, it shed much light on the notion of holography as applied to an entire universe. For instance, in Maldacena's work, the bulk description and the boundary description are on an absolutely equal footing. One is not primary and the other secondary. In much the same spirit as the relation between the five string theories, the bulk and boundary theories are translations of each other. The unusual feature of this particular translation, though, is that the bulk theory has more dimensions than the equivalent theory formulated on the boundary. Moreover, whereas the bulk theory includes gravity (since Maldacena formulated it using string theory), calculations show that the boundary theory doesn't. Nevertheless, any question asked or calculation done in one of the theories can be translated into an equivalent question or calculation in the other. While someone unfamiliar with the dictionary would think that the corresponding questions and calculations have absolutely nothing to do with each other (for example, since the boundary theory does not include gravity, questions involving gravity in the bulk theory are translated into very-different-sounding, gravity-less questions in the boundary theory), someone well versed in both languages—an expert on both theories—would recognize their relationship and realize that the answers to corresponding questions and the results of corresponding calculations must agree. Indeed, every calculation done to date, and there have been many, supports this assertion.

The details of all this are challenging to grasp fully, but don't let that obscure the main point. Maldacena's result is amazing. He found a concrete, albeit hypothetical, realization of holography within string theory. He showed that a particular quantum theory without gravity is a translation of—is indistinguishable from—another quantum theory that includes gravity but is formulated with one more space dimension. Vigorous research programs are under way to determine how these insights might apply to a more realistic universe, our universe, but progress is slow as the analysis is fraught with technical hurdles. (Maldacena chose the particular hypothetical example he did because it proved relatively easy to analyze mathematically; more realistic examples are much harder to deal with.) Nevertheless, we now know that string theory, at least in certain contexts, has the capacity to support the concept of holography. And, as with the case of geometric translations described earlier, this provides yet another hint that spacetime is not fundamental. Not only can the size and shape of spacetime change in translation from one formulation of a theory to another, equivalent form, but the *number* of space dimensions can change, too.

More and more, these clues point toward the conclusion that the form of spacetime is an adorning detail that varies from one formulation of a physical theory to the next, rather than being a fundamental element of reality. Much as the number of letters, syllables, and vowels in the word *cat* differ from those in *gato*, its Spanish translation, the form of space-time—its shape, its size, and even the number of its dimensions—also changes in translation. To any given observer who is using one theory to think about the universe, spacetime may seem real and indispensable. But should that observer change the formulation of the theory he or she uses to an equivalent, translated version, what once seemed real and indispensable necessarily changes, too. Thus, if these ideas are right—and I should emphasize that they have yet to be rigorously proven even though theorists have amassed a great deal of supporting evidence—they strongly challenge the primacy of space and time.

Of all the clues discussed here, I'd pick the holographic principle as the one most likely to play a dominant role in future research. It emerges from a basic feature of black holes—their entropy—the understanding of which, many physicists agree, rests on firm theoretical foundations. Even if the details of our theories should change, we expect that any sensible description of gravity will allow for black holes, and hence the entropy bounds driving this discussion will persist and holography will apply. That string theory naturally incorporates the holographic principle—at least in examples amenable to mathematical analysis—is another strong piece of evidence suggesting the principle's validity. I expect that regardless of where the search for the foundations of space and time may take us, regardless of modifications to string/M-theory that may be waiting for us around the bend, holography will continue to be a guiding concept.

The Constituents of Spacetime

Throughout this book we have periodically alluded to the ultramicro-scopic constituents of spacetime, but although we've given indirect arguments for their existence we've yet to say anything about what these constituents might actually be. And for good reason. We really have no idea what they are. Or, perhaps I should say, when it comes to identifying spacetime's elemental ingredients, we have no ideas about which we're really confident. This is a major gap in our understanding, but it's worthwhile to see the problem in its historical context.

Were you to have polled scientists in the late nineteenth century about their views on matter's elementary constituents, you wouldn't have found universal agreement. A mere century ago, the atomic hypothesis was controversial; there were well-known scientists—Ernst Mach was one—who thought it wrong. Moreover, ever since the atomic hypothesis received widespread acceptance in the early part of the twentieth century, scientists have been continuously updating the picture it supplies with what are believed to be ever more elementary ingredients (for example, first protons and neutrons, then quarks). String theory is the latest step along this path, but because it has yet to be confirmed experimentally (and even if it were, that wouldn't preclude the existence of a yet more refined theory awaiting development), we must forthrightly acknowledge that the search for nature's basic material constituents continues.

The incorporation of space and time into a modern scientific context goes back to Newton in the 1600s, but serious thought regarding their microscopic makeup required the twentieth-century discoveries of general relativity and quantum mechanics. Thus, on historical time scales, we've really only just begun to analyze spacetime, so the lack of a definitive proposal for its "atoms"—spacetime's most elementary constituents—is not a black mark on the subject. Far from it. That we've gotten as far as we have—that we've revealed numerous features of space and time vastly beyond common experience—attests to progress unfathomable a century ago. The search for the most fundamental of nature's ingredients, whether of matter or of spacetime, is a formidable challenge that will likely occupy us for some time to come.

For spacetime, there are currently two promising directions in the search for elementary constituents. One proposal comes from string theory and the other from a theory known as *loop quantum gravity*.

String theory's proposal, depending on how hard you think about it, is either intuitively pleasing or thoroughly baffling. Since we speak of the "fabric" of spacetime, the suggestion goes, maybe spacetime is stitched out of strings much as a shirt is stitched out of thread. That is, much as joining numerous threads together in an appropriate pattern produces a shirt's fabric, maybe joining numerous strings together in an appropriate pattern produces what we commonly call spacetime's fabric. Matter, like you and me, would then amount to additional agglomerations of vibrating strings—like sonorous music played over a muted din, or an elaborate pattern embroidered on a plain piece of material—moving within the context stitched together by the strings of spacetime.

I find this an attractive and compelling proposal, but as yet no one has turned these words into a precise mathematical statement. As far as I can tell, the obstacles to doing so are far from trifling. For instance, if your shirt completely unraveled you'd be left with a pile of thread—an outcome that, depending on circumstances, you might find embarrassing or irritating, although probably not deeply mysterious. But it thoroughly taxes the mind (my mind, at least) to think about the analogous situation with strings—the threads of spacetime in this proposal. What would we make of a "pile" of strings that had unraveled from the spacetime fabric or, perhaps more to the point, had not yet even joined together to produce the spacetime fabric? The temptation might be to think of them much as we do the shirt's thread—as raw material that needs to be stitched together—but that glosses over an absolutely essential subtlety. We picture strings as vibrating in space and through time, but without the spacetime fabric that the strings are themselves imagined to yield through their orderly union, *there is no space or time.* In this proposal, the concepts of space and time fail to have meaning until innumerable strings weave together to produce them.

Thus, to make sense of this proposal, we would need a framework for describing strings that does not assume from the get-go that they are vibrating in a preexisting spacetime. We would need a fully spaceless and timeless formulation of string theory, in which spacetime emerges from the collective behavior of strings.

Although there has been progress toward this goal, no one has yet come up with such a spaceless and timeless formulation of string theory—something that physicists call a *background-independent* formulation (the term comes from the loose notion of spacetime as a backdrop against which physical phenomena take place). Instead, essentially all approaches envision strings as moving and vibrating through a spacetime that is inserted into the theory "by hand"; spacetime does not emerge from the theory, as physicists imagine it would in a background-independent framework, but is supplied to the theory by the theorist. Many researchers consider the development of a background-independent formulation to be the single greatest unsolved problem facing string theory. Not only would it give insight into the origin of spacetime, but a background-independent framework would likely be instrumental in resolving the major hang-up encountered at the end of Chapter 12—the theory's current inability to select the geometrical form of the extra dimensions. Once its basic mathematical formalism is disentangled from any particular spacetime, the reasoning goes, string theory should have the capacity to survey all possibilities and perhaps adjudicate among them.

Another difficulty facing the strings-as-threads-of-spacetime proposal is that, as we learned in Chapter 13, string theory has other ingredients besides strings. What role do these other components play in spacetime's fundamental makeup? This question is brought into especially sharp relief by the braneworld scenario. If the three-dimensional space we experience is a three-brane, is the brane itself indecomposable or is it made from combining the theory's other ingredients? Are branes, for example, made from strings, or are branes and strings both elementary? Or should we consider yet another possibility, that branes and strings might be made from some yet finer ingredients? These questions are at the forefront of current research, but since this final chapter is about hints and clues, let me note one relevant insight that has garnered much attention.

Earlier, we talked about the various branes one finds in string/M-theory: one-branes, two-branes, three-branes, four-branes, and so on. Although I didn't stress it earlier, the theory also contains *zero-branes*—ingredients that have no spatial extent, much like point particles. This might seem counter to the whole spirit of string/M-theory, which moved away from the point-particle framework in an effort to tame the wild undulations of quantum gravity. However, the zero-branes, just like their higher dimensional cousins in Figure 13.2, come with strings attached, literally, and hence their interactions are governed by strings. Not surprisingly, then, zero-branes behave very differently from conventional point particles, and, most important, they participate fully in the spreading out and lessening of ultramicroscopic spacetime jitters; zero-branes do not reintroduce the fatal flaws afflicting point-particle schemes that attempt to merge quantum mechanics and general relativity.

In fact, Tom Banks of Rutgers University and Willy Fischler of the University of Texas at Austin, together with Leonard Susskind and Stephen Shenker, both now at Stanford, have formulated a version of string/M-theory in which zero-branes are *the* fundamental ingredients that can be combined to generate strings and the other, higher dimensional branes. This proposal, known as *Matrix theory*—still another possible meaning for the "M" in "M-theory"—has generated an avalanche of follow-up research, but the difficult mathematics involved has so far prevented scientists from bringing the approach to completion. Nevertheless, the calculations that physicists have managed to carry out in this framework seem to support the proposal. If Matrix theory is true, it might mean that everything—strings, branes, and perhaps even space and time themselves—is composed of appropriate aggregates of zero-branes. It's an exciting prospect, and researchers are cautiously optimistic that progress over the next few years will shed much light on its validity.

We have so far surveyed the path string theorists have followed in the search for spacetime's ingredients, but as I mentioned, there is a second path coming from string theory's main competitor, loop quantum gravity. Loop quantum gravity dates from the mid-1980s and is another promising proposal for merging general relativity and quantum mechanics. I won't attempt a detailed description (if you're interested, take a look at Lee Smolin's excellent book *Three Roads to Quantum Gravity*), but will instead mention a few key points that are particularly illuminating for our current discussion.

String theory and loop quantum gravity both claim to have achieved the long-sought goal of providing a quantum theory of gravity, but they do so in very different ways. String theory grew out of the successful particle physics tradition that has for decades sought matter's elementary ingredients; to most early string researchers, gravity was a distant, secondary concern, at best. By contrast, loop quantum gravity grew out of a tradition tightly grounded in the general theory of relativity; to most practitioners of this approach, gravity has always been the main focus. A one-sentence comparison would hold that string theorists start with the small (quantum theory) and move to embrace the large (gravity), while adherents of loop quantum gravity start with the large (gravity) and move to embrace the small (quantum theory).[9] In fact, as we saw in Chapter 12, string theory was initially developed as a quantum theory of the strong nuclear force operating within atomic nuclei; it was realized only later, serendipitously, that the theory actually included gravity. Loop quantum gravity, on the other hand, takes Einstein's general relativity as its point of departure and seeks to incorporate quantum mechanics.

This starting at opposite ends of the spectrum is mirrored in the ways the two theories have so far developed. To some extent, the main achievements of each prove to be the failings of the other. For example, string theory merges all forces and all matter, including gravity (a complete unification that eludes the loop approach), by describing everything in the language of vibrating strings. The particle of gravity, the graviton, is but one particular string vibrational pattern, and hence the theory naturally describes how these elemental bundles of gravity move and interact quantum mechanically. However, as just noted, the main failing of current formulations of string theory is that they presuppose a background spacetime within which strings move and vibrate. By contrast, the main achievement of loop quantum gravity—an impressive one—is that it does *not* assume a background spacetime. Loop quantum gravity is a background-independent framework. However, extracting ordinary space and time, as well as the familiar and successful features of general relativity when applied on large distance scales (something easily done with current formulations of string theory) from this extraordinarily unfamiliar spaceless/timeless starting point, is a far from trivial problem, which researchers are still trying to solve. Moreover, in comparison to string theory, loop quantum gravity has made far less progress in understanding the dynamics of gravitons.

One harmonious possibility is that string enthusiasts and loop quantum gravity aficionados are actually constructing the same theory, but from vastly different starting points. That each theory involves loops—in string theory, these are string loops; in loop quantum gravity, they're harder to describe nonmathematically, but, roughly speaking, they're elementary loops of space—suggests there might be such a connection. This possibility is further supported by the fact that on the few problems accessible to both, such as black hole entropy, the two theories agree fully.[10] And, on the question of spacetime's constituents, both theories suggest that there is some kind of atomized structure. We've already seen the clues pointing toward this conclusion that arise from string theory; those coming from loop quantum gravity are compelling and even more explicit. Loop researchers have shown that numerous loops in loop quantum gravity can be interwoven, somewhat like tiny wool loops crocheted into a sweater, and produce structures that seem, on larger scales, to approximate regions of spacetime. Most convincing of all, loop researchers have calculated the allowed areas of such surfaces of space.

And just as you can have one electron or two electrons or 202 electrons, but you can't have 1.6 electrons or any other fraction, the calculations show that surfaces can have areas that are one square Planck-length, or two square Planck-lengths, or 202 square Planck-lengths, but no fractions are possible. Once again, this is a strong theoretical clue that space, like electrons, comes in discrete, indivisible chunks.[11]

If I were to hazard a guess on future developments, I'd imagine that the background-independent techniques developed by the loop quantum gravity community will be adapted to string theory, paving the way for a string formulation that is background independent. And that's the spark, I suspect, that will ignite a third superstring revolution in which, I'm optimistic, many of the remaining deep mysteries will be solved. Such developments would likely also bring spacetime's long story full circle. In earlier chapters, we followed the pendulum of opinion as it swung between relationist and absolutist positions on space, time, and spacetime. We asked: Is space a something, or isn't it? Is spacetime a something, or isn't it? And, over the course of a few centuries' thought, we encountered differing views. I believe that an experimentally confirmed, background-independent union between general relativity and quantum mechanics would yield a gratifying resolution to this issue. By virtue of the background independence, the theory's ingredients might stand in some relation to one another, but with the absence of a spacetime that is inserted into the theory from the outset, there'd be no background arena in which they were themselves embedded. Only relative relationships would matter, a solution much in the spirit of relationists like Leibniz and Mach. Then, as the theory's ingredients—be they strings, branes, loops, or something else discovered in the course of future research—coalesced to produce a familiar, large-scale spacetime (either our real spacetime or hypothetical examples useful for thought experiments), its being a "something" would be recovered, much as in our earlier discussion of general relativity: in an otherwise empty, flat, infinite spacetime (one of the useful hypothetical examples), the water in Newton's spinning bucket would take on a concave shape. The essential point would be that the distinction between spacetime and more tangible material entities would largely evaporate, as they would both emerge from appropriate aggregates of more basic ingredients in a theory that's fundamentally relational, spaceless, and timeless. If this is how it turns out, Leibniz, Newton, Mach, and Einstein could all claim a share of the victory.

Inner and Outer Space

Speculating about the future of science is an entertaining and constructive exercise. It places our current undertakings in a broader context, and emphasizes the overarching goals toward which we are slowly and deliberately working. But when such speculation turns to the future of spacetime itself, it takes on an almost mystical quality: we're considering the fate of the very things that dominate our sense of reality. Again, there is no question that regardless of future discoveries, space and time will continue to frame our individual experience; space and time, as far as everyday life goes, are here to stay. What will continue to change, and likely change drastically, is our understanding of the framework they provide—the arena, that is, of experiential reality. After centuries of thought, we still can only portray space and time as the most familiar of strangers. They unabashedly wend their way through our lives, but adroitly conceal their fundamental makeup from the very perceptions they so fully inform and influence.

Over the last century, we've become intimately acquainted with some previously hidden features of space and time through Einstein's two theories of relativity and through quantum mechanics. The slowing of time, the relativity of simultaneity, alternative slicings of spacetime, gravity as the warping and curving of space and time, the probabilistic nature of reality, and long-range quantum entanglement were not on the list of things that even the best of the world's nineteenth-century physicists would have expected to find just around the corner. And yet there they were, as attested to by both experimental results and theoretical explanations.

In our age, we've come upon our own panoply of unexpected ideas: Dark matter and dark energy that appear to be, far and away, the dominant constituents of the universe. Gravitational waves—ripples in the fabric of spacetime—which were predicted by Einstein's general relativity and may one day allow us to peek farther back in time than ever before. A Higgs ocean, which permeates all of space and which, if confirmed, will help us to understand how particles acquire mass. Inflationary expansion, which may explain the shape of the cosmos, resolve the puzzle of why it's so uniform on large scales, and set the direction to time's arrow. String theory, which posits loops and snippets of energy in place of point particles and promises a bold version of Einstein's dream in which all particles and all forces are combined into a single theory. Extra space dimensions, emerging from the mathematics of string theory, and possibly detectable in accelerator experiments during the next decade. A braneworld, in which our three space dimensions may be but one universe among many, floating in a higher-dimensional spacetime. And perhaps even emergent spacetime, in which the very fabric of space and time is composed of more fundamental spaceless and timeless entities.

During the next decade, ever more powerful accelerators will provide much-needed experimental input, and many physicists are confident that data gathered from the highly energetic collisions that are planned will confirm a number of these pivotal theoretical constructs. I share this enthusiasm and eagerly await the results. Until our theories make contact with observable, testable phenomena, they remain in limbo—they remain promising collections of ideas that may or may not have relevance for the real world. The new accelerators will advance the overlap between theory and experiment substantially, and, we physicists hope, will usher many of these ideas into the realm of established science.

But there is another approach that, while more of a long shot, fills me with incomparable wonderment. In Chapter 11 we discussed how the effects of tiny quantum jitters can be seen in any clear night sky since they're stretched enormously by cosmic expansion, resulting in clumps that seed the formation of stars and galaxies. (Recall the analogy of tiny scribbles, drawn on a balloon, that are stretched across its surface when the balloon is inflated.) This realization demonstrably gives access to quantum physics through astronomical observations. Perhaps it can be pushed even further. Perhaps cosmic expansion can stretch the imprints of even shorter-scale processes or features—the physics of strings, or quantum gravity more generally, or the atomized structure of ultramicroscopic spacetime itself—and spread their influence, in some subtle but observable manner, across the heavens. Maybe, that is, the universe has already drawn out the microscopic fibers of the fabric of the cosmos and unfurled them clear across the sky, and all we need do is learn how to recognize the pattern.

Assessing cutting-edge proposals for deep physical laws may well require the ferocious might of particle accelerators able to re-create violent conditions unseen since moments after the big bang. But for me, there would be nothing more poetic, no outcome more graceful, no unification more complete, than for us to confirm our theories of the ultra-small—our theories about the ultramicroscopic makeup of space, time, and matter—by turning our most powerful telescopes skyward and gazing silently at the stars.

Notes

Chapter 1

1. Lord Kelvin was quoted by the physicist Albert Michelson during his 1894 address at the dedication of the University of Chicago's Ryerson Laboratory (see D. Kleppner, *Physics Today*, November 1998).

2. Lord Kelvin, "Nineteenth Century Clouds over the Dynamical Theory of Heat and Light," *Phil. Mag.* Ii—6th series, 1 (1901).

3. A. Einstein, N. Rosen, and B. Podolsky, *Phys. Rev.* 47, 777 (1935).

4. Sir Arthur Eddington, *The Nature of the Physical World* (Cambridge, Eng.: Cambridge University Press, 1928).

5. As described more fully in note 2 of Chapter 6, this is an overstatement because there are examples, involving relatively esoteric particles (such as K-mesons and B-mesons), which show that the so-called weak nuclear force does not treat past and future fully symmetrically. However, in my view and that of many others who have thought about it, since these particles play essentially no role in determining the properties of everyday material objects, they are unlikely to be important in explaining the puzzle of time's arrow (although, I hasten to add, no one knows this for sure). Thus, while it is technically an overstatement, I will assume throughout that the error made in asserting that the laws treat past and future on equal footing is minimal—at least as far as explaining the puzzle of time's arrow is concerned.

6. Timothy Ferris, *Coming of Age in the Milky Way* (New York: Anchor, 1989).

Chapter 2

1. Isaac Newton, *Sir Isaac Newton's Mathematical Principle of Natural Philosophy and His System of the World*, trans. A. Motte and Florian Cajori (Berkeley: University of California Press, 1934), vol. 1, p. 10.

2. Ibid., p. 6.

3. Ibid.

4. Ibid., p. 12.

5. Albert Einstein, in Foreword to Max Jammer, *Concepts of Space: The Histories of Theories of Space in Physics* (New York: Dover, 1993).

6. A. Rupert Hall, *Isaac Newton, Adventurer in Thought* (Cambridge, Eng.: Cambridge University Press, 1992), p. 27.

7. Ibid.

8. H. G. Alexander, ed., *The Leibniz-Clarke Correspondence* (Manchester: Manchester University Press, 1956).

9. I am focusing on Leibniz as the representative of those who argued against assigning space an existence independent of the objects inhabiting it, but many others also strenuously defended this view, among them Christiaan Huygens and Bishop Berkeley.

10. See, for example, Max Jammer, p. 116.

11. V. I. Lenin, *Materialism and Empiriocriticism: Critical Comments on a Reactionary Philosophy* (New York: International Publications, 1909). Second English ed. of *Materializm' i Empiriokrititsizm': Kriticheskia Zametki ob' Odnoi Reaktsionnoi Filosofii* (Moscow: Zveno Press, 1909).

Chapter 3

1. For the mathematically trained reader, these four equations are

$$\nabla \cdot E = \rho/\varepsilon_0, \ \nabla \cdot B = 0, \ \nabla \times E + \partial B/\partial t = 0, \ \nabla \times B - \varepsilon_0\mu_0\partial E/\partial t = \mu_0 J,$$
$$\text{where } E, B, \rho, J, \varepsilon_0, \mu_0$$

denote the electric field, the magnetic field, the electric charge density, the electric current density, the permittivity of free space, and the permeability of free space, respectively. As you can see, Maxwell's equations relate the rate of change of the electromagnetic fields to the presence of electric charges and currents. It is not hard to show that these equations imply a speed for electromagnetic waves given by $1/\sqrt{\varepsilon_0\mu_0}$, which when evaluated is in fact the speed of light.

2. There is some controversy as to the role such experiments played in Einstein's development of special relativity. In his biography of Einstein, *Subtle Is the Lord: The Science and the Life of Albert Einstein* (Oxford: Oxford University Press, 1982), pp. 115–19, Abraham Pais has argued, using Einstein's own statements from his later years, that Einstein was aware of the Michelson-Morley results. Albrecht Fölsing in *Albert Einstein: A Biography* (New York: Viking, 1997), pp. 217–20, also argues that Einstein was aware of the Michelson-Morley result, as well as earlier experimental null results in searching for evidence of the aether, such as the work of Armand Fizeau. But Fölsing and many other historians of science have also argued that such experiments played, at best, a secondary role in Einstein's thinking. Einstein was primarily guided by considerations of mathematical symmetry, simplicity, and an uncanny physical intuition.

3. For us to see anything, light has to travel to our eyes; similarly, for us to see light, the light itself would have to make the same journey. So, when I speak of Bart's seeing light that is speeding away, it is shorthand. I am imagining that Bart has a small army of helpers, all moving at Bart's speed, but situated at various distances along the path that he and the light beam follow. These helpers give Bart updates on how far ahead the light has sped and the time at which the light reached such distant locations. Then, on the basis of this information, Bart can calculate how fast the light is speeding away from him.

4. There are many elementary mathematical derivations of Einstein's insights on space and time arising from special relativity. If you are interested, you can, for example, take a look at Chapter 2 of *The Elegant Universe* (together with mathematical details given in the endnotes to that chapter). A more technical but extremely lucid account is Edwin Taylor and John Archibald Wheeler, *Spacetime Physics: Introduction to Special Relativity* (New York, W. H. Freeman & Co., 1992).

5. The stopping of time at light speed is an interesting notion, but it is important not to read too much into it. Special relativity shows that no material object can ever attain light speed: the faster a material object travels, the harder we'd have to push it to further

increase its speed. Just shy of light speed, we'd have to give the object an essentially infinitely hard push for it to go any faster, and that's something we can't ever do. Thus, the "timeless" photon perspective is limited to *massless* objects (of which the photon is an example), and so "timelessness" is permanently beyond what all but a few types of particle species can ever attain. While it is an interesting and fruitful exercise to imagine how the universe would appear when moving at light speed, ultimately we need to focus on perspectives that material objects, such as ourselves, can reach, if we want to draw inferences about how special relativity affects our experiential conception of time.

6. See Abraham Pais, *Subtle Is the Lord*, pp. 113–14.

7. To be more precise, we *define* the water to be spinning if it takes on a concave shape, and not spinning if it doesn't. From a Machian perspective, in an empty universe there is no conception of spinning, so the water's surface would always be flat (or, to avoid issues of the lack of gravity pulling on the water, we can say that the tension on the rope tied between two rocks will always be slack). The statement here is that, by contrast, in special relativity there *is* a notion of spinning, even in an empty universe, so that the water's surface can be concave (and the tension on the rope tied between the rocks can be taut). In this sense, special relativity violates Mach's ideas.

8. Albrecht Fölsing, *Albert Einstein* (New York: Viking Press, 1997), pp. 208–10.

9. The mathematically inclined reader will note that if we choose units so that the speed of light takes the form of one space unit per one time unit (like one light-year per year or one light-second per second, where a light-year is about 6 trillion miles and a light-second is about 186,000 miles), then light moves through spacetime on 45-degree rays (because such diagonal lines are the ones which cover one space unit in one time unit, two space units in two time units, etc.). Since nothing can exceed the speed of light, any material object must cover less distance in space in a given interval of time than would a beam of light, and hence the path it follows through spacetime must make an angle with the centerline of the diagram (the line running through the center of the loaf from crust to crust) that is less than 45 degrees. Moreover, Einstein showed that the time slices for an observer moving with velocity v—all of space at one moment of such an observer's time—have an equation (assuming one space dimension for simplicity) given by $t_{moving} = \gamma(t_{stationary} - (v/c^2) x_{stationary})$, where $\gamma = (1 - v^2/c^2)^{-1/2}$, and c is the velocity of light. In units where $c = 1$, we note that $v < 1$ and hence a time slice for the moving observer—the locus where t_{moving} takes on a fixed value—is of the form $(t_{stationary} - vx_{stationary}) = $ constant. Such time slices are angled with respect to the stationary time slices (the loci of the form $t_{stationary} = $ constant), and because $v < 1$, the angle between them is less than 45 degrees.

10. For the mathematically inclined reader, the statement being made is that the geodesics of Minkowski's spacetime—the paths of extremal spacetime length between two given points—are geometrical entities that do not depend on any particular choice of coordinates or frame of reference. They are intrinsic, absolute, geometric spacetime features. Explicitly, using the standard Minkowski metric, the (timelike) geodesics are straight lines (whose angle with respect to the time axis is less than 45 degrees, since the speed involved is less than that of light).

11. There is something else of importance that all observers, regardless of their motion, also agree upon. It's implicit in what we've described, but it's worth stating directly. If one event is the cause of another (I shoot a pebble, causing a window to break), all observers agree that the cause happened *before* the effect (all observers agree that I shot the pebble *before* the window broke). For the mathematically inclined reader, it is actually not difficult to see this using our schematic depiction of spacetime. If event A is the cause

of event B, then a line drawn from A to B intersects each of the time slices (time slices of an observer at rest with respect to A) at an angle that is *greater* than 45 degrees (the angle between the space axes—axes that lie on any given time slice—and the line between A and B is greater than 45 degrees). For instance, if A and B take place at the same location in space (the rubber band wrapped around my finger [A] causes my finger to turn white [B]) then the line connecting A and B makes a 90-degree angle relative to the time slices. If A and B take place at different locations in space, whatever traveled from A to B to exert the influence (my pebble traveling from slingshot to window) did so at less than light speed, which means the angle differs from 90 degrees (the angle when no speed is involved) by less than 45 degrees—i.e. the angle with respect to the time slices (the space axes) is greater than 45 degrees. (Remember from endnote 9 of this chapter that light speed sets the limit and such motion traces out 45-degree lines.) Now, as in endnote 9, the different time slicings associated with an observer in motion are angled relative to those of an observer at rest, but the angle is always *less* than 45 degrees (since the relative motion between two material observers is always less than the speed of light). And since the angle associated with causally related events is always *greater* than 45 degrees, the time slices of an observer, who necessarily travels at less than light speed, cannot first encounter the effect and then later encounter the cause. To all observers, cause will precede effect.

12. The notion that causes precede their effects (see the preceding note) would, among other things, be challenged if influences could travel faster than the speed of light.

13. Isaac Newton, *Sir Isaac Newton's Mathematical Principles of Natural Philosophy and His System of the World*, trans. A. Motte and Florian Cajori (Berkeley: University of California Press, 1962), vol. 1, p. 634.

14. Because the gravitational pull of the earth differs from one location to another, a spatially extended, freely falling observer can still detect a residual gravitational influence. Namely, if the observer, while falling, releases two baseballs—one from his outstretched right arm and the other from his left—each will fall along a path toward the earth's center. So, from the observer's perspective, he will be falling straight down toward the earth's center, while the ball released from his right hand will travel downward and slightly toward the left, while the ball released from his left hand will travel downward and slightly toward the right. Through careful measurement, the observer will therefore see that the distance between the two baseballs slowly decreases; they move toward one another. Crucial to this effect, though, is that the baseballs were released in slightly different locations in space, so that their freely falling paths toward earth's center were slightly different as well. Thus, a more precise statement of Einstein's realization is that the smaller the spatial extent of an object, the more fully it can eliminate gravity by going into free fall. While an important point of principle, this complication can be safely ignored throughout the discussion.

15. For a more detailed, yet general-level, explanation of the warping of space and time according to general relativity, see, for example, Chapter 3 of *The Elegant Universe*.

16. For the mathematically trained reader, Einstein's equations are $G_{\mu\nu} = (8\pi G/c^4) T_{\mu\nu}$, where the left-hand side describes the curvature of spacetime using the Einstein tensor and the right-hand side describes the distribution of matter and energy in the universe using the energy-momentum tensor.

17. Charles Misner, Kip Thorne, and John Archibald Wheeler, *Gravitation* (San Francisco: W. H. Freeman and Co., 1973), pp. 544–45.

18. In 1954, Einstein wrote to a colleague: "As a matter of fact, one should no longer speak of Mach's principle at all" (as quoted in Abraham Pais, *Subtle Is the Lord*, p. 288).

19. As mentioned earlier, successive generations have attributed the following ideas to Mach even though his own writings do not phrase things explicitly in this manner.

20. One qualification here is that objects which are so distant that there hasn't been enough time since the beginning of the universe for their light—or gravitational influence—to yet reach us have no impact on the gravity we feel.

21. The expert reader will recognize that this statement is, technically speaking, too strong, as there are nontrivial (that is, non–Minkowski space) empty space solutions to general relativity. Here I am simply using the fact that special relativity can be thought of as a special case of general relativity in which gravity is ignored.

22. For balance, let me note that there are physicists and philosophers who do not agree with this conclusion. Even though Einstein gave up on Mach's principle, during the last thirty years it has taken on a life of its own. Various versions and interpretations of Mach's idea have been put forward, and, for example, some physicists have suggested that general relativity *does* fundamentally embrace Mach's ideas; it's just that some particular shapes that spacetime can have—such as the infinite flat spacetime of an empty universe—don't. Perhaps, they suggest, any spacetime that is remotely realistic—populated by stars and galaxies, and so forth—does satisfy Mach's principle. Others have offered reformulations of Mach's principle in which the issue is no longer how objects, such as rocks tied by a string or buckets filled with water, behave in an otherwise empty universe, but rather how the various time slicings—the various three-dimensional spatial geometries—relate to one another through time. An enlightening reference on modern thinking about these ideas is *Mach's Principle: From Newton's Bucket to Quantum Gravity*, Julian Barbour and Herbert Pfister, eds. (Berlin: Birkhäuser, 1995), which is a collection of essays on the subject. As an interesting aside, this reference contains a poll of roughly forty physicists and philosophers regarding their view on Mach's principle. Most (more than 90 percent) agreed that general relativity does not fully conform to Mach's ideas. Another excellent and extremely interesting discussion of these ideas, from a distinctly pro-Machian perspective and at a level suited to general readers, is Julian Barbour's book *The End of Time: The Next Revolution in Physics* (Oxford: Oxford University Press, 1999).

23. The mathematically inclined reader might find it enlightening to learn that Einstein believed that spacetime had no existence independent of its metric (the mathematical device that gives distance relations in spacetime), so that if one were to remove everything—including the metric—spacetime would *not* be a something. By "spacetime" I always mean a manifold together with a metric that solves the Einstein equations, and so the conclusion we've reached, in mathematical language, is that metrical spacetime is a something.

24. Max Jammer, *Concepts of Space*, p. xvii.

Chapter 4

1. More accurately, this appears to be a medieval conception with historical roots that go back to Aristotle.

2. As we will discuss later in the book, there are realms (such as the big bang and black holes) that still present many mysteries, at least in part owing to extremes of small size and huge densities that cause even Einstein's more refined theory to break down. So, the statement here applies to all but the extreme contexts in which the known laws themselves become suspect.

3. An early reader of this text, and one who, surprisingly, has a particular expertise in voodoo, has informed me that something *is* imagined to go from place to place to carry out the voodoo practitioner's intentions—namely, a spirit. So my example of a fanciful nonlocal process may, depending on your take on voodoo, be flawed. Nevertheless, the idea is clear.

4. To avoid any confusion, let me reemphasize at the outset that when I say, "The universe is not local," or "Something we do over here can be entwined with something over there," I am not referring to the ability to exert an instantaneous intentioned control over something distant. Instead, as will become clear, the effect I am referring to manifests itself as *correlations* between events taking place—usually, in the form of correlations between results of measurements—at distant locations (locations for which there would not be sufficient time for even light to travel from one to the other). Thus, I am referring to what physicists call *nonlocal correlations*. At first blush, such correlations may not strike you as particularly surprising. If someone sends you a box containing one member of a pair of gloves, and sends the other member of the pair to your friend thousands of miles away, there will be a correlation between the handedness of the glove each of you sees upon opening your respective box: if you see left, your friend will see right; if you see right, your friend will see left. And, clearly, nothing in these correlations is at all mysterious. But, as we will gradually describe, the correlations apparent in the quantum world seem to be of a very different character. It's as if you have a pair of "quantum gloves" in which each member can be either left-handed or right-handed, and commits to a definite handedness only when appropriately observed or interacted with. The weirdness arises because, although each glove seems to choose its handedness randomly when observed, the gloves work in tandem, even if widely separated: if one chooses left, the other chooses right, and vice versa.

5. Quantum mechanics makes predictions about the microworld that agree fantastically well with experimental observations. On this, there is universal agreement. Nevertheless, because the detailed features of quantum mechanics, as discussed in this chapter, differ significantly from those of common experience, and, relatedly, as there are different mathematical formulations of the theory (and different formulations of how the theory spans the gap between the microworld of phenomena and the macroworld of measured results), there isn't consensus on how to *interpret* various features of the theory (and various puzzling data which the theory, nevertheless, is able to explain mathematically), including issues of nonlocality. In this chapter, I have taken a particular point of view, the one I find most convincing based on current theoretical understanding and experimental results. But, I stress here that not everyone agrees with this view, and in a later endnote, after explaining this perspective more fully, I will briefly note some of the other perspectives and indicate where you can read more about them. Let me also stress, as we will discuss later, that the experiments contradict Einstein's belief that the data could be explained solely on the basis of particles always possessing definite, albeit hidden, properties *without any use or mention of nonlocal entanglement*. However, the failure of this perspective only rules out a local universe. It does not rule out the possibility that particles have such definite hidden features.

6. For the mathematically inclined reader, let me note one potentially misleading aspect of this description. For multiparticle systems, the probability wave (the wavefunction, in standard terminology) has essentially the same interpretation as just described, but is defined as a function on the *configuration space* of the particles (for a single particle, the configuration space is isomorphic to real space, but for an N-particle system it has 3N dimensions). This is important to bear in mind when thinking about the question of whether the wavefunction is a real physical entity or merely a mathematical device, since if one takes the former position, one would need to embrace the reality of configuration space as well—an interesting variation on the themes of Chapters 2 and 3. In relativistic quantum field theory, the fields can be defined in the usual four spacetime dimensions of common experience, but there are also somewhat less widely used formulations that

invoke generalized wavefunctions—so-called *wavefunctionals* defined on an even more abstract space, *field space*.

7. The experiments I am referring to here are those on the *photoelectric effect*, in which light shining on various metals causes electrons to be ejected from the metal's surface. Experimenters found that the greater the intensity of the light, the greater the number of electrons emitted. Moreover, the experiments revealed that the energy of each ejected electron was determined by the color—the frequency—of the light. This, as Einstein argued, is easy to understand if the light beam is composed of particles, since greater light intensity translates into more light particles (more photons) in the beam—and the more photons there are, the more electrons they will hit and hence eject from the metallic surface. Furthermore, the frequency of the light would determine the energy of each photon, and hence the energy of each electron ejected, precisely in keeping with the data. The particlelike properties of photons were finally confirmed by Arthur Compton in 1923 through experiments involving the elastic scattering of electrons and photons.

8. Institut International de Physique Solvay, *Rapport et discussions du 5ème Conseil* (Paris, 1928), pp. 253ff.

9. Irene Born, trans., *The Born-Einstein Letters* (New York: Walker, 1971), p. 223.

10. Henry Stapp, *Nuovo Cimento* 40B (1977), 191–204.

11. David Bohm is among the creative minds that worked on quantum mechanics during the twentieth century. He was born in Pennsylvania in 1917 and was a student of Robert Oppenheimer at Berkeley. While teaching at Princeton University, he was called to appear in front of the House Un-American Activities Committee, but refused to testify at the hearings. Instead, he departed the United States, becoming a professor at the University of São Paulo in Brazil, then at the Technion in Israel, and finally at Birkbeck College of the University of London. He lived in London until his death in 1992.

12. Certainly, if you wait long enough, what you do to one particle can, in principle, affect the other: one particle could send out a signal alerting the other that it had been subjected to a measurement, and this signal could affect the receiving particle. However, as no signal can travel faster than the speed of light, this kind of influence is not instantaneous. The key point in the present discussion is that at the very moment that we measure the spin of one particle about a chosen axis we learn the spin of the other particle about that axis. And so, any kind of "standard" communication between the particles—luminal or subluminal communication—is not relevant.

13. In this and the next section, the distillation of Bell's discovery which I am using is a "dramatization" inspired by David Mermin's wonderful papers: "Quantum Mysteries for Anyone," *Journal of Philosophy* 78, (1981), pp. 397–408; "Can You Help Your Team Tonight by Watching on TV?," in *Philosophical Consequences of Quantum Theory: Reflections on Bell's Theorem*, James T. Cushing and Ernan McMullin, eds. (University of Notre Dame Press, 1989); "Spooky Action at a Distance: Mysteries of the Quantum Theory," in *The Great Ideas Today* (Encyclopaedia Britannica, Inc., 1988), which are all collected in N. David Mermin, *Boojums All the Way Through* (Cambridge, Eng.: Cambridge University Press, 1990). For anyone interested in pursuing these ideas in a more technical manner, there is no better place to start than with Bell's own papers, many of which are collected in J. S. Bell, *Speakable and Unspeakable in Quantum Mechanics* (Cambridge, Eng.: Cambridge University Press, 1997).

14. While the locality assumption is critical to the argument of Einstein, Podolsky, and Rosen, researchers have tried to find fault with other elements of their reasoning in an attempt to avoid the conclusion that the universe admits nonlocal features. For example, it is sometimes claimed that all the data require is that we give up so-called realism—the

idea that objects possess the properties they are measured to have independent of the measurement process. In this context, though, such a claim misses the point. If the EPR reasoning had been confirmed by experiment, there would be nothing mysterious about the long-range correlations of quantum mechanics; they'd be no more surprising than classical long-range correlations, such as the way finding your left-handed glove over here ensures that its partner over there is a right-handed glove. But such reasoning is refuted by the Bell/Aspect results. Now, if in response to this refutation of EPR we give up realism—as we do in standard quantum mechanics—that does nothing to lessen the stunning weirdness of long-range correlations between widely separated *random* processes; when we relinquish realism, the gloves, as in endnote 4, become "quantum gloves." Giving up realism does not, by any means, make the observed nonlocal correlations any less bizarre. It is true that if, in light of the results of EPR, Bell, and Aspect, we try to maintain realism—for example, as in Bohm's theory discussed later in the chapter—the kind of nonlocality we require to be consistent with the data seems to be more severe, involving nonlocal interactions, not just nonlocal correlations. Many physicists have resisted this option and have thus relinquished realism.

15. See, for example, Murray Gell-Mann, *The Quark and the Jaguar* (New York: Freeman, 1994), and Huw Price, *Time's Arrow and Archimedes' Point* (Oxford: Oxford University Press, 1996).

16. Special relativity forbids anything that has ever traveled slower than light speed from crossing the speed-of-light barrier. But if something has *always* been traveling faster than the speed of light, it is not strictly ruled out by special relativity. Hypothetical particles of this sort are called *tachyons*. Most physicists believe tachyons don't exist, but others enjoy tinkering with the possibility that they do. So far, though, largely because of the strange features that such a faster-than-light particle would have according to the equations of special relativity, no one has found any particular use for them—even hypothetically speaking. In modern studies, a theory that gives rise to tachyons is generally viewed as suffering from an instability.

17. The mathematically inclined reader should note that, at its core, special relativity claims that the laws of physics must be Lorentz invariant, that is, invariant under $SO(3,1)$ coordinate transformations on Minkowski spacetime. The conclusion, then, is that quantum mechanics would be squared with special relativity if it could be formulated in a fully Lorentz-invariant manner. Now, relativistic quantum mechanics and relativistic quantum field theory have gone a long way toward this goal, but as yet there isn't full agreement regarding whether they have addressed the quantum measurement problem in a Lorentz-invariant framework. In relativistic quantum field theory, for example, it is straightforward to compute, in a completely Lorentz-invariant manner, the probability amplitudes and probabilities for outcomes of various experiments. But the standard treatments stop short of also describing the way in which one particular outcome or another emerges from the range of quantum possibilities—that is, what happens in the measurement process. This is a particularly important issue for entanglement, as the phenomenon hinges on the effect of what an experimenter does—the act of measuring one of the entangled particle's properties. For a more detailed discussion, see Tim Maudlin, *Quantum Non-locality and Relativity* (Oxford: Blackwell, 2002).

18. For the mathematically inclined reader, here is the quantum mechanical calculation that makes predictions in agreement with these experiments. Assume that the axes along which the detectors measure spin are vertical and 120 degrees clockwise and counterclockwise from vertical (like noon, four o'clock, and eight o'clock on two clocks, one for each detector, that are facing each other) and consider, for argument's sake, two electrons

emerging back to back and heading toward these detectors in the so-called singlet state. That is the state whose total spin is zero, ensuring that if one electron is found to be in the spin-up state, the other will be in the spin-down state, about a given axis, and vice versa. (Recall that for ease in the text, I've described the correlation between the electrons as ensuring that if one is spin-up so is the other, and if one is spin-down, so is the other; in point of fact, the correlation is one in which the spins point in opposite directions. To make contact with the main text, you can always imagine that the two detectors are calibrated oppositely, so that what one calls spin-up the other calls spin-down.) A standard result from elementary quantum mechanics shows that if the angle between the axes along which our two detectors measure the electron's spins is θ, then the probability that they will measure opposite spin values is $\cos^2 (\theta/2)$. Thus, if the detector axes are aligned ($\theta = 0$), they definitely measure opposite spin values (the analog of the detectors in the main text always measuring the same value when set to the same direction), and if they are set at either +120° or −120°, the probability that they measure opposite spins is $\cos^2 (+120°$ or $-120°) = \frac{1}{4}$. Now, if the detector axes are set randomly, $\frac{1}{3}$ of the time they will point in the same direction, and $\frac{2}{3}$ of the time they won't. Thus, over all runs, we expect to find opposite spins $(\frac{1}{3})(1) + (\frac{2}{3})(\frac{1}{4}) = \frac{1}{2}$ of the time, as found by the data.

You may find it odd that the assumption of locality yields a higher spin correlation (greater than 50 percent) than what we find with standard quantum mechanics (exactly 50 percent); the long-range entanglement of quantum mechanics, you'd think, should yield a greater correlation. In fact, it does. A way to think about it is this: With only a 50 percent correlation over all measurements, quantum mechanics yields 100 percent correlation for measurements in which the left and right detector axes are chosen to point in the same direction. In the local universe of Einstein, Podolsky, and Rosen, a greater than 55 percent correlation over all measurements is required to ensure 100 percent agreement when the same axes are chosen. Roughly, then, in a local universe, a 50 percent correlation over all measurements would entail *less* than a 100 percent correlation when the same axes are chosen—i.e., less of a correlation than what we find in our nonlocal quantum universe.

19. You might think that an instantaneous collapse would, from the get-go, fall afoul of the speed limit set by light and therefore ensure a conflict with special relativity. And if probability waves were indeed like water waves, you'd have an irrefutable point. That the value of a probability wave suddenly dropped to zero over a huge expanse would be far more shocking than all of the water in the Pacific Ocean's instantaneously becoming perfectly flat and ceasing to move. But, quantum mechanics practitioners argue, probability waves are *not* like water waves. A probability wave, although it describes matter, is not a material thing itself. And, such practitioners continue, the speed-of-light barrier applies only to material objects, things whose motion can be directly seen, felt, detected. If an electron's probability wave has dropped to zero in the Andromeda galaxy, an Andromedan physicist will merely fail, with 100 percent certainty, to detect the electron. Nothing in the Andromedan's observations reveals the sudden change in the probability wave associated with the successful detection, say, of the electron in New York City. As long as the electron itself does not travel from one place to another at greater than light speed, there is no conflict with special relativity. And, as you can see, all that has happened is that the electron was found to be in New York City and not anywhere else. Its speed never even entered the discussion. So, while the instantaneous collapse of probability is a framework that comes with puzzles and problems (discussed more fully in Chapter 7), it need not necessarily imply a conflict with special relativity.

20. For a discussion of some of these proposals, see Tim Maudlin, *Quantum Nonlocality and Relativity.*

Chapter 5

1. For the mathematically inclined reader, from the equation $t_{moving} = \gamma(t_{stationary} - (v/c^2) x_{stationary})$ (discussed in note 9 of Chapter 3) we find that Chewie's now-list at a given moment will contain events that observers on earth will claim happened $(v/c^2) x_{earth}$ earlier, where x_{earth} is Chewie's distance from earth. This assumes Chewie is moving away from earth. For motion toward earth, v has the opposite sign, so the earthbound observers will claim such events happened $(v/c^2)x_{earth}$ later. Setting $v = 10$ miles per hour and $x_{earth} = 10^{10}$ light-years, we find $(v/c^2) x_{earth}$ is about 150 years.

2. This number—and a similar number given in a few paragraphs further on describing Chewie's motion toward earth—were valid at the time of the book's publication. But as time goes by here on earth, they will be rendered slightly inaccurate.

3. The mathematically inclined reader should note that the metaphor of slicing the spacetime loaf at different angles is the usual concept of *spacetime diagrams* taught in courses on special relativity. In spacetime diagrams, all of three-dimensional space at a given moment of time, according to an observer who is considered stationary, is denoted by a horizontal line (or, in more elaborate diagrams, by a horizontal plane), while time is denoted by the vertical axis. (In our depiction, each "slice of bread"—a plane—represents all of space at one moment of time, while the axis running through the middle of the loaf, from crust to crust, is the time axis.) Spacetime diagrams provide an insightful way of illustrating the point being made about the now-slices of you and Chewie.

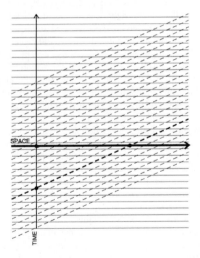

The light solid lines are equal time slices (now-slices) for observers at rest with respect to earth (for simplicity, we imagine that earth is not rotating or undergoing any acceleration, as these are irrelevant complications for the point being made), and the light dotted lines are equal time slices for observers moving away from earth at, say, 9.3 miles per hour. When Chewie is at rest relative to earth, the former represent his now-slices (and since you are at rest on earth throughout the story, these light solid lines always represent your now-slices), and the darkest solid line shows the now-slice containing you (the left dark dot), in earth's twenty-first century, and he (the right dark dot), both sitting still and reading. When Chewie is walking away from earth, the dotted lines represent his now-slices, and the darkest dotted line shows the now-slice containing Chewie (having just gotten up and started to walk) and John Wilkes Booth (the lower left dark dot). Note, too, that one of the subsequent dotted time slices will contain Chewie walking (if he is still around!) and you, in earth's twenty-first century, sitting still reading. Hence, a single moment for you will appear on two of Chewie's now-lists—one list of relevance before and one of relevance after he started to walk. This shows yet another way in which the simple intuitive notion of *now*—when envisioned as applying throughout space—is transformed by special relativity into a concept with highly unusual features. Furthermore, these now-lists do *not* encode causality: standard causality (note 11, Chapter 3) remains in full force. Chewie's now-lists jump because he jumps from one reference frame to another. But every observer—using a single, well-defined choice of spacetime coordinatization—will agree with every other regarding which events can affect which.

4. The expert reader will recognize that I am assuming spacetime is Minkowskian. A similar argument in other geometries will not necessarily yield the entire spacetime.

5. *Albert Einstein and Michele Besso: Correspondence 1903–1955*, P. Speziali, ed. (Paris: Hermann, 1972).

6. The discussion here is meant to give a qualitative sense of how an experience right *now*, together with memories that you have right *now*, forms the basis of your sense of having experienced a life in which you've lived out those memories. But, if, for example, your brain and body were somehow put into exactly the same state that they are right now, you would have the same sense of having lived the life that your memories attest to (assuming, as I do, that the basis of all experience can be found in the physical state of brain and body), even if those experiences never really happened, but were artificially imprinted into your brain state. One simplification in the discussion is the assumption that we can feel or experience things that happen at a single instant, when, in reality, processing time is required for the brain to recognize and interpret whatever stimuli it receives. While true, this is not of particular relevance to the point I'm making; it is an interesting but largely irrelevant complication arising from analyzing time in a manner directly tied to human experience. As we discussed earlier, human examples help make our discussion more grounded and visceral, but it does require us to tease out those aspects of the discussion that are more interesting from a biological as opposed to a physical perspective.

7. You might wonder how the discussion in this chapter relates to our description in Chapter 3 of objects "moving" through spacetime at the speed of light. For the mathematically disinclined reader, the rough answer is that the history of an object is represented by a curve in spacetime—a path through the spacetime loaf that highlights every place the object has been at the moment it was there (much as we see in Figure 5.1). The intuitive notion of "moving" through spacetime, then, can be expressed in "flowless" language by simply specifying this path (as opposed to imagining the path being traced out before your eyes). The "speed" associated with this path is then a measure of how long the path is (from one chosen point to another), divided by the time difference recorded on a watch carried by someone or something between the two chosen points on the path. This, again, is a conception that does not involve any time flow: you simply look at what the watch in question says at the two points of interest. It turns out that the speed found in this way, for *any* motion, is equal to the speed of light. The mathematically inclined reader will realize that the reason for this is immediate. In Minkowski spacetime the metric is $ds^2 = c^2 dt^2 - dx^2$

(where dx^2 is the Euclidean length $dx_1^2 + dx_2^2 + dx_3^2$), while the time carried by a clock ("proper" time) is given by $d\tau^2 = ds^2/c^2$. So, clearly, velocity through spacetime as just defined is given mathematically by $ds/d\tau$, which equals c.

8. Rudolf Carnap, "Autobiography," in *The Philosophy of Rudolf Carnap*, P. A. Schilpp, ed. (Chicago: Library of Living Philosophers, 1963), p. 37.

Chapter 6

1. Notice that the asymmetry being referred to—the arrow of time—arises from the order in which events take place *in* time. You could also wonder about asymmetries in time itself—for example, as we will see in later chapters, according to some cosmological theories time may have had a beginning but it may not have an end. These are distinct notions of temporal asymmetry, and our discussion here is focusing on the former. Even so, by the end of the chapter we will conclude that the temporal asymmetry of things in time relies on special conditions early on in the universe's history, and hence links the arrow of time to aspects of cosmology.

2. For the mathematically inclined reader, let me note more precisely what is meant by time-reversal symmetry and point out one intriguing exception whose significance for the issues we're discussing in this chapter has yet to be fully settled. The simplest notion of time-reversal symmetry is the statement that a set of laws of physics is time-reversal symmetric if given any solution to the equations, say $S(t)$, then $S(-t)$ is also a solution to the equations. For instance, in Newtonian mechanics, with forces that depend on particle positions, if $x(t) = (x_1(t), x_2(t), \ldots, x_{3n}(t))$ are the positions of n-particles in three space dimensions, then the fact that $x(t)$ solves $d^2x(t)/dt^2 = F(x(t))$ implies that $x(-t)$ is also a solution to Newton's equations, i.e. $d^2x(-t)/dt^2 = F(x(-t))$. Notice that $x(-t)$ represents particle motion that passes through the same positions as $x(t)$, but in reverse order, with reverse velocities.

More generally, a set of physical laws provides us with an algorithm for evolving an initial state of a physical system at time t_0 to some other time $t + t_0$. Concretely, this algorithm can be viewed as a map $U(t)$ which takes as input $S(t_0)$ and produces $S(t + t_0)$, that is: $S(t + t_0) = U(t)S(t_0)$. We say that the laws giving rise to $U(t)$ are time-reversal symmetric if there is a map T satisfying $U(-t) = T^{-1}U(t)T$. In English, this equation says that by a suitable manipulation of the state of the physical system at one moment (accomplished by T), evolution by an amount t forward in time according to the laws of the theory (accomplished by $U(t)$) is equivalent to having evolved the system t units of time backward in time (denoted by $U(-t)$). For instance, if we specify the state of a system of particles at one moment by their positions and velocities, then T would keep all particle positions fixed and reverse all velocities. Evolving such a configuration of particles forward in time by an amount t is equivalent to having evolved the original configuration of particles backward in time by an amount t. (The factor of T^{-1} undoes the velocity reversal so that, at the end, not only are the particle positions what they would have been t units of time previously, but so are their velocities.)

For certain sets of laws, the T operation is more complicated than it is for Newtonian mechanics. For example, if we study the motion of charged particles in the presence of an electromagnetic field, the reversal of particle velocities would be inadequate for the equations to yield an evolution in which the particles retrace their steps. Instead, the direction of the magnetic field must also be reversed. (This is required so that the $v \times B$ term in the Lorentz force law equation remains unchanged.) Thus, in this case, the T operation encompasses both of these transformations. The fact that we have to do more than just reverse all particle velocities has no impact on any of the discussion that follows in the text. All that matters is that particle motion in one direction is just as consistent with the physical laws as particle motion in the reverse direction. That we have to reverse any magnetic fields that happen to be present to accomplish this is of no particular relevance.

Where things get more subtle is the weak nuclear interactions. The weak interactions are described by a particular quantum field theory (discussed briefly in Chapter 9), and a general theorem shows that quantum field theories (so long as they are local, unitary, and Lorentz invariant—which are the ones of interest) are always symmetric under the *combined* operations of charge conjugation C (which replaces particles by their antiparticles), parity P (which inverts positions through the origin), and a bare-bones time-reversal operation T (which replaces t by $-t$). So, we could define a **T** operation to be the product *CPT*, but if **T** invariance absolutely requires the *CP* operation to be included, **T** would no longer be simply interpreted as particles retracing their steps (since, for example, particle identities would be changed by such **T**—particles would be replaced by their antiparticles—and hence it would not be the original particles retracing their steps). As it turns out, there are some exotic experimental situations in which we are forced into this corner. There are certain particle species (K-mesons, B-mesons) whose repertoire of behaviors is *CPT* invariant but is not invariant under *T* alone. This was established indirectly in 1964 by James Cronin, Val Fitch, and their collaborators (for which Cronin and Fitch received the 1980 Nobel Prize) by showing that the K-mesons violated *CP* symmetry (ensuring that they must violate *T* symmetry in order *not* to violate *CPT*). More recently, *T* symmetry violation has been directly established by the CPLEAR experiment at CERN and the KTEV experiment at Fermilab. Roughly speaking, these experiments show that if you were presented with a film of the recorded processes involving these meson particles, you'd be able to determine whether the film was being projected in the correct forward time direction, or in reverse. In other words, these particular particles can distinguish between past and future. What remains unclear, though, is whether this has any relevance for the arrow of time we experience in everyday contexts. After all, these are exotic particles that can be produced for fleeting moments in high-energy collisions, but they are not a constituent of familiar material objects. To many physicists, including me, it seems unlikely that the time nonreversal invariance evidenced by these particles plays a role in answering the puzzle of time's arrow, so we shall not discuss this exceptional example further. But the truth is that no one knows for sure.

3. I sometimes find that there is reluctance to accept the theoretical assertion that the eggshell pieces would really fuse back together into a pristine, uncracked shell. But the time-reversal symmetry of nature's laws, as elaborated with greater precision in the previous endnote, ensures that this is what would happen. Microscopically, the cracking of an egg is a physical process involving the various molecules that make up the shell. Cracks appear and the shell breaks apart because groups of molecules are forced to separate by the impact the egg experiences. If those molecular motions were to take place in reverse, the molecules would join back together, re-fusing the shell into its previous form.

4. To keep the focus on modern ways of thinking about these ideas, I am skipping over some very interesting history. Boltzmann's own thinking on the subject of entropy went through significant refinements during the 1870s and 1880s, during which time interactions and communications with physicists such as James Clerk Maxwell, Lord Kelvin, Josef Loschmidt, Josiah Willard Gibbs, Henri Poincaré, S. H. Burbury, and Ernest Zermelo were instrumental. In fact, Boltzmann initially thought he could prove that entropy would always and absolutely be nondecreasing for an isolated physical system, and not that it was merely highly unlikely for such entropy reduction to take place. But

objections raised by these and other physicists subsequently led Boltzmann to emphasize the statistical/probabilistic approach to the subject, the one that is still in use today.

5. I am imagining that we are using the Modern Library Classics edition of *War and Peace*, translated by Constance Garnett, with 1,386 text pages.

6. The mathematically inclined reader should note that because the numbers can get so large, entropy is actually defined as the logarithm of the number of possible arrangements, a detail that won't concern us here. However, as a point of principle, this is important because it is very convenient for entropy to be a so-called *extensive* quantity, which means that if you bring two systems together, the entropy of their union is the sum of their individual entropies. This holds true only for the logarithmic form of entropy, because the number of arrangements in such a situation is given by the product of the individual arrangements, so the logarithm of the number of arrangements is additive.

7. While we can, *in principle*, predict where each page will land, you might be concerned that there is an additional element that determines the page ordering: how you gather the pages together in a neat stack. This is not relevant to the physics being discussed, but in case it bothers you, imagine that we agree that you'll pick up the pages, one by one, starting with the one that's closest to you, and then picking up the page closest to that one, and so on. (And, for example, we can agree to measure distances from the nearest corner of the page in question.)

8. To succeed in calculating the motion of even a few pages with the accuracy required to predict their page ordering (after employing some algorithm for stacking them in a pile, such as in the previous note), is actually *extremely* optimistic. Depending on the flexibility and weight of the paper, such a comparatively "simple" calculation could still be beyond today's computational power.

9. You might worry that there is a fundamental difference between defining a notion of entropy for page orderings and defining one for a collection of molecules. After all, page orderings are discrete—you can count them, one by one, and so although the total number of possibilities might be large, it's finite. To the contrary, the motion and position of even a single molecule are continuous—you can't count them one by one, and so there is (at least according to classical physics) an infinite number of possibilities. So how can a precise counting of molecular rearrangements be carried out? Well, the short response is that this is a good question, but one that has been answered fully—so if that's enough to ease your worry, feel free to skip what follows. The longer response requires a bit of mathematics, so without background this may be tough to follow completely. Physicists describe a classical, many-particle system, by invoking *phase space*, a 6N-dimensional space (where N is the number of particles) in which each point denotes all particle positions and velocities (each such position requires three numbers, as does each velocity, accounting for the 6N dimensionality of phase space). The essential point is that phase space can be carved up into regions such that all points in a given region correspond to arrangements of the speeds and velocities of the molecules that have the same, overall, gross features and appearance. If the molecules' configuration were changed from one point in a given region of phase space to another point in the same region, a macroscopic assessment would find the two configurations indistinguishable. Now, rather than counting the number of points in a given region—the most direct analog of counting the number of different page rearrangements, but something that will surely result in an infinite answer—physicists define entropy in terms of the *volume* of each region in phase space. A larger volume means more points and hence higher entropy. And a region's volume, even a region in a higher-dimensional space, is something that can be given a rigorous mathematical definition. (Mathematically, it requires choosing something called a measure, and

for the mathematically inclined reader, I'll note that we usually choose the measure which is uniform over all microstates compatible with a given macrostate—that is, each microscopic configuration associated with a given set of macroscopic properties is assumed to be equally probable.)

10. Specifically, we know one way in which this *could* happen: if a few days earlier the CO_2 was initially in the bottle, then we know from our discussion above that if, right now, you were to simultaneously reverse the velocity of each and every CO_2 molecule, and that of every molecule and atom that has in any way interacted with the CO_2 molecules, and wait the same few days, the molecules *would* all group back together in the bottle. But this velocity reversal isn't something that can be accomplished in practice, let alone something that is likely to happen of its own accord. I might note, though, that one can prove mathematically that if you wait long enough, the CO_2 molecules will, of their own accord, *all find their way back into the bottle*. A result proven in the 1800s by the French mathematician Joseph Liouville can be used to establish what is known as the Poincaré recurrence theorem. This theorem shows that, if you wait long enough, a system with a finite energy and confined to a finite spatial volume (like CO_2 molecules in a closed room) *will return to a state arbitrarily close to its initial state* (in this case, CO_2 molecules all situated in the Coke bottle). The catch is how long you'd have to wait for this to happen. For systems with all but a small number of constituents, the theorem shows you'd typically have to wait far in excess of the age of the universe for the constituents to, of their own accord, regroup in their initial configuration. Nevertheless, as a point of principle, it is provocative to note that with endless patience and longevity, every spatially contained physical system will return to how it was initially configured.

11. You might wonder, then, why water ever turns into ice, since that results in the H_2O molecules becoming more ordered, that is, attaining lower, not higher, entropy. Well, the rough answer is that when liquid water turns into solid ice, it gives off energy to the environment (the opposite of what happens when ice melts, when it takes in energy from the environment), and that raises the environmental entropy. At low enough ambient temperatures, that is, below 0 degrees Celsius, the increase in environmental entropy exceeds the decrease in the water's entropy, so freezing becomes entropically favored. That's why ice forms in the cold of winter. Similarly, when ice cubes form in your refrigerator's freezer, their entropy goes down but the refrigerator itself pumps heat into the environment, and if that is taken account of, there is a total net increase of entropy. The more precise answer, for the mathematically inclined reader, is that spontaneous phenomena of the sort we're discussing are governed by what is known as *free energy*. Intuitively, free energy is that part of a system's energy that can be harnessed to do work. Mathematically, free energy, F, is defined by $F = U - TS$, where U stands for total energy, T stands for temperature, and S stands for entropy. A system will undergo a spontaneous change if that results in a decrease of its free energy. At low temperatures, the drop in U associated with liquid water turning into solid ice outweighs the decrease in S (outweighs the increase in $-TS$), and so will occur. At high temperatures (above 0 degrees Celsius), though, the change of ice to liquid water or gaseous steam is entropically favored (the increase in S outweighs changes to U) and so will occur.

12. For an early discussion of how a straightforward application of entropic reasoning would lead us to conclude that memories and historical records are not trustworthy accounts of the past, see C. F. von Weizsäcker in *The Unity of Nature* (New York: Farrar, Straus, and Giroux, 1980), 138–46, (originally published in *Annalen der Physik* 36 (1939). For an excellent recent discussion, see David Albert in *Time and Chance* (Cambridge, Mass.: Harvard University Press, 2000).

13. In fact, since the laws of physics don't distinguish between forward and backward in time, the explanation of having fully formed ice cubes a half hour earlier, at 10 p.m., would be *precisely* as absurd—entropically speaking—as predicting that by a half hour later, by 11:00 p.m., the little chunks of ice would have grown into fully formed ice cubes. To the contrary, the explanation of having liquid water at 10 p.m. that slowly forms small chunks of ice by 10:30 p.m. is *precisely* as sensible as predicting that by 11:00 p.m. the little chunks of ice will melt into liquid water, something that is familiar and totally expected. This latter explanation, from the perspective of the observation at 10:30 p.m., is perfectly temporally symmetric and, moreover, agrees with our subsequent observations.

14. The particularly careful reader might think that I've prejudiced the discussion with the phrase "early on" since that injects a temporal asymmetry. What I mean, in more precise language, is that we will need special conditions to prevail on (at least) one end of the temporal dimension. As will become clear, the special conditions amount to a low entropy boundary condition and I will call the "past" a direction in which this condition is satisfied.

15. The idea that time's arrow requires a low-entropy past has a long history, going back to Boltzmann and others; it was discussed in some detail in Hans Reichenbach, *The Direction of Time* (Mineola, N.Y.: Dover Publications, 1984), and was championed in a particularly interesting quantitative way in Roger Penrose, *The Emperor's New Mind* (New York: Oxford University Press, 1989), pp. 317ff.

16. Recall that our discussion in this chapter does not take account of quantum mechanics. As Stephen Hawking showed in the 1970s, when quantum effects are considered, black holes do allow a certain amount of radiation to seep out, but this does not affect their being the highest-entropy objects in the cosmos.

17. A natural question is how we know that there isn't some future constraint that also has an impact on entropy. The bottom line is that we don't, and some physicists have even suggested experiments to detect the possible influence that such a future constraint might have on things that we can observe today. For an interesting article discussing the possibility of future and past constraints on entropy, see Murray Gell-Mann and James Hartle, "Time Symmetry and Asymmetry in Quantum Mechanics and Quantum Cosmology," in *Physical Origins of Time Asymmetry*, J. J. Halliwell, J. Pérez-Mercader, W. H. Zurek, eds. (Cambridge, Eng.: Cambridge University Press, 1996), as well as other papers in Parts 4 and 5 of that collection.

18. Throughout this chapter, we've spoken of *the* arrow of time, referring to the apparent fact that there is an asymmetry along the time axis (any observer's time axis) of spacetime: a huge variety of sequences of events is arrayed in one order along the time axis, but the reverse ordering of such events seldom, if ever, occurs. Over the years, physicists and philosophers have divided these sequences of events into subcategories whose temporal asymmetries might, in principle, be subject to logically independent explanations. For example, heat flows from hot objects to cooler ones, but not from cool objects to hot ones; electromagnetic waves emanate outward from sources like stars and lightbulbs, but seem never to converge inward on such sources; the universe appears to be uniformly expanding, and not contracting; and we remember the past and not the future (these are called the thermodynamic, electromagnetic, cosmological, and psychological arrows of time, respectively). All of these are time-asymmetric phenomena, but they might, in principle, acquire their time asymmetry from completely different physical principles. My view, one that many share (but others don't), is that except possibly for the cosmological arrow, these temporally asymmetric phenomena are not fundamentally different, and ultimately are subject to the same explanation—the one we've described in this chapter. For

example, why does electromagnetic radiation travel in expanding outward waves but not contracting inward waves, even though both are perfectly good solutions to Maxwell's equations of electromagnetism? Well, because our universe has low-entropy, coherent, ordered sources for such outward waves—stars and lightbulbs, to name two—and the existence of these ordered sources derives from the even more ordered environment at the universe's inception, as discussed in the main text. The psychological arrow of time is harder to address since there is so much about the microphysical basis of human thought that we've yet to understand. But much progress has been made in understanding the arrow of time when it comes to computers—undertaking, completing, and then producing a record of a computation is a basic computational sequence whose entropic properties are well understood (as developed by Charles Bennett, Rolf Landauer, and others) and fit squarely within the second law of thermodynamics. Thus, if human thought can be likened to computational processes, a similar thermodynamic explanation may apply. Notice, too, that the asymmetry associated with the fact that the universe is expanding and not contracting is related to, but logically distinct from, the arrow of time we've been exploring. If the universe's expansion were to slow down, stop, and then turn into a contraction, the arrow of time would still point in the same direction. Physical processes (eggs breaking, people aging, and so on) would still happen in the usual direction, even though the universe's expansion had reversed.

19. For the mathematically inclined reader, notice that when we make this kind of probabilistic statement we are assuming a particular probability measure: the one that is uniform over all microstates compatible with what we see right *now*. There are, of course, other measures that we could invoke. For example, David Albert in *Time and Chance* has advocated using a probability measure that is uniform over all microstates compatible with what we see *now* and what he calls *the past hypothesis*—the apparent fact that the universe began in a low-entropy state. Using this measure, we eliminate consideration of all but those histories that are compatible with the low-entropy past attested to by our memories, records, and cosmological theories. In this way of thinking, there is no probabilistic puzzle about a universe with low entropy; it began that way, by assumption, with probability 1. There is still the same huge puzzle of *why* it began that way, even if it isn't phrased in a probabilistic context.

20. You might be tempted to argue that the known universe had low entropy early on simply because it was much smaller in size than it is today, and hence—like a book with fewer pages—allowed for far fewer rearrangements of its constituents. But, by itself, this doesn't do the trick. Even a small universe can have huge entropy. For example, one possible (although unlikely) fate for our universe is that the current expansion will one day halt, reverse, and the universe will implode, ending in the so-called big crunch. Calculations show that even though the size of the universe would decrease during the implosion phase, entropy would continue to rise, which demonstrates that small size does not ensure low entropy. In Chapter 11, though, we will see that the universe's small initial size does play a role in our current, best explanation of the low entropy beginning.

Chapter 7

1. It is well known that the equations of classical physics cannot be solved exactly if you are studying the motion of three or more mutually interacting bodies. So, even in classical physics, any actual prediction about the motion of a large set of particles will necessarily be approximate. The point, though, is that there is no fundamental limit to how good this approximation can be. If the world were governed by classical physics, then with

ever more powerful computers, and ever more precise initial data about positions and velocities, we would get ever closer to the exact answer.

2. At the end of Chapter 4, I noted that the results of Bell, Aspect, and others do not rule out the possibility that particles always have definite positions and velocities, even if we can't ever determine such features simultaneously. Moreover, Bohm's version of quantum mechanics explicitly realizes this possibility. Thus, although the widely held view that an electron doesn't have a position until measured is a standard feature of the conventional approach to quantum mechanics, it is, strictly speaking, too strong as a blanket statement. Bear in mind, though, that in Bohm's approach, as we will discuss later in this chapter, particles are "accompanied" by probability waves; that is, Bohm's theory always invokes particles *and* waves, whereas the standard approach envisions a complementarity that can roughly be summarized as particles *or* waves. Thus, the conclusion we're after— that the quantum mechanical description of the past would be thoroughly incomplete if we spoke exclusively about a particle's having passed through a unique point in space at each definite moment in time (what we *would* do in classical physics)—is true nevertheless. In the conventional approach to quantum mechanics, we must also include the wealth of other locations that a particle could have occupied at any given moment, while in Bohm's approach we must also include the "pilot" wave, an object that is also spread throughout a wealth of other locations. (The expert reader should note that the pilot wave is just the wavefunction of conventional quantum mechanics, although its incarnation in Bohm's theory is rather different.) To avoid endless qualifications, the discussion that follows will be from the perspective of conventional quantum mechanics (the approach most widely used), leaving remarks on Bohm's and other approaches to the last part of the chapter.

3. For a mathematical but highly pedagogical account see R. P. Feynman and A. R. Hibbs, *Quantum Mechanics and Path Integrals* (Burr Ridge, Ill.: McGraw-Hill Higher Education, 1965).

4. You might be tempted to invoke the discussion of Chapter 3, in which we learned that at light speed time slows to a halt, to argue that from the photon's perspective all moments are the same moment, so the photon "knows" how the detector switch is set when it passes the beam-splitter. However, these experiments can be carried out with other particle species, such as electrons, that travel slower than light, and the results are unchanged. Thus, this perspective does not illuminate the essential physics.

5. The experimental setup discussed, as well as the actual confirming experimental results, comes from Y. Kim, R. Yu, S. Kulik, Y. Shih, M. Scully, *Phys. Rev. Lett*, vol. 84, no. 1, pp. 1–5.

6. Quantum mechanics can also be based on an equivalent equation presented in a different form (known as matrix mechanics) by Werner Heisenberg in 1925. For the mathematically inclined reader, Schrödinger's equation is: $H\,\Psi(x,t) = i\hbar\,(d\Psi(x,t)/dt)$, where H stands for the Hamiltonian, Ψ stands for the wavefunction, and $\hbar$ is Planck's constant.

7. The expert reader will note that I am suppressing one subtle point here. Namely, we would have to take the complex conjugate of the particle's wavefunction to ensure that it solves the time-reversed version of Schrödinger's equation. That is, the T operation described in endnote 2 of Chapter 6 takes a wavefunction $\Psi(x,t)$ and maps it to $\Psi^*(x,-t)$. This has no significant impact on the discussion in the text.

8. Bohm actually rediscovered and further developed an approach that goes back to Prince Louis de Broglie, so this approach is sometimes called the de Broglie–Bohm approach.

9. For the mathematically inclined reader, Bohm's approach is local in *configuration* space but certainly *nonlocal* in real space. Changes to the wavefunction in one location in real space immediately exert an influence on particles located in other, distant locations.

10. For an exceptionally clear treatment of the Ghirardi-Rimini-Weber approach and its relevance to understanding quantum entanglement, see J. S. Bell, "Are There Quantum Jumps?" in *Speakable and Unspeakable in Quantum Mechanics* (Cambridge, Eng.: Cambridge University Press, 1993).

11. Some physicists consider the questions on this list to be irrelevant by-products of earlier confusions regarding quantum mechanics. The wavefunction, this view professes, is merely a theoretical tool for making (probabilistic) predictions and should not be accorded any but mathematical reality (a view sometimes called the "Shut up and calculate" approach, since it encourages one to use quantum mechanics and wavefunctions to make predictions, without thinking hard about what the wavefunctions actually mean and do). A variation on this theme argues that wavefunctions never actually collapse, but that interactions with the environment make it *seem* as if they do. (We will discuss a version of this approach shortly.) I am sympathetic to these ideas and, in fact, strongly believe that the notion of wavefunction collapse will ultimately be dispensed with. But I don't find the former approach satisfying, as I am not ready to give up on understanding what happens in the world when we are "not looking," and the latter—while, in my view, the right direction—needs further mathematical development. The bottom line is that measurement causes something that *is* or is *akin to* or *masquerades as* wavefunction collapse. Either through a better understanding of environmental influence or through some other approach yet to be suggested, this apparent effect needs to be addressed, not simply dismissed.

12. There are other controversial issues associated with the Many Worlds interpretation that go beyond its obvious extravagance. For example, there are technical challenges to define a notion of probability in a context that involves an infinite number of copies of each of the observers whose measurements are supposed to be subject to those probabilities. If a given observer is really one of many copies, in what sense can we say that he or she has a particular probability to measure this or that outcome? Who really is "he" or "she"? Each copy of the observer will measure—with probability 1—whatever outcome is slated for the particular copy of the universe in which he or she resides, so the whole probabilistic framework requires (and has been given, and continues to be given) careful scrutiny in the Many Worlds framework. Moreover, on a more technical note, the mathematically inclined reader will realize that, depending on how one precisely defines the Many Worlds, a preferred eigenbasis may need to be selected. But how should that eigenbasis be chosen? There has been a great deal of discussion and much written on all these questions, but to date there are no universally accepted resolutions. The approach based on decoherence, discussed shortly, has shed much light on these issues, and has offered particular insight into the issue of eigenbasis selection.

13. The Bohm or de Broglie–Bohm approach has never received wide attention. Perhaps one reason for this, as pointed out by John Bell in his article "The Impossible Pilot Wave," collected in *Speakable and Unspeakable in Quantum Mechanics,* is that neither de Broglie nor Bohm was particularly fond of what he himself had developed. But, again as Bell points out, the de Broglie–Bohm approach does away with much of the vagueness and subjectivity of the more standard approach. If for no other reason, even if the approach is wrong, it is worth knowing that particles can have definite positions and definite velocities at all times (ones beyond our ability, even in principle, to measure), and still

conform fully to the predictions of standard quantum mechanics—uncertainty and all. Another argument against Bohm's approach is that the nonlocality in this framework is more "severe" than that of standard quantum mechanics. By this it is meant that Bohm's approach has nonlocal interactions (between the wavefunction and particles) as a central element of the theory from the outset, while in quantum mechanics the nonlocality is more deeply buried and arises only through nonlocal correlations between widely separated measurements. But, as supporters of this approach have argued, because something is hidden does not make it any less present, and, moreover, as the standard approach is vague regarding the quantum measurement problem—the very place where nonlocality makes itself apparent—once that issue is fully resolved, the nonlocality may not be so hidden after all. Others have argued that there are obstacles to making a relativistic version of the Bohm approach, although progress has been made on this front as well (see, for example, John Bell Beables *for Quantum Field Theory*, in the collected volume indicated above). And so, it is definitely worth keeping this alternative approach in mind, even if only as a foil against rash conclusions about what quantum mechanics unavoidably implies. For the mathematically inclined reader, a very nice treatment of Bohm's theory and issues of quantum entanglement can be found in Tim Maudlin, *Quantum Nonlocality and Relativity* (Malden, Mass.: Blackwell, 2002).

14. For an in-depth, though technical, discussion of time's arrow in general, and the role of decoherence in particular, see H. D. Zeh, *The Physical Basis of the Direction of Time* (Heidelberg: Springer, 2001).

15. Just to give you a sense of how quickly decoherence takes place—how quickly environmental influence suppresses quantum interference and thereby turns quantum probabilities into familiar classical ones—here are a few examples. The numbers are approximate, but the point they convey is clear. The wavefunction of a grain of dust floating in your living room, bombarded by jittering air molecules, will decohere in about a billionth of a billionth of a billionth of a billionth (10^{-36}) of a second. If the grain of dust is kept in a perfect vacuum chamber and subject only to interactions with sunlight, its wavefunction will decohere a bit more slowly, taking a thousandth of a billionth of a billionth (10^{-21}) of a second. And if the grain of dust is floating in the darkest depths of empty space and subject only to interactions with the relic microwave photons from the big bang, its wavefunction will decohere in about a millionth of a second. These numbers are extremely small, which shows that decoherence for something even as tiny as a grain of dust happens very quickly. For larger objects, decoherence happens faster still. It is no wonder that, even though ours is a quantum universe, the world around us looks like it does. (See, for example, E. Joos, "Elements of Environmental Decoherence," in *Decoherence: Theoretical, Experimental, and Conceptual Problems*, Ph. Blanchard, D. Giulini, E. Joos, C. Kiefer, I.-O. Stamatescu, eds. [Berlin: Springer, 2000]).

Chapter 8

1. To be more precise, the symmetry between the laws in Connecticut and the laws in New York makes use of both translational symmetry *and* rotational symmetry. When you perform in New York, not only will you have changed location from Connecticut, but more than likely you will undertake your routines while facing in a somewhat different direction (east versus north, perhaps) than during practice.

2. Newton's laws of motion are usually described as being relevant for "inertial observers," but when one looks closely at how such observers are specified, it sounds circular: inertial observers are those observers for whom Newton's laws hold. A good way to

think about what's really going on is that Newton's laws draw our attention to a large and particularly useful class of observers: those whose description of motion fits completely and quantitatively within Newton's framework. By definition, these are inertial observers. Operationally, inertial observers are those on whom no forces of any kind are acting—observers, that is, who experience no accelerations. Einstein's general relativity, by contrast, applies to all observers, regardless of their state of motion.

3. If we lived in an era during which *all* change stopped, we'd experience no passage of time (all body and brain functions would be frozen as well). But whether this would mean that the spacetime block in Figure 5.1 came to an end, or, instead, carried on with no change along the time axis—that is, whether time would come to an end or would still exist in some kind of formal, overarching sense—is a hypothetical question that's both difficult to answer and largely irrelevant for anything we might measure or experience. Note that this hypothetical situation is different from a state of maximal disorder in which entropy can't further increase, but microscopic change, like gas molecules going this way and that, still takes place.

4. The cosmic microwave radiation was discovered in 1964 by the Bell Laboratory scientists Arno Penzias and Robert Wilson, while testing a large antenna intended for use in satellite communications. Penzias and Wilson encountered background noise that proved impossible to remove (even after they scraped bird droppings—"white noise"—from the inside of the antenna) and, with the key insights of Robert Dicke at Princeton and his students Peter Roll and David Wilkinson, together with Jim Peebles, it was ultimately realized that the antenna was picking up microwave radiation that originated with the big bang. (Important work in cosmology that set the stage for this discovery was carried out earlier by George Gamow, Ralph Alpher, and Robert Herman.) As we discuss further in later chapters, the radiation gives us an unadulterated picture of the universe when it was about 300,000 years old. That's when electrically charged particles like electrons and protons, which disrupt the motion of light beams, combined to form electrically neutral atoms, which, by and large, allow light to travel freely. Ever since, such ancient light—produced in the early stages of the universe—has traveled unimpeded, and today, suffuses all of space with microwave photons.

5. The physical phenomenon involved here, as discussed in Chapter 11, is known as *redshift*. Common atoms such as hydrogen and oxygen emit light at wavelengths that have been well documented through laboratory experiments. When such substances are constituents of galaxies that are rushing away, the light they emit is elongated, much as the siren of a police car that's racing away is also elongated, making the pitch drop. Because red is the longest wavelength of light that can be seen with the unaided eye, this stretching of light is called the redshift effect. The amount of redshift grows with increasing recessional speed, and hence by measuring the received wavelengths of light and comparing with laboratory results, the speed of distant objects can be determined. (This is actually one kind of redshift, akin to the Doppler effect. Redshifting can also be caused by gravity: photons elongate as they climb out of a gravitational field.)

6. More precisely, the mathematically inclined reader will note that a particle of mass m, sitting on the surface of a ball of radius R and mass density ρ, experiences an acceleration, d^2R/dt^2 given by $(4\pi/3)R^3G\rho/R^2$, and so $(1/R)\,d^2R/dt^2 = (4\pi/3)G\rho$. If we formally identify R with the radius of the universe, and ρ with the mass density of the universe, this is Einstein's equation for how the size of the universe evolves (assuming the absence of pressure).

7. See P.J.E. Peebles, *Principles of Physical Cosmology* (Princeton: Princeton University Press, 1993), p. 81.

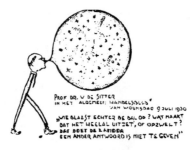

PROF DR. V DE SITTER
IN HET ALGEMEEN: HANDELSBLAD
VAN WOENSDAG 9 JULI 1930

"WIE BLAAST ECHTER DE BAL OP ? WAT MAAKT
DAT HET HEELAL UITZET, OF OPZWELT ?
DAT DOET DE LAMBDA
EEN ANDER ANTWOORD IS NIET TE GEVEN"

The caption reads: "But who is really blowing up this ball? What makes it so that the universe expands or inflates? A Lambda does the job! Another answer cannot be given." (Translation by Koenraad Schalm.) Lambda refers to something known as the cosmological constant, an idea we will encounter in Chapter 10.

8. To avoid confusion, let me note that one drawback of the penny model is that every penny is essentially identical to every other, while that is certainly not true of galaxies. But the point is that on the largest of scales—scales on the order of 100 million light-years—the individual differences between galaxies are believed to average out so that, when one analyzes huge volumes of space, the overall properties of each such volume are extremely similar to the properties of any other such volume.

9. You could also travel to just outside the edge of a black hole, and remain there, engines firing away to avoid being pulled in. The black hole's strong gravitational field manifests itself as a severe warping of spacetime, and that results in your clock's ticking far slower than it would in a more ordinary location in the galaxy (as in a relatively empty spatial expanse). Again, the time duration measured by your clock is perfectly valid. But, as in zipping around at high speed, it is a completely individualistic perspective. When analyzing features of the universe as a whole, it is more useful to have a widely applicable and agreed upon notion of elapsed time, and that's what is provided by clocks that move along with the cosmic flow of spatial expansion and that are subject to a far more mild, far more average gravitational field.

10. The mathematically inclined reader will note that light travels along null geodesics of the spacetime metric, which, for definiteness, we can take to be $ds^2 = dt^2 - a^2(t)(dx^2)$, where $dx^2 = dx_1^2 + dx_2^2 + dx_3^2$, and the x_i are comoving coordinates. Setting $ds^2 = 0$, as appropriate for a null geodesic, we can write $\int_t^{t_0} (dt/a(t))$ for the total comoving distance light emitted at time t can travel by time t_0. If we multiply this by the value of scale factor $a(t_0)$ at time t_0, then we will have calculated the physical distance that the light has traveled in this time interval. This algorithm can be widely used to calculate how far light can travel in any given time interval, revealing whether two points in space, for example, are in causal contact. As you can see, for accelerated expansion, even for arbitrarily large t_0, the integral is bounded, showing that the light will never reach arbitrarily distant comoving locations. Thus, in a universe with accelerated expansion, there are locations with which we can never communicate, and conversely, regions that can never communicate with us. Such regions are said to be beyond our cosmic horizon.

11. When analyzing geometrical shapes, mathematicians and physicists use a quantitative approach to curvature developed in the nineteenth century, which today is part of a mathematical body of knowledge known as differential geometry. One nontechnical way of thinking about this measure of curvature is to study triangles drawn on or within the

shape of interest. If the triangle's angles add up to 180 degrees, as they do when it is drawn on a flat tabletop, we say the shape is flat. But if the angles add up to more or less than 180 degrees, as they do when the triangle is drawn on the surface of a sphere (the outward bloating of a sphere causes the sum of the angles to exceed 180 degrees) or the surface of a saddle (the inward shrinking of a saddle's shape causes the sum of the angles to be less than 180 degrees), we say the shape is curved. This is illustrated in Figure 8.6.

12. If you were to glue the opposite vertical edges of a torus together (which is reasonable to do, since they are identified—when you pass through one edge you immediately reappear on the other) you'd get a cylinder. And then, if you did the same for the upper and lower edges (which would now be in the shape of circles), you'd get a doughnut. Thus, a doughnut is another way of thinking about or representing a torus. One complication of this representation is that the doughnut no longer looks flat! However, it actually is. Using the notion of curvature given in the previous endnote, you'd find that all triangles drawn on the surface of the doughnut have angles that add up to 180 degrees. The fact that the doughnut looks curved is an artifact of how we've embedded a two-dimensional shape in our three-dimensional world. For this reason, in the current context it is more useful to use the manifestly uncurved representations of the two- and three-dimensional tori, as discussed in the text.

13. Notice that we've been loose in distinguishing the concepts of shape and curvature. There are three types of *curvatures* for completely symmetric space: positive, zero, and negative. But two shapes can have the same curvature and yet not be identical, with the simplest example being the flat video screen and the flat infinite tabletop. Thus, symmetry allows us to narrow down the curvature of space to three possibilities, but there are somewhat more than three shapes for space (differing in what mathematicians call their global properties) that realize these three curvatures.

14. So far, we've focused exclusively on the curvature of three-dimensional space— the curvature of the spatial slices in the spacetime loaf. However, although it's hard to picture, in all three cases of spatial curvature (positive, zero, negative), the whole four-dimensional spacetime is curved, with the degree of curvature becoming ever larger as we examine the universe ever closer to the big bang. In fact, near the moment of the big bang, the four-dimensional curvature of spacetime grows so large that Einstein's equations break down. We will discuss this further in later chapters.

Chapter 9

1. If you raised the temperature much higher, you'd find a fourth state of matter known as a *plasma*, in which atoms disintegrate into their component particles.

2. There are curious substances, such as Rochelle salts, which become less ordered at high temperatures, and more ordered at low temperatures—the reverse of what we normally expect.

3. One difference between force and matter fields is expressed by Wolfgang Pauli's *exclusion principle*. This principle shows that whereas a huge number of force particles (like photons) can combine to produce fields accessible to a prequantum physicist such as Maxwell, fields that you see every time you enter a dark room and turn on a light, matter particles are generally excluded by the laws of quantum physics from cooperating in such a coherent, organized manner. (More precisely, two particles of the same species, such as two electrons, are excluded from occupying the same state, whereas there is no such restriction for photons. Thus, matter fields do not generally have a macroscopic, classical-like manifestation.)

4. In the framework of quantum field theory, every known particle is viewed as an excitation of an underlying field associated with the species of which that particle is a member. Photons are excitations of the photon field—that is, the electromagnetic field; an up-quark is an excitation of the up-quark field; an electron is an excitation of the electron field, and so on. In this way, all matter and all forces are described in a uniform quantum mechanical language. A key problem is that it has proved very difficult to describe all the quantum features of gravity in this language, an issue we will discuss in Chapter 12.

5. Although the Higgs field is named after Peter Higgs, a number of other physicists—Thomas Kibble, Philip Anderson, R. Brout, and François Englert, among others—played a vital part in its introduction into physics and its theoretical development.

6. Bear in mind that the field's *value* is given by its distance from the bowl's center, so even though the field has zero *energy* when its value is in the bowl's valley (since the height above the valley denotes the field's energy), its value is not zero.

7. In the text's description, the value of the Higgs field is given by its distance from the bowl's center, and so you may be wondering how points on the bowl's circular valley—which are all the same distance from the bowl's center—give rise to any but the *same* Higgs value. The answer, for the mathematically inclined reader, is that different points in the valley represent Higgs field values with the same magnitude but different phases (the Higgs field value is a complex number).

8. In principle, there are two concepts of mass that enter into physics. One is the concept described in the text: mass as that property of an object which resists acceleration. Sometimes, this notion of mass is called *inertial mass*. The second concept of mass is the one relevant for gravity: mass as that property of an object which determines how strongly it will be pulled by a gravitational field of a specified strength (such as the earth's). Sometimes this notion of mass is called *gravitational mass*. At first glance, the Higgs field is relevant only for an understanding of inertial mass. However, the equivalence principle of general relativity asserts that the force felt from accelerated motion and from a gravitational field are indistinguishable—they are equivalent. And that implies an equivalence between the concepts of inertial mass and gravitational mass. Thus, the Higgs field is relevant for both kinds of mass we've mentioned since, according to Einstein, they are the same.

9. I thank Raphael Kasper for pointing out that this description is a variation on the prize-winning metaphor of Professor David Miller, submitted in response to British Science Minister William Waldegrave's challenge in 1993 to the British physics community to explain why taxpayer money should be spent on searching for the Higgs particle.

10. The mathematically inclined reader should note that the photons and W and Z bosons are described in the electroweak theory as lying in the adjoint representation of the group $SU(2) \times U(1)$, and hence are interchanged by the action of this group. Moreover, the equations of the electroweak theory possess complete symmetry under this group action and it is in this sense that we describe the force particles as being interrelated. More precisely, in the electroweak theory, the photon is a particular mixture of the gauge boson of the manifest $U(1)$ symmetry and the $U(1)$ subgroup of $SU(2)$; it is thus tightly related to the weak gauge bosons. However, because of the symmetry group's product structure, the four bosons (there are actually two W bosons with opposite electric charges) do not fully mix under its action. In a sense, then, the weak and electromagnetic interactions are part of a single mathematical framework, but one that is not as fully unified as it might be. When one includes the strong interactions, the group is augmented by including an $SU(3)$ factor—"color" $SU(3)$—and this group's having *three* independent factors, $SU(3) \times SU(2) \times U(1)$, only highlights further the lack of complete unity. This is part of the moti-

vation for grand unification, discussed in the next section: grand unification seeks a single, semi-simple (Lie) group—a group with a single factor—that describes the forces at higher energy scales.

11. The mathematically inclined reader should note that Georgi and Glashow's grand unified theory was based on the group $SU(5)$, which includes $SU(3)$, the group associated with the strong nuclear force, and also $SU(2) \times U(1)$, the group associated with the electroweak force. Since then, physicists have studied the implications of other potential grand unified groups, such as $SO(10)$ and E_6.

Chapter 10

1. As we've seen, the big bang's bang is not an explosion that took place at one location in a preexisting spatial expanse, and that's why we've not also asked *where* it banged. The playful description of the big bang's deficiency we've used is due to Alan Guth; see, for example, his *The Inflationary Universe* (Reading, Eng.: Perseus Books, 1997), p. xiii.

2. The term "big bang" is sometimes used to denote the event that happened at time-zero itself, bringing the universe into existence. But since, as we'll discuss in the next chapter, the equations of general relativity break down at time-zero, no one has any understanding of what this event actually was. This omission is what we've meant by saying that the big bang theory leaves out the bang. In this chapter, we are restricting ourselves to realms in which the equations do not break down. Inflationary cosmology makes use of such well-behaved equations to reveal a brief explosive swelling of space that we naturally take to be the bang left out by the big bang theory. Certainly, though, this approach leaves unanswered the question of what happened at the initial moment of the universe's creation—if there actually was such a moment.

3. Abraham Pais, *Subtle Is the Lord* (Oxford: Oxford University Press, 1982), p. 253.

4. For the mathematically inclined reader: Einstein replaced the original equation $G_{\mu\nu} = 8\pi T_{\mu\nu}$ by $G_{\mu\nu} + \Lambda g_{\mu\nu} = 8\pi T_{\mu\nu}$ where Λ is a number denoting the size of the cosmological constant.

5. When I refer to an object's mass in this context, I am referring to the sum total mass of its particulate constituents. If a cube, say, were composed of 1,000 gold atoms, I'd be referring to 1,000 times the mass of a single such atom. This definition jibes with Newton's perspective. Newton's laws say that such a cube would have a mass that is 1,000 times that of a single gold atom, and that it would weigh 1,000 times as much as a single gold atom. According to Einstein, though, the weight of the cube also depends on the kinetic energy of the atoms (as well as all other contributions to the energy of the cube). This follows from $E=mc^2$: more energy (E), regardless of the source, translates into more mass (m). Thus, an equivalent way of expressing the point is that because Newton didn't know about $E=mc^2$, his law of gravity uses a definition of mass that misses various contributions to energy, such as energy associated with motion.

6. The discussion here is suggestive of the underlying physics but does not capture it fully. The pressure exerted by the compressed spring does indeed influence how strongly the box is pulled earthward. But this is because the compressed spring affects the total energy of the box and, as discussed in the previous paragraph, according to general relativity, the total energy is what's relevant. However, the point I'm explaining here is that pressure itself—not just through the contribution it makes to total energy—generates gravity, much as mass and energy do. According to general relativity, pressure gravitates. Also note that the repulsive gravity we are referring to is the *internal* gravitational field experienced

within a region of space suffused by something that has negative rather than positive pressure. In such a situation, negative pressure will contribute a repulsive gravitational field acting *within* the region.

7. Mathematically, the cosmological constant is represented by a number, usually denoted by Λ (see note 4). Einstein found that his equations made perfect sense regardless of whether Λ was chosen to be a positive or a negative number. The discussion in the text focuses on the case of particular interest to modern cosmology (and modern observations, as will be discussed) in which Λ is positive, since this gives rise to negative pressure and repulsive gravity. A negative value for Λ yields ordinary attractive gravity. Note, too, that since the pressure exerted by the cosmological constant is uniform, this pressure does not directly exert any force: only pressure differences, like what your ears feel when you're underwater, result in a pressure force. Instead, the force exerted by the cosmological constant is purely a gravitational force.

8. Familiar magnets always have both a north and a south pole. By contrast, grand unified theories suggest that there may be particles that are like a purely north or purely south magnetic pole. Such particles are called monopoles and they could have a major impact on standard big bang cosmology. They have never been observed.

9. Guth and Tye recognized that a supercooled Higgs field would act like a cosmological constant, a realization that had been made earlier by Martinus Veltman and others. In fact, Tye has told me that were it not for a page limit in *Physical Review Letters*, the journal to which he and Guth submitted their paper, they would not have struck a final sentence noting that their model would entail a period of exponential expansion. But Tye also notes that it was Guth's achievement to realize the important cosmological implications of a period of exponential expansion (to be discussed later in this and in the next chapter), and thereby put inflation front and center on cosmologists' maps.

In the sometimes convoluted history of discovery, the Russian physicist Alexei Starobinsky had, a few years earlier, found a different means of generating what we now call inflationary expansion, work described in a paper that was not widely known among western scientists. However, Starobinsky did not emphasize that a period of such rapid expansion would solve key cosmological problems (such as the horizon and flatness problems, to be discussed shortly), which explains, in part, why his work did not generate the enthusiastic response that Guth's received. In 1981, the Japanese physicist Katsuhiko Sato also developed a version of inflationary cosmology, and even earlier (in 1978), Russian physicists Gennady Chibisov and Andrei Linde hit upon the idea of inflation, but they realized that—when studied in detail—it suffered from a key problem (discussed in note 11) and hence did not publish their work.

The mathematically inclined reader should note that it is not difficult to see how accelerated expansion arises. One of Einstein's equations is $d^2a/dt^2/a = -4\pi/3(\rho + 3p)$ where a, ρ, and p are the scale factor of the universe (its "size"), the energy density, and the pressure density, respectively. Notice that if the righthand side of this equation is positive, the scale factor will grow at an increasing rate: the universe's rate of growth will accelerate with time. For a Higgs field perched on a plateau, its pressure density turns out to equal the negative of its energy density (the same is true for a cosmological constant), and so the righthand side is indeed positive.

10. The physics underlying these quantum jumps is the uncertainty principle, covered in Chapter 4. I will explicitly discuss the application of quantum uncertainty to fields in both Chapter 11 and Chapter 12, but to presage that material, briefly note the following. The value of a field at a given point in space, and the rate of change of the field's value at that point, play the same role for fields as position and velocity (momentum) play for a

particle. Thus, just as we can't ever know both a definite position and a definite velocity for a particle, a field can't have a definite value and a definite rate of change of that value, at any given point in space. The more definite the field's value is at one moment, the more uncertain is the rate of change of that value—that is, the more likely it is that the field's value will change a moment later. And such change, induced by quantum uncertainty, is what I mean when referring to quantum jumps in the field's value.

11. The contribution of Linde and of Albrecht and Steinhardt was absolutely crucial, because Guth's original model—now called *old inflation*—suffered from a pernicious flaw. Remember that the supercooled Higgs field (or, in the terminology we introduce shortly, the *inflaton* field) has a value that is perched on the bump in its energy bowl *uniformly* across space. And so, while I've described how quickly the supercooled inflaton field could take the jump to the lowest energy value, we need to ask whether this quantum-induced jump would happen everywhere in space at the same time. And the answer is that it wouldn't. Instead, as Guth argued, the relaxation of the inflaton field to a zero energy value takes place by a process called bubble nucleation: the inflaton drops to its zero energy value at one point in space, and this sparks an outward-spreading bubble, one whose walls move at light speed, in which the inflaton drops to the zero energy value with the passing of the bubble wall. Guth envisioned that many such bubbles, with random centers, would ultimately coalesce to give a universe with zero-energy inflaton field everywhere. The problem, though, as Guth himself realized, was that the space surrounding the bubbles was still infused with a non-zero-energy inflaton field, and so such regions would continue to undergo rapid inflationary expansion, driving the bubbles apart. Hence, there was no guarantee that the growing bubbles would find one another and coalesce into a large, homogeneous spatial expanse. Moreover, Guth argued that the inflaton field energy was not lost as it relaxed to zero energy, but was converted to ordinary particles of matter and radiation inhabiting the universe. To achieve a model compatible with observations, though, this conversion would have to yield a *uniform* distribution of matter and energy throughout space. In the mechanism Guth proposed, this conversion would happen through the collision of bubble walls, but calculations—carried out by Guth and Erick Weinberg of Columbia University, and also by Stephen Hawking, Ian Moss, and John Steward of Cambridge University—revealed that the resulting distribution of matter and energy was *not* uniform. Thus, Guth's original inflationary model ran into significant problems of detail.

The insights of Linde and of Albrecht and Steinhardt—now called *new inflation*—fixed these vexing problems. By changing the shape of the potential energy bowl to that in Figure 10.2, these researchers realized, the inflaton could relax to its zero energy value by "rolling" down the energy hill to the valley, a gradual and graceful process that had no need for the quantum jump of the original proposal. And, as their calculations showed, this somewhat more gradual rolling down the hill sufficiently prolonged the inflationary burst of space so that one single bubble easily grew large enough to encompass the entire observable universe. Thus, in this approach, there is no need to worry about coalescing bubbles. What was of equal importance, rather than converting the inflaton field's energy to that of ordinary particles and radiation through bubble collisions, in the new approach the inflaton gradually accomplished this energy conversion uniformly throughout space by a process akin to friction: as the field rolled down the energy hill—uniformly throughout space—it gave up its energy by "rubbing against" (interacting with) more familiar fields for particles and radiation. New inflation thus retained all the successes of Guth's approach, but patched up the significant problem it had encountered.

About a year after the important progress offered by new inflation, Andrei Linde had another breakthrough. For new inflation to occur successfully, a number of key elements must all fall into place: the potential energy bowl must have the right shape, the inflaton field's value must begin high up on the bowl (and, somewhat more technically, the inflaton field's value must itself be uniform over a sufficiently large spatial expanse). While it's possible for the universe to achieve such conditions, Linde found a way to generate an inflationary burst in a simpler, far less contrived setting. Linde realized that even with a simple potential energy bowl, such as that in Figure 9.1a, and even without finely arranging the inflaton field's initial value, inflation could still naturally take place. The idea is this. Imagine that in the very early universe, things were "chaotic"—for example, imagine that there was an inflaton field whose value randomly bounced around from one number to another. At some locations in space its value might have been small, at other locations its value might have been medium, and at yet other locations in space its value might have been high. Now, nothing particularly noteworthy would have happened in regions where the field value was small or medium. But Linde realized that something fantastically interesting would have taken place in regions where the inflaton field happened to have attained a high value (even if the region were tiny, a mere 10^{-33} centimeters across). When the inflaton field's value is high—when it is high up on the energy bowl in Figure 9.1a—a kind of cosmic friction sets in: the field's value tries to roll down the hill to lower potential energy, but its high value contributes to a resistive drag force, and so it rolls very slowly. Thus, the inflaton field's value would have been nearly constant and (much like an inflaton on the top of the potential energy hill in new inflation) would have contributed a nearly constant energy and a nearly constant negative pressure. As we are now very familiar, these are the conditions required to drive a burst of inflationary expansion. Thus, without invoking a particularly special potential energy bowl, and without setting up the inflaton field in a special configuration, the chaotic environment of the early universe could have naturally given rise to inflationary expansion. Not surprisingly, Linde had called this approach *chaotic inflation*. Many physicists consider it the most convincing realization of the inflationary paradigm.

12. Those familiar with the history of this subject will realize that the excitement over Guth's discovery was generated by its solutions to key cosmological problems, such as the horizon and flatness problems, as we describe shortly.

13. You might wonder whether the electroweak Higgs field, or the grand unified Higgs field, can do double duty—playing the role we described in Chapter 9, while also driving inflationary expansion at earlier times, before forming a Higgs ocean. Models of this sort have been proposed, but they typically suffer from technical problems. The most convincing realizations of inflationary expansion invoke a new Higgs field to play the role of the inflaton.

14. See note 11, this chapter.

15. For example, you can think of our horizon as a giant, imaginary sphere, with us at its center, that separates those things with which we could have communicated (the things within the sphere) from those things with which we couldn't have communicated (those things beyond the sphere), in the time since the bang. Today, the radius of our "horizon sphere" is roughly 14 billion light-years; early on in the history of the universe, its radius was much less, since there had been less time for light to travel. See also note 10 from Chapter 8.

16. While this is the essence of how inflationary cosmology solves the horizon problem, to avoid confusion let me highlight a key element of the solution. If one night you and a friend are standing on a large field happily exchanging light signals by turning flashlights on and off, notice that no matter how fast you then turn and run from each other,

you will *always* be able subsequently to exchange light signals. Why? Well, to avoid receiving the light your friend shines your way, or for your friend to avoid receiving the light you send her way, you'd need to run from each other at faster than light speed, and that's impossible. So, how is it possible for regions of space that were able to exchange light signals early on in the universe's history (and hence come to the same temperature, for example) to now find themselves beyond each other's communicative range? As the flashlight example makes clear, it must be that they've rushed apart at faster than the speed of light. And, indeed, the colossal outward push of repulsive gravity during the inflationary phase *did* drive every region of space away from every other at much faster than the speed of light. Again, this offers no contradiction with special relativity, since the speed limit set by light refers to motion through space, not motion from the swelling of space itself. So a novel and important feature of inflationary cosmology is that it involves a short period in which there is superluminal expansion of space.

17. Note that the numerical value of the critical density decreases as the universe expands. But the point is that if the actual mass/energy density of the universe is equal to the critical density at one time, it will decrease in exactly the same way and maintain equality with the critical density at all times.

18. The mathematically inclined reader should note that during the inflationary phase, the size of our cosmic horizon stayed fixed while space swelled enormously (as can easily be seen by taking an exponential form for the scale factor in note 10 of Chapter 8). That is the sense in which our observable universe is a tiny speck in a gigantic cosmos, in the inflationary framework.

19. R. Preston, *First Light* (New York: Random House Trade Paperbacks, 1996), p. 118.

20. For an excellent general-level account of dark matter, see L. Krauss, *Quintessence: The Mystery of Missing Mass in the Universe* (New York: Basic Books, 2000).

21. The expert reader will recognize that I am not distinguishing between the various dark matter problems that emerge on different scales of observation (galactic, cosmic) as the contribution of dark matter to the cosmic mass density is my only concern here.

22. There is actually some controversy as to whether this is the mechanism behind all type Ia supernovae (I thank D. Spergel for pointing this out to me), but the uniformity of these events—which is what we need for the discussion—is on a strong observational footing.

23. It's interesting to note that, years before the supernova results, prescient theoretical works by Jim Peebles at Princeton, and also by Lawrence Krauss of Case Western and Michael Turner of the University of Chicago, and Gary Steigman of Ohio State, had suggested that the universe might have a small nonzero cosmological constant. At the time, most physicists did not take this suggestion too seriously, but now, with the supernova data, the attitude has changed significantly. Also note that earlier in the chapter we saw that the outward push of a cosmological constant can be mimicked by a Higgs field that, like the frog on the plateau, is perched above its minimum energy configuration. So, while a cosmological constant fits the data well, a more precise statement is that the supernova researchers concluded that space must be filled with something *like* a cosmological constant that generates an outward push. (There are ways in which a Higgs field can be made to generate a long-lasting outward push, as opposed to the brief outward burst in the early moments of inflationary cosmology. We will discuss this in Chapter 14, when we consider the question of whether the data do indeed require a cosmological constant, or whether some other entity with similar gravitational consequences can fit the bill.) Researchers often use the term "dark energy" as a catchall phrase for an ingredient in the universe that

is invisible to the eye but causes every region of space to push, rather than pull, on every other.

24. Dark energy is the most widely accepted explanation for the observed accelerated expansion, but other theories have been put forward. For instance, some have suggested that the data can be explained if the force of gravity deviates from the usual strength predicted by Newtonian and Einsteinian physics when the distance scales involved are extremely large—of cosmological size. Others are not yet convinced that the data show cosmic acceleration, and are waiting for more precise measurements to be carried out. It is important to bear these alternative ideas in mind, especially should future observations yield results that strain the current explanations. But currently, there is widespread consensus that the theoretical explanations described in the main text are the most convincing.

Chapter 11

1. Among the leaders in the early 1980s in determining how quantum fluctuations would yield inhomogeneities were Stephen Hawking, Alexei Starobinsky, Alan Guth, So-Young Pi, James Bardeen, Paul Steinhardt, Michael Turner, Viatcheslav Mukhanov, and Gennady Chibisov.

2. Even with the discussion in the main text, you may still be puzzled regarding how a tiny amount of mass/energy in an inflaton nugget can yield the huge amount of mass/energy constituting the observable universe. How can you wind up with more mass/energy than you begin with? Well, as explained in the main text, the inflaton field, by virtue of its negative pressure, "mines" energy from gravity. This means that as the energy in the inflaton field increases, the energy in the gravitational field decreases. The special feature of the gravitational field, known since the days of Newton, is that its energy can become arbitrarily negative. Thus, gravity is like a bank that is willing to lend unlimited amounts of money—gravity embodies an essentially limitless supply of energy, which the inflaton field extracts as space expands.

The particular mass and size of the initial nugget of uniform inflaton field depend on the details of the model of inflationary cosmology one studies (most notably, on the precise details of the inflaton field's potential energy bowl). In the text, I've imagined that the initial inflaton field's energy density was about 10^{82} grams per cubic centimeter, so that a volume of $(10^{-26} \text{ centimeters})^3 = 10^{-78}$ cubic centimeters would have total mass of about 10 kilograms, i.e., about 20 pounds. These values are typical to a fairly conventional class of inflationary models, but are only meant to give you a rough sense of the numbers involved. To give a flavor of the range of possibilities, let me note that in Andrei Linde's chaotic models of inflation (see note 11 of Chapter 10), our observable universe would have emerged from an initial nugget of even smaller size, 10^{-33} centimeters across (the so-called Planck length), whose energy density was even higher, about 10^{94} grams per cubic centimeter, combining to give a lower total mass of about 10^{-5} grams (the so-called Planck mass). In these realizations of inflation, the initial nugget would have weighed about as much as a grain of dust.

3. See Paul Davies, "Inflation and Time Asymmetry in the Universe," in *Nature*, vol. 301, p. 398; Don Page, "Inflation Does Not Explain Time Asymmetry," in *Nature*, vol. 304, p. 39; and Paul Davies, "Inflation in the Universe and Time Asymmetry," in *Nature*, vol. 312, p. 524.

4. To explain the essential point, it is convenient to split entropy up into a part due to spacetime and gravity, and a remaining part due to everything else, as this captures intu-

itively the key ideas. However, I should note that it proves elusive to give a mathematically rigorous treatment in which the gravitational contribution to entropy is cleanly identified, separated off, and accounted for. Nevertheless, this doesn't compromise the qualitative conclusions we reach. In case you find this troublesome, note that the whole discussion can be rephrased largely without reference to gravitational entropy. As we emphasized in Chapter 6, when ordinary attractive gravity is relevant, matter falls together into clumps. In so doing, the matter converts gravitational potential energy into kinetic energy that, subsequently, is partially converted into radiation that emanates from the clump itself. This is an entropy-increasing sequence of events (larger average particle velocities increase the relevant phase space volume; the production of radiation through interactions increases the total number of particles—both of which increase overall entropy). In this way, what we refer to in the text as *gravitational entropy* can be rephrased as *matter entropy generated by the gravitational force*. When we say gravitational entropy is low, we mean that the gravitational force has the potential to generate significant quantities of entropy through matter clumping. In realizing such entropy potential, the clumps of matter create a non-uniform, non-homogeneous gravitational field—warps and ripples in spacetime—which, in the text, I've described as having higher entropy. But as this discussion makes clear, it really can be thought of as the clumpy matter (and radiation produced in the process) as having higher entropy (than when uniformly dispersed). This is good since the expert reader will note that if we view a classical gravitational background (a classical spacetime) as a coherent state of gravitons, it is an essentially unique state and hence has low entropy. Only by suitably coarse graining would an entropy assignment be possible. As this note emphasizes, though, this isn't particularly necessary. On the other hand, should the matter clump sufficiently to create black holes, then an unassailable entropy assignment becomes available: the area of the black hole's event horizon (as explained further in Chapter 16) is a measure of the black hole's entropy. And this entropy can unambiguously be called gravitational entropy.

5. Just as it is possible both for an egg to break and for broken eggshell pieces to reassemble into a pristine egg, it is possible for quantum-induced fluctuations to grow into larger inhomogeneities (as we've described) or for sufficiently correlated inhomogeneities to work in tandem to suppress such growth. Thus, the inflationary contribution to resolving time's arrow also requires sufficiently uncorrelated initial quantum fluctuations. Again, if we think in a Boltzmann-like manner, among all the fluctuations yielding conditions ripe for inflation, sooner or later there will be one that meets this condition as well, allowing the universe as we know it to initiate.

6. There are some physicists who would claim that the situation is better than described. For example, Andrei Linde argues that in chaotic inflation (see note 11, Chapter 10), the observable universe emerged from a Planck-sized nugget containing a uniform inflaton field with Planck scale energy density. Under certain assumptions, Linde further argues that the entropy of a *uniform* inflaton field in such a tiny nugget is roughly equal to the entropy of any other inflaton field configuration, and hence the conditions necessary for achieving inflation weren't special. The entropy of the Planck-sized nugget was small but on a par with the possible entropy that the Planck-sized nugget *could* have had. The ensuing inflationary burst then created, in a flash, a huge universe with an enormously higher entropy—but one that, because of its smooth, uniform distribution of matter, was also enormously far from the entropy that it could have. The arrow of time points in the direction in which this entropy gap is being lessened.

While I am partial to this optimistic vision, until we have a better grasp on the physics out of which inflation is supposed to have emerged, caution is warranted. For example, the expert reader will note that this approach makes favorable but unjustified assumptions about the high-energy (transplanckian) field modes—modes that can affect the onset of inflation and play a crucial role in structure formation.

Chapter 12

1. The circumstantial evidence I have in mind here relies on the fact that the strengths of all three nongravitational forces depend on the energy and temperature of the environment in which the forces act. At low energies and temperatures, such as those of our everyday environment, the strengths of all three forces are different. But there is indirect theoretical and experimental evidence that at very high temperatures, such as occurred in the earliest moments of the universe, the strengths of all three forces converge, indicating, albeit indirectly, that all three forces themselves may fundamentally be unified, and appear distinct only at low energies and temperatures. For a more detailed discussion see, for example, *The Elegant Universe*, Chapter 7.

2. Once we know that a field, like any of the known force fields, is an ingredient in the makeup of the cosmos, then we know that it exists everywhere—it is stitched into the fabric of the cosmos. It is impossible to excise the field, much as it is impossible to excise space itself. The nearest we can come to eliminating a field's presence, therefore, is to have it take on a value that minimizes its energy. For force fields, like the electromagnetic force, that value is zero, as discussed in the text. For fields like the inflaton or the standard-model Higgs field (which, for simplicity, we do not consider here), that value can be some nonzero number that depends on the field's precise potential energy shape, as we discussed in Chapters 9 and 10. As mentioned in the text, to keep the discussion streamlined we are only explicitly discussing quantum fluctuations of fields whose lowest energy state is achieved when their value is zero, although fluctuations associated with Higgs or inflaton fields require no modification of our conclusions.

3. Actually, the mathematically inclined reader should note that the uncertainty principle dictates that energy fluctuations are inversely proportional to the time resolution of our measurements, so the finer the time resolution with which we examine a field's energy, the more wildly the field will undulate.

4. In this experiment, Lamoreaux verified the Casimir force in a modified setup involving the attraction between a spherical lens and a quartz plate. More recently, Gianni Carugno, Roberto Onofrio, and their collaborators at the University of Padova have undertaken the more difficult experiment involving the original Casimir framework of two parallel plates. (Keeping the plates perfectly parallel is quite an experimental challenge.) So far, they have confirmed Casimir's predictions to a level of 15 percent.

5. In retrospect, these insights also show that if Einstein had not introduced the cosmological constant in 1917, quantum physicists would have introduced their own version a few decades later. As you will recall, the cosmological constant was an energy Einstein envisioned suffusing all of space, but whose origin he—and modern-day proponents of a cosmological constant—left unspecified. We now realize that quantum physics suffuses empty space with jittering fields, and as we directly see through Casimir's discovery, the resulting microscopic field frenzy fills space with energy. In fact, a major challenge facing theoretical physics is to show that the combined contribution of all field jitters yields a total energy in empty space—a total cosmological constant—that is within the observational limit currently determined by the supernova observations discussed in Chapter 10. So far, no one has been able to do this; carrying out the analysis exactly has proven to be beyond the capacity of current theoretical methods, and approximate calculations have

gotten answers *wildly* larger than observations allow, strongly suggesting that the approximations are way off. Many view explaining the value of the cosmological constant (whether it is zero, as long thought, or small and nonzero as suggested by the inflation and the supernova data) as one of the most important open problems in theoretical physics.

6. In this section, I describe one way of seeing the conflict between general relativity and quantum mechanics. But I should note, in keeping with our theme of seeking the true nature of space and time, that other, somewhat less tangible but potentially important puzzles arise in attempting to merge general relativity and quantum mechanics. One that's particularly tantalizing arises when the straightforward application of the procedure for transforming classical nongravitational theories (like Maxwell's electrodynamics) into a quantum theory is extended to classical general relativity (as shown by Bryce DeWitt in what is now called the Wheeler-DeWitt equation). In the central equation that emerges, it turns out that the time variable does not appear. So, rather than having an explicit mathematical embodiment of time—as is the case in every other fundamental theory—in this approach to quantizing gravity, temporal evolution must be kept track of by a physical feature of the universe (such as its density) that we expect to change in a regular manner. As yet, no one knows if this procedure for quantizing gravity is appropriate (although much progress in an offshoot of this formalism, called *loop quantum gravity*, has been recently achieved; see Chapter 16), so it is not clear whether the absence of an explicit time variable is hinting at something deep (time as an emergent concept?) or not. In this chapter we focus on a different approach for merging general relativity and quantum mechanics, *superstring theory*.

7. It is somewhat of a misnomer to speak of the "center" of a black hole as if it were a place in space. The reason, roughly speaking, is that when one crosses a black hole's event horizon—its outer edge—the roles of space and time are interchanged. In fact, just as you can't resist going from one second to the next in time, so you can't resist being pulled to the black hole's "center" once you've crossed the event horizon. It turns out that this analogy between heading forward in time and heading toward a black hole's center is strongly motivated by the mathematical description of black holes. Thus, rather than thinking of the black hole's center as a location in space, it is better to think of it as a location in time. Furthermore, since you can't go beyond the black hole's center, you might be tempted to think of it as a location in spacetime where time comes to an end. This may well be true. But since the standard general relativity equations break down under such extremes of huge mass density, our ability to make definite statements of this sort is compromised. Clearly, this suggests that if we had equations that don't break down deep inside a black hole, we might gain important insights into the nature of time. That is one of the goals of superstring theory.

8. As in earlier chapters, by "observable universe" I mean that part of the universe with which we could have had, at least in principle, communication during the time since the bang. In a universe that is infinite in spatial extent, as discussed in Chapter 8, all of space does *not* shrink to a point at the moment of the bang. Certainly, everything in the observable part of the universe will be squeezed into an ever smaller space as we head back to the beginning, but, although hard to picture, there are things—infinitely far away—that will forever remain separate from us, even as the density of matter and energy grows ever higher.

9. Leonard Susskind, in "The Elegant Universe," *NOVA*, three-hour PBS series first aired October 28 and November 4, 2003.

10. Indeed, the difficulty of designing experimental tests for superstring theory has been a crucial stumbling block, one that has substantially hindered the theory's accep-

tance. However, as we will see in later chapters, there has been much progress in this direction; string theorists have high hopes that upcoming accelerator and space-based experiments will provide at least circumstantial evidence in support of the theory, and with luck, maybe even more.

11. Although I haven't covered it explicitly in the text, note that every known particle has an *antiparticle*—a particle with the same mass but opposite force charges (like the opposite sign of electric charge). The electron's antiparticle is the positron; the up-quark's antiparticle is, not surprisingly, the anti-up-quark; and so on.

12. As we will see in Chapter 13, recent work in string theory has suggested that strings may be much larger than the Planck length, and this has a number of potentially critical implications—including the possibility of making the theory experimentally testable.

13. The existence of atoms was initially argued through indirect means (as an explanation of the particular ratios in which various chemical substances would combine, and later, through Brownian motion); the existence of the first black holes was confirmed (to many physicists' satisfaction) by seeing their effect on gas that falls toward them from nearby stars, instead of "seeing" them directly.

14. Since even a placidly vibrating string has *some* amount of energy, you might wonder how it's possible for a string vibrational pattern to yield a massless particle. The answer, once again, has to do with quantum uncertainty. No matter how placid a string is, quantum uncertainty implies that it has a minimal amount of jitter and jiggle. And, through the weirdness of quantum mechanics, these uncertainty-induced jitters have *negative* energy. When this is combined with the positive energy from the most gentle of ordinary string vibrations, the total mass/energy is zero.

15. For the mathematically inclined reader, the more precise statement is that the *square* of the masses of string vibrational modes are given by integer multiples of the square of the Planck mass. Even more precisely (and of relevance to recent developments covered in Chapter 13), the square of these masses are integer multiples of the *string scale* (which is proportional to the inverse square of the string length). In conventional formulations of string theory, the string scale and the Planck mass are close, which is why I've simplified the main text and only introduced the Planck mass. However, in Chapter 13 we will consider situations in which the string scale can be different from the Planck mass.

16. It's not too hard to understand, in rough terms, how the Planck length crept into Klein's analysis. General relativity and quantum mechanics invoke three fundamental constants of nature: c (the velocity of light), G (the basic strength of the gravitational force) and $\hbar$ (Planck's constant describing the size of quantum effects). These three constants can be combined to produce a quantity with units of length: $(\hbar G/c^3)^{1/2}$, which, by definition, is the Planck length. After substituting the numerical values of the three constants, one finds the Planck length to be about 1.616×10^{-33} centimeters. Thus, unless a dimensionless number with value differing substantially from 1 should emerge from the theory—something that doesn't often happen in a simple, well-formulated physical theory—we expect the Planck length to be the characteristic size of lengths, such as the length of the curled-up spatial dimension. Nevertheless, do note that this does not rule out the possibility that dimensions can be larger than the Planck length, and in Chapter 13 we will see interesting recent work that has investigated this possibility vigorously.

17. Incorporating a particle with the electron's charge, and with its relatively tiny mass, proved a formidable challenge.

18. Note that the uniform symmetry requirement that we used in Chapter 8 to narrow down the shape of the universe was motivated by astronomical observations (such as

those of the microwave background radiation) within the *three large dimensions*. These symmetry constraints have no bearing on the shape of the possible six tiny extra space dimensions. Figure 12.9a is based on an image created by Andrew Hanson.

19. You might wonder about whether there might not only be extra space dimensions, but also extra time dimensions. Researchers (such as Itzhak Bars at the University of Southern California) have investigated this possibility, and shown that it is at least possible to formulate theories with a second time dimension that seem to be physically reasonable. But whether this second time dimension is really on a par with the ordinary time dimension or is just a mathematical device has never been settled fully; the general feeling is more toward the latter than the former. By contrast, the most straightforward reading of string theory says that the extra space dimensions are every bit as real as the three we know about.

20. String theory experts (and those who have read *The Elegant Universe*, Chapter 12) will recognize that the more precise statement is that certain formulations of string theory (discussed in Chapter 13 of this book) admit limits involving eleven spacetime dimensions. There is still debate as to whether string theory is best thought of as fundamentally being an eleven spacetime dimensional theory, or whether the eleven dimensional formulation should be viewed as a particular limit (e.g., when the string coupling constant is taken large in the Type IIA formulation), on a par with other limits. As this distinction does not have much impact on our general-level discussion, I have chosen the former viewpoint, largely for the linguistic ease of having a fixed and uniform total number of dimensions.

Chapter 13

1. For the mathematically inclined reader: I am here referring to *conformal* symmetry—symmetry under arbitrary angle-preserving transformations on the volume in spacetime swept out by the proposed fundamental constituent. Strings sweep out two-spacetime-dimensional surfaces, and the equations of string theory are invariant under the two-dimensional conformal group, which is an *infinite* dimensional symmetry group. By contrast, in other numbers of space dimensions, associated with objects that are not themselves one-dimensional, the conformal group is finite-dimensional.

2. Many physicists contributed significantly to these developments, both by laying the groundwork and through follow-up discoveries: Michael Duff, Paul Howe, Takeo Inami, Kelley Stelle, Eric Bergshoeff, Ergin Szegin, Paul Townsend, Chris Hull, Chris Pope, John Schwarz, Ashoke Sen, Andrew Strominger, Curtis Callan, Joe Polchinski, Petr Hořava, J. Dai, Robert Leigh, Hermann Nicolai, and Bernard deWit, among many others.

3. In fact, as explained in Chapter 12 of *The Elegant Universe*, there is an even tighter connection between the overlooked tenth spatial dimension and p-branes. As you increase the size of the tenth spatial dimension in, say, the type IIA formulation, one-dimensional strings stretch into two-dimensional inner-tube-like membranes. If you assume the tenth dimension is very small, as had always been implicitly done prior to these discoveries, the inner tubes look and behave like strings. As is the case for strings, the question of whether these newly found branes are indivisible or, instead, are made of yet finer constituents, remains unanswered. Researchers are open to the possibility that the ingredients so far identified in string/M-theory will not bring to a close the search for *the* elementary constituents of the universe. However, it's also possible that they will. Since much of what follows is insensitive to this issue, we'll adopt the simplest perspective and imagine that all the ingredients—strings and branes of various dimensions—are fundamental. And

what of the earlier reasoning, which suggested that fundamental higher dimensional objects could not be incorporated into a physically sensible framework? Well, that reasoning was itself rooted in another quantum mechanical approximation scheme—one that is standard and fully battle tested but that, like any approximation, has limitations. Although researchers have yet to figure out all the subtleties associated with incorporating higher-dimensional objects into a quantum theory, these ingredients fit so perfectly and consistently within all five string formulations that almost everyone believes that the feared violations of basic and sacred physical principles are absent.

4. In fact, we could be living on an even higher-dimensional brane (a four-brane, a five-brane . . .) three of whose dimensions fill ordinary space, and whose other dimensions fill some of the smaller, extra dimensions the theory requires.

5. The mathematically inclined reader should note that for many years string theorists have known that closed strings respect something called T-duality (as explained further in Chapter 16, and in Chapter 10 of *The Elegant Universe*). Basically, T-duality is the statement that if an extra dimension should be in the shape of a circle, string theory is completely insensitive to whether the circle's radius is R or $1/R$. The reason is that strings can move around the circle ("momentum modes") and/or wrap around the circle ("winding modes") and, under the replacement of R with $1/R$, physicists have realized that the roles of these two modes simply interchange, keeping the overall physical properties of the theory unchanged. Essential to this reasoning is that the strings are closed loops, since if they are open there is no topologically stable notion of their winding around a circular dimension. So, at first blush, it seems that open and closed strings behave completely differently under T-duality. With closer inspection, and by making use of the Dirichlet boundary conditions for open strings (the "D" in D-branes), Polchinski, Dai, Leigh, as well as Hořava, Green, and other researchers resolved this puzzle.

6. Proposals that have tried to circumvent the introduction of dark matter or dark energy have suggested that even the accepted behavior of gravity on large scales may differ from what Newton or Einstein would have thought, and in that way attempt to account for gravitational effects incompatible with solely the material we can see. As yet, these proposals are highly speculative and have little support, either experimental or theoretical.

7. The physicists who introduced this idea are S. Giddings and S. Thomas, and S. Dimopoulus and G. Landsberg.

8. Notice that the contraction phase of such a bouncing universe is not the same as the expansion phase run in reverse. Physical processes such as eggs splattering and candles melting would happen in the usual "forward" time direction during the expansion phases and would continue to do so during the subsequent contraction phase. That's why entropy would increase during both phases.

9. The expert reader will note that the cyclic model can be phrased in the language of four-dimensional effective field theory on one of the three-branes, and in this form it shares many features with more familiar scalar-field-driven inflationary models. When I say "radically new mechanism," I am referring to the conceptual description in terms of colliding branes, which in and of itself is a striking new way of thinking about cosmology.

10. Don't get confused on dimension counting. The two three-branes, together with the space interval between them, have four dimensions. Time brings it to five. That leaves six more for the Calabi-Yau space.

11. An important exception, mentioned at the end of this chapter and discussed in further detail in Chapter 14, has to do with inhomogeneities in the gravitational field, so-called primordial gravitational waves. Inflationary cosmology and the cyclic model differ

in this regard, one way in which there is a chance that they may be distinguished experimentally.

12. Quantum mechanics ensures that there is always a nonzero probability that a chance fluctuation will disrupt the cyclic process (e.g., one brane twists relative to the other), causing the model to grind to a halt. Even if the probability is minuscule, sooner or later it will surely come to pass, and hence the cycles cannot continue indefinitely.

Chapter 14

1. A. Einstein, "Vierteljahrschrift für gerichtliche Medizin und öffentliches Sanitätswesen" 44 37 (1912). D. Brill and J. Cohen, *Phys. Rev.* vol. 143, no. 4, 1011 (1966); H. Pfister and K. Braun, *Class. Quantum Grav.* 2, 909 (1985).

2. In the four decades since the initial proposal of Schiff and Pugh, other tests of frame dragging have been undertaken. These experiments (carried out by, among others, Bruno Bertotti, Ignazio Ciufolini, and Peter Bender; and I. I. Shapiro, R. D. Reasenberg, J. F. Chandler, and R. W. Babcock) have studied the motion of the moon as well as satellites orbiting the earth, and found some evidence for frame dragging effects. One major advantage of Gravity Probe B is that it is the first fully contained experiment, one that is under complete control of the experimenters, and so should give the most precise and most direct evidence for frame dragging.

3. Although they are effective at giving a feel for Einstein's discovery, another limitation of the standard images of warped space is that they don't illustrate the warping of time. This is important because general relativity shows that for an ordinary object like the sun, as opposed to something extreme like a black hole, the warping of time (the closer you are to the sun, the slower your clocks will run) is far more pronounced than the warping of space. It's subtler to depict the warping of time graphically and it's harder to convey how warped time contributes to curved spatial trajectories such as the earth's elliptical orbit around the sun, and that's why Figure 3.10 (and just about every attempt to visualize general relativity I've ever seen) focuses solely on warped space. But it's good to bear in mind that in many common astrophysical environments, it's the warping of time that is dominant.

4. In 1974, Russell Hulse and Joseph Taylor discovered a binary pulsar system—two pulsars (rapidly spinning neutron stars) orbiting one another. Because the pulsars move very quickly and are very close together, Einstein's general relativity predicts that they will emit copious amounts of gravitational radiation. Although it is quite a challenge to detect this radiation directly, general relativity shows that the radiation should reveal itself indirectly through other means: the energy emitted via the radiation should cause the orbital period of the two pulsars to gradually decrease. The pulsars have been observed continuously since their discovery, and indeed, their orbital period has decreased—and in a manner that agrees with the prediction of general relativity to about one part in a thousand. Thus, even without direct detection of the emitted gravitational radiation, this provides strong evidence for its existence. For their discovery, Hulse and Taylor were awarded the 1993 Nobel Prize in Physics.

5. However, see note 4, above.

6. From the viewpoint of energetics, therefore, cosmic rays provide a naturally occurring accelerator that is far more powerful than any we have or will construct in the foreseeable future. The drawback is that although the particles in cosmic rays can have extremely high energies, we have no control over what slams into what—when it comes to cosmic ray collisions, we are passive observers. Furthermore, the number of cosmic ray particles with a given energy drops quickly as the energy level increases. While about 10

billion cosmic ray particles with an energy equivalent to the mass of a proton (about one-thousandth of the design capacity of the Large Hadron Collider) strike each square kilometer of earth's surface every second (and quite a few pass through your body every second as well), only about *one* of the most energetic particles (about 100 billion times the mass of a proton) would strike a given square kilometer of earth's surface each *century*. Finally, accelerators can slam particles together by making them move quickly, in opposite directions, thereby creating a large center of mass energy. Cosmic ray particles, by contrast, slam into the relatively slow moving particles in the atmosphere. Nevertheless, these drawbacks are not insurmountable. Over the course of many decades, experimenters have learned quite a lot from studying the more plentiful, lower-energy cosmic ray data, and, to deal with the paucity of high-energy collisions, experimenters have built huge arrays of detectors to catch as many particles as possible.

7. The expert reader will realize that conservation of energy in a theory with dynamic spacetime is a subtle issue. Certainly, the stress tensor of all sources for the Einstein equations is covariantly conserved. But this does not necessarily translate into a global conservation law for energy. And with good reason. The stress tensor does not take account of gravitational energy—a notoriously difficult notion in general relativity. Over short enough distance and time scales—such as occur in accelerator experiments—local energy conservation is valid, but statements about global conservation have to be treated with greater care.

8. This is true of the simplest inflationary models. Researchers have found that more complicated realizations of inflation can suppress the production of gravitational waves.

9. A viable dark matter candidate must be a stable, or very long-lived, particle—one that does not disintegrate into other particles. This is expected to be true of the lightest of the supersymmetric partner particles, and hence the more precise statement is that the lightest of the zino, higgsino, or photino is a suitable dark matter candidate.

10. Not too long ago, a joint Italian-Chinese research group known as the Dark Matter Experiment (DAMA), working out of the Gran Sasso Laboratory in Italy, made the exciting announcement that they had achieved the first direct detection of dark matter. So far, however, no other group has been able to verify the claim. In fact, another experiment, Cryogenic Dark Matter Search (CDMS), based at Stanford and involving researchers from the United States and Russia, has amassed data that many believe rule out the DAMA results to a high degree of confidence. In addition to these dark matter searches, many others are under way. To read about some of these, take a look at *http://hepwww.rl. ac.uk/ukdmc/dark_matter/other_searches.html.*

Chapter 15

1. This statement ignores hidden-variable approaches, such as Bohm's. But even in such approaches, we'd want to teleport an object's quantum state (its wavefunction), so a mere measurement of position or velocity would be inadequate.

2. Zeilinger's research group also included Dick Bouwmeester, Jian-Wi Pan, Klaus Mattle, Manfred Eibl, and Harald Weinfurter, and De Martini's has included S. Giacomini, G. Milani, F. Sciarrino, and E. Lombardi.

3. For the reader who has some familiarity with the formalism of quantum mechanics, here are the essential steps in quantum teleportation. Imagine that the initial state of a photon I have in New York is given by $|\Psi\rangle_1 = \alpha|0\rangle_1 + \beta|1\rangle_1$ where $|0\rangle$ and $|1\rangle$ are the two photon polarization states, and we allow for definite, normalized, but arbitrary values of the coefficients. My goal is to give Nicholas enough information so that he can produce a photon in London in exactly the same quantum state. To do so, Nicholas and I first

acquire a pair of entangled photons in the state, say $|\Psi\rangle_{23} = (1/\sqrt{2}) |0_2 0_3\rangle - (1/\sqrt{2})|1_2 1_3\rangle$. The initial state of the three-photon system is thus $|\Psi\rangle_{123} = (\alpha/\sqrt{2}) \{|0_1 0_2 0_3\rangle - |0_1 1_2 1_3\rangle\} + (\beta/\sqrt{2}) \{|1_1 0_2 0_3\rangle - |1_1 1_2 1_3\rangle\}$. When I perform a Bell-state measurement on Photons 1 and 2, I project this part of the system onto one of four states: $|\Phi\rangle_{\pm} = (1/\sqrt{2}) \{|0_1 0_2\rangle \pm |1_1 1_2\rangle\}$ and $|\Omega\rangle_{\pm} = (1/\sqrt{2}) \{|0_1 1_2\rangle \pm |1_1 0_2\rangle\}$. Now, if we re-express the initial state using this basis of eigenstates for Particles 1 and 2, we find: $|\Psi\rangle_{123} = \frac{1}{2}\{|\Phi\rangle_{+}(\alpha|0_3\rangle - \beta|1_3\rangle) + |\Phi\rangle_{-} (\alpha|0_3\rangle + \beta|1_3\rangle) + |\Omega\rangle_{+} (-\alpha|1_3\rangle + \beta|0_3\rangle) + |\Omega\rangle_{-} (-\alpha|1_3\rangle - \beta|0_3\rangle)\}$. Thus, after performing my measurement, I will "collapse" the system onto one of these four summands. Once I communicate to Nicholas (via ordinary means), which summand I find, he knows how to manipulate Photon 3 to reproduce the original state of Photon 1. For instance, if I find that my measurement yields state $|\Phi\rangle_{-}$, then Nicholas does not need to do anything to Photon 3, since, as above, it is already in the original state of Photon 1. If I find any other result, Nicholas will have to perform a suitable rotation (dictated, as you can see, by which result I find), to put Photon 3 into the desired state.

4. In fact, the mathematically inclined reader will note that it is not hard to prove the so-called no-quantum-cloning theorem. Imagine we have a unitary cloning operator U that takes any given state as input and produces two copies of it as output (U maps $|\alpha\rangle \rightarrow |\alpha\rangle|\alpha\rangle$, for any input state $|\alpha\rangle$). Note that U acting on a state like $(|\alpha\rangle + |\beta\rangle)$ yields $(|\alpha\rangle|\alpha\rangle + |\beta\rangle|\beta\rangle)$, which is not a two-fold copy of the original state $(|\alpha\rangle + |\beta\rangle)(|\alpha\rangle + |\beta\rangle)$, and hence no such operator U exists to carry out quantum cloning. (This was first shown by Wootters and Zurek in the early 1980s.)

5. Many researchers have been involved in developing both the theory and the experimental realization of quantum teleportation. In addition to those discussed in the text, the work of Sandu Popescu while at Cambridge University played an important part in the Rome experiments, and Jeffrey Kimble's group at the California Institute of Technology has pioneered the teleportation of continuous features of a quantum state, to name a few.

6. For extremely interesting progress on entangling many-particle systems, see, for example, B. Julsgaard, A. Kozhekin, and E. S. Polzik, "Experimental long-lived entanglement of two macroscopic objects," Nature 413 (Sept. 2001), 400–403.

7. One of the most exciting and active areas of research making use of quantum entanglement and quantum teleportation is the field of quantum computing. For recent general-level presentations of quantum computing, see Tom Siegfried, The Bit and the Pendulum (New York: John Wiley, 2000), and George Johnson, A Shortcut Through Time (New York: Knopf, 2003).

8. One aspect of the slowing of time at increasing velocity, which we did not discuss in Chapter 3 but will play a role in this chapter, is the so-called twin paradox. The issue is simple to state: if you and I are moving relative to one another at constant velocity, I will think your clock is running slow relative to mine. But since you are as justified as I in claiming to be at rest, you will think that mine is the moving clock and hence is the one that is running slow. That each of us thinks the other's clock is running slow may seem paradoxical, but it's not. At constant velocity, our clocks will continue to get farther apart and hence they don't allow for a direct, face-to-face comparison to determine which is "really" running slow. And all other indirect comparisons (for instance, we compare the times on our clocks by cell phone communication) occur with some elapsed time over some spatial separation, necessarily bringing into play the complications of different observers' notions of now, as in Chapters 3 and 5. I won't go through it here, but when these special relativistic complications are folded into the analysis, there is no contradiction between each of us declaring that the other's clock is running slow (see, e.g., E. Taylor and J. A. Wheeler, Spacetime Physics, for a complete, technical, but elementary discus-

sion). Where things appear to get more puzzling is if, for example, you slow down, stop, turn around, and head back toward me so that we can compare our clocks face to face, eliminating the complications of different notions of now. Upon our meeting, whose clock will be ahead of whose? This is the so-called twin paradox: if you and I are twins, when we meet again, will we be the same age, or will one of us look older? The answer is that my clock will be ahead of yours—if we are twins, I will look older. There are many ways to explain why, but the simplest to note is that when you change your velocity and experience an acceleration, the symmetry between our perspectives is lost—you can definitively claim that *you* were moving (since, for example, you *felt it*—or, using the discussion of Chapter 3, unlike mine, your journey through spacetime has not been along a straight line) and hence that your clock ran slow relative to mine. Less time elapsed for you than for me.

9. John Wheeler, among others, has suggested a possible central role for observers in a quantum universe, summed up in one of his famous aphorisms: "No elementary phenomenon is a phenomenon until it is an observed phenomenon." You can read more about Wheeler's fascinating life in physics in John Archibald Wheeler and Kenneth Ford, *Geons, Black Holes, and Quantum Foam: A Life in Physics* (New York: Norton, 1998). Roger Penrose has also studied the relation between quantum physics and the mind in his *The Emperor's New Mind*, and also in *Shadows of the Mind: A Search for the Missing Science of Consciousness* (Oxford: Oxford University Press, 1994).

10. See, for example, "Reply to Criticisms" in *Albert Einstein*, vol. 7 of *The Library of Living Philosophers*, P. A. Schilpp, ed. (New York: MJF Books, 2001).

11. W. J. van Stockum, *Proc. R. Soc. Edin.* A 57 (1937), 135.

12. The expert reader will recognize that I am simplifying. In 1966, Robert Geroch, who was a student of John Wheeler, showed that it is at least possible, in principle, to construct a wormhole without ripping space. But unlike the more intuitive, space-tearing approach to building wormholes in which the mere existence of the wormhole does not entail time travel, in Geroch's approach the construction phase itself would necessarily require that time become so distorted that one could freely travel backward and forward in time (but no farther back than the initiation of the construction itself).

13. Roughly speaking, if you passed through a region containing such exotic matter at nearly the speed of light and took the average of all your measurements of the energy density you detected, the answer you'd find would be negative. Physicists say that such exotic matter violates the so-called averaged weak energy condition.

14. The simplest realization of exotic matter comes from the vacuum fluctuations of the electromagnetic field between the parallel plates in the Casimir experiment, discussed in Chapter 12. Calculations show that the decrease in quantum fluctuations between the plates, relative to empty space, entails negative averaged energy density (as well as negative pressure).

15. For a pedagogical but technical account of wormholes, see Matt Visser, *Lorentzian Wormholes: From Einstein to Hawking* (New York: American Institute of Physics Press, 1996).

Chapter 16

1. For the mathematically inclined reader, recall from note 6 of Chapter 6 that entropy is defined as the *logarithm* of the number of rearrangements (or states), and that's important to get the right answer in this example. When you join two Tupperware containers together, the various states of the air molecules can be described by giving the state of the air molecules in the first container, and then by giving the state of those in the sec-

ond. Thus, the number of arrangements for the joined containers is the square of the number of arrangements of either separately. After taking the logarithm, this tells us that the entropy has doubled.

2. You will note that it doesn't really make much sense to compare a volume with an area, as they have different units. What I really mean here, as indicated by the text, is that the rate at which volume grows with radius is much faster than the rate at which surface area grows. Thus, since entropy is proportional to surface area and not volume, it grows more slowly with the size of a region than it would were it proportional to volume.

3. While this captures the spirit of the entropy bound, the expert reader will recognize that I am simplifying. The more precise bound, as proposed by Raphael Bousso, states that the entropy flux through a null hypersurface (with everywhere non-positive focusing parameter Θ) is bounded by A/4, where A is the area of a spacelike cross-section of the null hypersurface (the "light-sheet").

4. More precisely, the entropy of a black hole is the area of its event horizon, expressed in Planck units, divided by 4, and multiplied by Boltzmann's constant.

5. The mathematically inclined reader may recall from the endnotes to Chapter 8 that there is another notion of horizon—a cosmic horizon—which is the dividing surface between those things with which an observer can and cannot be in causal contact. Such horizons are also believed to support entropy, again proportional to their surface area.

6. In 1971, the Hungarian-born physicist Dennis Gabor was awarded the Nobel Prize for the discovery of something called *holography*. Initially motivated by the goal of improving the resolving power of electron microscopes, Gabor worked in the 1940s on finding ways to capture more of the information encoded in the light waves that bounce off an object. A camera, for example, records the intensity of such light waves; places where the intensity is high yield brighter regions of the photograph, and places where it's low are darker. Gabor and many others realized, though, that intensity is only part of the information that light waves carry. We saw this, for example, in Figure 4.2b: while the interference pattern is affected by the intensity (the amplitude) of the light (higher-amplitude waves yield an overall brighter pattern), the pattern itself arises because the overlapping waves emerging from each of the slits reach their peak, their trough, and various intermediate wave heights at different locations along the detector screen. The latter information is called *phase information*: two light waves at a given point are said to be *in phase* if they reinforce each other (they each reach a peak or trough at the same time), *out of phase* if they cancel each other (one reaches a peak while the other reaches a trough), and, more generally, they have phase relations intermediate between these two extremes at points where they partially reinforce or partially cancel. An interference pattern thus records phase information of the interfering light waves.

Gabor developed a means for recording, on specially designed film, both the intensity and the phase information of light that bounces off an object. Translated into modern language, his approach is closely akin to the experimental setup of Figure 7.1, except that one of the two laser beams is made to bounce off the object of interest on its way to the detector screen. If the screen is outfitted with film containing appropriate photographic emulsion, it will record an interference pattern—in the form of minute, etched lines on the film's surface—between the unfettered beam and the one that has reflected off the object. The interference pattern will encode both the intensity of the reflected light and phase relations between the two light beams. The ramifications of Gabor's insight for science have been substantial, allowing for vast improvements in a wide range of measurement techniques. But for the public at large, the most prominent impact has been the artistic and commercial development of holograms.

Ordinary photographs look flat because they record only light intensity. To get depth, you need phase information. The reason is that as a light wave travels, it cycles from peak to trough to peak again, and so phase information—or, more precisely, phase differences between light beams that reflect off nearby parts of an object—encodes differences in how far the light rays have traveled. For example, if you look at a cat straight on, its eyes are a little farther away than its nose and this depth difference is encoded in the phase difference between the light beams' reflecting off each facial element. By shining a laser through a hologram, we are able to exploit the phase information the hologram records, and thereby add depth to the image. We've all seen the results: stunning three-dimensional projections generated from two-dimensional pieces of plastic. Note, though, that your eyes do not use phase information to see depth. Instead, your eyes use parallax: the slight difference in the angles at which light from a given point travels to reach your left eye and your right eye supplies information that your brain decodes into the point's distance. That's why, for example, if you lose sight in one eye (or just keep it closed for a while), your depth perception is compromised.

7. For the mathematically inclined reader, the statement here is that a beam of light, or massless particles more generally, can travel from any point in the interior of anti-deSitter space to spatial infinity and back, in finite time.

8. For the mathematically inclined reader, Maldacena worked in the context of $AdS_5 \times S^5$, with the boundary theory arising from the boundary of AdS_5.

9. This statement is more one of sociology than of physics. String theory grew out of the tradition of quantum particle physics, while loop quantum gravity grew out of the tradition of general relativity. However, it is important to note that, as of today, only string theory can make contact with the successful predictions of general relativity, since only string theory convincingly reduces to general relativity on large distance scales. Loop quantum gravity is understood well in the quantum domain, but bridging the gap to large-scale phenomena has proven difficult.

10. More precisely, as discussed further in Chapter 13 of *The Elegant Universe*, we have known how much entropy black holes contain since the work of Bekenstein and Hawking in the 1970s. However, the approach those researchers used was rather indirect, and never identified microscopic rearrangements—as in Chapter 6—that would account for the entropy they found. In the mid-1990s, this gap was filled by two string theorists, Andrew Strominger and Cumrun Vafa, who cleverly found a relation between black holes and certain configurations of branes in string/M-theory. Roughly, they were able to establish that certain special black holes would admit exactly the same number of rearrangements of their basic ingredients (whatever those ingredients might be) as do particular, special combinations of branes. When they counted the number of such brane rearrangements (and took the logarithm) the answer they found was the area of the corresponding black hole, in Planck units, divided by 4—exactly the answer for black hole entropy that had been found years before. In loop quantum gravity, researchers have also been able to show that the entropy of a black hole is proportional to its surface area, but getting the exact answer (surface area in Planck units divided by 4) has proven more of a challenge. If a particular parameter, known as the Immirzi parameter, is chosen appropriately, then indeed the exact black hole entropy emerges from the mathematics of loop quantum gravity, but as yet there is no universally accepted fundamental explanation, within the theory itself, of what sets the correct value of this parameter.

11. As I have throughout the chapter, I am suppressing quantitatively important but conceptually irrelevant numerical parameters.

Glossary

absolute space: Newton's view of space; envisions space as unchanging and independent of its contents.

absolute spacetime: View of space emerging from special relativity; envisions space through the entirety of time, from any perspective, as unchanging and independent of its contents.

absolutist: Perspective holding that space is absolute.

acceleration: Motion that involves a change in speed and/or direction.

accelerator, atom smasher: Research tool of particle physics that collides particles together at high speed.

aether, luminiferous aether: Hypothetical substance filling space that provides the medium for light to propogate; discredited.

arrow of time: Direction in which time seems to point—from past to future.

background independence: Property of a physical theory in which space and time emerge from a more fundamental concept, rather than being inserted axiomatically.

big bang theory/standard big bang theory: Theory describing a hot, expanding universe from a moment after its birth.

big crunch: One possible end to the universe, analogous to a reverse of the big bang in which space collapses in on itself.

black hole: An object whose immense gravitational field traps anything, even light, that gets too close (closer than the black hole's event horizon).

braneworld scenario: Possibility within string/M-theory that our familiar three-spatial dimensions are a three-brane.

Casimir force: Quantum mechanical force exerted by an imbalance of vacuum field fluctuations.

classical physics: As used in this book, the physical laws of Newton and Maxwell. More generally, often used to refer to all nonquantum laws of physics, including special and general relativity.

closed strings: Filaments of energy in string theory, in the shape of loops.

collapse of probability wave, collapse of wavefunction: Hypothetical development in which a probability wave (a wavefunction) goes from a spread-out to a spiked shape.

Copenhagen interpretation: Interpretation of quantum mechanics that envisions large objects as being subject to classical laws and small objects as being subject to quantum laws.

cosmic microwave background radiation: Remnant electromagnetic radiation (photons) from the early universe, which permeates space.

cosmic horizon, horizon: Locations in space beyond which light has not had time to reach us, since the beginning of the universe.

cosmological constant: A hypothetical energy and pressure, uniformly filling space; origin and composition unknown.

cosmology: Study of origin and evolution of the universe.

critical density: Amount of mass/energy density required for space to be flat; about 10^{-23} grams per cubic meter.

D-branes, Dirichlet-p-branes: A p-brane that is "sticky"; a p-brane to which open string endpoints are attached.

dark energy: A hypothetical energy and pressure, uniformly filling space; more general notion than a cosmological constant as its energy/pressure can vary with time.

dark matter: Matter suffused through space, exerting gravity but not emitting light.

electromagnetic field: The field which exerts the electromagnetic force.

electromagnetic force: One of nature's four forces; acts on particles that have electric charge.

electron field: The field for which the electron particle is the smallest bundle or constituent.

electroweak theory: The theory unifying the electromagnetic and the weak nuclear forces into the electroweak force.

electroweak Higgs field: Field that acquires a nonzero value in cold, empty space; gives rise to masses for fundamental particles.

energy bowl: See *potential energy bowl*.

entropy: A measure of the disorder of a physical system; the number of rearrangements of a system's fundamental constituents that leave its gross, overall appearance unchanged.

entanglement, quantum entanglement: Quantum phenomenon in which spatially distant particles have correlated properties.

event horizon: Imaginary sphere surrounding a black hole delineating the points of no return; anything crossing the event horizon cannot escape the black hole's gravity.

field: A "mist" or "essence" permeating space; can convey a force or describe the presence/motion of particles. Mathematically, involves a number or collection of numbers at each point in space, signifying the field's value.

flat space: Possible shape of the spatial universe having no curvature.

flatness problem: Challenge for cosmological theories to explain observed flatness of space.

general relativity: Einstein's theory of gravity; invokes curvature of space and time.

gluons: Messenger particles of the strong nuclear force.

gravitons: Hypothetical messenger particles of the gravitational force.

grand unification: Theory attempting to unify the strong, weak, and electromagnetic forces.

Higgs field: See *electroweak Higgs field*.

Higgs field vacuum expectation value: Situation in which a Higgs field acquires a nonzero value in empty space; a Higgs ocean.

Higgs ocean: Shorthand, peculiar to this book, for a Higgs field vacuum expectation value.

Higgs particles: Finest quantum constituents of a Higgs field.

horizon problem: Challenge for cosmological theories to explain how regions of space, beyond each other's cosmological horizon, have nearly identical properties.

inertia: Property of an object that resists its being accelerated.

inflationary cosmology: Cosmological theory incorporating a brief but enormous burst of spatial expansion in the early universe.

inflaton field: The field whose energy and negative pressure drives inflationary expansion.

interference: Phenomenon in which overlapping waves create a distinctive pattern; in quantum mechanics, involves seemingly exclusive alternatives combining together.

Kaluza-Klein theory: Theory of universe involving more than three spatial dimensions.

Kelvin: Scale in which temperatures are quoted relative to absolute zero (the lowest possible temperature, $-273°$ on the Celsius scale).

luminiferous aether: See *aether.*

M-theory: Currently incomplete theory unifying all five versions of string theory; a fully quantum mechanical theory of all forces and all matter.

Mach's principle: Principle that all motion is relative and that the standard of rest is provided by average mass distribution in the universe.

Many Worlds interpretation: Interpretation of quantum mechanics in which all potentialities embodied by a probability wave are realized in separate universes.

messenger particle: Smallest "packet" or "bundle" of a force, which communicates the forces' influence.

microwave background radiation: See *cosmic microwave background radiation.*

negative curvature: Shape of space containing less than the critical density; saddle-shaped.

observable universe: Part of universe within our cosmic horizon; part of universe close enough so that light it emitted can have reached us by today; part of universe we can see.

open strings: Filaments of energy in string theory, in the shape of snippets.

***p*-brane:** Ingredient of string/M-theory with p-spatial dimensions. See also D-brane.

Planck length: Size (10^{-33} centimeters) below which the conflict between quantum mechanics and general relativity becomes manifest; size below which conventional notion of space breaks down.

Planck mass: Mass (10^{-5} grams, mass of a grain of dust; ten billion billion times the proton mass); typical mass of a vibrating string.

Planck time: Time (10^{-43} seconds) it takes light to traverse one Planck length; time interval below which conventional notion of time breaks down.

phase transition: Qualitative change in a physical system when its temperature is varied through a sufficiently wide range.

photon: Messenger particle of the electromagnetic force; a "bundle" of light.

potential energy: Energy stored in a field or object.

potential energy bowl: Shape describing the energy a field contains for a given field value; technically called the field's potential energy.

probability wave: Wave in quantum mechanics that encodes the probability that a particle will be found at a given location.

quantum chromodynamics: Quantum mechanical theory of the strong nuclear force.

quantum fluctuations, quantum jitters: The unavoidable, rapid variations in the value of a field on small scales, arising from quantum uncertainty.

quantum measurement problem: Problem of explaining how the myriad possibilities encoded in a probability wave give way to a single outcome when measured.

quantum mechanics: Theory, developed in the 1920s and 1930s, for describing the realm of atoms and subatomic particles.

quarks: Elementary particles subject to the strong nuclear force; there are six varieties (up, down, strange, charm, top, bottom).

relationist: Perspective holding that all motion is relative and space is not absolute.

rotational invariance, rotational symmetry: Characteristic of a physical system, or of a theoretical law, of being unaffected by a rotation.

second law of thermodynamics: Law that says that, on average, the entropy of a physical system will tend to rise from any given moment.

spacetime: The union of space and time first articulated by special relativity.

special relativity: Einstein's theory in which space and time are not individually absolute, but instead depend upon the relative motion between distinct observers.

spin: Quantum mechanical property of elementary particles in which, somewhat like a top, they undergo rotational motion (they have intrinsic angular momentum).

spontaneous symmetry breaking: Technical name for the formation of a Higgs ocean; process by which a previously manifest symmetry is hidden or spoiled.

standard candles: Objects of a known intrinsic brightness that are useful for measuring astronomical distances.

standard model: Quantum mechanical theory composed of quantum chromodynamics and the electroweak theory; describes all matter and forces, except for gravity. Based on conception of point particles.

strong nuclear force: Force of nature that influences quarks; holds quarks together inside protons and neutrons.

string theory: Theory based on one-dimensional vibrating filaments of energy (see superstring theory), but which does not necessarily incorporate supersymmetry. Sometimes used as shorthand for superstring theory.

superstring theory: Theory in which fundamental ingredients are one-dimensional loops (closed strings) or snippets (open strings) of vibrating energy, which unites general relativity and quantum mechanics; incorporates supersymmetry.

supersymmetry: A symmetry in which laws are unchanged when particles with a whole number amount of spin (force particles) are interchanged with particles that have half of a whole number amount of spin (matter particles).

symmetry: A transformation on a physical system that leaves the system's appearance unchanged (e.g., a rotation of a perfect sphere about its center leaves the sphere unchanged); a transformation of a physical system that has no effect on the laws describing the system.

time-reversal symmetry: Property of the accepted laws of nature in which laws make no distinction between one direction in time and the other. From any given moment, the laws treat past and future in exactly the same way.

time slice: All of space at one moment of time; a single slice through the spacetime block or loaf.

translational invariance, translational symmetry: Property of accepted laws of nature in which the laws are applicable at any location in space.

uncertainty principle: Property of quantum mechanics in which there is a fundamental limit on how precisely certain complementary physical features can be measured or specified.

unified theory: A theory that describes all forces and all matter in a single theoretical structure.

vacuum: The emptiest that a region can be; the state of lowest energy.

vacuum field fluctuations: See *quantum fluctuations*.

velocity: The speed and direction of an object's motion.

W and Z particles: The messenger particles of the weak nuclear force.

wavefunction: See *probability wave*.

weak nuclear force: Force of nature, acting on subatomic scales, and responsible for phenomena such as radioactive decay.

which-path information: Quantum mechanical information delineating the path a particle took in going from source to detector.

Suggestions for
Further Reading

The general and technical literature on space and time is vast. The references below, mostly suited to a general reader but a few requiring more advanced training, have proven helpful to me and are a good start for the reader who wants to explore further various developments addressed in this book.

Albert, David. *Quantum Mechanics and Experience.* Cambridge, Mass.: Harvard University Press, 1994.
———. *Time and Chance.* Cambridge, Mass.: Harvard University Press, 2000.
Alexander, H. G. *The Leibniz-Clarke Correspondence.* Manchester, Eng.: Manchester University Press, 1956.
Barbour, Julian. *The End of Time.* Oxford: Oxford University Press, 2000.
——— and Herbert Pfister. *Mach's Principle.* Boston: Birkhäuser, 1995.
Barrow, John. *The Book of Nothing.* New York: Pantheon, 2000.
Bartusiak, Marcia. *Einstein's Unfinished Symphony.* Washington, D.C.: Joseph Henry Press, 2000.
Bell, John. *Speakable and Unspeakable in Quantum Mechanics.* Cambridge, Eng.: Cambridge University Press, 1993.
Blanchard, Ph., and D. Giulini, E. Joos, C. Kiefer, I.-O Stamatescu. *Decoherence: Theoretical, Experimental and Conceptual Problems.* Berlin: Springer, 2000.
Callender, Craig, and Nick Hugget. *Physics Meets Philosophy at the Planck Scale.* Cambridge, Eng.: Cambridge University Press, 2001.
Cole, K. C. *The Hole in the Universe.* New York: Harcourt, 2001.
Crease, Robert, and Charles Mann. *The Second Creation.* New Brunswick, N.J.: Rutgers University Press, 1996.
Davies, Paul. *About Time.* New York: Simon & Schuster, 1995.
———. *How to Build a Time Machine.* New York: Allen Lane, 2001.
———. *Space and Time in the Modern Universe.* Cambridge, Eng.: Cambridge University Press, 1977.
D'Espagnat, Bernard. *Veiled Reality.* Reading, Mass.: Addison-Wesley, 1995.
Deutsch, David. *The Fabric of Reality.* New York: Allen Lane, 1997.
Ferris, Timothy. *Coming of Age in the Milky Way.* New York: Anchor, 1989.
———. *The Whole Shebang.* New York: Simon & Schuster, 1997.
Feynman, Richard. *QED.* Princeton: Princeton University Press, 1985.
Fölsing, Albrecht. *Albert Einstein.* New York: Viking, 1997.
Gell-Mann, Murray. *The Quark and the Jaguar.* New York: W. H. Freeman, 1994.
Gleick, James. *Isaac Newton.* New York: Pantheon, 2003.

Gott, J. Richard. *Time Travel in Einstein's Universe.* Boston: Houghton Mifflin, 2001.

Guth, Alan. *The Inflationary Universe.* Reading, Mass.: Perseus, 1997.

Greene, Brian. *The Elegant Universe.* New York: Vintage, 2000.

Gribbin, John. *Schrödinger's Kittens and the Search for Reality.* Boston: Little, Brown, 1995.

Hall, A. Rupert. *Isaac Newton.* Cambridge, Eng.: Cambridge University Press, 1992.

Halliwell, J. J., J. Pérez-Mercader, and W. H. Zurek. *Physical Origins of Time Asymmetry.* Cambridge, Eng.: Cambridge University Press, 1994.

Hawking, Stephen. *The Universe in a Nutshell.* New York: Bantam, 2001.

—— and Roger Penrose. *The Nature of Space and Time.* Princeton: Princeton University Press, 1996.

——, Kip Thorne, Igor Novikov, Timothy Ferris, and Alan Lightman. *The Future of Spacetime.* New York: Norton, 2002.

Jammer, Max. *Concepts of Space.* New York: Dover, 1993.

Johnson, George. *A Shortcut Through Time.* New York: Knopf, 2003.

Kaku, Michio. *Hyperspace.* New York: Oxford University Press, 1994.

Kirschner, Robert. *The Extravagant Universe.* Princeton: Princeton University Press, 2002.

Krauss, Lawrence. *Quintessence.* New York: Perseus, 2000.

Lindley, David. *Boltzmann's Atom.* New York: Free Press, 2001.

——. *Where Does the Weirdness Go?* New York: Basic Books, 1996.

Mach, Ernst. *The Science of Mechanics.* La Salle, Ill.: Open Court, 1989.

Maudlin, Tim. *Quantum Non-locality and Relativity.* Malden, Mass.: Blackwell, 2002.

Mermin, N. David. *Boojums All the Way Through.* New York: Cambridge University Press, 1990.

Overbye, Dennis. *Lonely Hearts of the Cosmos.* New York: HarperCollins, 1991.

Pais, Abraham. *Subtle Is the Lord.* Oxford: Oxford University Press, 1982.

Penrose, Roger. *The Emperor's New Mind.* New York: Oxford University Press, 1989.

Price, Huw. *Time's Arrow and Archimedes' Point.* New York: Oxford University Press, 1996.

Rees, Martin. *Before the Beginning.* Reading, Mass.: Addison-Wesley, 1997.

——. *Just Six Numbers.* New York: Basic Books, 2001.

Reichenbach, Hans. *The Direction of Time.* Mineola, N.Y.: Dover, 1956.

——. *The Philosophy of Space and Time.* New York: Dover, 1958.

Savitt, Steven. *Time's Arrows Today.* Cambridge, Eng.: Cambridge University Press, 2000.

Schrödinger, Erwin. *What Is Life?* Cambridge, Eng.: Canto, 2000.

Siegfried, Tom. *The Bit and the Pendulum.* New York: John Wiley, 2000.

Sklar, Lawrence. *Space, Time, and Spacetime.* Berkeley: University of California Press, 1977.

Smolin, Lee. *Three Roads to Quantum Gravity.* New York: Basic Books, 2001.

Stenger, Victor. *Timeless Reality.* Amherst, N.Y.: Prometheus Books, 2000.

Thorne, Kip. *Black Holes and Time Warps.* New York: W. W. Norton, 1994.

von Weizsäcker, Carl Friedrich. *The Unity of Nature.* New York: Farrar, Straus, and Giroux, 1980.

Weinberg, Steven. *Dreams of a Final Theory.* New York: Pantheon, 1992.

——. *The First Three Minutes.* New York: Basic Books, 1993.

Wilczek, Frank, and Betsy Devine. *Longing for the Harmonies.* New York: Norton, 1988.

Zeh, H. D. *The Physical Basis of the Direction of Time.* Berlin: Springer, 2001.

Index

Page numbers in *italics* refer to illustrations and tables.

ALSO BY BRIAN GREENE

"*[Greene] develops one fresh new insight after another. . . . In the great tradition of physicists writing for the masses,* The Elegant Universe *sets a standard that will be hard to beat.*"
—*George Johnson,* The New York Times Book Review

THE ELEGANT UNIVERSE

Superstrings, Hidden Dimensions, and the Quest for the Ultimate Theory

In a rare blend of scientific insight and writing as elegant as the theories it explains, Brian Greene, one of the world's leading string theorists, peels away the layers of mystery surrounding string theory to reveal a universe that consists of eleven dimensions in which the fabric of space tears and repairs itself, and all matter—from the smallest quarks to the most gargantuan supernovas—is generated by the vibrations of microscopically tiny loops of energy.

Greene uses everything from an amusement park ride to ants on a garden hose to illustrate the beautiful yet bizarre realities that modern physics is unveiling. Dazzling in its brilliance, unprecedented in its ability to both illuminate and entertain, *The Elegant Universe* is a tour de force of science writing—a delightful and lucid voyage through modern physics that brings us closer than ever to understanding how the universe works.

Science/Physics/0-375-70811-1

VINTAGE BOOKS
Available at your local bookstore, or call toll-free to order:
1-800-793-2665 (credit cards only)